VLADIMIR I. ARNOLD

Collected Works

Springer

Vladimir I. Arnold, drawing, 1968.
Photo by Dmitry Arnold

VLADIMIR I. ARNOLD

Collected Works

VLADIMIR I. ARNOLD

Collected Works

VOLUME II

Hydrodynamics, Bifurcation Theory, and Algebraic Geometry 1965-1972

VLADIMIR I. ARNOLD

Collected Works

VOLUME II

Hydrodynamics, Bifurcation Theory, and Algebraic Geometry 1965-1972

Edited by

Alexander B. Givental
Boris A. Khesin
Alexander N. Varchenko
Victor A. Vassiliev
Oleg Ya. Viro

Springer

Vladimir I. Arnold

June 12, 1937 – June 3, 2010

Editors

Alexander B. Givental
Department of Mathematics
University of California
Berkeley, CA, USA

Boris A. Khesin
Department of Mathematics
University of Toronto
Toronto, ON, Canada

Alexander N. Varchenko
Department of Mathematics
University of North Carolina
Chapel Hill, NC, USA

Victor A. Vassiliev
Steklov Mathematical Institute
Russian Academy of Sciences
Moscow, Russia

Oleg Ya. Viro
Institute for Mathematical Sciences
Stony Brook University
Stony Brook, NY, USA

ISBN 978-3-642-31030-0   ISBN 978-3-642-31031-7 (ebook)
DOI 10.1007/978-3-642-31031-7

Library of Congress Control Number: 2013937321

Cover design: WMXDesign GmbH, Heidelberg

Printed on acid-free paper

Springer is part of Springer Science+Business Media (www.springer.com)

Preface

This volume of the Collected Works appears in print after Vladimir Arnold's untimely death in June 2010. His passing was a terrible loss for mathematics and science in general.

We hope that this project of Collected Works, which is needed now more than ever, will contribute to establishing the tremendous legacy of V.I. Arnold, a remarkable mathematician and human being. Some memories of V.I. Arnold can be found in the recent March and April 2012 issues of the Notices of the AMS.

Our Editorial team has also suffered an unprecedented blow since Volume I was published in 2009. Jerry Marsden passed away in September 2010, and Vladimir Zakalyukin passed away in December 2011. We dedicate this volume to their memory.

This Volume II of the Collected Works includes papers written by V.I. Arnold mostly during the period from 1965 to 1972. This was an amazingly productive period, starting with a year Arnold spent in Paris. During this period he made fundamental contributions to the fields of hydrodynamics, algebraic geometry, singularity and bifurcation theories, and dynamical systems. We have also added later papers by Arnold on related topics so as to make this volume more complete and comprehensive.

Several papers were translated from Russian specifically for this volume. Unfortunately it was not possible to replace certain previously existing translations, completed in the era of the Iron Curtain, when it could not be expected that translators understood the subject. As an alarming example, we refer the reader to the paper "Local problems of analysis", and to the editors' comment therein. As a counter-example, see the paper "On the arrangements of ovals...", which has been translated once again.

For the same reason some of the titles in the reprinted translations were incorrect. Since it was not possible to correct them in the reprinted articles, we have given their correct versions in the Contents along with the versions used in the reprinted contributions.

November 2012

Alexander Givental
Boris Khesin
Alexander Varchenko
Victor Vassiliev
Oleg Viro

Acknowledgements

The Editors thank the Göttingen State and University Library for providing the original articles for this edition, as well as D. Auroux, A. Chenciner, G. Gould and R. Montgomery for the translation and editing of several papers in this volume. They also thank the Springer office in Heidelberg, and in particular Ruth Allewelt and Martin Peters, for their extensive help and tireless support with this project.

Contents

Books or book prefaces by V. Arnold written in 1965-1972, but not included in the "Collected Works"

1. (with A. Avez): "Problèmes ergodiques de la mècanique classique" (in French), Monographies Internationales de Mathèmatiques Modernes, No. 9, Gauthier-Villars, Paris, 1967, ii+243pp.; English translation: "Ergodic problems of classical mechanics", W. A. Benjamin, Inc., New York-Amsterdam, 1968, ix+286pp.

2. "Obyknovennye differentsialnye uravneniya" (in Russian) Nauka, Moscow, 1971, 239pp.; English translation: "Ordinary differential equations", The M.I.T. Press, Cambridge, 1973, ix+280pp.; Other translations/editions of this book are French (1984), Russian (2nd edition, unrevised, 1975), Russian (3rd edition, revised and extended, 1984), Polish (1975, 1983), Portuguese (1985), German (1980), etc.

3. Editor's preface to a collection of papers "Singularities of differentiable mappings", Mir, Moscow, 1968.

4. Translator's preface to F. Pham's book "Introduction to the topological study of Landau's singularities", Mir, Moscow, 1970.

5. Translator's preface to J. Milnor's "Singular points of complex hypersurfaces", Mir, Moscow, 1971.

VARIATIONAL PRINCIPLE FOR THREE-DIMENSIONAL STEADY-STATE FLOWS OF AN IDEAL FLUID

(VARIATSIONNYI PRINTSIP DLIA TREKHMERNYKH STATSIONARNYKH TECHENII IDEAL'NOI ZHIDKOSTI)

PMM Vol.29, № 5, 1965; pp. 846-851

V.I. ARNOL'D
(Moscow)

(Received June 14, 1965)

It is proved that a steady-state flow has an extremal kinetic energy in comparison with "equivorticity" flows. This result is applied to investigate the stability of steady-state flows: if the extremum is a minimum or a maximum, then the flow is stable, i.e. a small change in the initial velocity field causes only a small change in the velocity field for all time. To determine the nature of the extremum (maximum, minimum, etc.) a second variation is explicitly calculated. For the case of plane flows, sufficient conditions of the stability with respect to small finite perturbations are found. These conditions are close to the necessary ones.

1. Finite-dimensional model. We shall show that the equations of the three-dimensional hydrodynamics of an ideal fluid are infinite-dimensional analog to the following finite-dimensional situation. In the space $\mathbf{x} = x_1, \ldots, x_n$ let there be given a system of ordinary differential equations

$$\mathbf{x}^{\cdot} = f(\mathbf{x}) \tag{1.1}$$

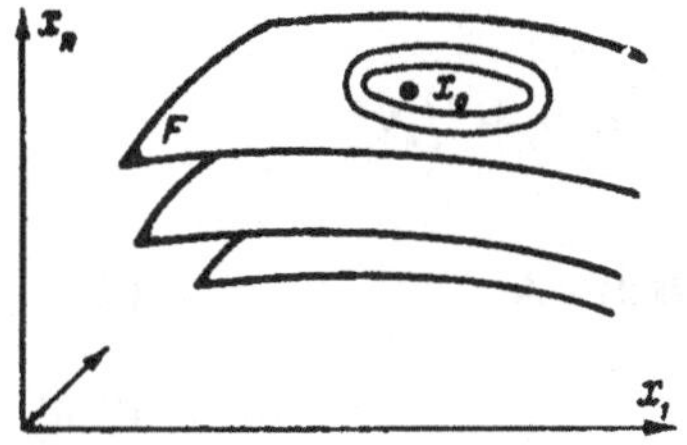

Fig. 1

In addition, we shall assume that a "k-dimensional structure" is given in the space $\mathbf{x}$ (Fig.1), i.e. that the space is decomposed into k-dimensional "sheets" (in the figure: $n = 3$, $k = 2$). We shall assume that the structure is invariant with respect to the system (1.1), i.e. that a trajectory $\mathbf{x}(t)$ which begins on the sheet F still remains on it. We shall call a point $\mathbf{x}$ of the sheet F regular, if in the neighborhood of this point there exists a system of coordinates $y_1, \ldots, y_n$ in which the sheets are given by the equations $y_{k+1} = c_{k+1}, \ldots, y_n = c_n$. In the whole, however, the sheets need not be given by equations (for example, a sheet may be everywhere dense).

We shall assume, finally, that the system (1.1) has a first integral $E(\mathbf{x})$.

Originally publ. in: Prikl. Mat. Mekh. 29:5, 846-851, © Nauka, Moscow 1965
English transl.: J. Appl. Math. Mech. 29:5, 1002-1008, © Elsevier, Oxford, 1965
Reprint by Springer-Verlag Berlin Heidelberg 2014 with kind permission, DOI 10.1007/978-3-642-31031-7_1

We shall consider a local conditional extremum of the function E on the sheet F. We shall assume that it occurs at a regular point $\mathbf{x}_0$ and that the quadratic form d^2E is nonsingular on the sheet F. The following three theorems are easily proved (cf.[1]).

Theorem 1.1. The point $\mathbf{x}_0$ is the equilibrium position of the system (1.1)

$$f(\mathbf{x}_0) = 0$$

Theorem 1.2. If the extremum is a maximum or a minimum, then the equilibrium position $\mathbf{x}_0$ is stable with respect to small finite perturbations.

Theorem 1.3. The spectrum of the problem of small oscillations corresponding to (1.1) $A\xi = \lambda\xi$ $(A = \partial\mathbf{f} / \partial\mathbf{x}$ in $\mathbf{x}_0)$ is symmetric with respect to the real and imaginary axes of λ.

The hydrodynamic analog of these theorems will be formulated below. They are, in fact, corollaries of the general theorems on Li geodesic groups provided partially with an invariant metric (cf.[2]). However, an independent proof is given here which makes use neither of Li groups, nor even of the existence and uniqueness theorems which correspond to the partial differential equations. From a mathematical point of view, these theorems will be "a priori" equalities and estimates.

2. Notation. Let D be a domain, bounded by the fixed surface Γ, in a three-dimensional Euclidean space; let $\mathbf{v}$ be the velocity field of an ideal fluid (incompressible, with density equal to 1, inviscid, exterior to a nonpotential mass force field) which fills the volume D; and let p be the pressure.

The Euler equation

$$\frac{\partial\mathbf{v}}{\partial t} + (\mathbf{v}\cdot\nabla)\,\mathbf{v} = -\operatorname{grad} p, \qquad \operatorname{div}\mathbf{v} = 0, \qquad \mathbf{v}\cdot\mathbf{n} = 0 \quad \text{on } \Gamma \tag{2.1}$$

has as a corollary the Bernoulli equation

$$\frac{\partial\mathbf{v}}{\partial t} = \mathbf{v}\times\mathbf{r} - \operatorname{grad}\lambda, \qquad \mathbf{r} = \operatorname{rot}\mathbf{v}, \qquad \lambda = p + {}^1/_2\mathbf{v}^2 \tag{2.2}$$

Hence, in view of the identity

$$\operatorname{rot}(\mathbf{A}\times\mathbf{B}) = \{\mathbf{AB}\} + \mathbf{A}\operatorname{div}\mathbf{B} - \mathbf{B}\operatorname{div}\mathbf{A} \tag{2.3}$$

there follows

$$\partial\mathbf{r} / \partial t = \{\mathbf{vr}\} \tag{2.4}$$

Here $\{\mathbf{AB}\}$ is the Poisson bracket of the vector fields

$$\{\mathbf{AB}\}_i = \sum (\partial A_i / \partial x_j)\, B_j - (\partial B_i / \partial x_j)\, A_j$$

If g is a smooth mapping of $x \to g(x)$, then we shall denote by g^* the corresponding mapping of the vectors

$$(g^*\xi)_i = \sum (\partial g_i / \partial x_j)\, \xi_j$$

3. Equivorticity fields. In order to formulate the law of coservation of vorticity in a form suitable for later use, we shall consider two vector fields $\mathbf{v}$ and $\mathbf{v}'$ in D

$$\operatorname{div}\mathbf{v}=0,\quad \operatorname{div}\mathbf{v}'=0,\quad (\mathbf{v}\cdot\mathbf{n})=0,\quad \mathbf{v}'\cdot\mathbf{n}=0 \quad \text{on } \Gamma$$

D e f i n i t i o n 3.1. The fields $\mathbf{v}$ and $\mathbf{v}'$ are equivorticity fields if there exists a smooth, volume preserving, mapping g of the domain D into itself such that (*)

$$\oint_{\gamma} \mathbf{v}\,dx = \oint_{g\gamma} \mathbf{v}'\,dx \tag{3.1}$$

for every closed contour γ in the domain D.

The law of conservation of vorticity now takes the following form. Let $\mathbf{v}(\mathbf{x},t)$ be the velocity field of the ideal fluid of (2.1).

T h e o r e m 3.1. The fields $\mathbf{v}(\mathbf{x},0)$ and $\mathbf{v}(\mathbf{x},t)$ are fields of equivorticity. In fact, let $\mathbf{x}(t)$ be the trajectory of a fluid particle. The mapping g is then that which transforms $\mathbf{x}(0)$ into $\mathbf{x}(t)$.

We shall now consider the Euler equation (2.1) as the system (1.1) in an infinite-dimensional space of the vector fields $\mathbf{v}(\mathbf{x})$ (where $\operatorname{div}\mathbf{v}=0$ and $\mathbf{v}\cdot\mathbf{n}=0$ on Γ). We shall show that this system has the characteristics of the system (1.1). In the space of the fields $\mathbf{v}(\mathbf{x})$ the following structure is specified: two fields belong to the same sheet if they are equivorticity fields. According to Theorem 3.1, this structure is invariant. Steady-state flows are "equilibrium position" of the system. Finally, the Euler equation (2.1) has a first integral of the energy

$$2E = \iiint \mathbf{v}^2\,dx$$

In order to transfer the results of Section 1 to the hydrodynamic equations (2.1) it is necessary to calculate the first and second variations of the function E on the sheet F.

4. Variational principle (**). The following fundamental Theorem is the analog of Theorem 1.1.

T h e o r e m 4.1. The steady-state flow $\mathbf{v}(\mathbf{x})$ has an extremal energy in comparison with all close equivorticity flows $\mathbf{v}'(\mathbf{x})$.

By closeness here is meant closeness "with respect to the sheet", i.e.

---

*) The mapping g transforms the vorticity of $\mathbf{v}$ into vorticity of $\mathbf{v}'$

$$g^*\operatorname{rot}\mathbf{v} = \operatorname{rot}\mathbf{v}' \tag{3.2}$$

Actually, if ξ, η is an infinitesimal parallelogram, then, since $\det g^* = 1$,

$$\xi\times\eta\cdot\operatorname{rot}\mathbf{v} = (g^*\xi)\times(g^*\eta)\cdot(g^*\operatorname{rot}\mathbf{v})$$

and corresponding to (3.1)

$$(\xi\times\eta\cdot\operatorname{rot}\mathbf{v}) = (g^*\xi)\times(g^*\eta)\cdot\operatorname{rot}\mathbf{v}'$$

If the domain D is not simply connected, then condition (3.1) is stronger than (3.2).

**) Another variational principle relating to unsteady flows has been determined and applied in the investigation of stability y Fjørtoft [3].

$\mathbf{v}'(\mathbf{x})$ is considered to be close to $\mathbf{v}(\mathbf{x})$ if the corresponding mapping g in (3.1) is close to the identity mapping. To determine the closeness of g to the identity mapping we shall introduce the "coordinates" f into the space g in the following way.

Let $\mathbf{f}(\mathbf{x})$ be a vector field in D such that

$$\operatorname{div} \mathbf{f} = 0, \qquad \mathbf{f}\cdot\mathbf{n} = 0 \quad \text{on } \Gamma$$

D e f i n i t i o n 4.1. L t $g_t = \exp ft$ be a mapping of D into itself, determined by the solutions $\mathbf{x}(t)$ of the ordinary differential equations $\dot{\mathbf{x}} = \mathbf{f}(\mathbf{x})$ according to Formula $g(\mathbf{x}(0)) = \mathbf{x}(t)$.

The field $\mathbf{v}'$ will be considered to be close to $\mathbf{v}$ if the "coordinates" f of the transformation g in (3.1) are small. In this case, the velocity perturbation $\delta\mathbf{v} = \mathbf{v}' - \mathbf{v}$. is also small. The relation between $\delta\mathbf{v}$ and $\mathbf{f}$ is given by the following Formula (4.2).

L e m m a 4.1. If for every closed contour γ in D

$$\oint\limits_{g_{-t}\gamma} \mathbf{v}\,d\mathbf{x} = \oint\limits_{\gamma} \mathbf{v}'\,d\mathbf{x}, \qquad \operatorname{div}\mathbf{v} = 0, \qquad \operatorname{div}\mathbf{v}' = 0 \tag{4.1}$$

then

$$\mathbf{v}' - \mathbf{v} = t(\mathbf{f}\times\mathbf{r}) + {}^1\!/_2 t^2\,\mathbf{f}\times\{\mathbf{f}\mathbf{r}\} + O(t^3) + \operatorname{grad}\alpha \tag{4.2}$$

where α is a single-valued function and $\mathbf{r} = \operatorname{rot}\mathbf{v}$.

P r o o f o f t h e L e m m a . According to the Stokes formula

$$\frac{d}{dt}\oint\limits_{g_{-t}\gamma} \mathbf{v}\,d\mathbf{x} = -\frac{1}{dt}\iint \mathbf{r}\,ds = \oint\limits_{g_{-t}\gamma} \mathbf{f}\times\mathbf{r}\,d\mathbf{x} \tag{4.3}$$

Since the Jacobian of $g_{-t}{}^*$ is equal to unity, we then have

$$\oint\limits_{g_{-t}\gamma} \mathbf{f}\times\mathbf{r}\,d\mathbf{x} = \int\limits_{\gamma} g_t{}^*\mathbf{f}(g_{-t}\mathbf{x})\times g_t{}^*\mathbf{r}(g_{-t}\mathbf{x})\,d\mathbf{x} \tag{4.4}$$

But, according to the definition of g_t we have $g_t{}^*\mathbf{f}(g_{-t}\mathbf{x}) = \mathbf{f}(\mathbf{x})$. Therefore, (4.3) and (4.4) give

$$\frac{d}{dt}\oint\limits_{g_{-t}\gamma} \mathbf{v}\,d\mathbf{x} = \oint\limits_{\gamma} \mathbf{f}\times\mathbf{r}(t)\,d\mathbf{x} \tag{4.5}$$

The field $\mathbf{r}(t)$ here is defined by Formula

$$\mathbf{r}(\mathbf{x}, t) = g_t{}^*\mathbf{r}(g_{-t}\mathbf{x}) \tag{4.6}$$

Differentiating (4.6), we find

$$\left.\frac{d\mathbf{r}}{dt}\right|_{t=0} = \{\mathbf{f}\mathbf{r}\}, \qquad \mathbf{r}(t) = \mathbf{r} + \{\mathbf{f}\mathbf{r}\}t + O(t^2) \tag{4.7}$$

But, according to condition (4.1),

$$\frac{d}{dt}\oint\limits_{g_{-t}\gamma} \mathbf{v}\,d\mathbf{x} = \oint\limits_{\gamma} \frac{\partial\mathbf{v}'}{\partial t}\,d\mathbf{x}, \qquad \mathbf{v}'|_{t=0} = \mathbf{v} \tag{4.8}$$

Integrating (4.5) and (4.6) with respect to t , we find from (4.6)

$$\oint\limits_{\gamma} (\mathbf{v}' - \mathbf{v})\,d\mathbf{x} = \oint\limits_{\gamma}\int\limits_0^t \mathbf{f}\times[\mathbf{r} + \{\mathbf{f}, \mathbf{r}\}t + O(t^2)]\,dt\,d\mathbf{x}$$

which is equivalent to (4.2).

P r o o f o f t h e b a s i c T h e o r e m . If $\mathbf{v}'$ is a equivorticity flow close to the steady-state flow $\mathbf{v}$, then according to (4.2) the first variation is

$$\delta\mathbf{v} = \mathbf{f}\times\mathbf{r} + \operatorname{grad}\alpha$$

Therefore,

$$\delta E = \iiint \mathbf{v}\cdot\delta\mathbf{v}\,d\mathbf{x} = \iiint \mathbf{v}\cdot(f\times r + \operatorname{grad}\alpha)\,d\mathbf{x} = \iiint [\mathbf{f}\cdot(\mathbf{r}\times\mathbf{v}) + \mathbf{v}\cdot\operatorname{grad}\alpha]\,d\mathbf{x}$$

For a steady-state flow, according to (2.2),

$$\mathbf{r}\times\mathbf{v} = -\operatorname{grad}\lambda$$

Therefore,

$$\delta E = \iiint (\mathbf{v}\cdot\operatorname{grad}\alpha - \mathbf{f}\cdot\operatorname{grad}\lambda)\,d\mathbf{x} = 0$$

This result is obtained by integrating by parts, taking into account Equalities $\operatorname{div}\mathbf{v} = 0$, $\operatorname{div}\mathbf{f} = 0$; $(\mathbf{v}\cdot\mathbf{n})|_\Gamma = 0$, $(f, n)|_\Gamma = 0$

The Theorem is proved.

5. Formula for the second variation. According to Lemma 4.1

$$\mathbf{v}' = \mathbf{v} + \delta\mathbf{v} + \delta^2\mathbf{v} + O(f^3)$$

where

$$\delta\mathbf{v} = \mathbf{f}\times\mathbf{r} + \operatorname{grad}\alpha_1, \quad \delta^2\mathbf{v} = \tfrac{1}{2}[\mathbf{f}\times\{\mathbf{fr}\}] + \operatorname{grad}\alpha_2$$

Correspondingly,

$$2\delta^2 E = \iiint [(\delta\mathbf{v})^2 + 2(\mathbf{v}\cdot\delta^2\mathbf{v})]\,dx = \iiint [(\delta\mathbf{v})^2 + \mathbf{v}\cdot\mathbf{f}\times\{\mathbf{fr}\} + 2\mathbf{v}\cdot\operatorname{grad}\alpha_2]\,dx$$

Integrating the last term by parts, we obtain the following form for the second variation of the energy E on the "sheet" F of the fields having equivorticity with $\mathbf{v}$, in terms of the variables $\mathbf{f}$ introduced in Section 4 :

$$2\delta^2 E = \iiint (\delta\mathbf{v})^2 + \mathbf{v}\times\mathbf{f}\{\mathbf{f}\cdot\mathbf{r}\}\,dx \tag{5.1}$$

This expression is quadratic with respect to $\mathbf{f}$, since $\delta\mathbf{v}$, linearly expressed in terms of $\mathbf{f}$, is

$$\delta\mathbf{v} = \mathbf{f}\times\mathbf{r} + \operatorname{grad}\alpha_1$$

where α_1 is determined from $\operatorname{div}\delta\mathbf{v} = 0$ and $(\delta\mathbf{v}\cdot\mathbf{n})|_\Gamma = 0$ and, therefore, is linearly dependent on $\mathbf{f}$. We also observe that according to Formula (2.3) $\{\mathbf{fr}\} = \delta\mathbf{r}$.

The following theorem is the analog of Theorem 1.2.

T h e o r e m 5.1. If the quadratic form (5.1) is of fixed sign, then the flow $\mathbf{v}$ is stable with respect to small finite perturbations. By a small perturbation here is meant one of which $\delta\mathbf{v}$ and $\mathbf{f}$, i.e. $\delta\mathbf{v}$ and $\delta\mathbf{r}$ or form $|\delta^2 E|$, are small.

The form (5.1) represents the first integral to the linear problem of small oscillations close to a steady-state flow $\mathbf{v}$. In accordance with Theorem 1.3, the spectrum of this problem is symmetric with respect to both axes. Hence —

T h e o r e m 5.2. If some perturbation of the steady-state flow **v** is damped, then some other perturbation is amplified and the flow **v** is unstable.

The author was not able to find a flow **v** for which the quadratic form $\delta^2 E$ was of fixed sign for three-dimensional perturbations. However, in specifically symmetric cases Theorem 5.1 gives simple stability criteria.

6. Supplementary integrals. Generalizing Theorem 1.2, we shall assume that the Euler equation (1.2) has a first integral M such that for a steady-state flow **v**

$$\delta M = \iiint \mathbf{A}\cdot\delta\mathbf{v}\,dx \qquad (\mathbf{A}\times\mathrm{rot}\,\mathbf{v} = \mathrm{grad}\,\alpha) \tag{6.1}$$

The assumption (6.1) is satisfied in the following three examples.

E x a m p l e 6.1. For the energy integral $M_1 = E$ we have

$$\mathbf{A} = \mathbf{v} = [\mathbf{v}\times\mathrm{rot}\ \mathbf{v}] = \mathrm{grad}\ \lambda$$

according to (2.2).

E x a m p l e 6.2. If the domain D and the flow **v** are invariant with respect to displacements along the x-axis, the integral

$$M_2 = \iiint \mathbf{v}\cdot\mathbf{e}_x dx\,dy\,dz$$

is then preserved.

For it $A = \mathbf{e}_x$ and $\mathbf{A}\times\mathrm{rot}\ \mathbf{v} = \mathrm{grad}\ (\mathbf{v}\cdot\mathbf{e}_x)$.

E x a m p l e 6.3. If the domain D and the flow **v** are invariant with respect to rotations about the z-axis, then

$$M_3 = \iiint (\mathbf{v}\times\mathbf{R}\cdot\mathbf{e}_z)\,dx\,dy\,dz$$

is preserved, where **R** is the radius vector of the point x, y, z. In this case

$$\mathbf{A} = \mathbf{R}\times\mathbf{e}_z, \qquad \mathbf{A}\times\mathrm{rot}\,\mathbf{v} = \mathrm{grad}\,(\mathbf{v}\times\mathbf{R}\cdot\mathbf{e}_z)$$

T h e o r e m 6.1. The value of the integral M over the velocity field of a steady-state flow **v** is an extremal in comparison with the values over close equivorticity fields , provided that M satisfies condition (6.1).

The proof is identical to the proof of Theorem 1.2. The corresponding formulas of the second variation have the form

$$2\delta^2 M_2 = \iiint (\mathbf{e}_x\times\mathbf{f})\,\{\mathbf{fr}\}\,dx\,dy\,dz \tag{6.2}$$

$$2\delta^2 M_3 = \iiint (R\times\mathbf{e}_z\times\mathbf{f})\,\{\mathbf{fr}\}\,dx\,dy\,dz \tag{6.3}$$

Fixed sign behavior of some linear combination

$$\lambda_1\delta^2 M_1 + \lambda_2\delta^2 M_2 + \lambda_3\delta^2 M_3$$

is sufficient for the stability of **v** .

We shall illustrate the application of Theorem 5.1, Formulas (5.1),(6.2) and (6.3) in an example of plane flows.

7. Plane perturbations of plane flows. Let the flow **v** have a stream function $\psi(x,y)$ such that

$$\mathbf{v} = \psi_y, \quad -\psi_x, \quad 0; \qquad \mathbf{r} = 0, \quad 0, \quad -\Delta\psi \tag{7.1}$$

Substituting (7.1) into (5.1), we obtain after a brief calculation, taking into consideration that $\{\mathbf{fr}\} = \delta\mathbf{r}$, the formula given in [1]

$$2\delta^2 E = \iint \left[(\delta\mathbf{v})^2 + \frac{\nabla\psi}{\nabla\Delta\psi} (\delta\mathbf{r})^2 \right] dx\, dy \tag{7.2}$$

From (7.2) and Theorem 5.1 there follows

Corollary 7.1 (cf.[1]). In any domain plane flows with a concave velocity profile $(\nabla\psi / \nabla\Delta\psi > 0)$ are stable with respect to finite plane perturbations.

We shall refer to specifically symmetric flows. The case of plane-parallel flows (the Rayleigh theorem) is considered in detail in [1]. We shall consider the flow in the annulus between concentric circles. After a brief calculation, Formula (6.3) is reduced in the plane case to the form

$$2\delta^2 M_3 = \frac{{}^1\!/_2 \nabla R^2}{\nabla\Delta\psi} (\delta\mathbf{r})^2 \tag{7.3}$$

From (7.3) and Theorem 5.1 there follows

Corollary 7.2. A plane circular flow in a circular annulus is stable with respect to small finite plane perturbations if the vorticity varies monotonously with the radius.

Actually, if the sign of $\nabla R^2 / \nabla\Delta\psi$ is preserved, then the form

$$\delta^2 H = \delta^2 E + \lambda\delta^2 M_3$$

is positive definite for suitable λ .

Finally, we note that the investigation of parallel flows with a single inflection point carried out in [1], owing to Formula (7.3), remains in effect for the case of circular flows.

BIBLIOGRAPHY

1. Arnol'd, V.I., Ob usloviiakh nelineinoi ustoichivosti ploskikh statsionarnykh krivolineinykh techenii ideal'noi zhidkosti (On nonlinear stability conditions of plane steady-state curvilinear flows of an ideal fluid . Dokl.Akad.Nauk SSSR, Vol.162, № 5, 1965.
2. Moreau, G.G., Méchanique des Fluides – une méthode de "Cinématique Fonctionnelle" et hydrodinamique. C.r.hebd.Séanc.Acad.Sci., Paris, Vol. 249, № 21, p.2156, 1959.
3. Fjørtoft, R., Application of integral theorems in deriving criteria of stability for laminar flows and for the baroclinic circular vortex. Geofys.Publ., Oslo, Vol.17, № 6, p.52, 1960.

Translated by R.D.C.

On the Riemann curvature of diffeomorphism groups

Note by Mr. **Vladimir Arnold**, presented by Mr. Jean Leray
Translated by Denis Auroux

Abstract One calculates explicitly the sectional curvatures of certain infinite-dimensional Lie groups equipped with left-invariant metrics whose geodesics correspond to flows of an ideal fluid. The sectional curvature turns out to be negative in certain two-dimensional directions.

1. Below I present an explicit expression (9) for the Riemann curvature of a Lie group equipped with a left-invariant metric. More generally, I will also call Riemann curvature the same expression for an infinite-dimensional group. In particular, I calculate the curvature of the group of area-preserving diffeomorphisms of the torus T^2, see (14). In that example, certain sectional curvatures (13) turn out to be negative. It has been known since Hadamard [1] that the sign of curvature influences the behavior of geodesics: negatively curved manifolds have unstable geodesics.

The interest in geodesics on the manifold underlying a Lie group is naturally justified by the following examples:

(a) For SO(3) they represent rotations of a rigid body in the three-dimensional Euclidean space E^3.
(b) For the group $S\,\mathrm{Diff}\,\mathscr{D}$ of volume-preserving diffeomorphisms of a Riemannian domain $\mathscr{D}$, they represent flows of an ideal fluid filling $\mathscr{D}$ [3, 4, 5].
(c) The group of positive dilations and translations of $\mathbb{R}^n$ gives rise to the geodesic flow on the space of constant negative curvature, while nilpotent groups give rise to "nilflows" [6].

2. Notations Given a Riemannian space M, denote by TM_x the tangent space at $x \in M$, and by $\langle\ ,\ \rangle$ the scalar product defined by the metric. Given $x \in M$ and

Translation of C.R. Acad. Sci. Paris 260 (1965), 5668-5671

Originally publ. in: C.R. Acad. Sc. Paris 260, 5668-5671, © French Acad. Sciences, Paris 1965
Translated by D. Auroux, Springer-Verlag Berlin Heidelberg 2014 with kind permission, DOI 10.1007/978-3-642-31031-7_2

$\xi \in TM_x$, denote by $\gamma(x,\xi,t) = \gamma(\xi,t) = \gamma(t) = \gamma$ the geodesic through $x = \gamma(0)$ with tangent vector $\xi = \dot{\gamma}(0)$, parametrized by time t.

The parallel transport along γ of a vector $\eta \in TM_x$ yields a vector $P_\gamma \eta$ which can be defined by the following construction. Set

$$\Pi_{\gamma(\xi,t)}\eta = \frac{1}{t}\frac{d}{d\tau}\bigg|_{\tau=0} \gamma(x,\xi+\tau\eta,t) \in TM_{\gamma(t)}. \tag{1}$$

Then $P_{\gamma(t)}\eta - \Pi_{\gamma(t)}\eta = O(t^2)$ as $t \to 0$. The covariant derivative $\nabla_\xi \eta$ of η along ξ is, by definition:

$$\nabla_\xi \eta = \frac{d}{dt}\bigg|_{t=0} P^{-1}_{\gamma(t)}\eta(\gamma(\xi,t)) = \frac{d}{dt}\bigg|_{t=0} \Pi^{-1}_{\gamma(t)}\eta(\gamma(\xi,t)). \tag{2}$$

Let ξ and η be two orthonormal vectors in TM_x. The sectional curvature $R_{\xi\eta}$ of M at x in the 2-plane defined by ξ and η is, by definition [7],

$$R_{\xi\eta} = -\langle \nabla_\xi \nabla_\eta \xi, \eta\rangle + \langle \nabla_\eta \nabla_\xi \xi, \eta\rangle + \langle \nabla_{[\xi,\eta]}\xi, \eta\rangle, \tag{3}$$

where $[\xi,\eta]$ is the Lie bracket of the vector fields ξ and η, whose restrictions at x are respectively $\xi(x)$ and $\eta(x)$.

Let G be a real Lie group, and $\mathfrak{A} = TG_e$ its Lie algebra equipped with the Lie bracket $[\,,\,]$. The group exponential map[1] $\exp : \mathfrak{A} \to G$ makes it possible to interpret the Lie algebra as a chart of G in a neighborhood of the identity element e.

For $a \in \mathfrak{A}$, I will denote $\exp a = \mathbf{a}$. Denote by

$$L_{\mathbf{g}} : TG_{\mathbf{h}} \to TG_{\mathbf{gh}}, \qquad L_g : T\mathfrak{A}_h \to T\mathfrak{A}_{gh}$$

the maps of tangent spaces induced by left translations.

It is easy to see that the map

$$L_a : \mathfrak{A} = T\mathfrak{A}_0 \to T\mathfrak{A}_a = \mathfrak{A}$$

from $\mathfrak{A}$ to itself is given by the formula

$$L_a\xi = \xi + \frac{1}{2}[a,\xi] + O(a^2), \qquad \text{where } a,\xi \in \mathfrak{A},\ |a| \ll 1. \tag{4}$$

Let $\langle a,b\rangle$ be any scalar product on the algebra $\mathfrak{A}$. A left-invariant metric on G is defined in terms of $\langle\ ,\ \rangle$ by the scalar product

$$\langle a,b\rangle_{\mathbf{g}} = \langle L_{\mathbf{g}^{-1}}a, L_{\mathbf{g}^{-1}}b\rangle, \qquad \text{where } a,b \in TG_{\mathbf{g}}.$$

From now on, all metric notions such as geodesics, curvature, etc., will be taken with respect to this metric.

---

[1] In general different from the geodesic exponential map.

The curvature of G will be expressed (9) in terms of the operation B defined as follows:

Let $a, c \in \mathfrak{A}$, then the formula

$$\langle [a,b], c\rangle = \langle B(c,a), b\rangle, \qquad \text{for all } b \text{ in } \mathfrak{A}, \tag{5}$$

defines a bilinear map $B : \mathfrak{A} \times \mathfrak{A} \to \mathfrak{A}$.

Example 1. Let $\langle a,b\rangle = (Aa,b)$, where $(\; , \;)$ is a bi-invariant scalar product, and A is a symmetric operator. Then $B(c,a) = A^{-1}[Ac,a]$.

Another example is given in section 4.

3. *The results*

Let $\gamma(\gamma_0, \dot\gamma, t)$ be a geodesic on G. Consider[2] the velocity vector $\dot\gamma \in TG_\gamma$ transported at e:

$$\xi(t) = L_{\gamma^{-1}}\dot\gamma \in \mathfrak{A}.$$

One has

Lemma 1. *The vector $\xi(t)$ satisfies Euler's equation [2]:*

$$\dot\xi = B(\xi,\xi). \tag{6}$$

For the proof, it suffices to consider the case of $\gamma_0 = e$.

Then one writes the Euler-Lagrange equations for the Lagrangian $L = \frac{1}{2}\langle \xi,\xi\rangle$ and uses (4).

Expressed differently, one has

Lemma 2. *The image of $\gamma(e,\xi,t)$ in $\mathfrak{A}$ is*

$$\gamma(\xi,t) = \xi t + B(\xi,\xi)\frac{t^2}{2} + O(t^3), \qquad t \to 0. \tag{7}$$

Let $\xi(g)$ and $\eta(g)$ be two left-invariant tangent vector fields on G. Then $\nabla_{\xi(g)}\eta(g)$ is also left-invariant. I denote by $\nabla_\xi \eta$ its value at $g = e$. Let us compute this value.

Lemma 3. *The covariant derivative is given by*

$$2\nabla_\xi \eta = [\xi,\eta] - B(\xi,\eta) - B(\eta,\xi). \tag{8}$$

It follows from (1) and (7) that

$$\Pi_{\gamma(\xi,t)}\eta = \eta + \delta t + O(t^2), \qquad t \to 0,$$

where

$$2\delta = B(\xi,\eta) + B(\eta,\xi).$$

[2] In dynamics of a rigid body, $\xi(t)$ is called the vector of angular velocity in the body. In hydrodynamics, it is the velocity field at time t.

Using (2) and (4), one obtains (8).
Next, using (8) and (3) one derives

Theorem 1. *The sectional curvature of G at e is*

$$R_{\xi\eta} = \langle\delta,\delta\rangle + 2\langle\alpha,\beta\rangle - 3\langle\alpha,\alpha\rangle - 4\langle B_\xi, B_\eta\rangle, \tag{9}$$

where

$$2\alpha = [\xi,\eta], \qquad 2\beta = B(\xi,\eta) - B(\eta,\xi),$$
$$2\delta = B(\xi,\eta) + B(\eta,\xi), \qquad 2B_\xi = B(\xi,\xi), \qquad 2B_\eta = B(\eta,\eta)$$

and the operation B is defined by (5).

4. Applications to diffeomorphism groups

Let $G = \mathrm{S\,Diff}\,\mathscr{D}$ be the group of diffeomorphisms of a Riemannian domain $\mathscr{D}$ which preserve the volume element. The algebra $\mathfrak{A}$ consists of vector fields $\vec{v}$ in $\mathscr{D}$ such that $\operatorname{div}\vec{v} = 0$, and satisfying $\vec{v}\cdot\vec{n} = 0$ on the boundary $\partial\mathscr{D}$ of $\mathscr{D}$. Let us define [3] a metric on $\mathfrak{A}$ by

$$\langle\vec{u},\vec{v}\rangle = \int_{\mathscr{D}} \vec{u}\cdot\vec{v}\,dx,$$

where dx is the volume element of $\mathscr{D}$.

In order to calculate the curvature (9), let us write down an explicit expression for the operator B. The expression for B is particularly simple if $\mathscr{D}$ is a domain in the Euclidean space E^3.

Denote by $\vec{u}\cdot\vec{v}$ the dot product, and by $\vec{u}\wedge\vec{v}$ the vector cross-product.

Theorem 2. *Let* $\vec{u},\vec{v}\in\mathfrak{A}$, *then*

$$\vec{B}(\vec{u},\vec{v}) = (\overrightarrow{\operatorname{curl}}\,\vec{u})\wedge\vec{v} + \overrightarrow{\operatorname{grad}}\,\alpha, \tag{10}$$

where α *is the function determined by the conditions:* $\operatorname{div}\vec{B} = 0$ *in* $\mathscr{D}$, $\vec{B}\cdot\vec{n} = 0$ *on* $\partial\mathscr{D}$.

If $\mathscr{D}$ is a domain in the (x,y)-plane, one can identify the algebra $\mathfrak{A}$ of vector fields $\vec{v}$ with the algebra of stream functions $\psi(x,y)$:

$$v_1 = \frac{\partial\psi}{\partial y}, \qquad v_2 = -\frac{\partial\psi}{\partial x} \tag{11}$$

with $[\psi_1,\psi_2]$ the Jacobian of the functions ψ_1 and ψ_2.

Using these notations, according to (10) the field $\vec{B}(\psi_1,\psi_2)\in\mathfrak{A}$ is given by

$$\vec{B}(\psi_1,\psi_2) = -\Delta\psi_1\,\overrightarrow{\operatorname{grad}}\,\psi_2 + \overrightarrow{\operatorname{grad}}\,\alpha, \tag{12}$$

---

[3] It follows from the principles of mechanics that geodesics on G describe the flow of an ideal fluid in $\mathscr{D}$.

where

$$\Delta = \frac{\partial^2}{\partial x^2} + \frac{\partial^2}{\partial y^2}.$$

Example 2. Let $\mathscr{D} = \mathrm{T}^2$ be any torus equipped with a flat metric. Consider at the identity of $G = \mathrm{S\,Diff}\,\mathscr{D}$ the two-dimensional plane spanned by the two stream functions

$$\psi_1 = \cos(\vec{k}_1 \cdot \vec{x}), \qquad \psi_2 = \cos(\vec{k}_2 \cdot \vec{x}), \tag{13}$$

where $\vec{k}_1$ and $\vec{k}_2$ are the "wave vectors".

Using (9), (11) and (12), one obtains

Theorem 3. *The sectional curvature of the group* $\mathrm{S\,Diff\,T}^2$ *at* e *in the direction of the plane* (13) *is*

$$R = -\frac{k_1^2 + k_2^2}{4S} \sin^2 \varphi \sin^2 \varphi', \tag{14}$$

where S *is the area of the torus,* φ *is the angle between* $\vec{k}_1$ *and* $\vec{k}_2$*, and* φ' *is the angle between* $\vec{k}_1 + \vec{k}_2$ *and* $\vec{k}_1 - \vec{k}_2$*.*

In particular, if $\mathrm{T}^2 = \{x \mod 2\pi, y \mod 2\pi\}$*, then the curvature of* $\mathrm{S\,Diff\,T}^2$ *in the two-dimensional direction defined by the vector fields* $\vec{u}$ *with components* $\sin y$ *and* 0 *and* $\vec{v}$ *with components* 0 *and* $\sin x$ *is* $R = -1/8\pi^2$*.*

References

[1] I. HADAMARD, *J. Math. pures et appl.*, 5e série, **4**, 1898, p. 27–73.
[2] L. EULER, *Theoria motus corporum solidorum seu rigidorum*, 1765.
[3] J.-J. MOREAU, *Comptes rendus*, **249**, 1959, p. 2156.
[4] V. JUDOVIČ, *C. R. Acad. U. R. S. S.*, **136**, 1961, p. 564.
[5] V. ARNOLD, *Journal de Mécanique* (in press).
[6] L. AUSLANDER, L. GREEN and F. HAHN, *Ann. Math. Studies*, **53**, 1963.
[7] J. MILNOR, *Ann. Math. Studies*, **51**, 1963.

TOPOLOGIE. — *Sur la topologie des écoulements stationnaires des fluides parfaits.* Note (*) de M. **Vladimir Arnold**, présentée par M. Jean Leray.

On considère les écoulements stationnaires d'un fluide parfait, incompressible et non visqueux, dans un domaine borné D. On suppose que le vecteur vitesse n'est pas partout colinéaire au vecteur rotation. On démontre alors que le domaine D est divisé, par certaines surfaces et courbes, en un nombre fini de « cellules » ouvertes, fibrées en tores ou en cylindres engendrés par des lignes de courant. Les lignes de courant sont fermées sur les cylindres, fermées ou denses sur les tores.

1. Soit M une variété riemannienne analytique réelle, orientée, compacte, connexe, à trois dimensions, de premier nombre de Betti nul :

$$b_1(M) = \dim H^1(M, \mathbf{R}) = 0. \tag{1}$$

Soient a et b deux champs de vecteurs tangents à M, analytiques, et de divergence nulle :

$$\operatorname{div} a = 0, \qquad \operatorname{div} b = 0. \tag{2}$$

Théorème 1. — *Si les champs de vecteurs a et b, à divergence nulle, commutent :*

$$[a, b] = 0, \tag{3}$$

et ne sont pas partout colinéaires :

$$a \wedge b \not\equiv 0, \tag{4}$$

alors, presque toutes les trajectoires du champ a sont fermées ou partout denses sur des tores T^2 *analytiques plongés dans* M. *Les autres trajectoires forment un vrai sous-ensemble analytique compact de* M.

Précisons les notations utilisées. La trajectoire du champ a issue de $x \in M$ est, par définition, la courbe : $T_a(x, t) = T_a(t) = T(t)$, $t \in \mathbf{R}$, définie par

$$\left.\frac{dT}{dt}\right|_{t=0} = a(x), \qquad T_a(x, 0) = x. \tag{5}$$

A chaque vecteur tangent a on peut associer un opérateur différentiel θ_a, une 1-forme ω_a^1, et une 2-forme ω_a^2 :

$$\theta_a \varphi = \left.\frac{d}{dt}\right|_{t=0} \varphi(T_a(t)); \qquad \omega_a^1 \xi = \langle a, \xi \rangle; \qquad \omega_a^2(\xi, \eta) = \langle a, \xi \wedge \eta \rangle, \tag{6}$$

où φ est une fonction dérivable arbitraire sur M, ξ un vecteur tangent arbitraire, (ξ, η) un bivecteur tangent arbitraire. La métrique riemannienne de M définit un produit scalaire $\langle \xi, \eta \rangle$ sur l'espace tangent à M, et, avec l'orientation de M, un produit vectoriel $\xi \wedge \eta$ sur l'espace tangent.

Originally published in: C. R. Acad. Sc. Paris 261, 17-20, © French Acad. Sciences, Paris, 1965
Reprint by Springer-Verlag Berlin Heidelberg 2014 with kind permission, DOI 10.1007/978-3-642-31031-7_3

Avec les notations (6), on peut définir les notions de crochet de Poisson [,], gradient, rotationnel, divergence :

$$(7)\qquad \begin{cases} \theta_{[a,b]} = \theta_a\theta_b - \theta_b\theta_a, & \langle \operatorname{grad}\varphi;\, b\rangle = \theta_b\varphi, \\ \omega^2_{\operatorname{rot} a} = d\omega^1_a, & (\operatorname{div} a)\,\tau = d\omega^2_a, \end{cases}$$

où τ est l'élément de volume de M.

Nous utiliserons l'identité classique :

$$(8)\qquad \operatorname{rot}(a \wedge b) = [b, a] + a.\operatorname{div} b - b.\operatorname{div} a.$$

2. *Démonstration du théorème* 1. — Les deux champs commutant a et b déterminent l'action du groupe $\mathbf{R}^2$ sur M. Étudions les orbites F de ce groupe. Nous allons montrer qu'elles sont compactes, et par conséquent toriques. D'après (2), (3), (8), $\operatorname{rot} a \wedge b = 0$. Il existe donc une fonction α analytique, univalente [d'après (1)], non constante [d'après (4)], telle que :

$$(9)\qquad a \wedge b = \operatorname{grad} \alpha.$$

Ainsi, chaque orbite F est contenue dans une surface de niveau de la fonction α. Puisque M est compacte, α analytique, α n'a qu'un nombre fini de valeurs critiques. D'après (4) les « surfaces » de niveau critiques forment un vrai sous-ensemble analytique de M. Soit C une composante connexe d'une surface de niveau non critique de α. Alors, sur C, $\operatorname{grad}\alpha \neq 0$; par suite, d'après (9), les champs de vecteurs a et b sont linéairement indépendants sur C. Il en résulte que C = F, qui est donc une orbite non dégénérée du groupe $\mathbf{R}^2$. Comme surface de niveau, C = F est compacte; c'est donc un tore.

Les trajectoires du champ a sont les orbites de $R \subset \mathbf{R}^2$. Cela achève la démonstration car on sait que ces orbites sont fermées ou partout denses sur le tore.

3. *Extension aux variétés à bord.* — Le théorème 1 est encore vrai si M est une variété à bord ∂M analytique, et si le champ a est tangent à ∂M.

Preuve. — Soit α_0 une valeur critique telle qu'il existe un point x dans M où $\operatorname{grad}\alpha = 0$ et $\alpha(x) = \alpha_0$, ou telle qu'il existe un point x sur ∂M où $\operatorname{grad}\alpha$ est orthogonal à ∂M et $\alpha(x) = \alpha_0$. On sait qu'il n'existe qu'un nombre fini de valeurs critiques α_0. Les surfaces de niveau non critiques sont maintenant des sous-variétés (éventuellement à bord) de M. Si C est une composante connexe de surface de niveau non critique, de bord ∂C, alors ∂C est formé d'un nombre fini de trajectoires fermées sur ∂M. Soit $x \in \partial$C,

$$(10)\qquad T_a(x, t+u) \equiv T_a(x, t)$$

la composante de ∂C contenant x. Sur $C \cup \partial C$, les vecteurs a et b sont partout linéairement indépendants. Chaque point z de C est donc de la forme : $z = T_{ta+sb}(x, 1)$.

D'après (3), (10) :

$$(11)\qquad T_a(z, u) = T_{(t+u)a+sb}(x, 1) = T_{ta+sb}(T_a(x, u), 1) = z.$$

La trajectoire issue de z est donc fermée, ce qui démontre le théorème. La formule (11) montre aussi que C est un cylindre : $\{(t, s)\ (t \bmod u, s \in [0, S)\}$.

4. *Applications aux fluides parfaits.* — Soit D un domaine riemannien, compact, connexe, à trois dimensions, de bord ∂D. Par exemple, une partie bornée de l'espace euclidien E^3.

THÉORÈME 2. — *Soit v le champ de vitesse d'un écoulement stationnaire d'un fluide parfait dans* D(v, D, ∂D *sont analytiques réels*). *Si v n'est pas partout colinéaire au vecteur rotation, alors presque toutes les lignes de courant sont fermées ou partout denses sur des tores analytiques réels plongés dans* D; *les autres lignes de courant forment un vrai sous-ensemble analytique compact de* D.

Remarque. — Il est probable que les écoulements tels que $\operatorname{rot} v = \lambda v$, $\lambda = \mathrm{Cte}$ [3], ont des lignes de courant à la topologie compliquée. De telles complications interviennent en mécanique céleste [*voir* [1], *fig.* 6]. La topologie des lignes de courant des écoulements stationnaires des fluides visqueux peut être semblable à celle de [1], *fig.* 6.

Démonstration du théorème 2. — Le champ de vitesse d'un fluide parfait de E^3 vérifie l'équation d'Euler-Newton :

$$(12)\qquad \frac{dv}{dt} = -\operatorname{grad} p, \qquad \operatorname{div} v = 0, \qquad \text{où} \quad \frac{dv}{dt} = \frac{\partial v}{\partial t} + \frac{\partial v}{\partial x} v, \quad p \text{ pression.}$$

Dans E^3, (12) équivaut à :

$$(13)\qquad \frac{\partial v}{\partial t} = v \wedge \operatorname{rot} v - \operatorname{grad} \alpha, \qquad \operatorname{div} v = 0, \qquad \text{où} \quad \alpha = p + \frac{v^2}{2},$$

que j'appelle équation de Bernoulli. L'équation de Bernoulli (13) est encore valide pour les écoulements dans un espace de Riemann D [*voir* [2]] : par définition, les écoulements d'un fluide parfait dans D sont les géodésiques du groupe S Diff D des difféomorphismes de D conservant l'élément de volume.

Pour les écoulements stationnaires (13) se réduit à

$$(14)\qquad v \wedge \operatorname{rot} v = \operatorname{grad} \alpha, \qquad \operatorname{div} v = 0.$$

En utilisant (8), on voit que :

$$(15)\qquad [v, \operatorname{rot} v] = 0.$$

Le théorème 2 se ramène donc au théorème 1 appliqué à v et rot v. Remarquons que la restriction topologique (1) n'est pas nécessaire pour le théorème 2, elle est remplacée par (14).

(*) Séance du 28 juin 1965.

(1) V. Arnold, *Russian mathematical surveys*, 18, n° 6, 1963, p. 91-192.

(2) V. Arnold, *Comptes rendus*, 260, 1965, p. 5668.

(3) Par exemple, sur $D = T^3 = \{x, y, z, \bmod 2\pi\}$:

$$\dot{x} = A \sin z + C \cos y, \qquad \dot{y} = B \sin x + A \cos z, \qquad \dot{z} = C \sin y + B \cos x.$$

(*Institut H. Poincaré*, 11, *rue Pierre-Curie, Paris*, 5e
et *Université B*-234, *Moscou*.)

Doklady 1965
Tom 162, No. 5

CONDITIONS FOR NONLINEAR STABILITY OF STATIONARY PLANE CURVILINEAR FLOWS OF AN IDEAL FLUID

V. I. ARNOL'D

1. The present article demonstrates some hydrodynamical corollaries of three simple theorems in the theory of ordinary differential equations. In particular, we shall derive sufficient conditions for the stability of stationary, plane, curvilinear flows of an inviscid incompressible fluid with respect to finite perturbations.

The stability condition for infinitely small perturbations of plane-paralled flows was derived by Rayleigh: it suffices that the velocity profile have no points of inflection [1–3]. It turns out that Rayleigh's condition is sufficient also for stability with respect to finite perturbations. But, in addition to that, our results prove the stability of certain flows which have one inflection point.

The term perturbation refers here to a small change of the initial velocity field, consistent with the condition of incompressibility. All functions occurring in our discussion, including the perturbations, are assumed to be differentiable as many times as necessary. For the sake of simplicity we shall consider only perturbations which do not change the value of velocity circulation along each boundary.*

By stability we mean "Ljapunov stability". In other words, we shall prove that if the perturbation $\delta\psi$ is small at the initial time, then the velocity field of the perturbed motion is near to the unperturbed motion at all times ("nearness" is understood here in the sense of metric (11)).

2. **Conditional extrema and states of equilibrium.** The basis of Theorems 1 and 2 is the well-known argument which proves the stability of Eulerian rotation of a solid body around its large or small axis of inertia (cf. [4]). Let the system of differential equations

$$\dot{x} = f(x) \quad (x = x_1, \ldots, x_n) \tag{1}$$

have single-valued first integrals $E(x);\ F_1(x), \cdots, F_k(x)\ (1 \leq k < n)$. Consider the level set

$$F_i(x) = c_i \quad (i = 1, \ldots, k). \tag{2}$$

Let x_0 be the point of conditional extremum of the function E under conditions (2), i.e. with the appropriate Lagrange multipliers we have at the point x_0

$$dH = dE + \lambda_1 dF_1 + \ldots + \lambda_k dF_k = 0. \tag{3}$$

Theorem 1. *The point x_0 is a position of equilibrium of system (1) if one of the following two conditions holds:*

1) *x_0 is a point of conditional maximum or minimum (possibly only local),*

---

*In the case of plane-parallel flows it can easily be shown that stability with respect to such perturbations leads to stability with respect to all perturbations. In fact, a perturbation which changes the value of circulation can be regarded as a perturbation which does not change the circulation of some neighboring stationary flow. In an analogous manner, the restriction on total flow in Example 2 of §6 is immaterial.

Originally publ. in: Dokl. Akad. Nauk SSSR, 162:5, 975-978, © Russ. Acad. Sciences 1965
English transl.: Sov. Math. Dokl. 162, No. 5, 773-777, © American Math. Soc., Providence, RI, 1965
Reprint by Springer-Verlag Berlin Heidelberg 2014 with kind permission, DOI 10.1007/978-3-642-31031-7_4

2) *the extremum is nondegenerate, i.e. the quadratic form*

$$d^2H = d^2E + \lambda_1 d^2F_1 + \ldots + \lambda_k d^2F_k \tag{4}$$

is nondegenerate in the subspace $dF_1 = 0, \cdots, dF_k = 0$.

By Theorem 1, the following *variational principle* holds: a stationary flow has extremal energy with respect to flows with equimeasurable vorticity.

Theorem 2. *If both conditions of Theorem* 1 *are satisfied, then the position of equilibrium* $x = x_0$ *is stable.*

The following simple theorem is a generalization of the theorem of Poincaré-Ljapunov on the eigenvalues of Hamiltonian systems [5, 6].

Theorem 3. *If the linear real system* $\dot{x} = Ax$ *has as a first integral a non-degenerate quadratic from* (Bx, x), *then the set of the eigenvalues* λ *of the operator* A *is symmetric (taking account of multiplicity) with respect to the real and imaginary axes. The number of points* λ *which lie strictly in the left half-plane* $\mathrm{Re}\,\lambda < 0$ *does not exceed the smallest of the indices of inertia* ν_+, ν_- *of the quadratic form* (Bx, x).

3. **Equations of motion and their first integrals.** Let $\psi(x, y; t)$ be the stream function, let $v_1 = \partial\psi/\partial y$, $v_2 = -\partial\psi/\partial x$ be the components of the velocity v, and let curl $v = -\Delta\psi$ be the vorticity.

The region D, filled with an ideal fluid, will be assumed to be an annulus bounded by two smooth stationary curves Γ_1, Γ_2 (Figure 3). The velocity at the boundary is assumed to be parallel to the boundary. Thus at each moment of time t the stream function ψ assumes at each boundary $\Gamma_{1,2}$ a constant value $\psi_1 = 0$, $\psi_2 = \phi(t)$.

Taking the curl of both sides of the $\dot{v} = -\mathrm{grad}\ p$, we obtain

$$\partial\Delta\psi / \partial t = -[\nabla\Delta\psi, \nabla\psi], \text{ where } [\xi, \eta] = \xi_1\eta_2 - \xi_2\eta_1. \tag{5}$$

Equation (5) defines a dynamical system (1) in the space of functions $\psi(x, y)$.*

Equation (5) represents the conservation of vorticity of a fluid particle, $d\Delta\psi/dt = 0$. In view of the incompressibility of the fluid, this leads to the "equimeasurability" of vorticity: the measure of the set of points in D at which $\Delta\psi < a$, does not change in time. Therefore, for every function $\Phi(\xi)$ the functional

$$F = \iint \Phi(\Delta\psi)\, dx\, dy \tag{6}$$

in the space of $\psi(x, y)$ is a first integral of system (5). Another integral is obtained from the law of conservation of the energy E, where

$$2E = \iint (\nabla\psi)^2\, dx\, dy. \tag{7}$$

The states of equilibrium of system (5) are the stationary flows $\psi(x, y)$. For such flows, according to (5), $\nabla\Delta\psi$ and $\nabla\psi$ are collinear, i.e.

$$\psi = \Psi(\Delta\psi), \tag{8}$$

---

*As the region D is doubly-connected, the vorticity equation (5) does not uniquely define a solution $\psi(x, y; t)$ from the initial condition $\psi(x, y)$ and boundary conditions $\psi|_{\Gamma_1} = 0$, $\psi|_{\Gamma_2} = \phi(t)$. Equation (5) must be supplemented by the law of conservation of circulation $\oint_{\Gamma_1} v\, ds$, which follows from the fact that the pressure p is single-valued. The value of the flow rate $\phi(t)$ does not enter the boundary conditions, and must be calculated from the solution.

if $\nabla\Delta\psi \neq 0$, i.e. if the vorticity does not have points of extremum inside D.*

4. **Calculation of the first and second variation.** Consider the class of functions $\psi(x, y)$ which are constant on each of the boundary curves Γ_1, Γ_2 and satisfy the conditions

$$\oint_{\Gamma_1} \frac{\partial\psi}{\partial n}\, ds = C_1, \quad \oint_{\Gamma_2} \frac{\partial\psi}{\partial n}\, ds = C_2. \tag{9}$$

The corresponding flows have fixed values of the velocity circulation $C_{1,2}$ on each component of the boundary. We shall seek a conditional extremum of E at fixed F according to Lagrange's formula

$$\delta H = \delta \iint {}^1/_2 (\nabla\psi)^2 + \lambda\Phi(\Delta\psi)\, dx\, dy = 0. \tag{10}$$

Integrating by parts, we obtain, for $u = \psi - \lambda\Phi'(\Delta\psi)$

$$\delta H = -\iint_D \Delta u \delta\psi\, dx\, dy + \oint_\Gamma \frac{\partial u}{\partial n} \delta\psi\, ds + \oint_\Gamma \lambda\Phi' \frac{\partial\delta\psi}{\partial n}\, ds.$$

When $u \equiv 0$, and taking account of (9), we see that the right-hand side vanishes. This proves

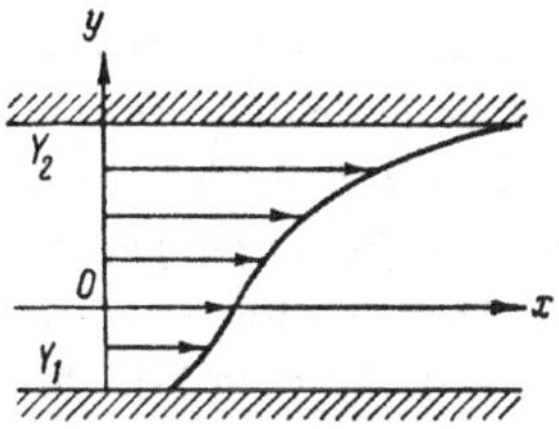

Figure 1

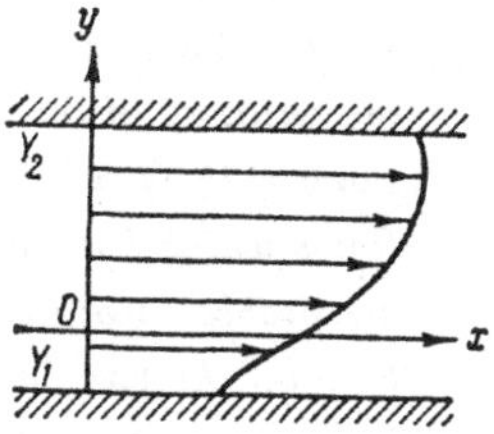

Figure 2

Theorem I. *The stationary flow* (8) *is a point of conditional extremum of E for fixed F and a point of absolute extremum of the functional $H = E + \lambda F$ defined by formulas* (6), (7), *where*

$$\lambda\Phi(\xi) = \int^{\xi} \Psi(\eta)\, d\eta.$$

Calculating the quadratic form $\delta^2 H$, we obtain the basic formula

$$2\delta^2 H = \iint (\nabla\delta\psi)^2 + \frac{\nabla\psi}{\nabla\Delta\psi} (\Delta\delta\psi)^2\, dx\, dy. \tag{11}$$

5. **Stability conditions.**

Theorem II. *The stationary flow defined by the stream function $\psi(x, y)$ is stable if form* (11) *is of fixed sign for $\delta\psi|_{\Gamma_1} = 0$, $\delta\psi|_{\Gamma_2} = C_0$.*

In fact, the first integral $H = E + \lambda F$ has ψ as its point of local maximum or minimum (in the class of smooth $\psi(x, y)$ with fixed boundary conditions (9)). This proves the stability of ψ with respect to smooth perturbations which preserve the value of circulation at the boundary.

If, without attempting a rigorous justification, we apply Theorem 3 to system (5) linearized about the stationary flow (8), we obtain

Theorem III. *The set of eigenvalues λ of the stability problem associated with a plane flow of an ideal fluid is symmetrical with respect to each of the axes: the real and the imaginary.*

---

*In the case of parallel flows this condition means that the velocity profile has no points of inflection.

Remark. In the case of a plane-parallel flow, this result can be proved by a straightforward analysis of the Orr-Sommerfeld equation for $\nu = 0$ (the fact that the equation is real corresponds to symmetry with respect to the imaginary axis of λ, and the invariance of the equation with respect to interchange of α and $-\alpha$ corresponds to symmetry with respect to $\lambda = 0$). It appears to the author that the spectrum is symmetric under conditions considerably more general than (8), e.g. in the case of three-dimensional flows of an inviscid fluid.

6. **Examples.** First we shall apply the results of §5 to flows parallel to the x axis in the strip $\{Y_1 \le y \le Y_2,\ x \bmod X\}$.* We then have

$$\psi = \psi(y), \quad \nabla\psi = v, \quad \Delta\psi = v', \ \nabla\Delta\psi = v''. \tag{12}$$

Example 1. Flows without points of inflection ($v'' \neq 0$).

Choose an inertial system of coordinates, in which the sign of v is everywhere equal to the sign of x''. In that case form (11) is positive definite. Thus *all flows which have no points of inflection are stable.*

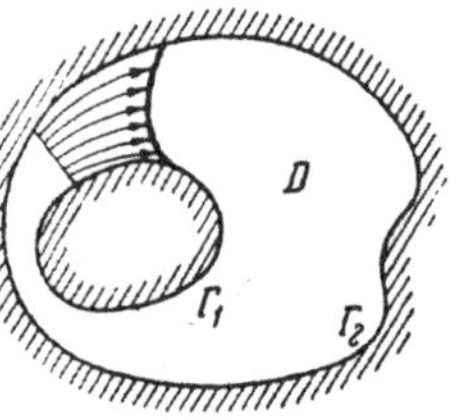

Figure 3

Example 2. Flows with one point of inflection ($v''(0) = 0$).

Assume that the velocity profile is symmetric with respect to the point of inflection. Choose an inertial system of coordinates, in which the velocity of the point of inflection is zero, so that $v(-y) = -v(y)$. Form (11) is positive definite if v and v'' are of the same sign. Thus *flows which have a velocity profile of the form represented in Figure 1 are stable.* For example, the flow with $v = a + by + cy^3$ is stable for $bc > 0$. As has been shown by Tollmien [7], this flow is unstable for $b = 0$, $Y_1 + Y_2 = 0$.

When v and v'' are of opposite sign, a sufficient condition for stability is that form (11) be negative definite. Consider those perturbations which also preserve, apart from circulation, the total flow rate of the unperturbed flow:** $\delta\psi|_\Gamma = 0$. With this boundary condition we have

$$\iint (\nabla\delta\psi)^2\,dx\,dy \leqslant \frac{(Y_2 - Y_1)^2}{\pi^2}\iint (\Delta\delta\psi)^2\,dx\,dy. \tag{13}$$

Inequality (13) leads to the stability of flows with

$$\left|\frac{v - v(0)}{v''}\right| > \frac{(Y_2 - Y_1)^2}{\pi^2}.$$

For example, *the flow with velocity* $v = a + b \sin y$ *is stable when* $Y_2 - Y_1 < \pi$ (Figure 2). As has been shown by Tollmien [7], this flow is unstable for $Y_2 - Y_1 > \pi$, $Y_2 + Y_1 = \pi$.

Example 3. Flow in a curvilinear annulus of general form.

Theorem II leads to the *stability of flows with a concave velocity profile* ($\nabla\psi/\nabla\Delta\psi > 0$, Figure 3).

When the velocity profile is convex ($\nabla\psi/\nabla\Delta\psi < 0$), then Theorem 3 and inequalities of type (13) lead to the *finiteness of the set of unstable eigenvalues* (Re $\lambda > 0$) *of the corresponding linear problem.*

---

* We consider the flow X to be periodic in x and the points $(x + X, y)$ as being identical with (x, y). The results are independent of X.

**It is clear from the law of conservation of momentum that in the case of such perturbation of the initial conditions the flow rate will remain equal to the flow rate of the unperturbed flow of all t.

The author expresses his sincere gratitude to A. L. Krylov and V. I. Judovič, for discussions which have led him to the present subject.*

Moscow State University Received 18/NOV/64

BIBLIOGRAPHY

[1] Lin Chia-chiao, *The theory of hydrodynamic stability*, Cambridge University Press, Cambridge, 1955; Russian transl., Moscow, 1958. MR 17, 1022.

[2] H. Schlichting, *Entstehung der Turbulenz*, Handbuch der Physik, Bd. 8/1, Strömungsmechanik I, Springer, Berlin, 1959, pp. 351–450; Russian transl., IL, Moscow, 1962. MR 21 #6836d; MR 26 #3319.

[3] Rayleigh, Scientific papers, Vol. 1, Cambridge University Press, Cambridge, 1880, p. 474.

[4] L. D. Landau and E. M. Livšic, *Mechanics*, Theoretical physics, Vol. I, Fizmatgiz, Moscow, 1958, Fig. 51; English transl., Pergamon Press, London and Addison-Wesley, Reading, Mass., 1960. MR 21 #985; MR 22 #11531.

[5] H. Poincaré, *Les méthodes nouvelles de la mécanique céleste*, Vol. 1, Paris, 1892, p. 193.

[6] A. M. Ljapunov, *General problem of stability of motion*, Kharkov, 1892, §51.

[7] W. Tollmien, Nachr. Akad. Wiss. Göttingen Math.-Phys. Kl. 50 (1935), 79.

[8] M. Rosenblatt and A. Simon, Phys. Fluids 7 (1964), no. 4, 557.

[9] R. Fjørtorf, Geofys. Publ. Norske Vid.-Akad. Oslo 17 (1950), no. 6. MR 14, 815.

Translated by:
A. Solan

Note added in proof. The author is grateful to L. A. Dikiĭ, who has brought to his attention two important papers [8, 9] which contain many results identical to the results of the present work.

ON THE TOPOLOGY OF THREE-DIMENSIONAL STEADY FLOWS OF AN IDEAL FLUID

(O TOPOLOGII TREKHMERNYKH STATSIONARNYKH TECHENII IDEAL'NOI ZHIDKOSTI)

PMM Vol. 30, No. 1, 1966, pp. 183-185

V.I. ARNOL'D
(Moscow)

(Received August 16, 1965)

We shall consider the rotational steady flows of an incompressible inviscid fluid in a bounded region D. It will be assumed that the vectors of velocity and vorticity are not everywhere colinear. It will be shown that the region of flow D is divided by the critical 'Bernoulli surfaces' into a finite number of cells, in each of which the streamlines are either closed, or else, everywhere they closely encircle toroidal surfaces.

1. The equations of motion. The Euler-Newton equation

$$\frac{d\mathbf{v}}{dt} = -\operatorname{grad} p, \quad \operatorname{div} v = 0 \qquad \left(\frac{dv}{dt} = \frac{\partial v}{\partial t} + \frac{\partial v}{\partial x} v\right) \tag{1.1}$$

is equivalent to the 'Bernoulli equation'

$$\frac{\partial \mathbf{v}}{\partial t} = [\mathbf{v}, \operatorname{curl} \mathbf{v}] - \operatorname{grad} \alpha, \quad \operatorname{div} v = 0 \qquad (\alpha = p + {}^1/{}_2 v^2) \tag{1.2}$$

For steady flow the Bernoulli equation takes the form

$$[v, \operatorname{curl} v] = \operatorname{grad} \alpha, \quad \operatorname{div} v = 0 \tag{1.3}$$

Let us make use of the well known identity of vector analysis

$$\operatorname{curl} [a, b] = \{b, a\} + a \operatorname{div} b - b \operatorname{div} a \tag{1.4}$$

Here $\{b, a\}$ is Poisson's bracket

$$\{b, a\}_i = \sum \frac{\partial a_i}{\partial x_j} b_j - \frac{\partial b_i}{\partial x_j} a_j$$

From the formulas (1.3) and (1.4) it follows that the velocity field of a steady flow commutates with its vorticity:

$$\{v, \operatorname{curl} v\} = 0 \tag{1.5}$$

We shall assume that the region of flow D is connected, finite and bounded by an

Originally publ. in: Prikl. Mat. Mekh. 30:1, 183-185, © Nauka, Moscow 1966
English transl.: J. Appl. Math. Mech. 30:1, 223-226, © Elsevier, Oxford, 1966
Reprint by Springer-Verlag Berlin Heidelberg 2014 with kind permission, DOI 10.1007/978-3-642-31031-7_5

analytical surface Γ; the boundary conditions are $(v, n)_\Gamma = 0$ (tangency).

2. Theorem. Let v be an analytic, steady velocity field, not everywhere colinear with its vorticity

$$[v, \text{curl } v] \not\equiv 0 \tag{2.1}$$

Then, almost all the streamlines are either closed or everywhere dense on two-dimensional toruses: all streamlines of other type fill a finite number of analytic sub-manifolds of D.

Note. To remove the condition (2.1) is probably impossible, since flows with curl $v \equiv \lambda v$ (λ = const) can evidently have streamlines with very complex topology, typical for the problems of celestial mechanics (see [1], fig. 6). Such intricate streamlines, however, can also exist in steady flows of a viscous fluid, closely resembling the flows of an ideal fluid. We notice, moreover, that formulas (1.1) to (1.5) and the theorem together with its proof are easily applicable to the case of flow of an ideal fluid in three-dimensional Riemann space (see [2]).

3. Proof. Let us consider the level surfaces of the function α (see (1.3)). The connected components of these surfaces will be called Bernoulli surfaces. The streamlines and lines of vorticity, according to (1.3), are orthogonal to grad α and therefore lie on the Bernoulli surfaces. We shall show that the majority of Bernoulli surfaces are toruses or rings.

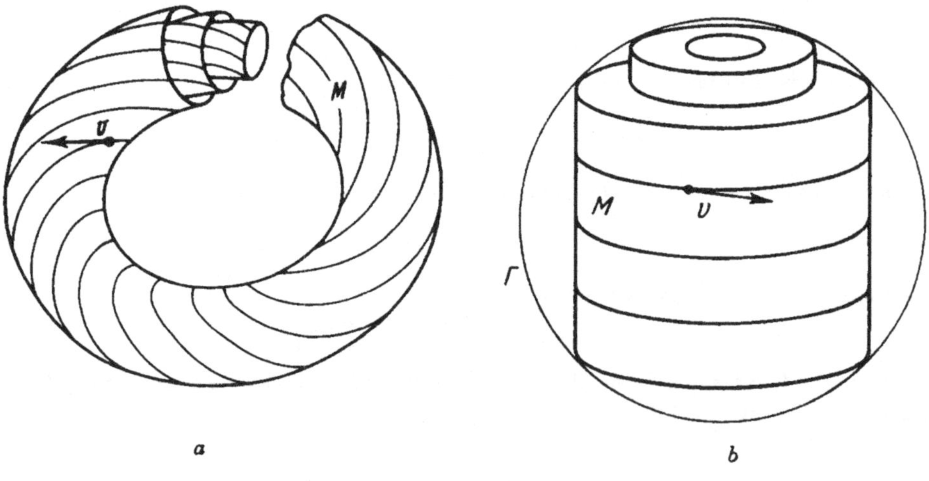

FIG. 1 *a*, *b*

We shall call the value α_0 *poor* if there exists a point x in the region D where grad $\alpha = 0$ and $\alpha(x) = \alpha_0$, or, if there exists a point x on the boundary Γ, at which grad α is orthogonal to Γ and $\alpha(x) = \alpha_0$. From the analycity of α and Γ it follows that poor values of α are finite in number. The points x at which the function α takes poor values form a finite number of analytic sub-manifolds of D of dimensionality not higher than 2 (since the function α is not constant, see (2.1)). These sub-manifolds can be called poor, whilst all the remaining Bernoulli surfaces are good.

The poor sub-manifolds divide the region D into cells, each of which is stratified by good Bernoulli surfaces. A good Bernoulli surface, not intersecting with the boundary of the region Γ, is a closed smooth two-dimensional surface, since grad $\alpha \neq 0$ on it. It turns out that this surface is a torus (see the case (1) and fig. 1*a*).

A good Bernoulli surface intersecting with the boundary of the region Γ intersects with it transversely (since on the boundary grad α is not orthogonal to Γ). Therefore such a surface is smooth, with a boundary consisting of a finite number of smooth closed curves lying on Γ. It turns out that this surface is a ring (see case (2) and fig. 1*b*).

Case (1). Let M be on unbounded Bernoulli surface. Let us construct on M a system of angular co-ordinates α, β (mod 2π) so that the streamlines would have the equation $d\alpha \,/\, d\beta = \lambda = \text{const}.$ This proves that M is a torus. But on the torus the lines $d\alpha \,/\, d\beta = \lambda$ are closed if λ is a rational number, and everywhere dense if λ is irrational. Therefore the theorem in case (1) is fully proved if the co-ordinate α, β can be constructed.

Let us consider a system of ordinary differential equations in $y\,(\tau,\, x,\, \sigma)$

$$\frac{dy}{d\tau} = s \operatorname{curl} v\,(y) + tv,\;\; y\,(0,\, x,\, \sigma) = x,\;\; \sigma = (s,\, t)$$

Here the parameter x is a point on the Bernoulli surface M, whilst σ is a point on the s, t-plane. Since the vectors v and curl v touch M, the point y lies on the same Bernoulli surface as x. When x is fixed, the formula

$$p_x(\sigma) = y\,(1,\, x,\, \sigma) \tag{3.1}$$

determines the mapping of the σ plane onto the Bernoulli surface M. From (1.5) the relation of commutativity follows

$$p_{p_x(\sigma)}\,(\sigma') = p_x\,(\sigma + \sigma') = p_{p_x(\sigma')}\,(\sigma) \tag{3.2}$$

Since the vectors v and curl v are linearly independent on M, the mapping (3.2) has a good overlap (i.e. the local value of σ can be taken as a co-ordinate on M). In fact, however, there are many points σ overlapping x. These points form, according to (3.2), a 'lattice' (if $p_x\;(\sigma) = p_x\;(\sigma') = x$, then also $p_x\;(\sigma + \sigma') = x$). From the commutativity of the Bernoulli surface M it follows, that this lattice has two generators σ_1 and σ_2 (two points of the plane such that any σ overlapping the point x has the form $m\sigma_1 + n\sigma_2$ with integral m and n). Let us make on the plane σ a linear substitution of the variables s and t by α and β, so that the co-ordinates of the points σ_1 and σ_2 would be $(2\pi, 0)$ and $(0, 2\pi)$. It is easy to see that α, β (mod 2π) are the required angular co-ordinates on the Bernoulli surface M. Hence the theorem is proved for the case (1).

Case (2). Let M be a Bernoulli surface with a boundary. The boundary of M consists of several closed streamlines lying on the boundary surface Γ (since the vector v is tangent to both M and Γ). Let x be a point on the boundary of M. Then, in the notation of (3.1), the closed streamline passing through x is

---

The above hypothesis was verified by M. Hennon by numerical experiment on the machine of the astrophysics institute in Paris.

$$p_x\ (0,\ t) = p_x\ (0,\ t+T) \quad (\infty < t < +\infty) \tag{3.3}$$

Let us put $z = p_x\ (s,\ t)$. Then, from the relations (3.2) and (3.3) it follows that

$$p_z\ (0,\ T) = p_x\ (s,\ t+T) = p_{p_x(0,T)}\ (s,\ t) = p_x\ (s,\ t) = z \tag{3.4}$$

i.e. the streamline passing through z is closed. But every point of M has the form $z = p_x\ (s,\ t)$ (in view of the linear independence of v and curl v and the connectedness of M). Therefore the formula (3.4) proves that all streamlines on M are closed. At the same time this formula introduces on M, the co-ordinates of the ring

$$t\ (\mathrm{mod}\ T),\ s,\quad 0 \leqslant s \leqslant S \ \text{ or } \ S \leqslant s \leqslant 0$$

Thus the proof of the theorem is completed.

BIBLIOGRAPHY

1. Arnol'd V.I. Malye znamenateli i problemy ustoichivosti dvizheniia v klassicheskoi i nebesnoi mekhanike (Small denominators and problems of stability of motion in classical and celestial mechanics). Uspekhi matem. nauk, Vol. 18, No. 6, pp 91-192, 1963.

2. Arnol'd V.I. Sur la topologie des écoulements stationaires des fluides parfaits. (Comptes Rendus Acad. Sci. Paris, Vol. 261, No. 1, 1965.

Translated by A.N.A.

ON AN APRIORI ESTIMATE IN THE THEORY OF HYDRODYNAMICAL STABILITY

UDC 517.917

V. I. ARNOL'D

This note contains a proof of the stability theorem stated in [1].

Let D be a domain on the (x, y)-plane bounded by the curves Γ_i. A solution $u(x, y; t)$ of the "vortex equation" $\partial\Delta u/\partial t = [\nabla u, \nabla\Delta u]$, where $[u, v] = u_1 v_2 - u_2 v_1$, with boundary conditions[1])

$$u|_{\Gamma_i} = c_i(t), \qquad c_1 \equiv 0, \qquad \frac{d}{dt}\oint_{\Gamma} \frac{\partial u}{\partial n}\, ds = 0.$$

is called the stream function of the flow of an ideal fluid in D.

Let $\psi(x, y)$ be the stream function of a stationary flow: $[\nabla\psi, \nabla\Delta\psi] = 0$. The vectors $\nabla\psi$ and $\nabla\Delta\psi$ are therefore collinear. We further assume that $\psi = \Psi(\Delta\psi)$; for this it is sufficient that $\nabla\Delta\psi \neq 0$. But $\psi = \Psi(\Delta\psi)$ in certain other cases also (see Example 2 in [1]). Let $u = \psi + \phi(x, y; t)$ be the stream function of another flow, where $\oint_{\Gamma_i}(\partial\phi/\partial n)\, ds = 0$ for $t = 0$; then by the law of conservation of circulations $\oint_{\Gamma_i}(\partial\phi/\partial n)\, ds = 0$ for all t. We assume, finally, that in D

$$c \leqslant \frac{\nabla\psi}{\nabla\Delta\psi} \leqslant C, \text{ where } 0 < c \leqslant C < \infty. \tag{1}$$

Theorem 1. *The perturbation $\phi(x, y; t)$ at any moment of time is estimated in terms of the initial perturbation $\phi_0 = \phi(x, y; 0)$ by*

$$\iint_D (\nabla\varphi)^2 + c\,(\Delta\varphi)^2\, dxdy \leqslant \iint_D (\nabla\varphi_0)^2 + C\,(\Delta\varphi_0)^2\, dxdy. \tag{2}$$

Proof. We set $\Phi = \int^{\xi}\Psi(\eta)\, d\eta$. Then $\Phi'' = \nabla\psi/\nabla\Delta\psi$, and thus for $\min\Delta\psi \le \xi \le \max\Delta\psi$ we have

$$c \le \Phi''(\xi) \le C. \tag{3}$$

We extend the definition of $\Phi(\xi)$ to cover the whole ξ axis subject to this inequality, and in what follows Φ denotes the function extended in this way. We form the functional

$$H_2(\varphi) = \iint_D \frac{(\nabla\varphi)^2}{2} + [\Phi(\Delta\psi + \Delta\varphi) - \Phi(\Delta\psi) - \Phi'(\Delta\psi)\,\Delta\varphi]\, dxdy. \tag{4}$$

---

1) The boundary conditions do not require the values of $c_i(t)$ but merely that c_i be independent of x and y.

Originally publ. in: Izv. Vyssh. Uchebn. Zaved. Mat. 5:54, 3-5, © Kazan State Univ. 1966
English transl.: Am. Math. Soc. Transl. (2) 79, 267-269, © American Math. Society, Providence, RI, 1969
Reprint by Springer-Verlag Berlin Heidelberg 2014 with kind permission, DOI 10.1007/978-3-642-31031-7_6

Lemma. *The functional H_2 is preserved:*

$$H_2(\varphi(x, y; t)) \equiv H_2(\varphi(x, y; 0)). \tag{5}$$

Proof. We consider the functional

$$H(u) = \iint_D \frac{1}{2}(\nabla u)^2 + \Phi(\Delta u)\, dxdy.$$

This functional is preserved by the laws of energy and vortex conservation: $H(u(x, y; t)) \equiv H(u(x, y; 0))$. Therefore $\hat{H}(\phi) = H(\psi + \phi) - H(\psi)$ is also a preserved functional:

$$\hat{H}(\varphi(x, y; t)) \equiv \hat{H}(\varphi(x, y; 0)). \tag{6}$$

We put $\hat{H}(\phi)$ in the form of a sum $\hat{H}(\phi) = H_1(\phi) + H_2(\phi)$, where

$$H_1(\varphi) = \iint_D \nabla\varphi\nabla\psi + \Phi'(\Delta\psi)\Delta\varphi\, dxdy,$$

$$H_2(\varphi) = \iint_D \frac{1}{2}(\nabla\varphi)^2 + [\Phi(\Delta\psi + \Delta\varphi) - \Phi(\Delta\psi) - \Phi'(\Delta\psi)\Delta\varphi]\, dxdy.$$

The first term is zero (this is Theorem 1 of [1]). For, on integrating by parts, we have

$$H_1(\varphi) = \iint_D (-\psi\Delta\varphi + \Phi'\Delta\varphi)\, dxdy + \oint_\Gamma \psi \frac{\partial\varphi}{\partial n}\, ds.$$

But $\Phi' = \Psi$, the $\psi|_{\Gamma_i}$ are constant and $\oint_{\Gamma_i}(\partial\phi/\partial n)\, ds = 0$. Hence $H_1(\phi) \equiv 0$. Therefore $\hat{H}(\phi) = H_2(\phi)$ and, in accordance with (6), H_2 is preserved. This proves the lemma.

Returning to the proof of the theorem, we note that it follows from (3) that for any h

$$c\frac{h^2}{2} \leqslant \Phi(\xi + h) - \Phi(\xi) - \Phi'(\xi)h \leqslant C\frac{h^2}{2}.$$

Hence

$$H_2(\varphi(t)) \geqslant \iint_D \frac{(\nabla\varphi)^2}{2} + c\frac{(\Delta\varphi)^2}{2}\, dxdy, \tag{7}$$

$$H_2(\varphi(0)) \leqslant \iint_D \frac{(\nabla\varphi_0)^2}{2} + C\frac{(\Delta\varphi_0)^2}{2}\, dxdy. \tag{8}$$

On comparing (5), (7) and (8) we obtain (2), which we needed to prove. Estimate (2) implies the stability of stationary flows with $\nabla\psi/\nabla\Delta\psi > 0$.

Now let a stationary flow be such that $c \leq -\nabla\psi/\nabla\Delta\psi \leq C$, $0 < c < C < \infty$.

Theorem 2. *The perturbation* $\phi(x, y; t)$ *is estimated in terms of* $\phi(x, y; 0)$ *by*

$$\iint_D c\,(\Delta\varphi)^2 - (\nabla\varphi)^2\, dxdy \leqslant \iint_D C\,(\Delta\varphi_0)^2 - (\nabla\varphi_0)^2\, dxdy. \tag{9}$$

Proof. Let $\Phi(\xi) = \int^{\xi}\Psi(\eta)\,d\eta$ be extended again over the whole ξ axis with $c < -\Phi'' < C$. Then in place of (7) and (8) we obtain

$$\left.\begin{aligned} -H_2(\varphi(t)) &\geqslant \iint_D c\,\frac{(\Delta\varphi)^2}{2} - \frac{(\nabla\varphi)^2}{2}\, dxdy,\\ -H_2(\varphi(0)) &\leqslant \iint_D C\,\frac{(\Delta\varphi_0)^2}{2} - \frac{(\nabla\varphi_0)^2}{2}\, dxdy, \end{aligned}\right\} \tag{10}$$

which, together with (5), gives (9). The theorem is proved.

If $\iint_D c(\Delta\phi)^2 - (\nabla\phi)^2 dxdy$ is positive definite, for a certain $\theta > 0$ we shall have

$$\iint_D c\,(\Delta\varphi)^2 - (\nabla\varphi)^2\, dxdy \geqslant \theta c \iint_D (\Delta\varphi)^2\, dxdy.$$

Therefore it follows from (9) that

$$\iint_D (\Delta\varphi)^2\, dxdy \leqslant \frac{C}{c\theta} \iint_D (\Delta\varphi_0)^2\, dxdy,$$

which expreses the stability of the stationary flow ψ.

Remark. The statement of Theorem II in [1] contains an error. It asserts that a sufficient condition for stability is that

$$\iint_D (\nabla\varphi)^2 + \frac{\nabla\psi}{\nabla\Delta\psi}\,(\Delta\varphi)^2\, dxdy \tag{11}$$

should be *of fixed sign* with respect to ϕ (where $\phi|_{\Gamma_i} = c_i$ and $\oint_{\Gamma_i}(\partial\phi/\partial n)\, ds = 0$). Stability is in fact proved when (11) is *positive definite* (Theorem 1 above) and when

$$\iint_D (\nabla\varphi)^2 + \left(\max\frac{\nabla\psi}{\nabla\Delta\psi}\right)(\Delta\varphi)^2\, dxdy$$

is negative definite (Theorem 2 above). In specific examples (§6 of [1]) Theorem II was used in this form. If, however, (11) is negative definite we have stability in the linear approximation.

BIBLIOGRAPHY

[1] V. I. Arnol'd, *On conditions for non-linear stability of plane stationary curvilinear flows of an ideal fluid*, Dokl. Akad. Nauk SSSR 162 (1965), 975–978 = Soviet Math. Dokl. 6 (1965), 773–777. MR 31 #4288.

Translated by T. Garrett

On the differential geometry of infinite-dimensional Lie groups and its applications to the hydrodynamics of perfect fluids*

V. Arnold
Translated by Alain Chenciner

In the year 1765, L. Euler [8] published the equations of rigid body motion which bear his name. It does not seem useless to mark the 200th anniversary of Euler's equations by a modern exposition of the question.

The eulerian motions of a rigid body are the geodesics on the group of rotations of three dimensional euclidean space endowed with a left invariant metric. Basically, Euler's theory makes use of nothing but this circumstance; hence Euler's equations still hold for an arbitrary group. For the other groups, one obtains the "Euler equations" of rigid body motion in the n-dimensional space, the equations of the hydrodynamics of ideal fluids, etc.

Euler's theorem on the stability of the rotations around the longest and shortest axes of the inertia ellipsoid also has analogues in the case of an arbitrary group. In the case of hydrodynamics, this analogy is an extension of Rayleigh's theorem on the stability of flows whose velocity profile is inflexion free (see §10).

As another application of Euler's theory, we prove in §8, the explicit formula of the riemannian curvature of a group endowed with a left invariant metric. In §11, this formula is used in the study of the curvature of the group of diffeomorphisms, whose geodesics are ideal fluid flows.

In what follows, I tried, following the call of Bourbaki [6], to always substitute blind computations for Euler's lucid ideas.

1 Modern notations

Let G be a Lie group, $\mathfrak{U}$ its Lie algebra. A curve $g(t)$ is a mapping $g : \mathbb{R} \to G$. The velocity vector $\dot{g} = dg/dt$ belongs to the tangent space to G at the point g; we shall denote this tangent space TG_g. Obviously, $TG_e = \mathfrak{U}$.

Annales de l'Institut Fourier, tome 16 $n^o 1$ (1966), p. 319–361

[1] Most of the results have been announced in [1, 2, 3, 4].

Originally publ. in: Annales de L'Institut Fourier, Vol. 16, No. 1, 319-361, © Institut Fourier, Grenoble 1966
Translated by A. Chenciner. Springer-Verlag Berlin Heidelberg 2014 with kind permission, DOI 10.1007/978-3-642-31031-7_7

Let $g \in G$. The left and right translations L_g and R_g are mappings from the group G into G, defined by

$$L_g h = gh, \quad R_g h = hg, \quad (h \in G). \tag{1}$$

The mappings induced on the tangent spaces will be denoted

$$\tilde{L}_g : TG_h \to TG_{gh}, \quad \tilde{R}_g : TG_h \to TG_{hg}. \tag{2}$$

We shall denote Ad_g the mapping from the algebra $\mathfrak{U}$ to itself

$$Ad_g \xi = \tilde{R}_{g^{-1}\tilde{L}_g \xi}, \quad (\xi \in \mathfrak{U}). \tag{3}$$

We shall denote $exp. : \mathfrak{U} \to G$ the natural map from the algebra into the group;

$$\exp(ft) = g(t), \quad (t \in \mathbb{R},\ f \in \mathfrak{U},\ g \in G) \tag{4}$$

is a one parameter group whose velocity vector is $dg/dt|_{t=0} = f$. We shall denote, for $\xi \in \mathfrak{U}, \eta \in \mathfrak{U}$

$$[\xi, \eta] \in \mathfrak{U}$$

the algebra commutation operation, which is bilinear and defined by

$$\exp[\xi, \eta] = \exp\xi \exp\eta \exp(-\xi)\exp(-\eta) + O(\xi^2) + O(\eta^2).$$

From the definition of the commutator and from (3) it follows that:

$$Ad_{\exp(ft)}\xi = \xi + t[f, \xi] + O(t^2), \quad (t \to 0). \tag{5}$$

The commutator is antisymmetric and satisfies the Jacobi identity:

$$[\xi + t\eta, \zeta] = [\xi, \eta] + t[\eta, \zeta], \quad [\xi, \eta] = -[\eta, \xi]; \quad (\xi, \eta, \zeta \in \mathfrak{U}),$$

$$[[\xi, \eta], \zeta] + [[\eta, \zeta], \xi] + [[\zeta, \xi], \eta] = 0 \tag{6}$$

The operator Ad_g is an algebra automorphism; for g variable, the operators Ad_g form an antirepresentation of the group:

$$Ad_g[\xi, \eta] = [Ad_g\xi, Ad_g\eta]; \quad Ad_{gh} = Ad_h Ad_g. \tag{7}$$

The dual of the tangent space TG_g will be called the cotangent space and denoted T^*G_g. An element $\xi \in T^*G_g$ is a linear form on TG_g; its value on η, will be denoted

$$(\xi, \eta) \in \mathbb{R}, \quad \xi \in T^*G_g, \quad \eta \in TG_g.$$

The conjugate operators of $\tilde{L}_g, \tilde{R}_g$ will be denoted

$$\tilde{L}_g^* : T^*G_{gh} \to T^*G_h, \quad \tilde{R}_g^* : T^*G_{hg} \to T^*G_h.$$

These operators are defined by

$$(\tilde{L}_g^*\xi,\eta) = (\xi,\tilde{L}_g\eta), \quad (\tilde{R}_g^*\xi,\eta) = (\xi,\tilde{R}_g\eta).$$

Finally, the operator $Ad_g^* : \mathfrak{U}^* \to \mathfrak{U}^*$ is defined by

$$(Ad_g^*\xi,\eta) = (\xi,Ad_g\eta). \tag{8}$$

Let $\mathfrak{L} : \mathfrak{U}^* \to \mathfrak{U}^*$ be a symmetric operator:

$$(\mathfrak{L}\xi,\eta) = (\mathfrak{L}\eta,\xi). \quad \xi,\eta \in \mathfrak{U}.$$

Let us define the symmetric operator $\mathfrak{L}_g : TG_g \to T^*G_g$ by left translation:

$$\mathfrak{L}_g\xi = \tilde{L}^*_{g^{-1}}\mathfrak{L}\tilde{L}_{g^{-1}}\xi.$$

The operator $\mathfrak{L}$ will be supposed to be positive definite, which means that the scalar product of $\xi,\eta \in TG_g$

$$\langle\xi,\eta\rangle_g = (\mathfrak{L}_g\xi,\eta) = (\mathfrak{L}_g\eta,\xi) = \langle\eta,\xi\rangle_g \tag{9}$$

is a positive definite form. The product $\langle\,,\,\rangle_g$ defines on G a left invariant riemannian metric.

The product $\langle\,,\,\rangle_g$ in the algebra will be denoted $\langle\,,\,\rangle$. Let us introduce the operation $B : \mathfrak{U} \to \mathfrak{U}$ defined by

$$\boxed{\langle[a,b],c\rangle = \langle B(c,a),b\rangle \quad \text{for all } b \in \mathfrak{U}.} \tag{10}$$

Of course, $B(c,a)$ depends bilinearly on c and a and, for fixed c, $\langle B(c,a),b\rangle$ is antisymmetric in a and b:

$$\langle B(c,a),b\rangle + \langle B(c,b),a\rangle = 0 \tag{11}$$

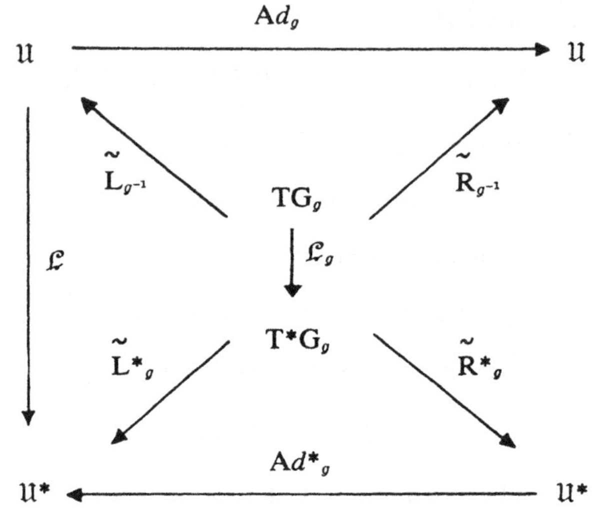

2 Classical notations

In rigid body mechanics, the objects defined above have the following names:

$G = SO(3)$	the configuration space;
g	rigid body configuration;
$g(t)$	rigid body motion;
$\dot{g}$	rotational velocity;
$\mathfrak{U} = \mathbb{R}^3$	space of rotation vectors;
$[\xi, \eta]$	vector product in $\mathbb{R}^3$;
$\omega_c = \tilde{L}_{g^{-1}}\dot{g} \in \mathfrak{U}$	rotation vector w.r.t. the body;
$\omega_s = \tilde{R}_{g^{-1}}\dot{g} = Ad_g\omega_c$	rotation vector w.r.t. space;
$\mathfrak{L}_g$	inertia operator;
$M = \mathfrak{L}_g\dot{g}$	angular momentum;
$\mathfrak{U}^*$	space of angular momenta;
$M_c = \tilde{L}_g^*M = \mathscr{L}\omega_c$	angular momentum w.r.t. the body;
$M_s = \tilde{R}_g^*M = Ad^*_{g^{-1}}M_c$	angular momentum w.r.t. space;
$T = \frac{1}{2}\langle\dot{g},\dot{g}\rangle_g = \frac{1}{2}\langle\omega_c,\omega_c\rangle = \frac{1}{2}\langle M_c,\omega_c\rangle = \frac{1}{2}(M,\dot{g})$	the kinetic energy.

The principle of least action asserts that the motions of a rigid body around a fixed point are (in the absence of exterior forces) the geodesics[2] of the group G endowed with a left invariant metric (9). To each geodesic $g(t)$ correspond 4 curves

$$\omega_c(t) \in \mathfrak{A}, \quad \omega_s(t) \in \mathfrak{A}, \quad M_c(t) \in \mathfrak{A}^*, \quad M_s(t) \in \mathfrak{A}^*. \tag{12}$$

3 Euler's theorems

For the curves (12) and $G = SO(3)$, Euler proved the following two theorems

Theorem 1. *The curves (12) satisfy the Euler equation*

$$\boxed{\frac{dM_c}{dt} = [M_c, \omega_c]} \tag{13}$$

Theorem 2. *The angular momentum with respect to space is invariant:*

$$\boxed{\frac{dM_s}{dt} = 0.} \tag{14}$$

With the notations of §1, one can write formula (13) in the form

$$\boxed{\frac{d\omega_c}{dt} = B(\omega_c, \omega_c)} \quad \omega_c = \tilde{L}_{g^{-1}}\dot{g}. \tag{15}$$

We shall prove formulæ (14) and (15) for the case of an arbitrary Lie group.

[2] A curve $g(t)$ in the riemannian space is a geodesic, if it is an extremal of the "action" integral $\frac{1}{2}\int \langle \dot{g}, \dot{g} \rangle dt$

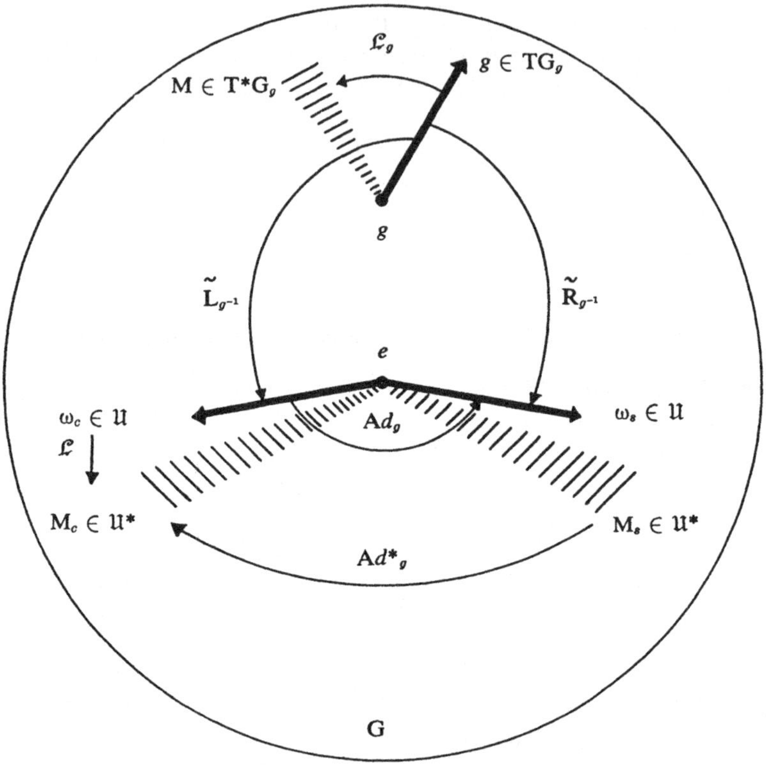

The tangent vectors to G are represented by straight arrows; the cotangent vectors are represented by series of parallel hatchings which represent the level planes of a corresponding 1-form on the tangent space.

4 Proof of Euler's first theorem

The left translate of a geodesic of a left invariant metric is also a geodesic. Hence, the derivative $d\omega_c/dt$ depends only on ω_c and not on g:

$$\frac{d\omega_c}{dt} = F(\omega_c).$$

In order to find the form of this universal function $F(\omega_c)$, it is sufficient to consider the geodesic $g(t)$ with $g(0) = e$, $\dot{g}(0) = \omega_c$.

In order to study this geodesic, let us write the Euler-Lagrange equation in a special coordinate system around the point $e \in G$.
In particular, the mapping $\exp : \mathfrak{U} \to G$ allows us to consider the neighborhood of 0 in $\mathfrak{U}$ as a chart of the neighborhood of e in G. The tangent space to $\mathfrak{U}$ at a, $T\mathfrak{U}_a$, is naturally identified with $\mathfrak{U}$.

Lemma 1. *Let us consider the operator $\tilde{L}_a$, induced by the left translation $L_{\exp a}$:*

$$\tilde{L}_a : \mathfrak{U} = T\mathfrak{U}_0 \to T\mathfrak{U}_a = \mathfrak{U}. \tag{16}$$

For this operator

$$\boxed{\tilde{L}_a \xi = \xi + \frac{1}{2}[a,\xi] + O(a^2)} \quad |a| \to 0. \tag{17}$$

Proof of lemma 1. It is known that all computations on Lie groups and Lie algebras may be done as if on a matrix group. For example, for any $a, b \in \mathfrak{U}$

$$(\exp a)(\exp b) = \exp\left(a + b + \frac{1}{2}[a,b] + O(a^2) + O(b^2)\right) \tag{18}$$

$$(a, b \to 0).$$

Setting $b = \xi t, t \to 0, |a| \to 0$ in (18) one finds

$$(\exp a)(\exp \xi t) = \exp\left[a + \left(\xi + \frac{1}{2}[a,\xi] + O(a^2)\right)t + O(t^2)\right], \ (t \to 0)$$

which is equivalent to (17). Lemma 1 is proved.

Proof of formula (15). Now, let q be cartesian coordinates in the algebra $\mathfrak{U}$, endowed with a metric $\langle,\rangle$. In a neighborhood of $e \in G$, on can use q as local coordinates. Let $g(t)$ be a curve in G. According to lemma 1, the coordinates $\dot{q} \in \mathfrak{U}$ of the velocity vector $\dot{g}$ are related to the vector[3] $\omega = \tilde{L}_{g^{-1}}\dot{g}$ by the formula

$$\omega = \dot{q} - \frac{1}{2}[q,\dot{q}] + O(q^2) \qquad |q| \to 0. \tag{19}$$

Hence, the Lagrange function $L(q,\dot{q})$ is given by the formula

$$2L = \langle \omega, \omega \rangle = \langle \dot{q}, \dot{q} \rangle - \langle \dot{q}, [q,\dot{q}] \rangle + O(q^2), \quad (|q| \to 0) \tag{20}$$

Using (10), (19) in order to compute the partial derivatives of L, one finds the impulse $p = \partial L / \partial \dot{q}$:

$$\begin{aligned} p &= \dot{q} - \frac{1}{2}[q,\dot{q}] - \frac{1}{2}B(\dot{q},q) + O(q^2) \\ &= \omega - \frac{1}{2}B(\omega,q) + O(q^2). \end{aligned} \tag{21}$$

According to (19), (20), (21), the Euler-Lagrange equation

$$\dot{p} = \frac{\partial L}{\partial q},$$

[3] Here and in the following computation, the index c is omitted in order to simplify the formulæ.

that is

$$\dot{p} = \frac{1}{2}B(\dot{q},\dot{q}) + O(q) = \frac{1}{2}B(\omega,\omega) + Oq) \tag{22}$$

becomes

$$\dot{\omega} - \frac{1}{2}B(\omega,\omega) = \frac{1}{2}B(\omega,\omega) + O(q). \tag{23}$$

When $q = 0$, (23) becomes (15). Theorem 1 is proved.

Remark. It follows also from (21), (22) that for a geodesic $q(t)$ with $q(0) = 0$, one has

$$\ddot{q}(0) = B(\dot{q}(0),\dot{q}(0)) = B(\omega,\omega). \tag{24}$$

5 Proof of Euler's second theorem

Euler's second theorem is a special case of "Noether's theorem". But, for what follows, it is useful to compute independently dM_s/dt. Here, in accordance with §2, the vector $M_s \in \mathfrak{U}^*$ is

$$M_s(t) = Ad_{g^{-1}(t)}\mathfrak{L}\omega_c(t).$$

Lemma 2. *Let $g(t)$ be a curve in the group G and ζ be an element of the algebra $\mathfrak{U}$. In the algebra, let us consider the curve defined by*

$$A(t) = Ad_{g^{-1}(t)}\zeta \tag{25}$$

Then

$$\frac{d}{dt}A(t) = [A(t),\xi(t)] \tag{26}$$

where

$$\xi(t) = \tilde{L}_{g^{-1}}\frac{dg}{dt}. \tag{27}$$

Proof of lemma 2. According to (27) one has

$$g(t+dt) = g(t)exp\big(\xi(t)dt + O(dt^2)\big).$$

Hence, substituting (25) in (3) and neglecting $O(dt^2)$, one finds

$$\begin{aligned} A(t+dt) = A_{g^{-1}(t+dt)}\zeta &= \tilde{R}_{exp(\xi dt)}\tilde{L}_{exp(-\xi dt)}\tilde{R}_g\tilde{L}_{g^{-1}}\zeta = \\ &= \tilde{R}_{exp(\xi dt)}\tilde{L}_{exp(-\xi dt)}A(t) = Ad_{exp(-\xi dt)}A(t). \end{aligned}$$

According to (5), this last expression is

$$Ad_{exp(-\xi dt)}A(t) = A(t) + [A,\xi]dt + O(dt^2)$$

which proves lemma 2.

Lemma 3. *If, in lemma 2, $g(t) = e^{ft}$, then*

$$A(t) = \zeta + t[\zeta, f] + \frac{t^2}{2}[[\zeta, f], f] + \cdots \tag{28}$$

Proof of lemma 3. The vector (27), $\xi(t) = f$ does not depend on t for $g(t) = e^{ft}$. Hence (26) gives

$$\frac{dA}{dt} = [A, f], \; \frac{d^2A}{dt^2} = [[A, f], f], \cdots$$

But as $A(0) = \zeta$, (28) is proved.

Remark. If $A(t) = Ad_{g(t)}\zeta$, then

$$\frac{d}{dt}A(t) = [A(t), \eta(t)], \tag{29}$$

where

$$\eta(t) = -\tilde{R}_{g^{-1}}\frac{dg}{dt}. \tag{30}$$

Because, in accordance with lemma 2 (29) holds with

$$\eta(t) = \tilde{L}_g\frac{d}{dt}g^{-1}(t), \quad \text{et} \quad \frac{d}{dt}g^{-1}(t) = -\tilde{R}_{g^{-1}}\tilde{L}_{g^{-1}}\frac{dg}{dt}.$$

Proof of formula (14). By definition,

$$M_s(t) = Ad^*_{g^{-1}(t)}\mathfrak{L}\omega_c(t) \in \mathfrak{U}^*. \tag{31}$$

Let us consider $\zeta \in \mathfrak{U}$. According to (8), (31), (9),

$$(M_s, \zeta) = (\mathfrak{L}\omega_c, Ad_{g^{-1}}\zeta) = \langle \omega_c, Ad_{g^{-1}}\zeta \rangle.$$

Now, theorem 1 and lemma 2 give

$$\frac{d(M_s, \zeta)}{dt} = \langle B(\omega_c, \omega_c), A \rangle + \langle \omega_c, [A, \omega_c] \rangle$$

with $A = Ad_{g^{-1}}\zeta$. Hence, by definition of the operation B (10),

$$d(M_s, \zeta)/dt = 0$$

for all ζ, which proves Euler's second theorem.

6 Stationary motions

Now, let us consider more carefully the system of Euler's equations

$$\boxed{\frac{d\omega}{dt} = B(\omega, \omega)} \tag{32}$$

which determine the evolution of the vector ω in the algebra $\mathfrak{U}$. Let v be a stationary point of system (32):

$$\boxed{B(v, v) = 0.} \tag{33}$$

Then, *the geodesic $g(t)$, originating from the point e with initial velocity $\dot{g}(0) = v$, is a subroup: $g(t) = expvt$.* Because according to (15), (33), for all t :

$$\tilde{L}_{g^{-1}} \frac{dg}{dt} = v = \tilde{L}_{exp(-vt)} \frac{dexp(vt)}{dt}.$$

Definition 1. The geodesic $g(t)$ is called a "stationary motion", if it is a one parameter subgroup.

We just proved that *the stationary motions are the only motions for which the "rotation vector with respect to the body" $\omega = \tilde{L}_g^{-1} \dot{g}$ is constant: $\dot{\omega} = 0$.*

For the classical rigid body ($G = SO(3)$) the stationary motions are the rotations around the great, the middle, or the small axis of the "inertia ellipsoid". In the general case, the stationary motions possess an interestsing property of extremality. In order to describe this property, we start with some remarks.

Lemma 4. *The system (32) admits the energy first integral*

$$E = \frac{1}{2}\langle \omega, \omega \rangle. \tag{34}$$

Proof. According to (32), (10)

$$\dot{E} = \langle \omega, \dot{\omega} \rangle = \langle \omega, B(\omega, \omega) \rangle = \langle [\omega, \omega], \omega \rangle = 0, \quad \text{C.Q.F.D.}$$

Definition 2. Under the name "foliation", I shall include every decomposition into equivalence classes of an arbitrary equivalence relation. I shall say that a foliation of the algebra $\mathfrak{U}$ is invariant, if with a point $\omega(0)$, the whole trajectory $\omega(t)$ of the system (32) belongs to one and the same leaf.

Example. According to lemma 4, the decomposition of $\mathfrak{U}$ into level surfaces of E is an invariant foliation. The leaves are the spheres $\langle \omega, \omega \rangle = 2E > 0$ and the point $\omega = 0$.

Euler's second theorem provides another example of an invariant foliation.

Definition 3. The vectors $\omega \in \mathfrak{U}$ and $\omega' \in \mathfrak{U}$ are "isorotating",

$$\omega \sim \omega'$$

if there exists $g \in G$ such that

$$Ad_g^* \mathfrak{L}\omega = \mathfrak{L}\omega'.^4 \tag{35}$$

Lemma 5. *The relation $\sim$ defines an invariant foliation in $\mathfrak{U}$.*

Proof. Obviously, $\omega \sim \omega$. Equation (7) implies the relations $(\omega \sim \omega') \Rightarrow (\omega' \sim \omega)$ and $(\omega \sim \omega' \sim \omega'') \Rightarrow (\omega \sim \omega'')$. Hence $\sim$ is an equivalence relation. The invariance of the leaves results from (14), (31), (35). Lemma 5 is proved.

In what follows, the words leaf, foliation, will indicate the foliation of the algebra $\mathfrak{U}$ into leaves of isorotating ω. Obviously, these leaves are the linear images of orbits of the adjoint action Ad_ω of the group in the algebra. [5]

Examples. For $G = SO(3)$ (rigid body) the leaves are concentric ellipsoids and the point

$$\langle\langle M_s(\omega), M_s(\omega)\rangle\rangle = C$$

where $\langle\langle,\rangle\rangle$ is the bi-invariant product in $\mathfrak{U}^*$. For $G = SO(n)$, the leaves are the "flag manifolds". For the group of homotheties and translations of a line, the leaves are two half-planes and all the points of the line which separates them.

Theorem 3. *The stationary points ν of the system (32) are the conditional extrema ω of the energy E on the leaves made of the ω's isorotating with ν.*

For the proof of theorem 3, we shall use some computations.

Lemma 6. *Let $\omega \in \mathfrak{U}, f \in \mathfrak{U}$, and*

$$\omega(t) = \mathfrak{L}^{-1} Ad_{e^{ft}}^* \mathfrak{L}\omega \tag{36}$$

Then :

$$\frac{d\omega}{dt} = B(\omega, f), \tag{37}$$

$$\omega(t) = \omega + tB(\omega, f) + \frac{t^2}{2} B(B(\omega, f), f) + \cdots \tag{38}$$

Proof. For any $\zeta \in \mathfrak{U}$, using (36), (9), (8), (9), one finds :

$$\begin{aligned} \langle \omega(t), \zeta\rangle = \langle \mathfrak{L}^{-1} Ad_{e^{ft}}^* \mathfrak{L}\omega, \zeta\rangle = (Ad_{e^{ft}}^* \mathfrak{L}\omega, \zeta) = \\ = (\mathfrak{L}\omega, Ad_{e^{ft}}\zeta) = \langle \omega, Ad_{e^{ft}}\zeta\rangle. \end{aligned} \tag{39}$$

For every $\zeta \in \mathfrak{U}$, according to (39), (29) and (10), one has for $t = 0$

$$\langle \frac{d\omega}{dt}, \zeta\rangle = \langle \omega, \frac{dA}{dt}\rangle = \langle \omega, [f, A]\rangle = \langle B(\omega, f), \zeta\rangle, \tag{40}$$

[4] In rigid body dynamics, the isorotating ω correspond to the same angular momentum. In hydrodynamics, they are the "isorotational" velocity fields (see[1]).

[5] transl. note: Nowadays these 'isorotating leaves' are usually viewed in the dual Lie algebra where they form the famous co-adjoint orbits

where $A(t) = Ad_{e^{ft}}\zeta$. As ζ is arbitrary, formula (40) proves (37) for $t = 0$. Formula (36) implies the identity

$$\omega(t_1 + t_2, \omega(0)) = \omega(t_1, \omega(t_2, \omega(0))).$$

Hence, formula (37) is proved for all t. Hence the derivatives

$$\frac{d^2\omega}{dt^2} = B(B(\omega, f), f); \cdots ; \frac{d^n\omega}{dt^n} = B(\cdots(B(\omega, f), \cdots, f); \cdots$$

which is equivalent to (38). Lemma 6 is proved.

Proof of theorem 3. Let ξ be a vector tangent to a leaf at the point v. As the leaves are orbits of a differentiable action of the group G (see (35)), there exists $f(\xi) \in \mathfrak{U}$ such that

$$v + \xi t = \omega(t) + O(t^2) \quad t \to 0$$

where the function $\omega(t)$ is defined by (36) with $\omega(0) = v$.

According to (37) one has

$$\xi = B(v, f). \tag{41}$$

In accordance with (34), (41), (11), the total differential of the energy on the leaf is

$$dE = \langle v, \xi \rangle = \langle v, B(v, f) \rangle = -\langle B(v, v), f \rangle. \tag{42}$$

Formula (42) shows that $dE = 0$ for all ξ if and only if $B(v, v) = 0$. C.q.f.d., according to (33).

7 Stability of stationary motions

Consider a system of ordinary differential equations

$$\dot{x} = f(x). \quad x \in \mathbb{R}^n. \tag{43}$$

Definition 4. A point x_0 (with $f(x_0) = 0$) is stable if

$$\forall \varepsilon > 0, \exists \delta > 0 : (|x(0) - x_0| < \delta) \Rightarrow (|x(t) - x_0| < \varepsilon \, \forall t > 0).$$

Suppose now that a foliation invariant with respect to the system (43) is defined in the space $\mathbb{R}^n$. The point x_0 is a *regular point*, if there exists, on a neighborhood of x_0, a system of differentiable local coordinates y ($y(x_0) = 0$) such that the connected components of the intersections of the leaves with the neighborhood are parallel planes of a fixed dimension k for the coordinates y.

Let E be a first integral of system (43). Suppose that
1) x_0 is a conditional extremum point of E on the leaf;
2) x_0 is a regular point and that
3) the second derivative of E on the leaf at x_0 is a non degenerate quadratic form.

One proves easily a lemma well known since Lagrange:

Lemma 7. *A point x_0 satisfying conditions 1), 2), 3) is a stationary point of system (43). If, moreover, the second derivative of E on the leaf is a positive-definite (or negative-definite) quadratic form, the stationary point x_0 is stable.*

Indeed, in coordinates y,

$$E(y) = \frac{1}{2}(E_2 y, y) + O(y^3) \quad \text{où} \quad (E_2 y, z) = (y, E_2 z),$$
$$\dot{E} = (E_2 y, \dot{y}) + O(y^2)\dot{y}. \qquad (y \to 0).$$

If $\dot{y} \neq 0$ for $y = 0$, then for y small enough $(E_2 y, y) \neq 0$ and E is not a first integral. Hence $\dot{y} = 0$ and x_0 is a stationary point. The claim of stability is obvious (Lagrange, Dirichlet, etc.).

Let us apply lemma 7 to system (32).

Let v be a stationary point of Euler's system (32). Suppose that v is a regular point, i.e. that v belongs to a leaf of maximal dimension. Consider the quadratic form

$$\boxed{2d^2E = \langle B(v,f), B(v,f)\rangle + \langle [f,v], B(v,f)\rangle.} \tag{44}$$

Theorem 4. *If the quadratic form (44) is positive-definite or negative-definite, the stationary point v is stable.*

Here the form d^2E is considered as a quadratic form on the vector space tangent to the leaf

$$\xi = B(v,f), \quad f \in \mathfrak{U}.$$

Proof of theorem 4. According to lemma 7, it is enough to prove that the second derivative of the energy on the leaf has the form (44). Formula (38) defines an algebra mapping on the leaf, which transforms

$$f \to \omega = v + \xi_1 + \xi_2 + O(f^3),$$
$$\xi_1 = B(v,f), \qquad 2\xi_2 = B(B(v,f)).$$

Substituting ω in (34) and taking (42), (33) into account, one finds

$$2d^2E = \langle \xi_1, \xi_1 \rangle + 2\langle\rangle$$

which is equivalent to (44).

It follows from the definition of d^2E that the form (44) does not depend on f, but rather only on $\xi_1 = B(v,f)$. In fact, one can check this by a small computation:

Lemma 8. *The bilinear form $C(x,y) = \langle [x,v], B(v,y)\rangle$ is symmetric:*

$$C(x,y) = C(y,x). \tag{45}$$

Proof. According to (11), (10), (6), (10), (33)

$$\begin{aligned}\langle [x,v],B(v,y)\rangle &= \langle B(v,[v,x]),y\rangle = \langle [[v,x],y],v\rangle \\ &= \langle [v,[x,y]],v\rangle + \langle [x,[y,v]],v\rangle = \\ &= \langle B(v,v),[x,y]\rangle + \langle B(v,x),[y,v]\rangle = \\ &= \langle [y,v],B(v,x)\rangle. \qquad\qquad C.Q.F.D.\end{aligned}$$

Let now $B(v,f_1) = B(v,f_2)$. Let us show that the corresponding values of d^2E coincide. We set $x = f_1 - f_2$, $y = f_1$. Then $B(v,x) = 0$. According to (44), (45)

$$\begin{aligned}2(d^2E(f_1) - d^2E(f_2)) &= \langle [x,v],B(v,y)\rangle = \\ &= \langle [y,v],B(v,x)\rangle = 0.\end{aligned}$$

Hence the quadratic form (44) depends only on $\xi = B(v,f)$. Theorem 4 now follows immediately from lemma 7.

Remark. Consider the "variational equations" of Euler's equations (32) in the neighborhood of v:

$$\dot{\xi} = B(v,\xi) + B(\xi,v). \tag{46}$$

As the leaves and the energy are invariant, the form (44) must be a first integral of system (46). Actually, one can check the conservation of d^2E by a direct computation.

Indeed, according to (44), (45), (41)

$$\frac{d}{dt}d^2E = \langle \xi,\dot{\xi}\rangle + \langle [f,v],\dot{\xi}\rangle.$$

Substituting $\dot{\xi}$ from (46), one gets by using successively (11), (45), (41), (10), (45), (11):

$$\frac{d}{dt}d^2E =$$

$$\begin{aligned}&\langle \xi,B(v,\xi)\rangle + \langle \xi,B(\xi,v)\rangle + \langle [f,v],B(v,\xi)\rangle + \langle [f,v],B(\xi,v)\rangle = \\ &\langle \xi,B(\xi,v)\rangle + \langle [\xi,v],\xi\rangle\rangle + \langle [v,[f,v]],\xi\rangle\rangle = \\ &= \langle [v,[f,v]],B(v,f)\rangle = -\langle [f,v],B(v,[f,v])\rangle = 0, \qquad C.Q.F.D.\end{aligned}$$

8 The riemannian curvature of Lie groups

As an example of application of Euler's theorems, I shall compute the connexion (55) and the riemannian curvature (53) of a Lie group, endowed with a left invariant metric (compare to [3]).

Let M be a riemannian manifold, $x \in M$ and $\xi \in TM_x$ a tangent vector to M at x. We shall denote

$$\gamma(x,\xi,t) = \gamma(\xi,t) = \gamma(t) = \gamma$$

the geodesic[6] on M, originating from $x = \gamma(0)$ with initial velocity $\xi = \dot{\gamma}(0)$.

One determines the *parallel transport along* γ

$$\pi_{\gamma(t)} : TM_{\gamma(0)} \to TM_{\gamma(t)}$$

with the help of the following construction.

Let

$$P_{\gamma(\xi,t}\eta = \frac{1}{t}\frac{d}{d\tau}_{|\tau=0}\gamma(x,\xi+\eta\tau,t) \in TM_{\gamma(t)}. \tag{47}$$

Then

$$\pi_{\gamma(t)}\eta = P_{\gamma(t)}\eta + O(t) \qquad (t \to 0).$$

The *covariant derivative* $\nabla_\xi \eta$ of a tangent vector-field η along the direction of the tangent vector $\xi \in TM_x$ is, by definition, the vector-field

$$\nabla_\xi \eta = \frac{d}{dt}_{|t=0} P^{-1}_{\gamma(t)}\eta(\gamma(\xi,t)) = \frac{d}{dt}_{|t=0} \pi^{-1}_{\gamma(t)}\eta(\gamma(\xi,t)) \in TM_x. \tag{48}$$

Let ξ, η be two orthogonal unit vectors in TM_x. One calls[7] the *riemannian sectional curvature* of M along ξ, η the real number

$$C_{\xi\eta} = \langle -\nabla_\xi \nabla_\eta \xi, \eta\rangle + \langle \nabla_\eta \nabla_\xi \xi, \eta\rangle - \langle \nabla_{\{\xi,\eta\}}\xi, \eta\rangle, \tag{49}$$

where $\langle , \rangle$ is the scalar product determining the metric, and ξ, η denote two vector-fields. Finally, the vector-field $\{\xi, \eta\}$ denotes the *Poisson bracket* [8] of the two vector-fields ξ, η.

With regard to Poisson bracket, we shall use the following known things. To each vector-field ζ on M, corresponds a one-parameter group of diffeomorphisms of M

$$\exp\xi t : x \mapsto z(x,t) \in M; \quad z(x,0) = x, \ \frac{dz}{dt} = \zeta.$$

The Poisson bracket of the vector-fields ξ, η tangent to M is defined as the commutator in the Lie algebra[9]: the vector-field $\{\xi, \eta\}$ depends bilinearly on ξ, η and satisfies the equation:

$$\exp\xi t \exp\eta s \exp(-\xi t)\exp(-\eta s) = \exp\{\xi,\eta\}st + O(s^2) + O(t^2). \tag{50}$$

In a system of local coordinates x on M the Poisson bracket is given by the formula:

$$\{\xi,\eta\} = \frac{\partial \xi}{\partial x}\eta - \frac{\partial \eta}{\partial x}\xi. \tag{51}$$

---

[6] That is the extremal of the "least action principle" $\delta \int \dot{\gamma}^2 dt = 0$.

[7] See for example [13].

[8] transl. note: Nowadays called the *Lie bracket*

[9] And differs by the sign from [13].

Indeed, computing the left hand side of (50), one finds

$$\exp \xi t : x \mapsto x + \xi t + O(t^2),$$

$$\exp \xi t \exp \eta s : x \mapsto x + \xi t + \eta s + \frac{\partial \xi}{\partial x} \eta st + O(s^2) + O(t^2),$$

$$\exp \xi t \exp \eta s \exp(-\xi t) \exp(-\eta s) :$$

$$x \mapsto x + \left(\frac{\partial \xi}{\partial x} \eta - \frac{\partial \eta}{\partial x} \xi \right) st + O(s^2) + O(t^2),$$

which proves formula (51).

Let now M be a Lie group G, $\mathfrak{U}$ its Lie algebra, ξ and η two *right*-invariant vector-fields. Then the vector-field $\{\xi, \eta\}$ also is right-invariant. A right-invariant vector-field is well-defined by its value at the neutral element, $\xi_e \in \mathfrak{U}$. *The Poisson bracket of the right-invariant vector-fields coincide at the point e with the commutator of the Lie algebra:*

$$\{\xi, \eta\}_e = [\xi_e, \eta_e].$$

Indeed, let $\exp(\xi t)$ be a one-parameter group of diffeomorphisms of the group G, corresponding to the right-invariant vector-field ξ. Then, for $x \in G$

$$\exp(\xi t) : x \mapsto (exp \xi_e t)x$$

where $exp \xi_e t$ is the one-parameter subgroup of the group G. Hence

$$\exp \xi t \exp \eta s \exp(-\xi t) \exp(-\eta s) : x \mapsto \exp \xi_e t \exp \eta_e s \exp(-\xi_e t) \exp(-\eta_e s) x.$$

C.Q.F.D.

Finally, let ξ and η be two *left*-invariant vector-fields on G. *The value at e of the Poisson bracket of left-invariant vector-fields differs by the sign of the commutator in the Lie algebra :*

$$\{\xi, \eta\}_e = -[\xi_e, \eta_e]. \tag{52}$$

Indeed, a left-invariant vector-field on G is a right-invariant vector-field on the group G' with the operation $g_1 \star g_2 = g_2 g_1$. But the commutator of the Lie algebra $\mathfrak{U}'$ of the group G' differs by the sign from the commutator in the algebra $\mathfrak{U}$.

Now, let as above G be a Lie group, endowed with a left-invariant metric, $\mathfrak{U}$ its Lie algebra, $\langle , \rangle$ the scalar product in the algebra, which determines the metric, $[,]$ the Lie bracket in the algebra, and B the operation defined in §1.

Theorem 5. *The sectional curvature of the group G at the point e for the plane defined by the two orthogonal and unitary vectors $\xi, \eta \in \mathfrak{U}$, is given by the formula*

$$\boxed{C_{\xi\eta} = \langle \delta, \delta \rangle + 2\langle \alpha, \beta \rangle - 3\langle \alpha, \alpha \rangle - 4\langle B_\xi, B_\eta \rangle} \tag{53}$$

where
$$\left.\begin{aligned} &2\delta = B(\xi,\eta)+B(\eta,\xi), \qquad 2\beta = B(\xi,\eta)-B(\eta,\xi),\\ &2\alpha = [\xi,\eta], \qquad 2B_\xi = B(\xi,\xi), \qquad 2B_\eta = B(\eta,\eta). \end{aligned}\right\} \tag{54}$$

Remark. For a bi-invariant metric, one has $B(\xi,\eta) = [\xi,\eta]$, hence $B_\xi = B_\eta = \delta = 0$, $\beta = 2\alpha$. Thus, (53) becomes the known formula

$$C_{\xi\eta} = \frac{1}{4}\langle[\xi,\eta],[\xi,\eta]\rangle.$$

In order to prove theorem 5, we shall at first express the covariant derivative in terms of B operations.

Lemma 9. *Let ξ,η be two left-invariant vector-fields on the group G. Then at the point e, the left-invariant vector-field $\nabla_\xi\eta$ is given by the formula*

$$\boxed{\nabla_\xi\eta = \frac{1}{2}([\xi,\eta]-B(\xi,\eta)-B(\eta,\xi)).} \tag{55}$$

Proof of lemma 9. One first computes the parallel transport (47) in algebra coordinates (§4). According to (24), the geodesics of the group have the following expression in the algebra

$$\gamma(0,\xi,t) = t\xi + \frac{t^2}{2}B(\xi,\xi)+O(t^3) \qquad t\to 0. \tag{56}$$

It follows from (56) that the vector

$$P_{\gamma(\xi,t)}\eta = \frac{1}{t}\frac{d}{d\tau}_{|\tau=0}\gamma(0,\xi+\eta\tau,t)\in T\mathfrak{U}_\gamma = \mathfrak{U}$$

has coordinates

$$P_{\gamma(\xi,t)}\eta = \eta+\frac{t}{2}(B(\xi,\eta)+B(\eta,\xi))+O(t^2). \tag{57}$$

By definition (48), for a left-invariant vector-field η

$$\nabla_\xi\eta = \frac{d}{dt}_{|t=0}P_{\gamma^{-1}(\xi,t)}\tilde{L}_{\gamma(\xi,t)}\eta. \tag{58}$$

Substituting in (58) the expressions (17) for $\tilde{L}_{\gamma(\xi,t)}\eta$ and (57) for P_γ, one finds successively

$$\begin{aligned}\nabla_\xi\eta &= \frac{d}{dt}_{|t=0}P_{\gamma^{-1}(\xi,t)}\left(\eta+\frac{t}{2}[\xi,\eta]+O(t^2)\right) =\\ &= \frac{d}{dt}_{|t=0}\eta+\frac{t}{2}([\xi,\eta]-B(\xi,\eta)-B(\eta,\xi))+O(t^2) \quad C.Q.F.D.\end{aligned}$$

Remark. Parallel transport π_γ on a riemannian manifold preserves the scalar product $\langle a,b\rangle$. But the left-invariant product of left-invariant vector-fields $\langle a,b\rangle$ is constant. Hence ∇_e is an antisymmetric operator :

$$\langle \nabla_e a, b\rangle + \langle \nabla_e b, a\rangle = 0. \tag{59}$$

One checks easily that the expression (55) satisfies the antisymmetry relation (59).

Proof of theorem 5. Let ξ,η be two left-invariant vector-fields. Then the vector-fields $[\xi,\eta]$, $\nabla_\xi\eta$ and $\nabla_\eta\xi$ also are left-invariant. According to (54) and (55), their values at point e are

$$\begin{aligned} \nabla_\xi\eta &= \alpha-\delta, \quad & \nabla_\eta\xi &= -(\alpha+\delta),\\ \nabla_\xi\xi &= -2B_\xi, \quad & \nabla_\eta\eta &= -2B_\eta \end{aligned} \tag{60}$$

Substituting (60) in (49) and using (59), one finds

$$\langle -\nabla_\xi\nabla_\eta\xi,\eta\rangle = \langle \nabla_\xi\eta,\nabla_\eta\xi\rangle = -\langle \alpha-\delta,\alpha+\delta\rangle, \tag{61}$$

$$\langle \nabla_\eta\nabla_\xi\xi,\eta\rangle = \langle -\nabla_\xi\xi,\nabla_\eta\eta\rangle = -4\langle B_\xi, B\eta\rangle. \tag{62}$$

It follows from (55) that

$$\begin{aligned} -\langle \nabla_{\{\xi,\eta\}}\xi,\eta\rangle = \langle \nabla_{[\xi,\eta]}\xi,\eta\rangle = \frac{1}{2}\langle [[\xi,\eta],\xi],\eta\rangle &- \frac{1}{2}\langle B([\xi,\eta],\xi),\eta\rangle\\ &- \frac{1}{2}\langle B(\xi,[\xi,\eta]),\eta\rangle. \end{aligned} \tag{63}$$

Taking into account (10), (11), (54), one finds

$$\left.\begin{aligned} \langle [[\xi,\eta],\xi],\eta\rangle &= \langle -B(\eta,\xi),[\xi,\eta]\rangle,\\ \langle B([\xi,\eta],\xi),\eta\rangle &= \langle [\xi,\eta],[\xi,\eta]\rangle,\\ \langle B(\xi,[\xi,\eta]),\eta\rangle &= \langle -B(\xi,\eta),[\xi,\eta]\rangle. \end{aligned}\right\} \tag{64}$$

Substituting (64) in (63), one finds in accordance with (54)

$$-\langle \nabla_{\{\xi,\eta\}}\xi,\eta\rangle = -2\langle\alpha,\alpha\rangle + 2\langle\alpha,\beta\rangle. \tag{65}$$

Finally, substituting (61), (62), (65) in (49), one finds

$$C_{\xi,\eta} = -\langle \alpha-\delta,\alpha+\delta\rangle - 4\langle B_\xi,B_\eta\rangle - 2\langle\alpha,\alpha\rangle + 2\langle\alpha,\beta\rangle,$$

which is equivalent to (53). Theorem 5 is proved.

9 The case of a diffeomorphism group

Let D be a bounded domain in a Riemann space. We shall denote SDiff D *the group of volume preserving diffeomorphisms of the domain D.* Even though SDiff D is not a Lie group, it is interesting to see the assertions which the formulas of the previous paragraphs become transformed into, in this case. Such a "heuristic" study leads to propositions which can be rigorously checked afterwards (see [2]). In accordance with the heuristic character of this and the following paragraph, we leave aside differentiability problems and other similar details.

The algebra $\mathfrak{U}$ corresponding to the group SDiff D *is made of the vector-fields v tangent to D such that*

$$div\, v = 0, \qquad \text{and} \qquad (v, n) = 0 \qquad \text{on} \qquad \partial D. \tag{66}$$

One defines the scalar product in the algebra $\mathfrak{U}$ by the formula

$$\langle v_1, v_2 \rangle = \int_D (v_1 \cdot v_2) dx, \tag{67}$$

where $(\cdot)$ is the scalar product which determines the metric in D, and dx is the riamannian volume element on D.

In order to write the formulas of the previous paragraphs for the group SDiff D, we shall use some notions from vector analysis. According to (50), the Lie commutator $[v_1, v_2]$ of two vector-fields coincide with their Poisson bracket $\{v_1, v_2\}$.

Let ω^k be a differential form of degree k on D, ξ^k a k-vector $\xi^k = \xi_1 \wedge \cdots \wedge \xi_k$. The value of the form ω^k on the polyvector ξ^k will be denoted by $\omega^k | \xi^k$. The convolution operation

$$i(\xi^k) : \omega^l \to \omega^{l-k}$$

is defined by the formula

$$i(\xi^k)\omega^l | \eta^{l-k} = \omega^l | \xi^k \wedge \eta^{l-k} \tag{68}$$

for any $(l-k)$-vector η^{l-k}.

The riemannian structure associates with each tangent vector c a 1-form ω_c^1, namely

$$\omega_c^1 | \xi = i(\xi)\omega_c^1 = (c, \xi) \tag{69}$$

for all tangent vectors ξ.

Theorem 6. *Let a and c be vector fields in $\mathfrak{U}$. The vector-field $B(c,a) \in \mathfrak{U}$ is given by the formula (see (10)):*

$$\boxed{\omega_B^1 = i(a) d\omega_c^1 + d\alpha} \tag{70}$$

where α is the univalent function on D defined by (70) and $B \in \mathfrak{U}$ (see (66)).

Proof of theorem 6. – According to H. Weyl's orthogonal decomposition theorem, each vector-field ξ on D admits a unique decomposition into the sum of two orthogonal vector-fields

$$\xi = \xi_1 + \xi_2 \quad \text{where} \quad \xi_1 \in \mathfrak{U},\ \xi_2 = \operatorname{grad}\alpha,$$

α being a univalent function on M. Applying the decomposition theorem to the vector-field which corresponds to the form $i(a)d\omega_c^1$, one finds a unique decomposition into ω_B^1 and $-d\alpha$. So that formula (70) determines a unique field $B \in \mathfrak{U}$.

Formula (10) determines the scalar products $\langle B, b\rangle$ for all $b \in \mathfrak{U}$:

$$\langle B(c,a), b\rangle = \langle [a,b], c\rangle \tag{10}$$

Hence, there cannot exist more than on vector $B(c,a)$ satisfying (10). Therefore, in order to prove theorem 6 it remains to prove that the function B, determined by (70), satisfies the identity (10) for all $b \in \mathfrak{U}$.

One shall make use of the known identities (71), (76) in vector analysis

$$d(\omega^k \wedge \omega^l) = d\omega^k \omega^l + (-1)^k \omega^k \wedge d\omega^l \tag{71}$$

$$\omega^k \wedge \omega^l = (-1)^{kl} \omega^l \wedge \omega^k, \tag{72}$$

$$i(a)(\omega^k \omega^l) = (i(a)\omega^k) \wedge \omega^l + (-1)^k \omega^k \wedge (i(a)\omega^l), \tag{73}$$

$$i(a \wedge b)\omega = i(b)i(a)\omega \quad \text{for all vectors } a, b \tag{74}$$

Let τ be the riemannian volume on D (an n-form, $n = \dim D$).
Then, for all tangent vector-fields a, b, one has

$$(a,b)\tau = \omega_a^1 \wedge (i(b)\tau), \tag{75}$$

$$\boxed{d(i(a \wedge b)\tau) = i(\{a,b\})\tau + i(a)di(b)\tau - i(b)di(a)\tau.} \tag{76}$$

It follows from (67), (50) and (75) that for all vector-fields a, b, c

$$\langle [a,b], c\rangle = \int_D (\{a,b\}, c)\tau = \int_D \omega_c^1 \wedge i(\{a,b\})\tau. \tag{77}$$

For two fields $a, b \in \mathfrak{U}$, on finds according to (66)

$$di(a)\tau = \tau \operatorname{div} a = 0, \quad di(b)\tau = \tau \operatorname{div} b = 0. \tag{78}$$

According to (78), it follows from (76) that for $a, b \in \mathfrak{U}$,

$$di(a \wedge b)\tau = i(\{a,b\})\tau. \tag{79}$$

It follows from (79), (71) and Stokes formula that

$$\int_D \omega_c^1 \wedge i(a\wedge b)\tau = \int_D d\omega_c^1 \wedge i(a\wedge b)\tau - \int_{\partial D} \omega_c^1 \wedge i(a\wedge b)\tau. \qquad (80)$$

According to (66), for the fields $a,b \in \mathfrak{U}$, the integral along ∂D is equal to 0. Indeed, let $\xi^{n-1} = \xi_1 \wedge \cdots \wedge \xi_{n-1}$ be a polyvector tangent to ∂D. According to (68),

$$\omega_c^1 \wedge i(a\wedge b)\tau|\xi = \sum_i \pm(\omega_c^1|\xi_i)(\tau|a\wedge b\wedge \xi_i')$$

where

$$\xi_i' = \xi_1 \wedge \cdots \xi_{i-1} \wedge \xi_{i+1} \wedge \cdots \wedge \xi_n.$$

But $\tau|a\wedge b\wedge \xi_i' = 0$, because the n-vector $a\wedge b\wedge \xi_i'$ is tangent to the manifold ∂D of dimension $n-1$. Hence, for $a,b \in \mathfrak{U}$

$$\int_{\partial D} \omega_c^1 \wedge i(a\wedge b)\tau = 0. \qquad (81)$$

It follows from (72), (74), that

$$d\omega_c^1 \wedge i(a\wedge b)\tau = (i(a\wedge b)\tau)\wedge d\omega_c^1 = (i(b)i(a)\tau)\wedge d\omega_c^1. \qquad (82)$$

Let us notice that, being of degree $n+1$, the form $(i(a)\tau)\wedge d\omega_c^1$ is equal to zero. Hence it follows from (73) that

$$(i(b)i(a)\tau)\wedge d\omega_c^1 = (-1)^n (i(a)\tau)\wedge (i(b)d\omega_c^1). \qquad (83)$$

The form $\tau \wedge i(b)d\omega_c^1$ is also of degree $n+1$, hence equal to zero. Therefore (73) gives

$$(i(a)\tau)\wedge (i(b)d\omega_c^1) = (-1)^{n+1}(\tau \wedge i(a)i(b)d\omega_c^1) = (-1)^n (i(b)i(a)d\omega_c^1)\tau. \qquad (84)$$

Comparing the formulas (77), (80), (81), (82), (83), (84), one finds

$$\langle [a,b],c\rangle = \int_D (i(b)i(a)d\omega_c^1)\tau. \qquad (85)$$

It follows from (70) and (69), that

$$i(b)i(a)d\omega_c^1 = i(b)(\omega_B^1 - d\alpha) = (b, B - \operatorname{grad}\alpha). \qquad (86)$$

According to the orthogonal decomposition theorem, the field $b \in \mathfrak{U}$ is orthogonal to $\operatorname{grad}\alpha$. Hence it follows from (85) and (86) :

$$\langle [a,b],c\rangle = \int_D (b, B - \operatorname{grad}\alpha)\tau = \langle B(c,a), b\rangle \qquad C.Q.F.D.$$

The tridimensional case. If $n=3$, the formula (70) takes the form

$$\boxed{B(c,a) = \operatorname{rot} c \wedge a + \operatorname{grad} \alpha} \tag{87}$$

where $\wedge$ is the vector product associated to the metric[10] of D. The proof of formula (87) is immediate.

The bidimensional case. Let D be a riemannian manifold of dimension 2, possibly with boundary ∂D. The vector-fields $v \in \mathfrak{U}$ admit "*stream functions*" ψ. Indeed, let $v \in \mathfrak{U}$, I the operator of right angle rotation "to the right". Then, for each contour which is homologous to zero, $\gamma = d\sigma$,

$$0 = \iint\limits_{\sigma} \operatorname{div} v\, dx = \oint\limits_{\partial\sigma} v\, dn = \oint\limits_{\partial\sigma} Iv\, dx$$

hence the 1-form $(Iv)dx$ is closed.

So

$$\boxed{v = I \operatorname{grad} \psi = \operatorname{rot} \psi} \tag{88}$$

where ψ is the, possibly multivalued, "*stream function*". it follows from (71), that the *stream function* $\psi_{\{v_1,v_2\}}$ of the commutator $\{v_1, v_2\}$ of the two fields $v_1, v_2 \in \mathfrak{U}$ *is the jacobian of the stream functions of these fields :*

$$\psi_{\{v_1,v_2\}} = J(\psi_{v_1}, \psi_{v_2}) = v_1 \wedge v_2. \tag{89}$$

Theorem 6 now becomes the following assertion: let ψ_c, ψ_a be the stream functions of the fields $c, a \in \mathfrak{U}$. Then the vector-field $B(c,a)$ is given by the formula

$$\boxed{B = -\Delta \psi_c \operatorname{grad} \psi_a + \operatorname{grad} \alpha,} \tag{90}$$

where $\Delta = \operatorname{div} \operatorname{grad} = -\operatorname{rot} \operatorname{rot}$

One deduces formula (90) from (87). Now $c = \operatorname{rot} \psi_c, a = \operatorname{rot} \psi_a, \operatorname{rot} c \wedge a = \operatorname{rot} \operatorname{rot} \psi_c \cdot \operatorname{grad} \psi_a$. In the special case of euclidean space x, y the formulas (88), (89), (90) become

$$v_x = \frac{\partial \psi}{\partial y}, v_y = -\frac{\partial \psi}{\partial x}, \psi_{\{v_1,v_2\}} = \frac{\partial \psi_{v_1}}{\partial x}\frac{\partial \psi_{v_2}}{\partial y} - \frac{\partial \psi_{v_1}}{\partial y}\frac{\partial \psi_{v_2}}{\partial x}, \Delta = \frac{\partial^2}{\partial x^2} + \frac{\partial^2}{\partial y^2}. \tag{91}$$

10 Applications to hydrodynamics

Let, as in §9, *SDiffD* be the group of those diffeomorphisms of a bounded riemannian domain D which preserve the volume element.

[10] Changing the orientation of D changes simultaneously the signs of rot and $\wedge$.

It follows from the principles of Mechanics that *the geodesics of the right invariant metric corresponding to* (67) *are the ideal fluid flows in D.*

Indeed, let us consider an ideal fluid (incompressible, inviscid, homogeneous), which fills the domain D. The transformation

$$g(t) : D \to D$$

which displaces each particle of the fluid from its position at time 0 to its position at time t preserves the volume element. Hence, $g(t) \in SDiffD$. According to the least action principle, the curve $g(t)$ is a geodesic of the group $SDiffD$, because it extremizes the integral $\int T\,dt$, where T is the kinetic energy:

$$T = \frac{1}{2}\int\limits_D v^2 dx$$

where v is the velocity field[11]. Now

$$g(t+dt) = (\exp(vdt))g(t).$$

Hence the velocity field is the right translation of the vector $\dot{g}$ in the algebra $\mathfrak{U}$:

$$v = \tilde{R}_{g^{-1}}\dot{g}.$$

Hence, $g(t)$ is the geodesic of a right invariant metric, corresponding to the form (67).

The formulas of the preceding paragraphs correspond to left invariant metrics. In order to obtain the corresponding formulas in the right invariant case, it is enough, by (52), to change the signs in front of all the commutators and the B operations; this does not change the quadratic forms (44), (45), (53).

In accordance with (32) and (52), the velocity field of the ideal fluid satisfies the equation

$$\frac{\partial v}{\partial t} = -B(v,v). \tag{92}$$

Comparing to (87), one finds the *equation of motion "in the form of Bernouilli"* for the ideal fluid in a riemannian domain D of dimension 1:

$$\frac{\partial v}{\partial t} = v \wedge \operatorname{rot} v + \operatorname{grad}\alpha,\ \operatorname{div} v = 0,\ (v,n) = 0 \text{ on } \partial D. \tag{93}$$

In the case of a steady flow, (93) and (76) imply

$$v \wedge \operatorname{rot} v = -\operatorname{grad}\alpha, \quad \{v, \operatorname{rot} v\} = 0. \tag{94}$$

[11] From the mathematician's point of view, one can consider this principle as a definition of the ideal fluids. See[14, 11].

Thus, *the velocity field of a steady flow commutes with its curl.* One immediately deduces from this a topological classification of analytic steady flows of ideal fluids in three-dimensional domains [4].

In particular, the domain D turns out to be divided by an analytic subspace into a finite number of open "cells", fibered by tori or cylinders generated by the stream lines[12]. The stream lines are closed on the cylinders, closed or dense on the tori (fig. 2, 3).

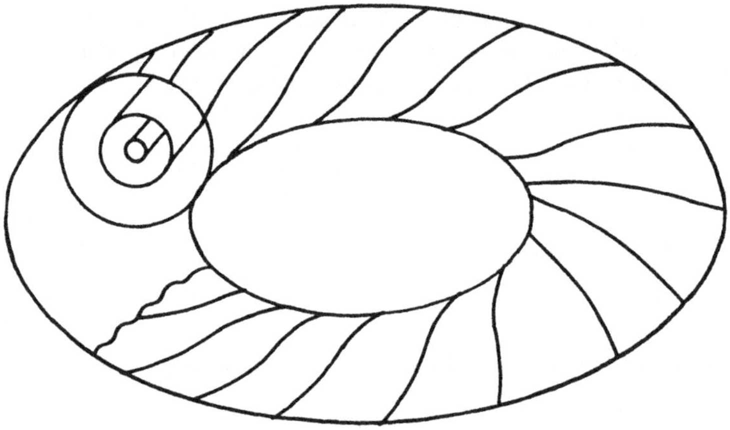

More precisely, the following theorem holds true

Theorem 7. *(see [4]). Let us suppose that*
1) The three-dimensional domain D is connected and compact.
2) The steady flow velocity field v, the domain D and the boundary ∂D are real analytic.
3) the velocity and the curl are not eveywhere collinear:

$$v \wedge rot\, v \not\equiv 0.$$

Then almost every stream line is either closed or everywhere dense on analytic tori analytically embedded in D: all the other stream lines make up a true compact analytic subset in D.

Remark. The analyticity condition is probably not very important. The non collinearity condition is probably important.

[12] The trajectories of the velocity field v.

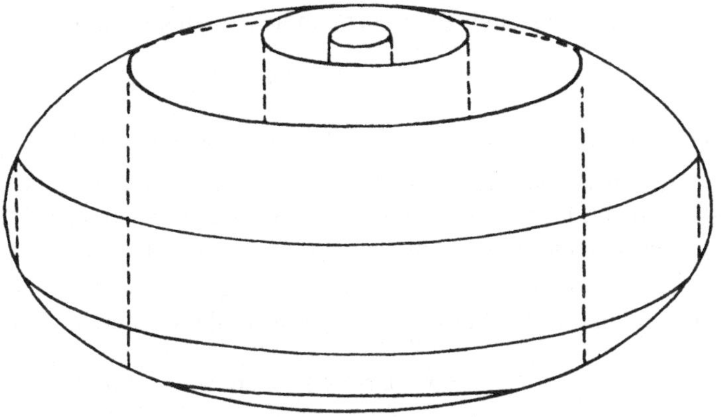

Because the flows with rot $v = \lambda v$[13] allow probably for stream lines with a topology as complicated as the one of orbits in celestial mechanics (see [5], fig. 6). In the same way, the stream lines of stationary flows of viscous fluids may possess a complicated topology.

Proof of theorem7 According to (94), the stream lines, as the curl lines, are orthogonal to grad α/ Hence they belong to the level surfaces $\alpha =$ constant. Let us study the topology of these surfaces.

We shal call α_0 a critical value if there exists a point x in D, where grad $\alpha(x) = 0$ and $\alpha(x) = \alpha_0$, or if there exists a point x in ∂D, where grad $\alpha(x)$ is orthogonal to ∂D and $\alpha(x) = \alpha_0$. From the analyticity of α and ∂D, it follows that there is only a finite number of critical values. As the function α is not constant (non collinearity), the points x such that $\alpha(x)$ is critical make up a finite number of analytic surfaces and curves in D

These surfaces divide $D \cup \partial D$ into "cells", in each of which grad $\alpha \neq 0$ and grad α is not orthogonal to ∂D. We shall show that in each cell, the stream lines are closed or everywhere dense on tori.

Let us consider a connected component C of a non-critical level surface $\alpha(x) = \alpha_1$, the value α_1 being non-critical. This component C, either
1) does not intersect ∂D, then C is a closed analytic surface in D, or
2) intersects ∂D, $C \cap \partial D = \partial C$ is made of a finite number of closed analytical curves (the intersection is transversal because grad α is not orthogonal to ∂D).

If C is a surface without boundary, the group $\mathbb{R}^2$ acts on C globally and without fixed point. Indeed, let $(s,t) \in \mathbb{R}^2, x \in C$. Consider the solution $y(s_0;s,t;\tau)$ of the system

$$\frac{dy}{d\tau} = sr + tv,\ y(x_0;s,t;0) = x_0,\ (r = \text{rot}\, v).$$

[13] Example: $D = \mathbb{T}^3$, $v_x = A\sin z + C\cos y$, $v_y = B\sin x + C\cos z$, $v_z = C\sin y + B\cos x$.

As the vector-fields r and v commute (94), the mapping

$$x : C \times \mathbb{R}^2 \to C$$

given by $x(x_0;s,t) = y(x_0;s,t;1)$, defines the action of $\mathbb{R}^2$ on C.

But α_1 is non-critical, hence $v \wedge r \neq 0$ everywhere on C, so that $\mathbb{R}^2$ acts without fixed point. Hence C is a torus, and the stream line $x(x_0;0,t)$ is the orbit of a one parameter subgroup; this proves theorem7 when $\partial C = 0$.

Suppose now that $\partial C \neq 0$. Then ∂C is composed of a finite number of closed stream lines in ∂D. Let $x(x_0;0,\tau)$ be one of these components of ∂C:

$$x(x_0;0,\tau+T) = x(x_0;0,\tau) \quad \text{for all } \tau.$$

Due connectedness of C and the condition of non-collinearity, each point of C can be written in the form

$$x_1 = x(x_0;s,t).$$

Hence the commutativity of $\mathbb{R}^2$ implies

$$x(x_1;0,T) = x_1$$

that is, the stream line through x_1 is closed. One sees also that C has the topology of the cylinder $S^1 \times I^1$. Theorem 7 is proved.

In hydrodynamics, the variational equation (46) in the neighborhood of the steady flow (94) corresponds to the Rayleigh equation (Orr-Sommerfeld equation for a non viscous fluid). The quadratic form (44) is a first integral of this linear equation. In accordance with theorem 4, it is sufficient for the stability of the steady flow that the quadratic form (44) be definite (positive or negative).

It seems that, for three-dimensional flows, the form (44) is never definite. It is only for two-dimensional flows that theorem 4 implies effective stability criteria.

Let, for example, D be a planar annulus (fig. 4).

The steady flow in D is stable if its stream function satisfies the concavity condition of its velocity profile.

$$\frac{\nabla \psi}{\nabla \Delta \psi} > 0. \tag{95}$$

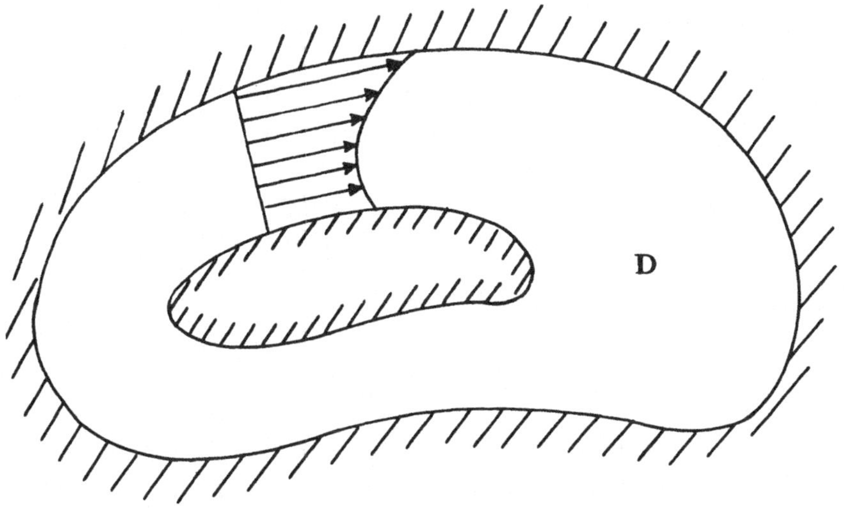

Indeed

Theorem 8. *The quadratic form* (44) *is*

$$\boxed{2\delta^2 E = \iint_D (\delta v)^2 + \frac{\nabla\psi}{\nabla\Delta\psi}(\delta r)^2 dxdy} \tag{96}$$

where $\delta r = \mathit{rot}\, v$.

Remark. In (95) and (96) the ratio of the vectors $\nabla\psi$ and $\nabla\Delta\psi$ is a scalar, because for the steady flow these vectors are collinear (see (94) and (91)).

The condition (95) implies: the form (96) is definite positive. Hence, according to theorem 4, the steady flow (95) is stable. One will find a rigourous proof of stability in [2].

Proof of theorem 8. According to formulæ (41), (44), (67), the quadratic form (44) is

$$2\delta^2 E = \iint_D \delta v^2 + \delta v\{f, v\} dxdy, \text{ where } \delta v = B(v, f) = r \wedge f + \operatorname{grad}\alpha, \quad r = -\Delta\psi. \tag{97}$$

Integrating by parts, one finds according to (76) and (71)

$$\iint_D \delta v\{f, v\} dxdy = \iint_D \delta v \operatorname{rot}(f \wedge v) dxdy = \iint_D (\delta r)(f \wedge v) dxdy, \tag{98}$$

with obvious notations.

Formula (88) implies

$$f \wedge v = f \wedge I \nabla \psi = -(f, \nabla \psi). \tag{99}$$

On the other hand, it follows from (97), (51) and (91), that

$$\begin{aligned} \delta r = \{r, f\} &= -\Delta \psi_x f_1 - \Delta \psi f_{1_x} - \Delta \psi_y f_2 - \Delta \psi f_{2_y} \\ &= -\Delta \psi (f_{1_x} + f_{2_y}) - \Delta \psi_x f_1 - \Delta \psi_y f_2 = -(f, \nabla \Delta \psi). \end{aligned} \tag{100}$$

Comparing formulæ (99) and (100), one finds

$$f \wedge v = \frac{\nabla \psi}{\nabla \Delta \psi} \delta r. \tag{101}$$

Formula (96) results directly from (97), (98) and (101). Theorem 8 is proved.

Remark. The steady flow may be stable even in case formula (96) is neither positive definite, nor negative definite.

Indeed, the circular flow in a circular annulus

$$(\psi = \psi(x^2 + y^2))$$

is stable, if $\frac{\nabla \psi}{\nabla \Delta \psi}$ does not change sign. For the proofs see [1]; one uses theorem 4 and the first integral of angular momentum.

In the same way, the planar-parallel flow

$$\vec{v} = v(y), 0; \quad Y_1 \leq y \leq Y_2; \quad x \bmod X$$

is stable, if $v_{yy} \neq 0$, that is if the velocity profile has no inflexion points (non linear analogue of Rayleigh's theorem, see [12, 9]). Moreover, a flow with a single inflexion point is stable in the case of figure 5 ($v/v_{yy} > 0$), while in the case of figure 6 ($v/\frac{1}{2}u_{yy} < 0$), it is stable in a sufficiently narrow band. For example, if

$$v(y) = \sin y, Y_1 + Y_2 = 0, \; X = 2\pi,$$

the flow is stable for $|Y_2 - Y_1| > \pi$. For the proofs, see [1] and [2]; one uses theorem 4 and the first integral of linear momentum.

The methods of [1, 2] also apply to the other two-dimensional motions. For example, in [7] , L.A. Dikii has studied spherical flows.

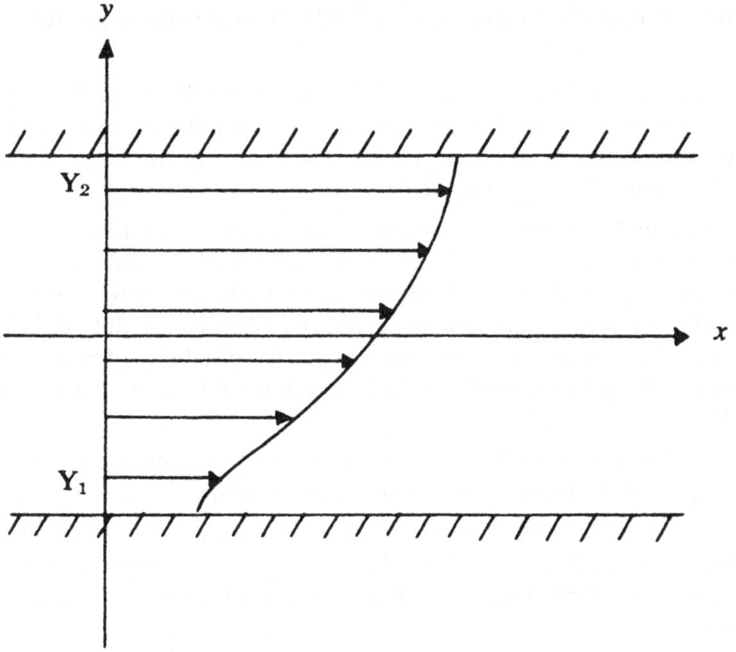

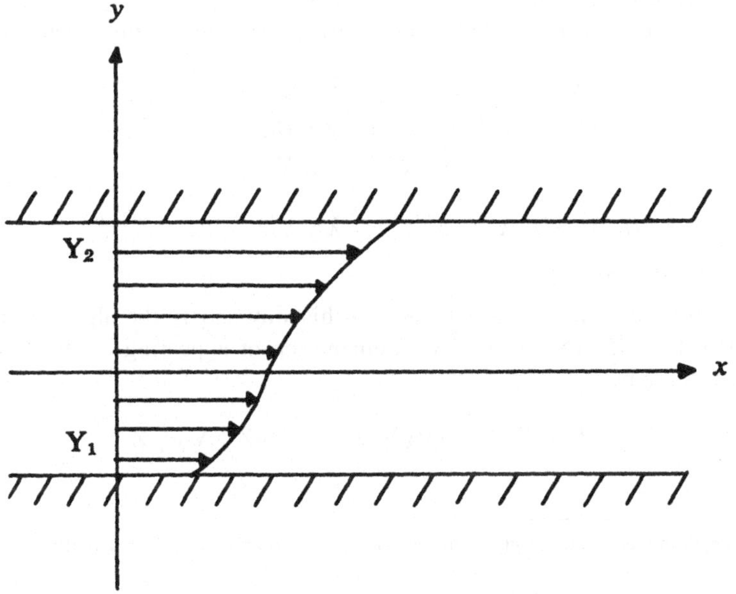

The problem: *does there exist a stable flow among the three-dimensional steady flows of an ideal fluid ?* remains open.

11 The riemannian curvature of diffeomorphisms groups

The expression (53) for the curvature of the Lie group endowed with a left invariant metric[14], is still meaningful in the case of the infinite dimensional group *SDiffD* of diffeomorphisms of a riemannian domain *D*. I call this same expression the "riemannian curvature" of the group *SDiffD*.

As an example, in this paragraph we shall compute the curvature of $SDiff\mathbb{T}^2$, the diffeomorphisms of a torus $\mathbb{T}^2$, endowed with a locally euclidean metric. The computations show that, in "good enough" sections, the curvature is *finite*. One finds many sections in which the curvature is *negative* (theorem 9). But one knows since Hadamard [10], that negative curvature implies the instability of geodesics. As the geodesics of the group *SDiffD* are ideal fluid flows, this shows the instability[15] of some flows.

On the other hand, one finds sections in which the curvature is *positive*. I do not know if there exists conjugate points on the group $SDiff\mathbb{T}^2$.

Let us first prove some auxiliary formulæ. As we saw in §9, the algebra $\mathfrak{U}$ of the group $SDiff\mathbb{T}^2$ may be considered as the algebra of stream functions $\psi(x,y)$. From now on we shall restrict ourselves to the subalgebra $\mathfrak{U}_0$ of *univalent* stream functions[16].

Let $\mathbb{C}\mathfrak{U}_0$ be the complex extension of the algebra $\mathfrak{U}_0$. The extension of the operations $[,], \langle,\rangle, B(,), \nabla, R$, defined in the former paragraphs, is immediate; one has

$$\begin{aligned}[]
[\lambda z, \lambda' z'] = \lambda\lambda'[z,z'];\ \langle \lambda z, \lambda z'\rangle = \lambda\lambda'\langle z,z'\rangle,\\
B(\lambda z, \lambda' z') = \lambda\lambda' B(z,z'),\\
\nabla_{\lambda z}\lambda' z' = \lambda\lambda'\nabla_z z';
\end{aligned} \tag{102}$$

$$\langle R(\lambda z, \lambda' z')\lambda'' z'', \lambda''' z'''\rangle = \lambda\lambda'\lambda''\lambda'''\langle R(z,z')z'', z'''\rangle, \tag{103}$$

for all $z \in \mathbb{C}\mathfrak{U}_0$ and all $\lambda \in \mathbb{C}$.

Obviously, the operations extended in this way satisfy the algebraic identities $(6), (10), (11), (45), (53), (59)$. The Riemann tensor appearing in (103) is defined in the real case by

$$R(X,Y)Z = -\nabla_X\nabla_Y Z + \nabla_Y\nabla_X Z + \nabla_{[X,Y]}Z. \tag{104}$$

[14] Of course, in the case of a right invariant metric, the curvature of a corresponding section is the same.

[15] here, the word instability has a meaning which differs from the one in par. 10: the question is about the proximity of the curves $g(t)$ in the group and not the the one of their images $v(t)$ in the algebra.

[16] This subalgebra is an ideal. The associated invariant subgroup (group of motions preserving the "center of mass") is a totally geodesic submanifold of $SDiff\mathbb{T}^2$, as results from the law of conservation of linear momentum. Hence, the curvature of the corresponding sections in the group $SDiff\mathbb{T}^2$ and in the subgroup are the same.

It is related to the the curvature (49) in the section defined by the two orthonormal vectors ξ, η, by the formula

$$C_{\xi\eta} = \langle R(\xi, \eta)\xi, \eta \rangle, \tag{105}$$

and it satisfies the symetry and Bianchi identities:

$$\langle R(X,Y)Z,W\rangle = -\langle R(X,Y)W,Z\rangle = \langle R(Z,W)X,Y\rangle \tag{106}$$

$$\langle R(X,Y)Z,W\rangle + \langle R(X,Z)W,Y\rangle + \langle R(X,W)Y,Z\rangle = 0. \tag{107}$$

The relations (106) and (107) remain valid in $\mathbb{C}\mathfrak{U}_0$.

Now, set

$$e_k = e^{i(k,x)},$$

a univalent function on the torus $\mathbb{T}^2 = \mathbb{R}^2/\Gamma$, $\mathbb{R}^2 = \{x\}$, k a vector in the euclidean plane $\mathbb{R}^2$. Obviously, the admissible vectors k form a lattice and the functions e_k, $k \neq 0$, form a basis in the space $\mathbb{C}\mathfrak{U}_0$. One shall compute the operations (102), (103) on the basis vectors.

Let S be the area of the torus $\mathbb{T}^2$, and $R_{k,l,m,n} = \langle R(e_k, e_l)e_m, e_n\rangle$.

Lemma 10. *The following formulae hold true:*

$$\langle e_k, e_l \rangle = 0 \quad \textit{for} \quad k+l \neq 0; \quad \langle e_k, e_{-k}\rangle = k^2 S; \tag{108}$$

$$[e_k, e_l] = (k \wedge l)e_{k+l}; \tag{109}$$

$$B(e_k, e_l) = b_{k,l}e_{k+l}, \quad b_{k,l} = (k \wedge l)\frac{k^2}{(k+l)^2}; \tag{110}$$

$$\nabla_{e_k} e_l = d_{l,k+l}e_{k+l}, \quad d_{u,v} = \frac{(v \wedge u)(u \cdot v)}{v^2} \tag{111}$$

$$\boxed{R_{k,l,m,n} = (a_{l,n}a_{k,m} - a_{l,m}a_{k,n})S,} \tag{112}$$

for k+l+m+n=0 one has

$$\boxed{a_{u,v} = \frac{(u \wedge v)^2}{|u+v|}}; \tag{113}$$

if $k+l+m+n \neq 0$, *then* $R_{k,l,m,n} = 0$.

Proof of lemma 10. Formula (108) is an immediate consequence of definition (67), and formula (109) is a consequence of formulæ(51) and (52).

Then, it follows from formulæ(10) and (109) that

$$\langle B(e_k, e_l), e_m\rangle = (l \wedge m)\langle e_{l+m}, e_k\rangle. \tag{114}$$

According to (108), formula (114) shows that $B(e_k, e_l)$ is orthogonal to e_m for $k+l+m \neq 0$. Hence $B(e_k, e_l) = b_{k,l} e_{k+l}$. One finds the expression (110) for $b_{k,l}$ by using (114) and (108) with $m = -k-l$.

According to (109), (110), it follows from (55) that

$$\nabla_{e_k} e_l = \frac{1}{2}(k \wedge l)\left(1 - \frac{k^2 - l^2}{(k+l)^2}\right) e_{k+l}. \tag{115}$$

But obviously

$$\frac{1}{2}(k \wedge l)\left(1 - \frac{k^2 - l^2}{(k+l)^2}\right) = \frac{\langle l, k+l \rangle}{(k+l)^2}. \tag{116}$$

Formulae (115) and (116) imply (111).

Then, substituting (111) into (104) , one finds according to (103)

$$\nabla_{e_k} \nabla_{e_l} e_m = d_{l+m,k+l+m} d_{m,l+m} e_{k+l+m}, \tag{117}$$

$$\nabla_{e_l} \nabla_{e_k} e_m = d_{k+m,k+l+m} d_{m,k+m} e_{k+l+m}, \tag{118}$$

$$\nabla_{[e_k, e_l]} e_m = (k \wedge l) d_{m,k+l+m} e_{k+l+m}. \tag{119}$$

Comparing formulæ (104), (117), (118), (119) and (108), one finds $R_{k,l,m,n} = 0$ for $k+l+m+n \neq 0$, and for $k+l+m+n = 0$, one finds

$$R_{k,l,m,n} = (d_{k+m,n} d_{m,k+m} - d_{l+m,n} d_{m,l+m} + (k \wedge l) d_{m,n}) n^2 S, \tag{120}$$

(one has used the fact that, according to (111), $d_{u,v}$ is symetric in u and in v)).

Now, let $k+l+m+n = 0$. According to (111), one has

$$\left.\begin{aligned}
-d_{k+m,n} &= \frac{l \wedge (k+m)}{n^2}(k+m, n); \\
-d_{m,l+m} &= \frac{m \wedge l}{(l+m)^2}(l+m, m); \\
-d_{m,k+m} &= \frac{m \wedge k}{(k+m)^2}(k+m, m); \\
-d_{m,n} &= \frac{(l+k) \wedge m}{n^2}(m, n). \\
-d_{l+m,n} &= \frac{k \wedge (l+m)}{n^2}(l+m, n);
\end{aligned}\right\} \tag{121}$$

It follows from (121) and (120) that:

$$S^{-1} R_{k,l,m,n} = A(l \wedge k)(m \wedge k) + (A-B)(l \wedge m)(m \wedge k) + B(k \wedge l)(l \wedge m), \tag{122}$$

where

$$A = \frac{(k+m,m)(k+m,n)}{(k+m)^2} - (m,n);$$
$$B = \frac{(l+m,m)(l+m,n)}{(l+m)^2} - (m,n);. \qquad (123)$$

Now one shall use the identiy

$$\frac{(a,b)(a,c)}{a^2} + \frac{(a \wedge b)(a \wedge c)}{a^2} = (b,c) \qquad (124)$$

where a, b, c are three arbitrary vectors in the euclidean plane.

(Formula (124) is best known in the form

$$\cos\alpha\cos\beta + \sin\alpha\sin\beta = \cos(\alpha - \beta)).$$

It follows from $(123), (124)$ and $k+l+m+n=0$, that

$$A = \frac{(k \wedge m)(l \wedge n)}{(k+m)^2}, \quad B = \frac{(l \wedge m)(k \wedge n)}{(l+m)^2}. \qquad (125)$$

Formula (122) is equivalent to

$$S^{-1}R_{k,l,m,n} = A(m \wedge k)(l \wedge (k+m)) + B(l \wedge m)(k \wedge (l+m)). \qquad (126)$$

Taking into account that $k+l+m+n=0$, (126) reduces to

$$S^{-1}R_{k,l,m,n} = A(m \wedge k)(n \wedge l) + B(l \wedge m)(n \wedge k). \qquad (127)$$

Comparing (127) and (125), one finds

$$S^{-1}R_{k,l,m,n} = \frac{(k \wedge m)^2 (l \wedge n)^2}{(k+m)^2} - \frac{(l \wedge m)^2 (k \wedge n)^2}{(l+m)^2}. \qquad (128)$$

Formula (128) is equivalent to $(112), (113)$, because $k+m = -(l+n)$ and $l+m = -(k+n)$. This proves lemma 10.

Theorem 9. *In each section which contains the vector $\xi \in \mathfrak{U}$, corresponding to the stream function* $\cos kx$, *the curvature of the group* $SDiff\,\mathbb{T}^2$ *is non-negative.*

Theorem 9 is an immediate consequence of formula (129) below.

Lemma 11. *Let*

$$2\xi = e_k + e_{-k}, \quad \eta = \sum x_l e_l$$

be two real vectors in $\mathfrak{U}$, *that is such that* $x_{-l} = \bar{x}_l$. *Then, with the notations of* (113),

$$\boxed{\langle R(\xi,\eta)\xi,\eta\rangle = -\frac{S}{4}\sum_l a_{k,l}^2 \,|x_l + x_{l+2k}|^2} \qquad (129)$$

We note a corollary of formula (129):

Corollary 1. *In the section defined by the stream functions* $\xi = \cos kx, \eta = \cos lx$, *the curvature of the group* $SDiff\mathbb{T}^2$ *is*

$$\boxed{C_{\xi,\eta} = -\frac{k^2+l^2}{4S}\sin^2\alpha\sin^2\beta} \tag{130}$$

where S *is the area of the torus,* α *the angle between* k *and* l, β *the angle between* $k+l$ *and* $k-l$.

For example, for $\mathbb{T}^2 = \{x,y \mod 2\pi\}$ the curvature in the section determined by the vector-fields $(\sin y, 0)$ and $(0, \sin x)$ is

$$\boxed{C_{\xi,\eta} = -\frac{1}{8\pi^2}.} \tag{131}$$

Formula (130) implies (131); here $k^2 = l^2 = 1$, $S = 4\pi^2$, $\alpha = \beta = \frac{\pi}{2}$.

Proof of formula (130). As the vectors ξ, η are orthogonal in $\mathfrak{U}$

$$\boxed{C_{\xi,\eta} = -\frac{\langle R(\xi,\eta)\xi,\eta\rangle}{\langle\xi,\xi\rangle\langle\eta,\eta\rangle}.} \tag{132}$$

According to (108),

$$\langle\xi,\xi\rangle = \frac{1}{2}k^2S, \quad \langle\eta,\eta\rangle = \frac{1}{2}l^2S. \tag{133}$$

Setting in (129), $x_l = x_{-l} = \frac{1}{2}$, one finds

$$\langle R(\xi,\eta)\xi,\eta\rangle = -\frac{S}{16}(a_{k,l}^2) + a_{k,l-2k}^2 + a_{k,-l}^2) + a_{k,-l-2k}^2. \tag{134}$$

In accordance with formula (113), $a_{k,l-2k}^2 = a_{k,-l}^2$ and $a_{k,-l-2k}^2 = a_{k,l}^2$. Hence (134) becomes

$$\langle R(\xi,\eta)\xi,\eta\rangle = -\frac{S}{8}(a_{k,l}^2 + a_{k,-l}^2). \tag{135}$$

It follows from (113) that

$$a_{k,l}^2 + a_{k,-l}^2 = (k\wedge l)^4\left(\frac{1}{h_+^2} - \frac{1}{h_-^2}\right), \quad h_+ = k+l,\ h_- = k-l. \tag{136}$$

One shall use the obvious identities

$$h_+^2 + h_-^2 = 2(k^2+l^2), \quad h_+\wedge h_- = -2(k\wedge l). \tag{137}$$

It follows from (136) and (137) that

$$a_{k,l}^2 + a_{k,-l}^2 = \frac{(k\wedge l)^2(h_+\wedge h_-)^2}{2h_+^2h_-^2}(k^2+l^2). \tag{138}$$

Comparing formulæ (138), (135), (133), (132) and taking into account that

$$(k\wedge l)^2 = k^2l^2\sin^2\alpha, \quad (h_+\wedge h_-)^2 = h_+^2h_-^2\sin^2\beta,$$

one gets (130). The corollary is proved.

Proof of lemma 11. According to (103),

$$\begin{aligned}\langle R(\xi,\eta)\xi,\eta\rangle = \frac{1}{4}\sum_l (&R_{k,l,k,-2k-l}x_lx_{-2k-l} + R_{-k,l,-k,2k-l}x_lx_{2k-l}\\ &+R_{k,l,-k,-l}x_lx_{-l} + R_{-k,l,k,-l}x_lx_{-l}).\end{aligned} \tag{139}$$

Using (112), one finds for the coefficients of the quadratic form (139) the values

$$R_{k,l,k,-2k-l} = R_{-k,l,k,-l} = -a_{k,l}^2S, \quad R_{k,l,-k,-l} = R_{-k,l,-k,2k-l} = -a_{k,-l}^2S.$$

Hence, the form (139) becomes

$$\langle R(\xi,\eta)\xi,\eta\rangle = -\frac{S}{4}\sum_l[a_{k,l}^2(x_lx_{-2k-l}+x_lx_{-l}) + a_{k,-l}^2(x_lx_{2k-l}+x_lx_{-l})]. \tag{140}$$

But it follows from formula (113) that

$$a_{k,-l}^2 = a_{k,l-2k}^2. \tag{141}$$

Using (141), one shows easily that

$$\sum_l a_{k,-l}^2(x_lx_{2k-l}+x_lx_{-l}) = \sum_l a_{k,l}^2(x_{l+2k}x_{-l}+x_{l+2k}x_{-l-2k}). \tag{142}$$

It follows from (140), (142) and the reality relation $x_{-j} = \bar{x}_j$, that

$$\langle R(\xi,\eta)\xi,\eta\rangle = -\frac{S}{4}\sum_l a_{k,l}^2(x_l\bar{x}_{l+2k} + x_l\bar{x}_l + x_{l+2k}\bar{x}_l + x_{l+2k}\bar{x}_{l+2k}),$$

which is equivalent to (129). Lemma 11 is proved.

Remark. There also exist sections in which the curvature of the group $SDiff\,\mathbb{T}^2$ is positive. Indeed, let us consider a section, defined by two stream lines

$$\begin{aligned}\xi &= \cos(3px-y)+\cos(3px+2y),\\ \eta &= \cos(px+y)+\cos(px-2y),\end{aligned} \tag{143}$$

on the torus $\mathbb{T}^2 = \{x, y \mod 2\pi\}$. The sectional curvature is

$$C_{\xi,\eta} = \frac{\langle R(\xi,\eta)\xi,\eta\rangle}{\langle\xi,\xi\rangle\langle\eta,\eta\rangle} \to \frac{9}{8\pi^2} \quad \text{when} \quad p \to \infty. \tag{144}$$

Proof of formula (144). Let k, l, m, n be integer planar vectors, $k + m = l + n$, and

$$\begin{aligned} \xi &= x_k \cos kx + x_m \cos mx, \\ \eta &= y_l \cos lx + y_n \cos nx. \end{aligned} \tag{145}$$

Then, according to (103), (106), (107), one finds

$$\begin{aligned} \langle R(\xi,\eta)\xi,\eta\rangle =& \frac{S}{16}(\alpha_{k,l}x_k^2x_l^2 + \alpha_{k,n}x_k^2y_n^2 + \alpha_{m,l}x_m^2y_l^2 \\ &+ \alpha_{m,n}x_m^2y_n^2 + \alpha x_k x_m y_l y_n), \end{aligned} \tag{146}$$

where, in accordance with formula (112),

$$\left.\begin{aligned} \alpha_{k,l} &= -2(a_{k,l}^2 + a_{k,-l}^2); & \alpha_{k,n} &= -2(a_{k,n}^2 + a_{k,-n}^2); \\ \alpha_{m,l} &= -2(a_{m,l}^2 + a_{m,-l}^2); & \alpha_{m,n} &= -2(a_{m,n}^2 + a_{m,-n}^2); \\ \alpha &= 4(2a_{k,m}a_{l,n} - a_{k,-n}a_{l,-m} - a_{k,-l}a_{m,-n}). \end{aligned}\right\} \tag{147}$$

The vectors (143) have the form (145) with $x_k = x_m = y_l = y_n = 1$, and

$$\left.\begin{aligned} k &= -3p, 1; & m &= 3p, 2; \\ l &= p, 1; & n &= -p, 2. \end{aligned}\right\} \tag{148}$$

It follows from expressions (148) that for $p \to \infty$, all the products $k \wedge l \ldots$ are of order p, with the sums and differences $|k \pm l| \ldots$, except

$$|k + m| = |l + n| = 3.$$

Hence, according to (113), all the $a_{k,l} \ldots$ in (147) are of order p, except

$$a_{k,m} = 27p^2, \qquad a_{l,n} = 3p^2.$$

Thus, one obtains for (146) the asymptotic expression:

$$\langle R(\xi,\eta)\xi,\eta\rangle \sim \frac{S\alpha}{16} \sim \frac{S \cdot 81p^4}{2}, \quad (p \to \infty). \tag{149}$$

According to (108), $\langle\xi,\xi\rangle = \frac{1}{2}(k^2 + m^2)S$; $\langle\eta,\eta\rangle = \frac{1}{2}(l^2 + n^2)S$. Hence it follows from (148) that

$$\langle\xi,\xi\rangle \sim 9p^2 S, \quad \langle\eta,\eta\rangle \sim p^2 S \quad (p \to \infty). \tag{150}$$

Given the area $S = 4\pi^2$ of the torus, formulæ(149) and (150) give (144). Q.E.D.

A more precise computation shows that $C_{\xi,\eta} > 0$ already for $p = 2$, that is in the section

$$\xi = \cos(6x - y) + \cos(6x + 2y), \quad \eta = \cos(2x + y) + \cos(2x - 2y).$$

References

[1] V. Arnold, Journal de Mécanique (to appear)
[2] V. Arnold, Doklady, 163 $n^0 2$ (1965)
[3] V. Arnold, C.R.A.S., 260 (1965), p. 5668
[4] V. Arnold, C.R.A.S., 261 (1965), p. 17
[5] V. Arnold, Russian mathematicla surveys, 18 $n^0 6$ (1963), 91-192
[6] N. Bourbaki, *l'Architecture de la Mathématique*, in N. Bourbaki, Istoria Matematiki, Moscow, Fizmatgiz, (1963).
[7] L.AL Dikii, *Sur la théorie non-linéaire de la stabilité hydrodynamique*, Prikladnaia Mathematiqua i Mecaniqua, 29 $n^0 5$ (1965), 852-855
[8] L. Euler, *Theoria motus corporum solidorum sei rigidorum*, (1765)
[9] R. Fjortioft, Geofysiske Publikasjoner, Oslo, 17 $n^0 6$ (1950)
[10] J. Hadamard, J. Math? Pures et Appl. 5e série, 4 (1898), 27-73
[11] V. Judovic, Doklady, 1366 (1961), p. 564
[12] C.C. Lin, *The theory of hydrodynamic stability*, Oxford (1953)
[13] J. Milnor, *Morse theory*, Ann. Math. Studies, 51 (1963)
[14] J.J. Moreau, C.R.A.S., 249 (1959), p. 2156

Manuscript received on October 13 1965

V. Arnold,
Faculty of Mechanics and Mathematics,
Moscow University, Moscow V-234 (S.S.S.R.)

On a variational principle for the steady flows of perfect fluids and its applications to problems of non-linear stability

V. Arnold[1]
Translated by Alain Chenciner

Summary One shows that a steady fluid flow has an extremal energy value among "isovorticed flows". If the extremum is maximum or minimum, the steady flow is stable with respect to a finite perturbation. In order to clarify the nature of the extremum, an explicit expression for the second variation of energy is given. This way one obtains sufficient conditions of flow stability. These conditions are close to be necessary ones, at least for planar flows.

1 A finite-dimensional model

We shall show that the system of hydrodynamical equations of an ideal fluid may be considered as an infinite-dimensional generalization of finite-dimensional systems of differential equations of a certain kind. Indeed, let

$$\dot{x} = f(x), \qquad x = (x_1, \cdots, x_n) \tag{1}$$

be a system of ordinary differential equations. We assume that the following hypotheses are satisfied:

(H_1) the space of x's is endowed with an "invariant foliated structure", i.e., the space is stratified by leaves F of dimension k (in Figure 1: $n = 3, k = 2$) in such a way that an integral curve $x(t)$ of (1) originating from a point in leaf F, remains forever in this leaf. We shall say that a point x in F is a "regular point", if in a neighbourhood of x there exists a coordinate system $y_1, \cdots, y_n$ such that the equations of leaves are $y_{k+1} = c_{k+1}, \cdots, y_n = c_n$.

(H_2) the system (1) admits a first integral $E(x)$.

Translation of Journal de Mécanique, Vol. 5 $n^o 1$ (March 1966), p. 29–43.

[1] Professor of Moscow University. The author thanks Professor P. Germain who helped him write this text in French.

Originally publ. in: Journal de Mécanique, Vol. 5, No. 1, 29-43, © Gauthier-Villars, Paris 1966
Translated by A. Chenciner. Springer-Verlag Berlin Heidelberg 2014 with kind permission, DOI 10.1007/978-3-642-31031-7_8

Now, let us consider a regular point x_0 at which the function E assumes an extremal value on the leaf F and let us suppose that at x_0 the quadratic form d^2E on F is non-degenerate. Then, one proves easily the following three theorems.

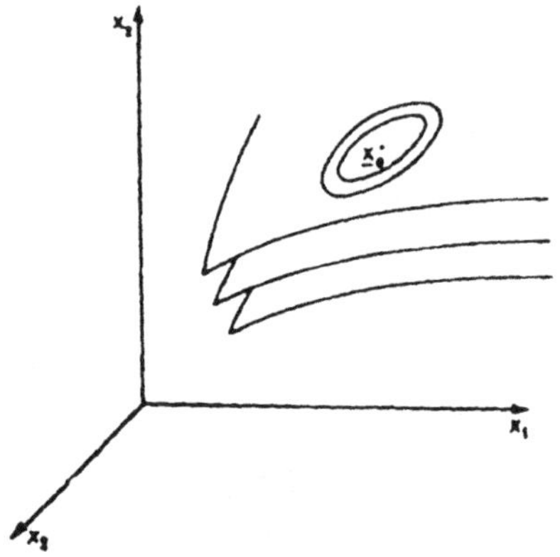

Fig. 1

Theorem 1. *The point x_0 is stationary: $f(x_0) = 0$.*

Theorem 2. *If the extremum is a maximum or a minimum, the point x_0 is stable with respect to finite perturbations.*

Theorem 3. *The spectrum $\{\lambda\}$ for the problem of "small motions" $A\xi = \lambda\xi$ associated to (1), where $A = |\frac{\partial f}{\partial x}|_{x_0}$, is symmetric with respect to both the real and imaginary axes of the λ plane.*

Therefore, if there exists a damped ($Re(\lambda) < 0$) motion of the linearized system, then there also necessarily exists a motion of this system whose amplitude increases without bound.

Theorem 3 is a corollary of the following proposition:

Let $\dot{x} = Ax$ be a system of linear differential equations which admits a non-degenerate quadratic first integral (Bx, x). Then the spectrum satisfies the symmetry conditions expressed in Theorem 3. The proof is immediate:

$$\det(A - \lambda E) = \det(BAB^{-1} - \lambda E) = (-1)^n \det(A' + \lambda E) = (-1)^n \det(A + \lambda E)$$

because $BA + A'B = 0$, where A' is the transpose of the matrix A.

Of course, the roots of $\det(A - \lambda E) = 0$ are either real or form an imaginary conjugate pair, while the above identity shows that if λ_1 is a root, so is $-\lambda_1$. This proves the proposition.

The hydrodynamical analogues of Theorems 1, 2, 3, are Theorems 5, 6, 7 which are proved below. These theorems are particular cases of general theorems on geodesics of Lie groups endowed with one-sided invariant metrics ([1, 2]). But here it is not necessary to use the theory of Lie groups or theory of partial differential equations. Hence, from the mathematical point of view, the results given below are *a priori* equalities and inequalities.

2 Notations

Let $\mathscr{D}$ be a domain in the standard space and with a fixed boundary Γ. Let v be the velocity field of an ideal fluid flow (incompressible, nonviscous, with density 1). The exterior forces are neglected. Let p be the pressure. The field v must satisfy the following equations (the Euler equation, the continuity equation, boundary conditions):

$$\frac{\partial v}{\partial t} + v \cdot \nabla v = -\operatorname{grad} p, \quad \operatorname{div} v = 0, \quad (v, n) = 0 \text{ on } \Gamma. \tag{2}$$

As a consequence, one can write the Bernoulli equation

$$\frac{\partial v}{\partial t} = v \wedge r - \operatorname{grad} h, \quad h = p + \frac{||v||^2}{2}, \quad r = \operatorname{curl} v. \tag{3}$$

Let us denote by $\{A, B\}$ the Poisson bracket

$$\{A, B\}_i = \frac{\partial A_i}{\partial x_j} B_j - \frac{\partial B_i}{\partial x_j} A_j.$$

Thanks to the identity

$$\operatorname{curl} A \wedge B = \{A, B\} + A \operatorname{div} B - B \operatorname{div} A$$

the equation of motion may be put in the form

$$\frac{\partial r}{\partial t} = \{v, r\}. \tag{4}$$

Let G be a differentiable mapping $x \mapsto G(x)$, and denote by G^* the associated linear mapping on the vectors ξ:

$$(G^* \xi)_i = \sum_j \frac{\partial G_i}{\partial x_j} \xi_j.$$

By hypothesis, the linear mapping G^* is non-singular.

By definition, we say that a field v defined on $\mathscr{D}$ is a kinematically admissible field, if $\operatorname{div} v = 0$ in $\mathscr{D}$ and $(v, n) = 0$ on the boundary Γ of $\mathscr{D}$.

3 Isovorticed flows

Let v and v' be two kinematically admissible fields defined on $\mathscr{D}$. We shall introduce the following definition.

Definition 1. The kinematically admissible fields v and v' are called isovorticed if there exists a differentiable mapping G from $\mathscr{D}$ to $\mathscr{D}$, which preserves the volume and satisfies

$$\oint_{\gamma} v dx = \oint_{G\gamma} v' dx \tag{5}$$

for every curve γ drawn in $\mathscr{D}$.

First let us show that the mapping G preserves the vorticity, that is one has

$$G^* \operatorname{curl} v = \operatorname{curl} v'. \tag{6}$$

Indeed, G is volume preserving, hence one can write the following identity for the mixed products:

$$(G^*\xi, G^*\eta, G^* \operatorname{curl} v) = (\xi, \eta, \operatorname{curl} v)$$

for any vectors ξ, η. On the other hand, according to (5) and the Stokes theorem one has:

$$(G^*\xi, G^*\eta, G^* \operatorname{curl} v') = (\xi, \eta, \operatorname{curl} v),$$

which proves (6). We note that in the case when $\mathscr{D}$ is multi-connected, condition (5) is stronger than condition (6).

Now, the theorem of conservation of vorticity assumes the following form.

Theorem 4. *Let $v(x,t)$ be a fluid velocity field at time t (this field, by hypothesis, satisfies equations* (2)*), then the fields $v(x,0)$ and $v(x,t)$ are isovorticed.*

Let us consider system of equations (2) as a system generalizing (1) to the (infinite-dimensional) functional space of kinematically admissible vector fields $v(x)$. The latter space is endowed with a foliated structure: two vector-fields belong to the same leaf is they are isovorticed. This foliated structure is invariant (Theorem 4). Steady flows are the stationary points of system (2). Finally, this system admits the energy integral $2E = \int v^2 dx$.

Hence the conditions of §1 are all satisfied.

4 The variational principle[2]

Theorem 5. *The energy of a steady flow v is extremal with respect to all sufficiently close kinematically admissible isovorticed flows v'.*

[2] Another variational principle, valid in the non-steady case, is given by Fjorthoft [3] who also applies it to stability problems, see also [4].

Here, "sufficiently close" means "sufficiently close in the leaf". In other words, two flows v and v' are close if the mapping G [see (5)] is close to the identity. It is easy to construct mappings close to the identity in the following way. Let $f(x)$ be a kinematically admissible field on $\mathcal{D}$ ($\operatorname{div} f = 0$ in $\mathcal{D}$, $(f,n) = 0$ on Γ). Let

$$G_t = \exp(ft) : x(0) \mapsto G_t x(0) = x(t)$$

be the mapping from $\mathcal{D}$ to $\mathcal{D}$ defined by solutions $x(t)$ of the system of differential equations $\dot{x} = f(x)$. Such a mapping preserves volumes. The mapping g_t is close to the identity if f is small. Hence one can use f as "coordinates" of v' in (5) on the leaf F. If f is small, so is $v' - v$. The precise expression of the difference $v' - v$ is given by the following lemma:

Lemma *If for every closed curve γ drawn in $\mathcal{D}$ and for every t*

$$\oint_{G_{-t}\gamma} v dx = \oint_{\gamma} v' dx, \quad (\operatorname{div} v = \operatorname{div} v' = 0) \tag{7}$$

then

$$v' - v = t f \wedge r + \frac{t^2}{2} f \wedge \{f \wedge r\} + \operatorname{grad} \alpha + O(t^3) \tag{8}$$

where α is a univalent function in $\mathcal{D}$ and $r = \operatorname{curl} v$.

Proof of the lemma. According to the Stokes formula and to the formula of Lagrangian derivative for the flux of a conservative field,

$$\frac{d}{dt} \oint_{G_{-t}\gamma} v dx = -\frac{d}{dt} \iint r \cdot ds = \oint_{G_{-t}\gamma} f \wedge r dx. \tag{9}$$

The Jacobian of G_{-t} is equal to 1, therefore

$$\oint_{G_{-t}\gamma} f \wedge r dx = \oint_{\gamma} G_t^* f(G_{-t}x) \wedge G_t^* r(G_{-t}x) dx. \tag{10}$$

But $G_t^* f(G_{-t}x) = f(x)$. From (9) and (10) one derives

$$\frac{d}{dt} \int_{G_{-t}\gamma} v dx = \int_{\gamma} f \wedge r(t) dx, \tag{11}$$

where $r(t)$ is the vector field

$$r(x,t) = G_t^* r(G_{-t}x). \tag{12}$$

By differentiating (12) one obtains

$$\frac{dr}{dt}\Big|_0 = \{f, r\}, \tag{13}$$

and hence

$$r(t) = r + t\{f,r\} + O(t^2).$$

But, according to (7),

$$\frac{d}{dt}\int\limits_{G_{-t}\gamma} v dx = \int\limits_{\gamma} \frac{\partial v'}{\partial t} dx$$

and moreover $v'|_{t=0} = v$, hence, integrating (11), (13) between moments 0 and t one gets

$$\oint\limits_{\gamma}(v' - v)dx = \oint\limits_{\gamma}\int\limits_{0}^{t} f \wedge (r + t\{f,r\} + O(t^2))dtdx,$$

which leads to the formula (8).

Proof of theorem 1. Let v' be a flow isovorticed with and sufficiently close to the steady flow v. According to (5) and (8), $\delta v = f \wedge r + \operatorname{grad}\alpha$. Therefore

$$\delta E = \iiint\limits_{\mathscr{D}} v\delta v d\mathrm{v} = \iiint\limits_{\mathscr{D}} v(f \wedge r + \operatorname{grad}\alpha)\, d\mathrm{v}$$

$$= \iiint\limits_{\mathscr{D}} (f \cdot r \wedge v + v \cdot \operatorname{grad}\alpha)\, d\mathrm{v}.$$

For a steady flow, according to (3),

$$r \wedge v = -- \operatorname{grad} h.$$

Hence

$$\delta E = \iiint\limits_{\mathscr{D}} (v \cdot \operatorname{grad}\alpha - f \cdot \operatorname{grad} h) d\mathrm{v}$$

$$= \iiint\limits_{\mathscr{D}} \operatorname{div}(\alpha v - hf) d\mathrm{v} = \iint\limits_{\Gamma} (\alpha v - hf) ds = 0.$$

This result is obtained by taking into account the equalities

$$\operatorname{div} v = \operatorname{div} f = 0 \qquad \text{and} \qquad (v,n) = (f,n) = 0 \quad \text{on } \Gamma.$$

5 The second variation

The second variation of energy on the leaf of fields isovorticed with v can be written in the following form (14), making use of the "coordinates" f introduced in the last section:

$$\delta^2 E = \iiint_{\mathscr{D}} \left((\delta v)^2 + \frac{1}{2}(v \wedge f) \cdot \{f, r\}\right) dv. \tag{14}$$

Proof. According to (8)

$$\delta v = f \wedge r + \operatorname{grad} \alpha_1,$$
$$\delta^2 v = \frac{1}{2} f \wedge \{f, r\} + \operatorname{grad} \alpha_2.$$

Then

$$2\delta^2 E = \iiint_{\mathscr{D}} 2((\delta v)^2 + v\delta^2 v) dv = \iiint_{\mathscr{D}} (2(\delta v)^2 + v \cdot (f \wedge \{f, r\})) dv$$

(in the above, the integral with grad α_2 is equal to 0). Formula (14) is proved.

Note that (14) is a quadratic form in f because δv is a linear homogeneous function in f. Indeed, $\delta v = f \wedge r + \operatorname{grad} \alpha$, where the function α is determined[3] by $\operatorname{div} \delta v = 0, (\delta v, n) = 0$ on Γ, and hence, α depends linearly on f.

The hydrodynamical analogues of Theorems 2, 3 are:

Theorem 6. *The steady flow v is stable with respect to a finite (but sufficiently small) perturbation provided that the quadratic form (14) is positive definite or negative definite.*

Theorem 7. *If some perturbation goes to 0, then another one increases and hence the steady flow is unstable.*

This statement is made more precise and proved in the appendix.

We did not find flows for which the quadratic form (14) is sign definite for all *three-dimensional* perturbations. But in some particular cases with symmetries Theorem 6 gives effective criteria (see §7).

6 Symmetric flows

To generalize Theorem 5, let us suppose that the Euler equation admits a first integral M such that for a steady flow v one has:

$$\delta M = \iiint_{\mathscr{D}} A \delta v \, dx dy dz, \qquad \text{where} \qquad A \wedge \operatorname{curl} v = \operatorname{grad} \alpha. \tag{15}$$

Condition (15) is satisfied in the following examples:

[3] The function α is a solution of the Poisson equation defined by a Neumann boundary condition. This solution exists and it is defined up to an additive constant due to the divergence theorem.

EXAMPLE 1. For the energy integral $M_1 = E$ one takes

$$A = v \qquad \text{and} \qquad A \wedge \operatorname{curl} v = \operatorname{grad} h$$

according to the Bernoulli equation (3).

EXAMPLE 2. Let a domain $\mathscr{D}$ and a steady flow v be invariant under translations parallel to the x axis. Then one can take

$$M = M_2 = \iiint_{\mathscr{D}} v \cdot e_x dxdydz.$$

One also has

$$A = e_x \qquad \text{and} \qquad \alpha = v \cdot e_x.$$

EXAMPLE 3. Let a domain $\mathscr{D}$ and a steady flow v be invariant under rotations around the z axis. Then one can take

$$M = M_3 = \iiint_{\mathscr{D}} (v \wedge R) \cdot e_z dxdydz.$$

One has

$$A = R \wedge e_z \qquad \text{and} \qquad \alpha = v \wedge R \cdot e_z,$$

where R is the vector (x,y,z).

Theorem 8. *The value of M for a steady flow v is extremal among the values corresponding to neighbouring fields v' isovorticed with v.*

The proof repeats the one of Theorem 5. The corresponding second variation formulæ are:

$$2\delta^2 M_2 = \iiint_{\mathscr{D}} e_x \wedge f \cdot \{f,r\} dv, \tag{16}$$

$$2\delta^2 M_3 = \iiint_{\mathscr{D}} (R \wedge e_z) \wedge f \cdot \{f,r\} dv. \tag{17}$$

Let us add that in (14), (16), (17)

$$\{f,r\} = \delta r = \operatorname{curl} \delta v$$

[see (4) and (8)].

Arguing as in Theorem 6, one obtains

Theorem 9. *A sufficient condition for stability of the field v is that a certain linear combination*

$$\delta^2 H = \lambda_1 \delta^2 M_1 + \lambda_2 \delta^2 M_2 + \lambda_3 \delta^2 M_3$$

is positive definite or negative definite.

7 Applications to steady flows (see also [3, 4])

Now suppose that the (unperturbed and perturbed) flows are planar. A planar flow admits a stream function $\psi(x,y)$:

$$v = (\psi_y, -\psi_x, 0); \quad \operatorname{curl} v = (0,0,-\Delta\psi). \tag{18}$$

For a stationary flow, one derives from (4):

$$\nabla\psi \wedge \nabla\Delta\psi = 0. \tag{19}$$

Hence, the vectors $\nabla\psi$ and $\nabla\Delta\psi$ are collinear. Substituting (18) in (14), (16), (17), and using (19), one obtains

Theorem 10. *In the planar case, the second variations (14), (16), (17) are*

$$2\delta^2 E = \iint\limits_{\mathscr{D}} \left(2(\delta v)^2 + \frac{\nabla\psi}{\nabla\Delta\psi}(\delta r)^2\right) dxdy, \tag{20}$$

$$2\delta^2 M_2 = \iint\limits_{\mathscr{D}} \frac{\nabla y}{\nabla\Delta\psi}(\delta r)^2 dxdy, \tag{21}$$

$$2\delta^2 M_3 = \iint\limits_{\mathscr{D}} \frac{\nabla(x^2+y^2)}{2\nabla\Delta\psi}(\delta r)^2 dxdy, \tag{22}$$

where $\delta r = curl\,\delta v$.

Here are some examples of application of theorems 9 and 10.

EXAMPLE 1. *A stationary flow in an arbitrary domain (Figure 2).*

Let us suppose that the velocity profile is concave $\left(\frac{\nabla\Delta\psi}{\nabla\psi} > 0\right)$, i.e., the form (20) is positive. Theorems 9, 10 give

Theorem 11. *All stationary flows with a concave velocity profile are stable.*

Recall the precise notion of non-linear stability used here.

For any $\varepsilon > 0$ there exists $\delta > 0$ such that, if one changes the velocity field v at instant $t = 0$ by a sufficiently small perturbation (i.e. $|v'(0) - v| < \delta$), the difference between the two solutions of the Euler equation remains small for all $t < +\infty$ (i.e. $|v'(t) - v| < \varepsilon$). Here

$$||\delta v||^2 = \iint (\delta v^2 + \delta r^2) dxdy, \quad \delta r = \operatorname{curl}\delta v.$$

EXAMPLE 2. *Flows parallels to the x axis: $\psi = \psi(y)$.*

In this case, one can use the integral M_2.

If the velocity profile has no inflection point ($\nabla\Delta\psi \neq 0$) (Figure 3), one can choose λ such that $\frac{\nabla(\psi+\lambda y)}{\nabla\Delta\psi} > 0$, and hence $\delta^2 E + \lambda\delta^2 M_2 > 0$, see (21).

Therefore, Theorems 9, 10 give

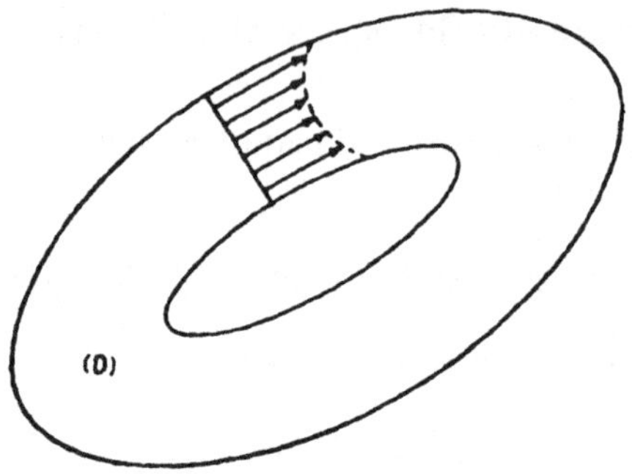

Fig. 2

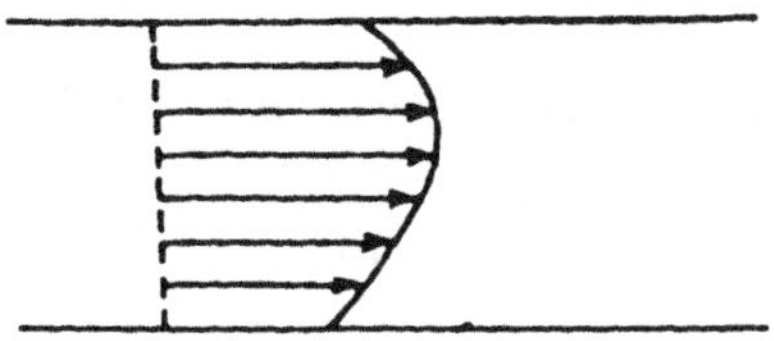

Fig. 3

Theorem 12. *The parallel flow is stable if the velocity profile has no inflection point.*

The stability under infinitesimally small perturbations of flows satisfying these conditions constitute the well known theorem of Rayleigh [5, 6].

EXAMPLE 3. *Flows with an inflection point (see also [7].)*

Suppose now that the velocity profile $v(y)$ possesses an inflection point y_0. Choose $\lambda = -v(y_0)$. Then:

$$\delta^2 E + \lambda \delta^2 M_1 = \iint \left(\delta v^2 + \frac{v(y) - v(y_0)}{v''(y)} \right) (\delta r)^2 dx dy > 0$$

if the velocity profile has the shape sketched in Figure 4. Hence, one obtains from Theorems 9, 10, the following

Corollary 1 *The flow whose velocity is given by*

$$v(y) = ay + by^3, \quad ab > 0$$

is stable.

If the velocity profile has the shape depicted in Figure 5, then

$$\frac{v(y) - v(y_0)}{v''(y)} < 0.$$

Therefore

$$\delta^2 E + \lambda \delta^2 M_1 < 0 \quad \text{for a sufficiently narrow strip } \mathscr{D}.$$

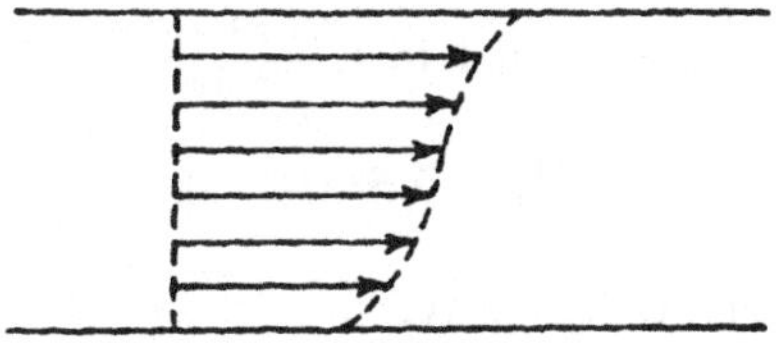

Fig. 4

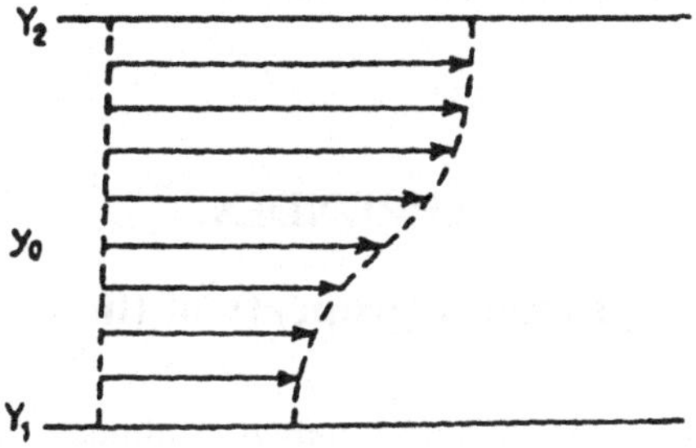

Fig. 5

For example, one gets from Theorems 9, 10,

Corollary 2 *The flow whose velocity field is given by* $v(y) = \sin y$ *in the strip* $Y_1 < y < Y_2$ *is stable if* $Y_1 - Y_2 < \pi$.

On the other hand, Tollmien [8] has shown that the same flow is unstable in the band $Y_1 < y < Y_2$, if $Y_2 - Y_1 > \pi$ and $Y_1 + Y_2 = \pi$.

EXAMPLE 4. *Circular flows:* $\psi = \psi(x^2 + y^2)$ *(Figure 6).*
In this case one can use the integral M_3.

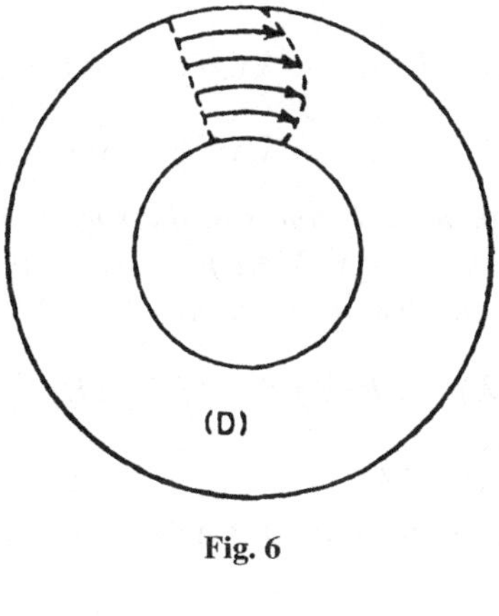

Fig. 6

If $\nabla\Delta\psi \neq 0$ one can choose λ such that $\frac{\nabla(\psi+\frac{\lambda}{2}R^2)}{\nabla\Delta\psi} > 0$, hence

$$\delta^2 E + \lambda\delta^2 M_3 > 0,$$

see (22). Therefore, theorems 9, 10 give:

Theorem 13. *The circular flow is stable if its rotation is monotone (i.e. if $\nabla\Delta\psi \neq 0$).*

One can also study inflection points (Example 3) for circular flows.

APPENDIX.

On the symmetry property of the spectrum

Let $< x,y >$ be a real non-degenerate symmetric bilinear form, and let A be a real linear operator satisfying

$$< Ax,y > + < x,Ay >= 0. \tag{1}$$

The proof of the symmetry of the operator A spectrum given in §1 uses determinants. Hence, in the infinite-dimensional case (where the spectrum may be continuous), the proof must be modified.

Extend the form $< x,y >$ and the operator A to the complex space X by setting

$$\begin{cases} A(x+iy) = Ax + iAy, \\ < x+iy, u+iv >=< x,u > + < y,v > +i(< y,u > - < x,v >). \end{cases} \tag{2}$$

So, for any x,y in X and any complex numbers λ,μ,

$$A(\lambda x) = \lambda Ax; \quad < \lambda x, \mu y >= \lambda\bar{\mu} < x,y >, \tag{3}$$

$$< x,y >= \overline{< y,x >} =< \bar{y},\bar{x} > . \tag{4}$$

Definition *A complex number ν belongs to the spectrum of the operator A, if the resolvent of A has a singularity at ν. More precisely, ν is in the spectrum of A if for some x,y the function of one complex variable λ*

$$R(\lambda) =< R_\lambda x,y >= \langle (A-\lambda E)^{-1}x,y\rangle$$

is not holomorphic at the point $\lambda = \nu$.

Theorem 7 *The spectrum of the operator A is symmetric with respect to both the real and imaginary axes.*

Proof. It follows from (4) and from the reality of A that

$$< R_{\bar{\lambda}}\bar{x},\bar{y} >= \overline{< R_\lambda x,y >}. \tag{5}$$

The formula (5) shows that if ν is a singularity of $< R_\lambda x, y >$, then $\bar{\nu}$ is a singularity of $< R_\lambda x, \bar{y} >$; hence the spectrum is symmetric with respect to the real axis.

Below one shall prove the formula

$$< R_\lambda x, y >= - < R_{-\lambda} \bar{y}, x > \tag{6}$$

using (1). It follows from formula (6) that if ν is a a singularity of $< R_\lambda x, y >$, then $-\nu$ is a singularity of $< R_\lambda \bar{y}, \bar{x} >$. Hence the spectrum is symmetric with respect to the origin. Together with the symmetry with respect to the real axis proved earlier, this completes the proof of the theorem.

Proof of formula (6). Let X^* be the dual space of X; the elements y of X^* are the linear forms (x, y) on X such that:

$$(\lambda x, \mu y) = \lambda \bar{\mu}(x, y) \quad \text{for } x \in X,\ y \in X^*. \tag{7}$$

The dual operator $A^* : X^* \to X^*$ of $A : X \to X$ is defined by

$$(Ax, y) = (x, A^* y) \quad \text{for } x \text{ in } X,\ y \text{ in } X^*. \tag{8}$$

For example, for the operator proportional to the identity

$$(\lambda E)^* = \bar{\lambda} E. \tag{9}$$

The form $< x, y >$ defines an operator $B : X \to X^*$ such that, according to (4),

$$< x, y >= (x, By) = (\bar{y}, B\bar{x}) = \overline{< y, x >} =< \bar{y}, \bar{x} > . \tag{10}$$

It follows from formulæ (1) and (2), that for any x, y in X

$$< Ax, y > + < x, Ay >= 0. \tag{11}$$

Formulæ(10) and (11) imply

$$(Ax, By) + (x, BAy) = 0. \tag{12}$$

It follows from (12) and (8) that for any x, y in X

$$(x, A^* By).(x, BAy) = 0,$$

and hence

$$A^* B + BA = 0. \tag{13}$$

Using formulæ(13) and (9), one finds

$$\begin{aligned} B(A - \lambda E)^{-1} B^{-1} &= \left(B(A - \lambda E) B^{-1} \right)^{-1} \\ &= \left((BA) B^{-1} - \lambda E \right)^{-1} = -(A^* + \lambda E)^{-1} = -(A + \bar{\lambda} E)^{-1^*}. \end{aligned}$$

In other words,

$$B(A-\lambda E)^{-1} = -(A+\bar{\lambda}E)^{-1^*}B. \tag{14}$$

Using consquentially formulæ(10), (14), (8) and again (10), one finds:

$$\begin{aligned}\langle (A-\lambda E)^{-1}x, y\rangle &= \overline{(y, B(A-\lambda E)^{-1}x)} = -\overline{(y, (A+\bar{\lambda}E)^{-1^*}Bx)}\\ &= \overline{((A+\bar{\lambda}E)^{-1}y, Bx)} = -\langle (A+\lambda E)^{-1}\bar{y}, \bar{x}\rangle.\end{aligned}$$

This formula is equaivalent to (6).

References

[1] V.I. Arnold, *C.R.A.S.*, 261 (1965), p. 17

[2] J.J. Moreau, *C.R.A.S.*, 249 (1959), p. 2156

[3] R.G. Fjortjoft, *Geofysiste Publikasjoner*, Oslo, vol. 17, $n^0 6$ (1950)

[4] V. Arnold, D.A.N. (1965)

[5] C.C. Lin, *The theory of hydrodynamical stability*, Cambridge (1955)

[6] H. Schlichting, *Boundary layer theory*, London (1955)

[7] M. Rosenblath & A. Simon, *The physics of fluids*, vol. 7, $N^0 4$, p. 557-558 (1964)

[8] C. Tollmien, *Ein allgemeines Kriterium der Instabilität laminärer Geschwindigkeitsverteilungen*, Nachr. Ges. Wiss. Göttingen; Math. phys., Klasse, Bd. 5o, S. 79-114 (1935)

CHARACTERISTIC CLASS ENTERING IN QUANTIZATION CONDITIONS

V. I. Arnol'd

Recently V. P. Maslov gave a mathematically rigorous treatment of multidimensional asymptotic methods of "quasiclassical" type in the large, i.e., for any number of conjugate points [1, 2]. It turned out that there appeared in the asymptotic formulas certain integers, reflecting homological properties of curves on surfaces of the phase space and closely related to the Morse indexes of the corresponding variational problems. In particular, Maslov defined a one-dimensional class of integer-valued cohomologies whose values on the basis cycles enter into the so-called "quantization conditions."

In this note we give new formulas for the calculation of this class of cohomologies. This class is characteristic in the category of real vector bundles, whose complexification is trivial and trivialized, and also in certain wider categories.

§ 1. NOTATION

1.1. Phase Space

Phase space will be 2n-dimensional real arithmetic space

$$\mathbf{R}^{2n} = \{x\}, \quad x = q,\ p; \quad q = q_1, \ldots, q_n; \quad p = p_1, \ldots, p_n.$$

In $\mathbf{R}^{2n}$ we shall consider the following three structures:

1. The Euclidean structure, given by the scalar quadratic

$$(x,\ x) = p^2 + q^2.$$

2. The complex structure, given by the operator

$$I\colon\ \mathbf{R}^{2n} \to \mathbf{R}^{2n},\ I(p,\ q) = (-q,\ p); \quad z = p + iq, \quad \mathbf{C}^n = \{z\}.$$

3. The simplectic structure, given by the skew-scalar product

$$[x,\ y] = (Ix,\ y) = -[y,\ x]. \tag{1}$$

The automorphism groups of $\mathbf{R}^{2n}$ preserving these structures are called the orthogonal group $\mathbf{O}(2n)$, the complex linear group $\mathbf{GL}(n, \mathbf{C})$, and the simplectic group $\mathbf{Sp}(n)$, respectively. From (1) there follows

LEMMA 1.1. An automorphism preserving two of these structures preserves the third also, so that

$$\mathbf{O}(2n) \cap \mathbf{GL}(n,\ \mathbf{C}) = \mathbf{GL}(n,\ \mathbf{C}) \cap \mathbf{Sp}(n) = \mathbf{Sp}(n) \cap \mathbf{O}(2n) = \mathbf{U}(n).$$

The automorphisms preserving two (and thus all three) structures form the unitary group $\mathbf{U}(n)$. The determinant det of a unitary automorphism is a complex number with modulus 1. Thus there arises a mapping of $\mathbf{U}(n)$ onto the circle

$$\mathbf{SU}(n) \to \mathbf{U}(n) \xrightarrow{\det} \mathbf{S}^1, \tag{2}$$

which is obviously a fibering (the fiber is the group $\mathbf{SU}(n)$ of unitary automorphisms with determinant 1).

1.2. The Lagrangian Grassmanian $\Lambda(n)$

We consider an n-dimensional plane $\mathbf{R}^n \subset \mathbf{R}^{2n}$. It is called Lagrangian if the skew-scalar product of any two vectors of $\mathbf{R}^n$ equals zero. For example, the planes $p = 0$ and $q = 0$ are Lagrangian.*

* The name comes from the "Lagrange brackets" in classical mechanics.

Moscow State University. Translated from Funktsional'nyi Analiz i Ego Prilozheniya, Vol. 1, No. 1, pp.1-14, January-March, 1967. Original article submitted September 28, 1966.

Originally publ. in: Funkts. Anal. Prilozh. 1:1, 1-14, © Russ. Acad. Sciences 1967
English transl.: Funct. Anal. Appl. 1, 1-13, © Consultants Bureau, New York 1967
Reprint by Springer-Verlag Berlin Heidelberg 2014 with kind permission, DOI 10.1007/978-3-642-31031-7_9

The manifold of all (nonoriented) Lagrangian subspaces of $\mathbf{R}^{2n}$ is called the Lagrangian Grassmanian $\Lambda(n)$.

From the complex point of view Lagrangian planes can be called real-similar, since there holds

LEMMA 1.2. The unitary group U(n) acts on $\Lambda(n)$ transitively with stationary subgroup O(n).

Proof. Let λ be a Lagrangian plane. By (1) this means that the plane $I\lambda$ is orthogonal to λ. Let $\lambda' \in \Lambda(n)$ and ξ, ξ' be orthogonal frames in λ, λ'. Then the automorphism of $\mathbf{R}^{2n}$ carrying ξ into ξ' and $I\xi$ into $I\xi'$ is unitary.

From this lemma it follows that $\Lambda(n)$ is a manifold, $\Lambda(n) = U(n)/O(n)$; thus there is a fibering

$$O(n) \to U(n) \to \Lambda(n). \tag{3}$$

1.3 The Mapping Det^2: $\Lambda(n) \to S^1$

The determinant of an orthogonal automorphism $A \in O(n) \subset U(n)$ equals ± 1. Therefore the square of the determinant of a unitary automorphism carrying the plane $p = 0$ into the Lagrangian plane λ depends only on λ. In this way a mapping is constructed

$$\mathrm{Det}^2\colon \Lambda(n) \to S^1.$$

Denote by $S\Lambda(n)$ the set of Lagrangian planes $\lambda \in \Lambda(n)$ with $\mathrm{Det}^2 \lambda = 1$. On this set the group SU(n) of unitary unimodular automorphisms acts transitively, and the stationary subgroup of any point is isomorphic to the rotation group SO(n). Therefore $S\Lambda(n) = SU(n)/SO(n)$ is a manifold.

Thus we obtain a diagram (obviously commutative) of six fiberings:

$$\begin{array}{ccccc} SO(n) & \to & O(n) & \xrightarrow{\det} & S^0, \\ \downarrow & & \downarrow & & \downarrow \\ SU(n) & \to & U(n) & \xrightarrow{\det} & S^1, \\ \downarrow & & \downarrow & & \downarrow z^2 \\ S\Lambda(n) & \to & \Lambda(n) & \xrightarrow{\mathrm{Det}^2} & S^1, \end{array}$$

where z^2 is the mapping of the circle ($z = e^{i\varphi} \to e^{2i\varphi} = z^2$).

1.4. The Cohomology Class $\alpha \in H^1(\Lambda(n), Z)$

LEMMA 1.4.1. The fundamental group $\Lambda(n)$ is free cyclic,

$$\pi_1(\Lambda(n)) = Z,$$

and its generator goes into the generator of S^1 under the mapping induced by Det^2.

The proof is obtained from the exact homotopy sequences of the left column and lower row of the diagram of section 1.3.

COROLLARY 1.4.2. The one-dimensional homology and cohomology groups of $\Lambda(n)$ are free cyclic:

$$H_1(\Lambda(n), Z) \simeq H^1(\Lambda(n), Z) \simeq \pi_1(\Lambda(n)) \simeq Z.$$

For the generator α of the cohomology group $H^1(\Lambda(n), Z)$ we take the number of rotations of Det^2, i.e., the cocycle whose value on a closed curve $\gamma\colon S^1 \to \Lambda(n)$ is equal to the degree of the composition

$$S^1 \xrightarrow{\gamma} \Lambda(n) \xrightarrow{\mathrm{Det}^2} S^1.$$

(Here S^1 is the circle $e^{i\varphi}$, oriented on the side of increasing φ.)

Example 1.4.3. Let λ be a Lagrangian plane: $\lambda \in \Lambda(n)$. Consider the automorphisms $e^{i\varphi}E \in U(n)$. The Lagrangian planes $e^{i\varphi}\lambda$ $(0 \le \varphi \le \pi)$ form a closed curve $\gamma\colon S^1 \to \Lambda(n)$, since $e^{i\pi}E = -E$.

The value of the class α on the curve γ equals n.

Indeed, $\det(e^{i\varphi}E) = e^{in\varphi}$, therefore $\mathrm{Det}^2 e^{i\varphi}\lambda = e^{2in\varphi}\mathrm{Det}^2\lambda$.

1.5. Lagrangian Manifolds

Let M be an n-dimensional submanifold of the phase space R^{2n}. The manifold M is called Lagrangian if its tangent plane at each point is Lagrangian. For example, in the case n = 1 every curve M on the phase plane R^2 is Lagrangian.

Let M be a Lagrangian manifold. We consider the tangential mapping

$$\tau:\ M \to \Lambda(n),$$

carrying each point $x \in M$ into the subspace $\tau x \in \Lambda(n)$ parallel to the tangent plane to M at x.

The cohomology class $\alpha \in H^1(\Lambda(n), \mathbf{Z})$ introduced above induces on M a one-dimensional cohomology class

$$\alpha^* = \tau^* \alpha \in H^1(M, \mathbf{Z}).$$

The value of α^* on an oriented closed curve γ: $S^1 \to M$ is defined as the number of rotations of the square of the determinant of the tangent plane, i.e., as the degree of the composition

$$S^1 \xrightarrow{\gamma} M \xrightarrow{\tau} \Lambda(n) \xrightarrow{\mathrm{Det}^2} S^1.$$

The aim of this note is the proof of the following assertion.

THEOREM 1.5. The cohomology class $\alpha^* \in H^1(M, Z)$ coincides with the "index of closed curves on the Lagrangian manifold M" introduced by Maslov in [1].

§ 2. PROOF OF THEOREM 1.5

The Maslov index is defined by him as the index of intersection of a certain two-sided (n-1)-dimensional cycle on M^n – the singular cycle.

2.1. The Singular Cycle

Let M be an n-dimensional Lagrangian manifold. Consider the projection f: M $\to \mathbf{R}^n$ of the manifold M onto the plane p = 0; $f(p, q) = q$. The set Σ of points of M where the rank of the differential of f is less than n is called the singularity of the mapping f. Regarding the singularity Σ, Maslov formulates the following assertions 1-5 (proofs are given below in § 3 and 4).

THEOREM 2.1. By an arbitrarily small unitary rotation the manifold M can be brought into "general position" relative to the projection f, so that the following assertions are valid:

ASSERTION 1. The singularity Σ consists of an open (n−1)-dimensional manifold Σ^1, where df has rank n−1, and the boundary $(\Sigma - \Sigma^1)$ dimension strictly less than n−2, so that Σ determines an (n−1)-dimensional (unoriented) cycle in M.

ASSERTION 2. This cycle is two-sided in M.

The choice of a positive side of Σ can be carried out in the following way.

ASSERTION 3. In a neighborhood of the point $x \in \Sigma^1$ the Lagrangian manifold M is given by n equations of the form

$$q_k = q_k(p_k, q_{\hat{k}}),\ p_{\hat{k}}(p_k, q_{\hat{k}}), \tag{4}$$

where $\hat{k} = 1, 2, \ldots, k-1, k+1, \ldots, n$ for some k, $1 \leq k \leq n$.

Obviously, in a neighborhood of such a point x the singularity Σ^1 is given by an equation $\dfrac{\partial q_k}{\partial p_k} = 0$.

ASSERTION 4. On passing through Σ^1 the quantity $\partial q_k/\partial p_k$ changes sign.

It turns out that for the positive side of Σ^1 we can take the one for which $\partial q_1/\partial p_1 > 0$.

ASSERTION 5. Such a definition of the positive side is correct, i.e., does not depend on which of the coordinate systems p_k, $q_{\hat{k}}$ (k = 1, . . ., n) we use.

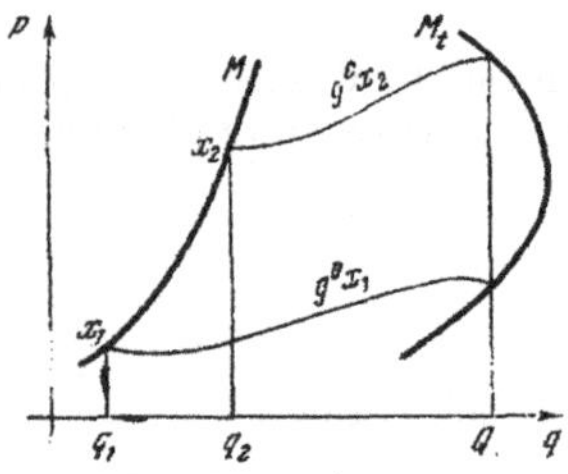

Fig. 1

2.2. The Maslov Index, ind$\in H^1(M, Z)$

Suppose there is given on the Lagrangian manifold M in "general position" in the sense of Theorem 2.1. a curve γ, transversal to the cycle Σ, with the initial and end points of γ nonsingular:

$$\partial\gamma = x_1 - x_0,\ x_1 \overline{\in} \Sigma,\ x_0 \overline{\in} \Sigma.$$

Maslov calls the index ind γ of the curve γ its index of intersection with Σ, i.e., the number ν_+ of points of passage from the negative side to the positive side minus the number ν_- of points of passage from the positive side to the negative:

$$\operatorname{ind}\gamma = \nu_+ - \nu_-.$$

Example. Let n = 1 and let M be a curve in the pq-plane (Fig. 1). For M in general position, Σ consists of separate points a, b, c, ... The indexes of the curves γ_i ($\partial\gamma_i = x_i - x_0$) are equal to 0, 1, 0, 1, 2. respectively.

THEOREM 2.2. The index of a closed oriented curve γ on the Lagrangian manifold in general position M depends only on the homology class of γ and is the value of a one-dimensional class of integer-valued cohomologies of M on the cycle γ: ind$\in H^1$ (M, Z).

2.3. Index of Curves on the Grassmanian $\Lambda(n)$

Proofs of the formulated theorems (1.5, 2.1, 2.2) are based on the following construction.

In the manifold of Lagrangian planes of $\Lambda(n)$ we single out the sets $\Lambda^k(n)$ of planes having a k-dimensional intersection with a fixed plane $\sigma \in \Lambda(n)$ (namely, the plane q = 0). It turns out that the closure $\overline{\Lambda^1(n)}$ determines a cycle (nonoriented) of codimension 1 (see 3.2.2).

In section 3.5 we prove

THE FUNDAMENTAL LEMMA. $\Lambda^1(n)$ is two-sidedly imbedded in $\Lambda(n)$, i.e., there exists a continuous vector field transversal to $\Lambda^1(n)$ and tangent to $\Lambda(n)$.

Such a vector field is constructed by means of the orbits of action of $S^1 = \{e^{i\theta}\}$ on $\Lambda(n)$. In § 3 we prove

LEMMA 3.5.1. Every circle

$$\theta \to e^{i\theta}\lambda, \quad 0 \leqslant \theta \leqslant \pi, \quad \lambda \in \Lambda(n), \tag{4}$$

is transversal to $\Lambda^1(n)$.

For the positive side of $\Lambda^1(n)$ we choose the one toward which the velocity vectors of the curves (4) are directed.

The two-sidedness of $\Lambda^1(n)$ allows us to define

$$\operatorname{Ind} \in H^1(\Lambda(n), Z)$$

as the index of intersection of oriented closed curves on $\Lambda(n)$ with $\overline{\Lambda^1(n)}$ (Definition 3.6.1).

The index Ind is connected with the Maslov index ind and the cohomology class α of section 1.4. Namely, it turns out that the choice of a positive side of $\Lambda^1(n)$ by means of the curves (4) agrees with the definition of the positive side of Σ^1 from section 1.2. In § 4 we prove

LEMMA 4.3.1. The index Ind generates the Maslov index ind under the tangential mapping τ: $M^n \to \Lambda(n)$; ind = τ^* Ind, i.e., for every curve γ: $S^1 \to M$ we have ind γ = Ind $\tau\gamma$.

Proof of Theorem 1.5. Calculation of the index of the curves (4) (see Example 3.6.2) gives Ind γ = n = $\alpha(\gamma)$ (Example 1.4.3). But $H^1(\Lambda(n), Z) = Z$ (Corollary 1.4.2). So, Ind = α. By Lemma 4.3.1 ind = τ^* Ind and by Definition 1.5. $\alpha^* = \tau^*\alpha$. Thus ind = α^*, which was to be proved.

3. PROOF OF THE FUNDAMENTAL LEMMA

In this section we prove the two-sidedness of the singular cycle $\overline{\Lambda^1(n)}$ and define the index $\mathrm{Ind} \in H^1(\Lambda(n), Z)$.

3.1. Generating Functions

Let M be a manifold in phase space which is given in a simply connected neighborhood of the point $q = q_0$, $p = p_0$ by an equation of the form $p = p(q)$.

LEMMA 3.1.1. The manifold M is Lagrangian if and only if there is a "generating function" s(q) such that

$$p = \frac{\partial s}{\partial q}. \tag{5}$$

Proof. Let $s(q) = \int_{q_0}^{q} p(q)dq$. Independence of this integral of path is equivalent to the differential $d(p\,dq) = dp \wedge dq$ being 0 on M. But the value of $dp \wedge dq$ on the bivector $\xi \wedge \eta$ is exactly equal to the skew-scalar product $[\xi, \eta]$, so that equality of $dp \wedge dq$ to zero on M is equivalent to M being Lagrangian. The function s(q) satisfies (5), proving the lemma.

Remark 3.1.2. The function s(q) is determined up to a constant summand. In the particular case where M is a subspace, this summand can be chosed so that s(q) is a quadratic form. From this there follows

COROLLARY 3.1.3. The set of Lagrangian subspaces of the form $p = p(q)$ (i.e., transversal to the plane $q = 0$) make up in the manifold $\Lambda(n)$ an open cell $\Lambda^0(n)$, diffeomorphic to the linear space D of all real symmetric matrices of order n. The diffeomorphism is given by the mapping

$$\varphi\colon\ D \to \Lambda^0(n), \quad \varphi(S) = \lambda_S \quad (S \in D,\ \lambda_S \in \Lambda^0(n)),$$

where λ_S denotes the plane $p = Sq$.

The proof is obtained from (5) by setting $s(q) = \frac{1}{2}(Sq, q)$.

The space of symmetric matrices D is $R^{n(n+1)/2}$. Thus we have proved

COROLLARY 3.1.4. The manifold $\Lambda(n)$ has dimension

$$\dim \Lambda(n) = n(n+1)/2.$$

3.2. The Singular Cycle $\overline{\Lambda^1(n)}$

Notation 3.2.0. Let σ be the Lagrangian plane $q = 0$. We denote by $\Lambda^k(n)$ the set of all Lagrangian planes $\lambda \in \Lambda(n)$ whose intersection with the plane σ is k-dimensional

LEMMA 3.2.1. The set $\Lambda^k(n)$ is an open manifold of codimension $k(k + 1)/2$ in the Lagrangian Grassmanian $\Lambda(n)$.

Proof. We compare with each plane $\lambda \in \Lambda^k(n)$ its intersection with the plane σ. There arises a mapping of $\Lambda^k(n)$ on the Grassman manifold $G_{n,k}$ of all k-dimensional subspaces of the n-dimensional space σ. It is easily verified that this mapping determines a fibering

$$\Lambda^0(n-k) \to \Lambda^k(n) \to G_{n,k}.$$

By Corollary 3.1.4 $\dim \Lambda^0(n-k) = \frac{(n-k)(n-k+1)}{2}$. Since $\dim G_{n,k} = k(n-k)$, we find

$$\dim \Lambda^k(n) = \frac{(n-k)(n-k+1)}{2} + k(n-k) = \frac{n(n+1)}{2} - \frac{k(k+1)}{2} = \dim \Lambda(n) - \frac{k(k+1)}{2},$$

which was to be proved.

COROLLARY 3.2.2. $\overline{\Lambda^1(n)}$ determines an (unoriented) cycle of codimension 1 in $\Lambda(n)$.

Proof. The manifold $\Lambda(n)$ can be considered algebraic. The closure $\overline{\Lambda^1(n)} = \bigcup_{k \geq 1} \Lambda^k(n)$ is an algebraic submanifold of codimension $1(k(k+1)/2 \geq 1$ for $k \geq 1)$. Therefore $\overline{\Lambda^1(n)}$ determines a (nonoriented) chain. The singularity of the algebraic manifold $\overline{\Lambda^1(n)}$ is the algebraic submanifold $\overline{\Lambda^2(n)} = \bigcup_{k \geq 2} \Lambda^k(n)$ of codimension 3 in $\Lambda(n)$, since $k(k+1)/2 \geq 3$ for $k \geq 2$. Thus the homological boundary of the chain $\overline{\Lambda^1(n)}$ equals 0, which was to be proved.

3.3. Coordinates on $\Lambda(n)$

We consider a Lagrangian plane $\lambda \in \Lambda(n)$. Let $\lambda \in \Lambda^k(n)$, i.e., let the intersection $\lambda \cap \sigma$ be k-dimensional. We introduce coordinates on $\Lambda(n)$ in a neighborhood of λ.

Notation 3.3.0. Let K be a subset of the set 1,2, . . ., n. Denote by σ_K the Lagrangian coordinate plane

$$\sigma_K = \{p, q: p_k = 0, \quad q_l = 0 \quad \forall k \in K, \ \forall l \bar{\in} K\}.$$

LEMMA 3.3.1. The plane $\lambda \in \Lambda^k(n)$ is transversal to one of the C_n^k coordinate planes σ_K, where K has k elements.

Proof. The intersection $\lambda \cap \sigma = \lambda_0$ is k-dimensional. Consequently, the plane λ_0 in σ is transversal to one of the C_n^k (n−k)-dimensional coordinate planes $\tau = \sigma_K \cap \sigma$, i.e., for some K we have $\lambda_0 \cap \sigma_K \cap \sigma = 0$. We shall show that the plane σ_K is transversal to λ: $\sigma_K \cap \lambda = 0$.

By the condition, $\lambda_0 + \tau = \sigma$. From the Lagrangian property of λ and σ_K it follows that $[\lambda, \lambda_0] = 0$ (since $\lambda_0 \subset \lambda$) and $[\sigma_K, \tau] = 0$ (since $\tau \subset \sigma_K$). Thus, $[\lambda \cap \sigma_K, \lambda_0 + \tau] = 0$, i.e., $[\lambda \cap \sigma_K, \sigma] = 0$. But the largest number of pairwise skew-orthogonal independent vectors in $\mathbf{R}^{2n}$ equals n. Therefore the n-dimensional plane σ is itself a maximal skew-orthogonal plane, thus $(\lambda \cap \sigma_K) \subset \sigma$. So, $(\lambda \cap \sigma_K) \subseteq (\lambda \cap \sigma \cap \sigma_K) = (\lambda_0 \cap \tau) = 0$, which was to be proved.

From the lemma just proved it follows that every plane $\lambda \in \Lambda(n)$ is transversal to one of the 2^n coordinate planes σ_K. This allows us to set up an atlas of $\Lambda(n)$ of 2^n charts.

One of the maps was constructed in section 3.1.1: the region $\Lambda^0(n)$ is diffeomorphic to the space of symmetric matrices $D = \mathbf{R}^{n(n+1)/2}$, where the diffeomorphism $\varphi: D \to \Lambda^0(n)$ is defined as

$$\varphi(S) = \lambda_S = \{p, q: p = Sq\} \ \forall S \in D.$$

Notation 3.3.2. We denote by I_K the operator of multiplication by i of the variables $z_\varkappa = p_\varkappa + iq_\varkappa$, $\varkappa \in K$:

$$I_K: \ \mathbf{R}^{2n} \to \mathbf{R}^{2n},$$

and for $\eta = I_K \xi$

$$q_\varkappa(\eta) = +p_\varkappa(\xi), \quad p_\varkappa(\eta) = -q_\varkappa(\xi), \quad \forall \varkappa \in K,$$

$$q_\nu(\eta) = q_\nu(\xi), \qquad p_\nu(\eta) = p_\nu(\xi), \qquad \forall \nu \bar{\in} K.$$

The operator I_K is unitary, therefore it carries Lagrangian planes into Lagrangian planes. In particular, $I_K\sigma = \sigma_K$. Thus I_K carries the set $\Lambda^0(n)$ of planes transversal to σ into the set of planes $I_K\Lambda^0(n)$ transversal to σ_K. Thus the formula

$$\varphi_K(S) = I_K\lambda_S \in \Lambda(n), \quad S \in D \tag{6}$$

gives a diffeomorphism φ_K: $D \to I_K\Lambda^0(n)$, where $I_K\Lambda^0(n)$ is the set of all Lagrangian planes transversal to σ_K.

By Lemma 3.3.1 the 2^n regions $I_K\Lambda^0(n)$ cover $\Lambda(n)$ entirely, so that formula (6) gives an atlas of $\Lambda(n)$ of 2^n charts.

LEMMA 3.3.3. The set $\Lambda^k(n)$ is covered by C_n^k charts $\varphi_K D$(K consisting of k elements) and in the coordinates $S = \varphi_K^{-1}\lambda$ is given by $k(k+1)/2$ linear equations $S_{\mu\nu} = 0$ $(\forall \mu \in K, \forall \nu \in K)$.

Proof. Let $\dim \lambda \cap \sigma = k$. By Lemma 3.3.1 $\lambda \cap \sigma_K = 0$ for some K of k elements. Therefore the plane $I_K\lambda = I_K^{-1}\lambda$ is transversal to σ and has an equation $p = Sq$. The intersection $(I_K\lambda) \cap \sigma_K = I_K(\lambda \cap \sigma)$ has dimension k. But on σ_K we have $q_l = 0$ $(\forall l \bar{\in} K)$, $p_m = 0$ $(\forall m \in K)$. Therefore on a k-dimensional subspace $q_l = 0$ $(l \bar{\in} K)$ of the plane $p = 0$, k of the functions p_μ $(p = Sq, \mu \in K)$ must vanish identically. This is equivalent to the equations $S_{\mu\nu} = 0$, as was to be proved.

3.4. Unitary Paramaterization

By means of the coordinates S introduced above it is possible to express the unitary transformations carrying the "purely imaginary" plane $p = 0$ into the plane $\lambda_S \in \Lambda^0(n)$.

It is obvious that we have

LEMMA 3.4.1. Let S, U be square $n \times n$ matrices with complex elements. Then

$$\left(U = \frac{E - iS}{E + iS}\right) \Leftrightarrow \left(S = -i\frac{E - U}{E + U}\right), \tag{7}$$

and for S, U, related by formulas (7),

S is self-adjoint if and only if U is unitary,

S is symmetric if and only if U is symmetric.

Thus, formulas (7) set up a diffeomorphism between the space D of real symmetric matrices S and the manifold of unitary symmetric nonsingular matrices U. (The unitary matrix U is nonsingular if -1 is not a proper value; for a real symmetric S we always have $\det(E + iS) \neq 0$).

It is always possible to take the square root of a nonsingular unitary matrix, defining it by continuity, beginning from $\sqrt{E} = E$.

LEMMA 3.4.2. Let $\lambda_S \in \Lambda^0(n)$ be a plane $p = Sq$. Then the matrix

$$\sqrt{U} = \frac{E - iS}{\sqrt{E + S^2}}$$

gives a unitary transformation carrying the plane $p = 0$ into the plane λ_S.

Proof. Since S is symmetric and real, $\sqrt{E + S^2}$ is real, and $\sqrt{U}$ carries the plane $p = 0$ into the same image as $E - iS$. The latter transformation carries the point

$$(0, q) \in \mathbf{R}^{2n}, \text{ i.e. }, \quad iq \in i\mathbf{R}^n \subset \mathbf{C}^n$$

into the point

$$iq + Sq \in \mathbf{C}^n, \quad \text{i.e.}, \quad (Sq, q) \in \lambda_S \subset \mathbf{R}^{2n},$$

which was to be proved.

COROLLARY 3.4.3. The mapping Det^2: $\Lambda(n) \to S^1$ of section 1.3 is given by the formula

$$\mathrm{Det}^2 \lambda_S = \det \frac{E - iS}{E + iS}.$$

COROLLARY 3.4.4. The symmetric nonsingular unitary matrix U for which $\sqrt{U}$ carries the plane $p = 0$ into the plane λ is uniquely determined by this plane $\lambda \in \Lambda^0(n)$.

In fact, by 3.4.1 U is uniquely determined by S, and by 3.4.2 S is uniquely determined by λ.

3.5 Two-Sidedness of the Singular Cycle

Let λ be a Lagrangian plane. Then each of the planes $e^{i\theta}\lambda$ is Lagrangian.

LEMMA 3.5.1. If $\lambda \in \Lambda^1(n)$, then the curve γ: $S^1 \to \Lambda(n), e^{i\theta} \to e^{i\theta}\lambda$ is transversal to the cycle $\Lambda^1(n)$ at the point $\theta = 0$.

Thus, the velocity vectors $v(\lambda) = \left.\frac{d}{d\theta}\right|_{\theta=0} (e^{i\theta}\lambda)$ form a transversal structure to $\Lambda^1(n)$, from which there follows the

FUNDAMENTAL LEMMA. The singular cycle $\overline{\Lambda^1(n)}$ is two-sidedly imbedded in $\Lambda(n)$.

We will carry out the proof of Lemma 3.5.1 in three stages.

A. First let $\lambda \in \Lambda^0(n)$, $\lambda = \lambda_S$, where $S \in D$ is a real symmetric matrix. We shall compute the coordinates of the velocity vector of the curve $e^{i\theta}\lambda$ in this coordinate system.

LEMMA 3.5.2. For any matrix $S \in D$

$$\left.\frac{d}{d\theta}\right|_{\theta=0} \varphi^{-1} e^{i\theta} \lambda_S = -(E + S^2).$$

7

Proof of Lemma 3.5.2. According to section 3.4 the plane λ is in a unique correspondence with a nonsingular unitary symmetric matrix U so that

$$\lambda = \lambda_{S(U)}, \quad S(U) = -i\frac{E-U}{E+U},$$

and $\sqrt{U}$ carries the plane $p = 0$ into λ.

Let $U(\theta) = e^{2i\theta} U$. The matrix $U(\theta)$ is unitary, symmetric, and for small $|\theta|$ nonsingular, so that $\sqrt{U(\theta)} = e^{i\theta}\sqrt{U}$. Therefore $\sqrt{U(\theta)}$ carries the plane $p = 0$ into $e^{i\theta}\lambda$, so that

$$\lambda_{S(U(\theta))} = e^{i\theta}\lambda_{S(U)}, \quad \text{or} \quad \varphi^{-1} e^{i\theta}\lambda_{S(U)} = S(U(\theta)). \tag{8}$$

The vector $\left.\frac{d}{d\theta}\right|_{\theta=0} \varphi^{-1} e^{i\theta}\lambda$ lies in the tangent space to the linear space of symmetric matrices D; this tangent space is naturally identified with D itself. With this identification, by the formula of 3.4.1,

$$-\left.\frac{d}{d\theta}\right|_{\theta=0} S(U(\theta)) = \left.\frac{d}{d\theta}\right|_{\theta=0} i\frac{E-e^{2i\theta}U}{E+e^{2i\theta}U} = \frac{4U}{(E+U)^2} = E + S^2,$$

which together with the formulas (8) proves Lemma 3.5.2.

B. Now let $\lambda \in \Lambda^1(n)$. By Lemma 3.3.3 the point $\lambda \in \Lambda^1(n)$ belongs to one of the n charts $\varphi_K D$, where K consists of one element $\varkappa$, $1 \le \varkappa \le n$. In other words, in the notation of 3.3.2

$$\lambda = I_K \lambda_S, \text{where } S \in D, \quad \lambda_S \in \Lambda^0(n).$$

It is easy to calculate the velocity vector of the curve γ: $S^1 \to \Lambda(n)$, $e^{i\theta} \to e^{i\theta}\lambda$ for $\theta = 0$ in the coordinate system $\varphi_K^{-1}\lambda = S$.

LEMMA 3.5.3. For any matrix $S \in D$, $\lambda = I_K \lambda_S$,

$$\left.\frac{d}{d\theta}\right|_{\theta=0} \varphi_K^{-1} e^{i\theta}\lambda = -(E + S^2).$$

Indeed, by Definition 3.3.2, $\varphi_K = I_K\varphi$, and I_K commutes with $e^{i\theta}$. Therefore

$$\varphi_K^{-1} e^{i\theta}\lambda = \varphi^{-1} I_K^{-1} e^{i\theta} I_K \lambda_S = \varphi^{-1} e^{i\theta}\lambda_S,$$

and Lemma 3.5.3 follows from Lemma 3.5.2.

C. The singular cycle $\Lambda^1(n)$ in the coordinates $S = \varphi_K^{-1}\lambda$ has an equation $S_{\varkappa\varkappa} = 0$ (Lemma 3.3.3). The velocity vector $v = \left.\frac{d}{d\theta}\right|_{\theta=0} \varphi_K^{-1} e^{i\theta}\lambda$ by Lemma 3.5.3 is a negative definite matrix, $-v_{\varkappa\varkappa} \ge 1$. Thus, v and $\Lambda^1(n)$ are transversal, which proves Lemma 3.5.1.

Remark 3.5.4. At the same time we have shown that the vector v is directed to the side of $\Lambda^1(n)$ where $S_{\varkappa\varkappa} > 0$.

3.6. The Index Ind of Curves on $\Lambda(n)$

Let γ be an oriented curve in $\Lambda(n)$, transversal to $\Lambda^1(n)$, and let $v(\lambda)$ be the velocity field of Lemma 3.5.1.

DEFINITION 3.6.1. By Ind γ we shall denote the index of intersection of the curve γ with the cycle $\Lambda^1(n)$, equipped with the field $v(\lambda)$.

In other words, Ind $\gamma = \nu_+ - \nu_-$, where ν_+ is the number of points of intersection of γ with $\Lambda^1(n)$ in which the vectors $\dot{\gamma}$ and v lie on the same side of $\Lambda^1(n)$, and ν_- is the number of those on which they lie on opposite sides.

The index of the closed curve γ, like every index of intersection, is determined by the homology class of γ and can be considered as a one-dimensional cohomology class

$$\text{Ind} \in H^1(\Lambda(n), \mathbb{Z}).$$

EXAMPLE 3.6.2. The index of the closed curve γ: $S^1 \to \Lambda(n)$, formed by $e^{i\theta}\lambda$, $0 \le \theta \le \pi$, is n:

$$\text{Ind}\,\gamma = n.$$

Proof. We have that dim $\Lambda^2(n) = \dim \Lambda^1(n)$ is 2 (Lemma 3.2.1). Therefore for almost all λ the curve $e^{i\theta}\lambda$ does not intersect $\overline{\Lambda^2(n)}$. Such a curve is transversal to $\Lambda^1(n)$ at every point of intersection (Lemma 3.5.1). In this case Ind γ is simply the number of these points of intersection (Definition 3.6.1).

Let $\lambda \in \Lambda^0(n)$. By Lemma 3.4.4 we have $\lambda = \lambda_{S(U)}$, where U is a unitary symmetric nonsingular matrix. We may consider the plane λ to be such that all proper vales of the matrix U

$$e^{i\alpha_k},\ 1 \leqslant k \leqslant n,\ \ |\alpha_k| < \pi,$$

are distinct.

But

$$e^{i\theta}\lambda = \lambda_{S(e^{2i\theta}U)}$$

by formula (8) of 3.5.2, and

$$(\lambda_{S(e^{2i\theta}U)} \overline{\in} \Lambda^0(n)) \Leftrightarrow (\det(E + e^{2i\theta}U) = 0)$$

by Lemma 3.4.2. In other words, in the points of intersection of γ with $\Lambda^1(n)$

$$\theta \equiv \frac{\pi - \alpha_k}{2} \pmod{\pi}.$$

There are precisely n such θ on the interval $0 \leq \theta \leq \pi$. Thus, Ind $\gamma = n$, as was to be proved.

§ 4. PROOFS OF THE THEOREMS ON GENERAL POSITION

Here we prove Theorems 2.1 and 2.2 of § 2.

4.1. Transversality

Notation 4.1.1. Let A be a smooth manifold, and let $a \in A$. By TA_a we denote the tangent space to A at the point a. If f: A → B is a smooth mapping, then by $f_*: TA_a \to TB_{f(a)}$ we denote the differential of f at a.

Let f: A → B, h: C → B be two smooth mappings. The mappings f, h are called transversal at the point $b \in B$ if

$$f_* TA_a + h_* TC_c = TB_b$$

for every pair of points $a \in A$, $c \in C$, for which $f(a) = h(c) = b$. The mappings f and h are transversal if they are transversal at every point $b \in B$.

In the particular case where f or h is an imbedding, we may speak of the transversality of a mapping to a submanifold or of the transversality of two submanifolds.

The concept of transversality extends also to the case where A is the union of several manifolds, $A = \cup A_k$ [for example, $\overline{\Lambda^1(n)} = \bigcup_{k \geqslant 1} \Lambda^k(n)$ in § 3] – in this case the restriction of f to each A_k must be transversal to h.

It is easy to prove (see, for example, [3]) the lemma of M. Morse and A. Sard

LEMMA 4.1.2. Let f: A → B be a smooth mapping. Then the measure of the set of points $b \in B$ not transversal to f equals 0 (the point $b \subset B$ is a zero-dimensional submanifold of B).

From Lemma 4.1.2 there follows (see, for example, [4])

LEMMA 4.1.3. Let B be a homogeneous space on which a Lie group G acts transitively ($\forall g \in G$, g: B → B is a diffeomorphism). Let $C \subseteq B$ be a smooth submanifold of B and let f: A → B be a smooth mapping. Then the measure of the set of points $g \in G$ for which the mapping

$$f_g\colon A \to B,\quad f_g(a) = gf(a)$$

is not transversal to C is zero.

For completeness we carry out the proof of Lemma 4.1.3.

Remark 4.1.4. Since the union of a countable number of sets of measure zero has measure zero, it

9

is sufficient to prove Lemma 4.1.3 for a neighborhood A_0 of a point $a_0 \in A$, a neighborhood C_0 of a point $c_0 \in C$ and a neighborhood of the identity e in the group G.

From the transitivity of the action of G there follows easily

ASSERTION 4.1.5. There exists a diffeomorphism of the product of spheres

$$u: D_1 \times D_2 \to G,$$

$$D_i = \{x \in R^{\nu_i}, |x| < 1\}, \quad \nu_1 = \dim B - \dim C, \quad \nu_2 = \dim G - \nu_1,$$

such that $u(0, 0) = e$, and the mapping

$$\beta: D_1 \times D_2 \times C_0 \to B \times D_2,$$

given by the formula

$$\beta(x, y, c) = (u(x, y)c, y), \forall x \in D_1, y \in D_2, c \in C_0,$$

is a diffeomorphism of $D_1 \times D_2 \times C_0$ onto some neighborhood E of the point $(c_0, 0)$ in $B \times D_2$.

Now define the projection of $E \subset B \times D_2$ on $D_1 \times D_2$

$$\Phi: E \to D_1 \times D_2 \text{ by the formula } \Phi(\beta(x, y, c)) = (x, y).$$

Further, define the mapping

$$\hat{f}: A \times D_2 \to B \times D_2 \text{ by the formula } \hat{f}(a, y) = (f(a), y).$$

We apply Lemma 4.1.2 to the composite mapping

$$\Theta = \Phi \circ \hat{f}: A_0 \times D_2 \to D_1 \times D_2.$$

ASSERTION 4.1.6. Let $x,y \in D_1 \times D_2$ be a point transversal to the mapping Θ. Then the mapping

$$f_g: A_0 \to B, \; g = (u(x, y))^{-1},$$

is transversal to the imbedding $C_0 \subset B$.

Proof of Assertion 4.1.6. Consider $\Phi^{-1}(x, y) (x \in D_1, y \in D_2)$. Obviously, $\Phi^{-1}(x, y) = (C_{x,y}, y)$, where $C_{x,y} = u(x, y)C_0 \subset B$. The kernel of the differential $\Phi_*: T(B \times D_2)_{b,y} \to T(D_1 \times D_2)_{x,y}$ is exactly the tangent space to $(C_{x,y}, y)$:

$$\ker \Phi_* = T(C_{x,y}, y)_{b,y}.$$

Therefore the transversality of the mapping $\Theta = \Phi \circ \hat{f}$ to the point x,y implies

$$\hat{f}_* T(A_0 \times D_2)_{a,y} + T(C_{x,y}, y)_{f(a),y} = T(B \times D_2)_{b,y}$$

for all $a \in A_0$, for which $f(a) = b \in C_{x,y}$. Thus, the mapping $f: A_0 \to B$ is transversal to the imbedding $C_{x,y} \subset B$. Applying the diffeomorphism $g = (u(x, y))^{-1} \in G$, we see that $gf: A_0 \to B$ is transversal to $gC_{x,y} = C_0$, as required.

Proof of Lemma 4.1.3. We apply Lemma 4.1.2 to Θ. The set of points $x, y \in D_1 \times D_2$ not transversal to Θ has measure zero. The corresponding set of points $g = (u(x, y))^{-1} \in G$ has measure zero in G. For the remaining g close to e the mapping f_g is transversal to C_0 by 4.1.6. This proves Lemma 4.1.3 according to remark 4.1.4.

4.2. Proof of Theorem 2.1.

We apply Lemma 4.1.3 to the case where A is a Lagrangian manifold M^n, B is the Lagrangian Grassmanian $\Lambda(n)$, f is the tangential mapping $\tau: M^n \to \Lambda(n)$, C is the submanifold $\Lambda^k(n) \subset \Lambda(n)$, and G is the unitary group $\mathbf{U}(n)$.

From Lemma 4.1.3 it follows that for almost all $u \in \mathbf{U}(n)$ the manifold uM^n is such that its tangential mapping τ is transversal to every $\Lambda^k(n) \subset \Lambda(n)$, $k = 1,2, \ldots$. Let us show that such a manifold uM^n is in "general position" in the sense of Theorem 2.1.

Assertion 1 of Theorem 2.1 follows from the implicit functions theorem and Lemmas 3.2.1 and 3.2.2. Assertion 2 follows from the fundamental lemma of section 3.5. Assertion 3 is deduced from Lemma 3.3.1. Assertion 4 is obtained from Lemma 3.3.3 for $k = 1$. Finally, Assertion 5 follows from Lemma 3.5.1 and remark 3.5.4. Theorem 2.1 is thus proved.

4.3. Proof of Theorem 2.2

Let M^n be a Lagrangian manifold in general position, and let $\gamma: S^1 \to M^n$ be an oriented closed curve, transversal to the singular cycle Σ.

LEMMA 4.3.1. Let $\tau\circ\gamma: S^1 \to \Lambda(n)$ be the tangential mapping of M^n onto the curve γ. Then

$$\operatorname{ind}\gamma = \operatorname{Ind}\tau\circ\gamma.$$

In fact, $\Sigma^1 = \tau^{-1}\Lambda^1(n)$ (definitions of sections 2.1 and 3.2). Further, the positive (in the sense of section 2.1) side of Σ^1 is carried under the mapping τ into the positive (in the sense of Definition 3.6.1) side of $\Lambda^1(n)$ — this follows from remark 3.5.4. Thus, each point of intersection of γ with Σ^1 gives to ind the same contribution as the corresponding point of intersection of $\tau\circ\gamma$ with $\Lambda^1(n)$ to Ind, which proves Lemma 4.3.1.

At the same time we have proved Theorem 2.2. also, since Ind $\tau\circ\gamma$ does not change under replacement of γ by a homologous curve γ' [this follows from the fact that $\dim \overline{\Lambda^2(n)} = \dim \Lambda^1(n) - 2$].

§ 5. A QUASICLASSICAL ASYMPTOTIC EXPRESSION

Here we give without proof the asymptotic formulas of Maslov, in which the index plays a role, for the simplest example.

5.1. Asymptotic Expression as h → 0 of the Solution of the Schrödinger Equation

$$ih\frac{\partial\psi}{\partial t} = -\frac{h^2}{2}\Delta\psi + U(q)\psi,\quad \psi = \psi(q, t),\ q\in\mathbf{R}^n \tag{9}$$

with the initial condition

$$\psi|_{t=0} = \varphi(q)e^{\frac{i}{h}f(q)},\ \text{where } \varphi(q) \text{ is a finite function.} \tag{10}$$

To the Schrödinger equation there corresponds the classical dynamical system given in the 2n-dimensional phase space $\mathbf{R}^{2n}$ by the Hamilton equations

$$\dot q = \frac{\partial H}{\partial p},\quad \dot p = -\frac{\partial H}{\partial q},\qquad H = \frac{1}{2}p^2 + U(q).$$

The solutions of the Hamilton equation define a one-parameter group of canonical* diffeomorphisms of the phase space — the phase flow $g^t: \mathbf{R}^{2n} \to \mathbf{R}^{2n}$.

To the initial condition (10) there corresponds a function $\varphi(q)$ on a surface M^n, given in the phase space $\mathbf{R}^{2n}$ by the equations

$$M = \left\{p, q:\ p(q) = \frac{\partial f}{\partial q}\right\}.$$

The surface M is projected uniquely onto the q-plane. It is Lagrangian by Lemma 3.1.1. The phase flow g^t carries M into another Lagrangian surface $g^t M = M_t$. The surface M_t is no longer necessarily projected uniquely onto the q-plane. There arises a mapping $Q(q) = q(g^t(p(q), q))$ (Fig. 2).

Let $x_j = (p_j, q_j)$ be points of M such that $g^t x_j = (P_j, Q) \in M_t$. Assume that the Jacobian $\left.\frac{DQ}{Dq}\right|_{q=q_j} \neq 0.$

Maslov proved the following asymptotic formula for the solution of equation (9) with the condition (10).

THEOREM 5.1. As $h \to 0$

*A diffeomorphism $g: \mathbf{R}^{2n} \to \mathbf{R}^{2n}$ is called canonical, if for every closed curve $\gamma: S^1 \to \mathbf{R}^{2n}$ we have $\oint_\gamma p\,dq = \oint_{g\gamma} p\,dq$. The differential g_* has then at each point a simplectic matrix: under a canonical transformation the 2-form $dp \wedge dq$ goes into itself. Therefore a Lagrangian manifold under a canonical diffeomorphism goes into a Lagrangian.

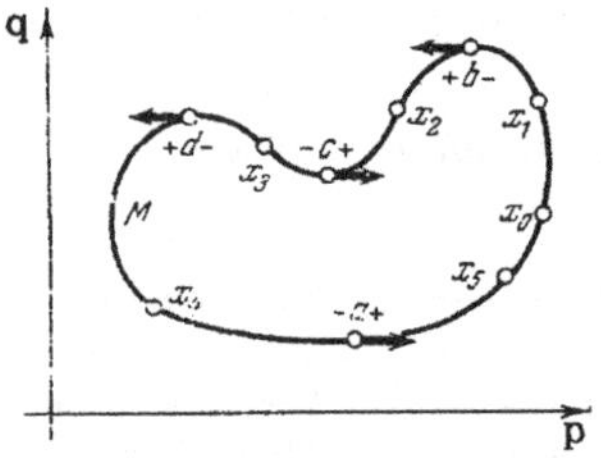

Fig. 2

$$\psi(Q,t)=\sum \varphi(q_j)\left|\frac{DQ}{Dq_j}\right|^{-1/2} e^{\frac{i}{h}S_j(Q,t)-\frac{i\pi}{2}\mu_j}+O(h),$$

where $S_j(Q, t)$ is the action along the trajectory $g^\theta x_j$:

$$S_j(Q,t)=f(q_j)+\int_0^t L\,d\theta,\quad L=\frac{q^2}{2}-U(q);\quad (p(\theta),\ q(\theta))=g^\theta x_j,$$

and μ_j is the Morse index of the trajectory $g^\theta x_j$, i.e., the number of points $q(g^\theta x_j)$, $0 < \theta < t$, focal to M.

5.2. Connection between the Maslov and Morse Indexes

In Theorem 5.1 there appeared the Morse index μ. The Morse index is a special case of the Maslov index; there holds

THEOREM 5.2. Consider in the $(2n + 2)$-dimensional phase space $\mathbf{R}^{2n+2}=\{\hat{p},\hat{q}\}$, $\hat{p}=p_0, p$; $\hat{q}=q_0, q$; $(p, q)\in\mathbf{R}^{2n}$, the $(n + 1)$-dimensional manifold $\hat{M}$: $(p, q)\in M_t$, $q_0=t$, $p_0=-H(p, q)$. Then the manifold $\hat{M}$ is Lagrangian, and the Morse index of the trajectory $g^\theta x$, $0 < \theta < t$, is the Maslov index of the curve $(\theta, -H, g^\theta x)$ on the manifold $\hat{M}$.

The proof follows easily from the definitions of the indexes μ and ind: since $\left(\frac{\partial^2 H}{\partial p^2}\xi, \xi\right) > 0$, a simple focal point gives a contribution of + 1 to ind.

COROLLARY. For any curve γ on M

$$\operatorname{ind} g^t\gamma-\operatorname{ind}\gamma=\mu(g^\theta\gamma^+)-\mu(g^\theta\gamma^-),$$

where $g^\theta\gamma^+$, $g^\theta\gamma^-$ $(0 \le\theta\le t)$ are trajectories with endpoints γ, $\partial\gamma = \gamma^+ - \gamma^-$.

For, the quadrilateral γ, $g^\theta\gamma^+$, $(g^t\gamma)^{-1}$, $(g^\theta\gamma^{-1})^{-1}$ on $\hat{M}$ is, obviously, homologous to zero; therefore its Maslov index equals zero (Theorem 2.2), which in view of Theorem 5.2 proves the desired relationship.

5.3. Quantization Conditions

In Theorem 5.1 there appear indexes of curves which are not closed. The indexes of closed curves enter into asymptotic formulas for stationary problems.

Let M be an invariant Lagrangian manifold of the phase flow g^t, lying on the hypersurface $H = E$ (such invariant manifolds exist not only for integrable systems: see [5]).

Maslov proved

THEOREM 5.3. The equation

$$\frac{1}{2}\Delta\psi=\lambda^2(U(q)-E)\psi$$

has a series of proper numbers $\lambda_N\to\infty$ with asymptotic expressions $\lambda_N=\mu_N+O(\mu_N^{-1})$ if for every $\gamma\in H_1(M, Z)$

$$\frac{2\mu_N}{\pi}\oint_\gamma p\,dq\equiv\operatorname{ind}\gamma\ (\mathrm{mod}\ 4). \tag{11}$$

In this case the characteristic functions ψ_N are also related to the manifold M (in a sense defined precisely in [1] and under assumptions of the type of a simple spectrum).

In the particular case $n = 1$ the index of the circle equals 2, and formula (11) becomes the classical "quantization condition"

$$\mu_N\oint_\gamma p\,dq=2\pi(N+{}^1/_2).$$

LITERATURE CITED

1. V. P. Maslov, Theory of Perturbations and Asymptotic Methods [in Russian], izd. MGU (1965).
2. V. P. Maslov, The WKB Method in the Multidimensional Case, Supplement to Heading's Book "An Introduction to Phase-Integral Methods " [in Russian], Biblioteka sb. "Matematika," Mir (1965).

3. L. S. Pontryagin, Smooth manifolds and their applications in homotopy theory, Trudy V. A. Steklov Matem. inst. AN SSSR, Vol. 45 (1955).
4. R. Abrakham, Supplement to Lang's Book "Differentiable Manifolds" [in Russian], Biblioteka sb. "Matematika," Mir (1966).
5. V. Arnold and A. Avez, Problèmes ergodiques de la mécanique classique, G.-V. (1966).

13

A NOTE ON WEIERSTRASS' AUXILIARY THEOREM

V. I. Arnol'd

Let there be given an analytic mapping, $y = f(x)$, $f(0) = 0$, in the neighborhood of zero in the n-dimensional complex space C^n_x to another complex space, C^n_y. Then each function $\varphi(y)$, analytic in the neighborhood of the point $y = 0$, can be considered as a function of x, $\varphi(f(x))$. In general, these functions do not exhaust the ring of all functions of x which are analytic in the neighborhood of $x = 0$.

The auxiliary theorem of Weierstrass asserts that, if the mapping f at 0 has finite multiplicity (i.e., there are only a finite number of proimages of points close to 0), then the ring of analytic functions of x forms a module with a finite number of generators on the ring of functions $\varphi(f(x))$ (see below, Theorem 1).

Thus, if we denote the generators of this module by $e_i(x)$, then each function, $\varphi(x)$, permits a "Weierstrass expansion"

$$\varphi(x) = \sum_{i=1}^{r} \varphi_i(y) e_i(x),$$

where $y = f(x)$ and the $\varphi_i(y)$ are analytic functions in the neighborhood of $y = 0$. More than that, this is a free module, i.e., the generators $e_i(x)$ can be so chosen that the expansion is unique.

There thus appears a linear operator which puts into correspondence with each function $\varphi(x)$ its collection of coefficients $(\varphi_i(y))$. For this operator, we shall obtain, in this note, a "bound without loss of domain," showing that its analytic properties are no worse than those of the operator of (r-1)-fold differentiation.

Such a bound is useful, for example, in studying the deformations of mappings: it is easy to derive from it the analytical stability of each infinitesimally stable shoot.*

§ 1. Notation

Let $C^n_x = \{x\}$ be an n-dimensional complex space, $x = x_1, \ldots, x_n$, $U \subset C^n_x$,

$$f: U \to C^n_y, \quad y = f(x),$$

a mapping which is analytic in neighborhood U of the point $x = 0$. Let $f(0) = 0$ and let 0 be an isolated point, the preimage $f^{-1}(0)$.

By the multiplicity, r, of mapping f at 0, we mean

$$r = \lim_{U \to 0} \max_{y} r(y, U),$$

where $r(y, U)$ is the number of points in the preimage, $f^{-1}(y)$ in the neighborhood U of point $x = 0$. It is known that the limit, r, exists and is finite, whereby, for sufficiently small U, almost every point y sufficiently close to 0 has, in U, exactly r preimages (see [1]).

By the local ring, Q, or mapping f at 0, we mean

$$Q = H_x/(f^*\mathfrak{M}_y) H_x,$$

*An analogous theorem of stability in the infinitely differentiable case was announced recently by Dzh. Mezer (but not yet published).

Moscow State University. Translated from Funktsional'nyi Analiz i Ego Prilozheniya, Vol. 1, No. 3, pp. 1-8, July-September, 1967. Original article submitted January 25, 1967.

Originally publ. in: Funkts. Anal. Prilozh. 1:3, 1-8, © Russ. Acad. Sciences 1967
English transl.: Funct. Anal. Appl. 1, 173-179, © Consultants Bureau, New York 1967
Reprint by Springer-Verlag Berlin Heidelberg 2014 with kind permission, DOI 10.1007/978-3-642-31031-7_10

where H_x (H_y) are rings of shoots of analytic functions of x (of y) at 0, $\mathfrak{M}_y$ is a maximal ideal of H_y, and f^* is the mapping induced by f: $f^*(\varphi(y))(x) = \varphi(f(x))$. One of the modern formulations of the so-called auxiliary theorem of Weierstrass is as follows:

THEOREM 1. As a C-linear space, ring Q has dimensionality r. If $e_i \in H_x$ ($i = 1, \ldots, r$) are some representatives of the generators of Q as a C-linear space, then each shoot, $\varphi(x) \in H_x$, is represented, in a single-valued manner, in the form

$$\varphi(x) = \sum_{i=1}^{r} \varphi_i(y) e_i(x), \tag{1}$$

where $\varphi_i \in H_y$, and $y = f(x)$ (i.e., $\varphi_i(y)$ means $(f^* \varphi_i)(x)$).

The existence of Expansion (1) was proven in [2], its uniqueness in [3]* (see also, [4]).

Expansion (1) defines the linear "Weierstrass operator"

$$W: H_x \to (H_y)^r, \quad W\varphi = (\varphi_1, \ldots, \varphi_r). \tag{2}$$

Example. Let $n = 1$, $f(x) = x^2$, $e_1 = 1$, $e_2 = x$, then

$$W[\varphi(x)](x^2) = \left(\frac{\varphi(x) + \varphi(-x)}{2}, \frac{\varphi(x) - \varphi(-x)}{2x}\right).$$

The aim of the present note is to study the analytic properties of the Weierstrass operator.

§ 2. Bounds Without Loss of Domain

We fix mapping f and generators e_i. Let V be a sufficiently small neighborhood of point $y = 0$ in C^n_y, let $U \subseteq f^{-1}(V)$ be a connected component of the null preimage of V in C^n_x, and let $\delta_0 > 0$ be a sufficiently small positive number. Then, Theorems 2 and 3 following are valid.

THEOREM 2. If $\varphi(x)$ is an analytic function in the neighborhood of region U, then the vector-function W_φ is analytic in the closure of V, and

$$\| (W\varphi)(y) \|_V \leqslant C_1 \frac{\| \varphi(x) \|_{U+\delta}}{\delta^{r-1}} \tag{3}$$

for any δ, $0 < \delta < \delta_0$, where the constant $C_1 > 0$ does not depend on φ, δ, U or V, but depends only on the original mapping, f, and the generators, e_i.

Here, $\|\psi(z)\|_G = \sup_{z \in G} |\psi(z)|$, $|\psi_1, \ldots, \psi_l| = \max_{1 \leqslant k \leqslant l} |\psi_k|$, and $U + \delta$ is a δ-neighborhood of U in the metric $|x| = \max |x_k|$.

Note. It is impossible to decrease the exponent $r - 1$ in (3); for example, in the case $n = 1$, $f(x) = x^r$; $e_1 = 1, \ldots, e_r = x^{r-1}$; $\varphi(x) = x^{r-1}$.

The proof of Theorem 2 is based on the construction of a certain representation of the ring of analytic functions of x in the ring of analytic matrices of y.

We denote by Q' the r-dimensional space spanned by the $e_i(x)$:

$$Q' = \{\alpha\}, \ \alpha = \sum_{i=1}^{r} \alpha_i e_i(x), \ \alpha_i \in C, \ \|\alpha\|^2 = \sum_{i=1}^{r} |\alpha_i|^2.$$

We make correspond to each function $\varphi(x)$, analytic in a neighborhood of region U, the endomorphism $\Phi(y)$: $Q' \to Q'$, analytically depending on y and acting as multiplication by $\varphi(x)$:

$$(\Phi(y)\alpha)(x) = \varphi(x)\alpha(x), \quad \text{if} \quad y = f(x), \ \alpha \in Q'. \tag{4}$$

*See [3], p. 299.

THEOREM 3. The endomorphism $\Phi(y): Q' \to Q'$, which satisfies Condition (4) and is continuous for $y \in V$, exists, is unique, is analytic in y in the closure of V, and admits the bound

$$\|\Phi(y)\|_V \leqslant C_2 \frac{\|\varphi(x)\|_{U+\delta}}{\delta^{r-1}} \tag{5}$$

for any δ, $0 < \delta < \delta_0$, where the constant $C_2 > 0$ does not depend on Φ, δ, U or V, but depends only on f and the e_i.

We note that Theorem 2 is a consequence of Theorem 3. Indeed, according to Theorem 1, in some neighborhood of the point x = 0 (it can be considered that this neighborhood contains $U + \delta_0$), there exists the expansion $1 = \sum_{i=1}^{r} \beta_i(f(x)) e_i(x)$. Consider the vector, depending on $y \in V$, $1_y = \sum_{i=1}^{r} \beta_i(y) e_i(x) \in Q'$. If $\Phi(y) 1_y = \sum_{i=1}^{r} \psi_i(y) e_i(x) \in Q'$, then, by virtue of (1) and (4), $(W\varphi)(y) = (\psi_1(y), \ldots, \psi_r(y))$ and, from the bound in (5) the bound of (3) follows. Thus, for the proof of Theorem 2, it suffices to construct the endomorphism $\Phi(y)$ satisfying Theorem 3, which we shall do in the following sections.

§ 3. The Basis Endomorphisms $X_i(y)$

Turning now to the construction of the representation of function $\varphi(x)$ by the endomorphisms $\Phi(y)$ in the entire domain $V = f(U)$: the existence of $\Phi(y)$ in some neighborhood of point y = 0 for each fixed analytic function $\varphi(y)$ follows directly from Theorem 1. In particular, for each of the coordinate functions x_i $(1 \le i \le n)$, we define the endomorphism $X_i(y): Q' \to Q'$ of its action on basis Q':

$$X_i(y) e_k = \sum_{l=1}^{r} \chi_{ikl}(y) e_l, \qquad 1 \leqslant k \leqslant r,$$

where the coefficients $\chi_{ikl}(y)$ are the same as in the rn expansions of the form of (1):

$$x_i e_k = \sum_{l=1}^{r} \chi_{ikl}(y) e_l(x), \quad y = f(x), \quad 1 \leqslant k \leqslant r, \quad 1 \leqslant i \leqslant n.$$

Then, $X_i(f(x))$ acts in Q' as multiplication by x_i, i.e.,

$$(X_i(f(x))\alpha)(x) = x_i \alpha(x), \qquad \forall \alpha \in Q'. \tag{6}$$

In a sufficiently small neighborhood, V_0, of the point y = 0, all the endomorphisms $X_i(y)$ are analytic and bounded:

$$\|X_i(y)\|_{V_0} \leqslant C_3, \qquad 1 \leqslant i \leqslant n. \tag{7}$$

§ 4. Eigenvalues

We now find the eigenvalues of the endomorphisms $X_i(y)$. We denote by U_0 the connected component of zero in the preimage $f^{-1}(V_0)$. If region V_0 is sufficiently small, then almost every point $y \in V_0$ has exactly r preimages in U_0. We shall call such points y points of general position.

Let $\xi^1, \ldots, \xi^r \in f^{-1}(y) \cap U_0$ be preimages of point of general position $y \in V_0$.

LEMMA 1. If endomorphism $\Phi(y): Q' \to \varphi'$ satisfies Relationship (4), then each of its r eigenvalues equals the value of φ at one of the points ξ^i, and the corresponding eigenvector is that function, $\delta_{\xi^i}(x)$, whose value at point ξ^i equals 1 and, at all other points ξ^k, equals zero:

$$\delta_{\xi^i}(\xi^k) = \delta_{ik} \text{ *)}, \tag{8}$$

$$\Phi(y)\delta_{\xi^i} = \varphi(\xi^i)\delta_{\xi^i}, \quad 1 \leqslant i \leqslant r. \tag{9}$$

* $\delta_{ik} = \begin{cases} 1 \text{ if } k = i, \\ 0 \text{ if } k \ne i. \end{cases}$

Proof. Relationship (9) follows immediately from (4) and (8). If, now, y is a point of general position, then the r vectors $\delta_\xi i (1 \le i \le r)$ form a basis in Q', which also proves the Lemma.

COROLLARY 1. For each point $y \in V_0$, the endomorphisms $X_i(y)$ $(1 \le i \le n)$ commute in pairs, and the eigenvalues of $X_i(y)$ equal the values of the coordinates x_i at the points $f^{-1}(y)$.

Indeed, in accordance with Relationship (6), Condition (4) is met (where $\Phi = X_i$, $\varphi = x_i$). For a point y of general position, the assertion follows from (9). But then, by the continuity of $X_i(y)$, it is proven for all $y \in V_0$.

COROLLARY 2. Endomorphisms $\Phi_1(y)$ and $\Phi_2(y)$, continuously depending on y and satisfying Condition (4), coincide.

Indeed, they coincide, according to (9), on the everywhere dense set of points y of general position.

Thus, the uniqueness of $\Phi(y)$ is proven. In order to construct $\Phi(y)$, we consider function ϕ and the endomorphisms X_i.

§ 5. Construction of $\Phi(y)$

Let V be an arbitrarily small neighborhood of the point y = 0, so small that its connected component in the preimage of V enters into U_0 in a δ_0-neighborhood. Let $\varphi(x)$ be an analytic function in $U + \delta$, where $0 < \delta < \delta_0$. We fix point $y_0 \in V$. Starting out from the basis endomorphisms $X_i(y)$, we provide ourselves with the endomorphism $\Phi(y)$: $Q' \to Q'$ in the neighborhood of point y_0 by the Cauchy integral

$$\Phi(y) = \left(\frac{1}{2\pi i}\right)^n \int_{\gamma(y_0)} \frac{\varphi(x)\, dx_1 \wedge \ldots \wedge dx_n}{\prod_{k=1}^{n} (x_k E - X_k(y))}. \tag{10}$$

The contour of integration, $\gamma(y_0)$, depends both on y_0 and on δ; it will be described in §6. We so choose it that $\Phi(y)$ will act on Q' as multiplication by Φ: Relationship (9) will hold for each point of general position $y \in V$.

For the endomorphism of (10), Relationship (9) is proven in §8, while the bound of (5) is obtained in §7.

§ 6. Choice of the Contour of Integration γ

We begin with certain notation related to the Cauchy integral.

We call the cross, K_ξ, of a point $\xi \in C^n$, the union of all the coordinate planes passing through ξ:

$$K_\xi = \left\{x : \prod_{k=1}^{n} (x_k - \xi_k) = 0\right\}.$$

We call a polycylinder in C^n a direct product of circles,

$$\omega = \{x : |x_k - x_k^0| \leqslant R_k,\ k = 1, \ldots, n\}.$$

We call the hull of a polycylinder the direct product of neighborhoods $\gamma = \{x : |x_k - x_k^0| = R_k,\ k = 1, \ldots, n\}$, oriented in the usual fashion.

Let g(x) be an analytic function in polycylinder ω. If the cross of point ξ does not intersect hull γ of polycylinder ω, we then define the Cauchy integral

$$G(\xi) = \left(\frac{1}{2\pi i}\right)^n \int_{\gamma} \frac{g(x)\, dx_1 \wedge \ldots \wedge dx_n}{\prod_{k=1}^{n} (x_k - \xi_k)}. \tag{11}$$

With this, if ξ lies within ω, then $G(\xi) = g(\xi)$, while if ξ is outside ω, then $G(\xi) = 0$.

LEMMA 2. Let $\xi^1, \ldots, \xi^r$ be points in C^n (not necessarily different), and let $\tau > 0$. There then exist $\rho \le r$ pairwise non-intersecting polycylinders $\omega^j (1 \le j \le \rho)$ such that:

1) each point ξ lies in one and only one of the polycylinders ω^j;

2) the hull of each of the polycylinders ω^j is at a distance of not less than τ from each of the r crosses $K\xi^i$;

3) the radius of none of the polycylinders ω^j exceeds $r\tau$.

The proof of Lemma 2 is given in §9.

Turning now to the construction of the contour of integration in Cauchy Integral (10), we apply Lemma 2 to the r (not necessarily different) preimages $\xi_0^j \in f^{-1}(y_0) \cap U$ of point $y_0 \in V$, setting $\tau = \delta/2r$. Then, all the polycylinders ω^j lie in $U + \delta$.

We form the contour of integration, $\gamma(y_0)$, in (10) from the hulls of the polycylinders $\omega^j (1 \leq j \leq \rho)$.

§ 7. Bound on the Cauchy Integral

The bound on Integral (10) is based on the following Lemma, proven in §10.

LEMMA 3. Let A_k: $C^r \to C^r (k = 1, \ldots, n)$ be commuting endomorphisms of C^r. If $\|A_k\| \leq a$ and all eigenvalues of all A_k are not less than ε in modulus, then

$$\left\| \prod_{k=1}^{n} A_k^{-1} \right\| \leqslant C_4 \varepsilon^{-(n+r-1)}, \text{ where } C_4(a, r, n) = r(ran)^{r-1}$$

(the norm is Hermitian).

Let $y \in V$ be a point of V sufficiently close to y_0. We denote its preimages (not necessarily different) by $\xi^j \in f^{-1}(y) \cap U (1 \leq j \leq r)$. Let $x = (x_1, \ldots, x_n) \in \gamma$. We set $A_k = x_k E - X_k(y)$. It follows from (7) that $\|A_k\| \leq 2C_3$. According to §4, the endomorphisms A_k commute, and their eigenvalues are $x_k - x_k(\xi^j)$ $(1 \leq j \leq r)$. According to the choice of γ, and by virtue of assertion 2) of Lemma 2, $|x_k - x_k(\xi_0^j)| \geq \tau$ so that, for a point y sufficiently close to y_0, we have

$$|x_k - x_k(\xi^j)| \geqslant \varepsilon = \tau/2 = \delta/4r.$$

Applying Lemma 3 to the A_k, we obtain

$$\left\| \prod_{k=1}^{n} (x_k E - X_k(y))^{-1} \right\| \leqslant C_5 \delta^{-(n+r-1)},$$

where $C_5 = C_4(2C_3, r, n)\,(5r)^{n+r-1}$. It follows from this that Integral (10) exists, is analytic with respect to y, and satisfies the bound

$$\|\Phi(y)\| \leqslant C_6 \delta^{-(r-1)} \|\varphi(x)\|_{U+\delta}, \tag{12}$$

where $C_6 = C_5 r$ because, according to assertion 3) of Lemma 2, the measure of γ does not exceed $r(2\pi\delta)^n$.

§ 8. Proof of Theorem 3

Let y be a point of general position, sufficiently close to y_0. We shall prove Relationship (9). We apply $\Phi(y)$ of (10) to vector δ of (8). Since, according to § 4, $X_k(y)\delta_{\xi j} = \xi_k^j \delta_{\xi j}$, then, using Cauchy Formula (11), we find that

$$\Phi(y)\delta_{\xi^j} = \sum_{l=1}^{\rho} \left(\frac{1}{2\pi i}\right)^n \int_{\gamma_l} \frac{\varphi(x)\, dx_1 \wedge \ldots \wedge dx_n}{\prod_{k=1}^{n} (x_k - \xi_k^j)} \delta_{\xi^j} = \sum_{l=1}^{\rho} \varphi(\xi^j)\delta_{\xi^j}\chi_l(\xi^j),$$

where γ_l is the hull of polycylinder ω^l, and χ_l is the characteristic function of ω^l.

In accordance with the choice of contour γ, and by virtue of assertion 1) of Lemma 2, each point ξ^j lies in exactly one of the polycylinders ω^l. Therefore, $\sum_{l=1}^{\rho} \chi_l(\xi^j) \equiv 1$, and Relationship (9) is proven.

It was shown in §5 that (4) follows from Relationship (9). Since the bound of (12) has the form of (5), Theorem 3 is proven.

§ 9. Proof of Lemma 2

Initially, let n = 1. We call a mapping, Ω, of a finite set, $\{\xi^j\}$, of points of C^1 into the set of circles in C^1 a τ-admissible system of circles if 1) the point ξ^i lies in a τ-neighborhood of circle $\Omega(\xi^i)$; 2) the radius of $\Omega(\xi^i)$ does not exceed $\nu\tau$, where ν is the number of points ξ^j for which $\Omega(\xi^i) = \Omega(\xi^j)$.

For example, the mapping Ω_0, making correspond to each point ξ^i, $1 \le i \le r$ the closure of its τ neighborhood, is a τ-admissible system of r circles.

If the circles of a τ-admissible system are pairwise non-intersecting, the Lemma is then proven. If, now, two circles, ω^1 and ω^2, of system Ω intersect, then the number of circles can be decreased, replacing ω^1 and ω^2 by the circumscribed circle ω, i.e., setting

$$\Omega'(\xi^i) = \begin{cases} \omega, & \text{if } \Omega(\xi^i) = \omega^1 \text{ or } \omega^2, \\ \Omega(\xi^i), & \text{if } \Omega(\xi^i) \neq \omega^1, \text{ nor } \omega^2. \end{cases}$$

It is easily verified that the system Ω' thus obtained will be τ-admissible system consisting of the least number of circles, the circles do not intersect and, for n = 1, Lemma 2 is proven.

Now, let n > 1. For given i $(1 \le i \le n)$, we consider the i'th coordinate of our r points $\xi_i^1, \ldots \xi_i^r$ C^1. Since the Lemma is proven for n = 1, there exists a τ-admissible system of $\rho(i)$ circles $\Omega_i(\xi_i^j)$ $(1 \le j \le r)$, pairwise non-intersecting. We cover each point $\xi^j \in C^n$ by the polycylinder corresponding to the circle covering its projection:

$$\omega(\xi^j) = \{x\colon x_i \in \Omega_i(\xi_i^j),\ i = 1, \ldots, n\}.$$

These polycylinders satisfy all the requirements of Lemma 2 (requirement 2) is satisfied because for any i, each of the $\rho(i)$ neighborhoods $\partial\Omega_i(\xi_i^j)$ is at a distance no less than τ from each of the r points $\xi_i^j \in C^1$). Lemma 2 is proven.

§ 10. Bounds on the Product of Commuting Matrices

LEMMA 4. Let T be an upper-triangular r × r matrix all of whose diagonal elements are not less than ε in modulus, and none of whose elements exceeds "a" in modulus. Then, the inverse matrix, T^{-1}, has the majorant

$$T^{-1} \ll \frac{1}{\varepsilon}\left(E + \frac{A}{\varepsilon} + \frac{A^2}{\varepsilon^2} + \ldots + \frac{A^{r-1}}{\varepsilon^{r-1}}\right), \tag{13}$$

where A is a nilpotent upper-triangular matrix all of whose $r(r-1)/2$ elements equal "a."

Proof. We expand T into a diagonal and a superdiagonal part: $T = \Lambda + \Delta = \Lambda(E + \Lambda^{-1}\Delta)$. Then,

$$T^{-1} = \Lambda^{-1} - \Lambda^{-1}\Delta\Lambda^{-1} + (\Lambda^{-1}\Delta)^2\Lambda^{-1} \pm \ldots \pm (\Lambda^{-1}\Delta)^{r-1}\Lambda^{-1}. \tag{14}$$

But, by hypothesis, $\Lambda^{-1} \ll \frac{E}{\varepsilon}$, $\Delta \ll A$, and (13) follows from (14).

Proof of Lemma 3. Since the A_k commute, their matrices have the triangular forms T_k in one and the same Hermitian orthonormal basis. With this, each element of matrix T_k is no larger than "a," and the diagonal elements are no smaller than ε.

By Lemma 4, the matrices T_k^{-1} have a common majorant, given in (13). Since $A^r = 0$,

$$\prod_{k=1}^{n} T_k^{-1} \ll \frac{1}{\varepsilon^n}\left(E + \frac{A}{\varepsilon} + \ldots + \frac{A^{r-1}}{\varepsilon^{r-1}}\right)^n \ll \frac{1}{\varepsilon^n}\left(E + \frac{nA}{\varepsilon} + \frac{n^2A^2}{\varepsilon^2} + \ldots + \frac{n^{r-1}A^{r-1}}{\varepsilon^{r-1}}\right).$$

But, obviously $\|A\| \le ra$, $\varepsilon \le a$, so that

$$\left\|\prod_{k=1}^{n} A_k^{-1}\right\| < \frac{1}{\varepsilon^n}\left(1 + \frac{nra}{\varepsilon} + \ldots + \left(\frac{nra}{\varepsilon}\right)^{r-1}\right) < \frac{r\,(nra)^{r-1}}{\varepsilon^{n+r-1}},$$

q.e.d.

LITERATURE CITED

1. M. Erve, Functions of Many Complex Variables [in Russian], Moscow, "Mir" (1965).
2. Séminaire H. Cartan, Paris, No. 18 (1960/1961).
3. G. Grauert, Complex Spaces [in Russian], Moscow, "Mir" (1965).
4. V. P. Palamodov, "On the multiplicity of holomorphic mappings," Funkts. Analiz, 1, No. 3, 54-65 (1967)

THE STABILITY PROBLEM AND ERGODIC PROPERTIES FOR CLASSICAL DYNAMICAL SYSTEMS

V. I. ARNOL'D

A *classical dynamical system* consists of a smooth manifold M and a one-parameter group g^t ($t \in R$ or Z) of diffeomorphisms $g^t\colon M \to M$.

The differential equations of classical mechanics (for example, the three body problem) provide us with many examples of systems (M, g^t). However, these systems belong to the narrower class of *Hamiltonian systems* (M, g^t, ω^1): the manifold $M = M^{2n}$ has an even number of dimensions, and on it there is defined a canonical structure, i.e. a fixed 1-form ω^1 (defined up to the total differential of a one-valued function) with nondegenerate derivative $d\omega^1 = \Sigma_{i=1}^n dp_i \wedge dq_i$ in suitable local coordinates p, q on M; the diffeomorphisms g^t are canonical; for an arbitrary closed curve γ

$$\oint_\gamma \omega^1 = \oint_{g^t\gamma} \omega^1.$$

Intermediate between the general and Hamiltonian systems are the "systems with invariant measure" (M, g^t, τ): on M there is a fixed nondegenerate differential form τ of maximal dimensionality and g^t preserves this form.

The theory of dynamical systems deals with the problem of studying the behavior of a typical orbit of a typical system (M, g^t), (M, g^t, τ) or (M, G^t, ω^1). The theory is completely different for the three types of systems. For general systems one of the typical possibilities consists of asymptotically stable motions, attracting the neighboring motions. But for systems with invariant measure asymptotically stable motions are impossible. Less well known are the special subsidiary properties of Hamiltonian systems: they show themselves, for example, in the conservative resistance of the evolution of the system.

At the present time there is no even remotely satisfactory general theory for any of the three cases. Only certain special systems have been carefully studied. Let us attempt to investigate the situation by transforming special systems into general ones by means of small perturbations.

1. **Theory of perturbations.** Let $M = T^k \times R^l$ be the direct product of the k-dimesnional torus T^k with the euclidean space R^l; let $\phi = \phi_1 \cdots \phi_k$ mod 2π be angular coordinates on T^k, and $I = I_1 \cdots I_l$ Cartesian coordinates on R^l. As an "unperturbed system" let us take the system defined by

Originally publ. in: Proc. Internat. Congr. Math., Moscow 1966, 387-392. © MIR, Moscow 1968
English transl.: Am. Math. Soc. Transl. (2) 70, 5-11, © American Math. Society, Providence, RI, 1969
Reprint by Springer-Verlag Berlin Heidelberg 2014 with kind permission, DOI 10.1007/978-3-642-31031-7_11

the differential equations

$$\frac{d\phi}{dt} = \omega(I), \qquad \frac{dI}{dt} = 0 \qquad (\omega = \omega_1, \cdots, \omega_k). \tag{1}$$

It is obvious that every torus $I = \text{const}$ is invariant. If the frequencies on it are incommensurable: $n_1\omega_1 + \cdots + n_k\omega_k \neq 0$ for integral n_i, $n_1^2 + \cdots + n_k^2 = 0$; then the orbit $\phi(t)$ is everywhere dense on the torus and the motion is said to be quasiperiodic, with k frequencies. If the frequencies are dependent, then the closure of the orbit is a torus of less than k dimensions (resonance).

For the investigation of the "perturbed system"

$$\frac{d\phi}{dt} = u(I) + \epsilon f(I, \phi), \qquad \frac{dI}{dt} = \epsilon F(I, \phi) \tag{2}$$

the classical (i.e. nonrigorous) theory of perturbations prescribes that we set up the mean value $\overline{F}(I) = (2\pi)^{-k} \oint_{T^k} F(I, \phi)\, d\phi$ and the "evolutionary equation"

$$\dot{J} = \epsilon \overline{F}(J). \tag{3}$$

It is assumed that for $\epsilon \ll 1$ the difference between the solutions $I(t)$ and $J(t)$ of the systems (2) and (3) with identical initial conditions remains small, at least for a long period of time $0 < t < 1/\epsilon$.

Problem 1. How are $I(t)$ and $J(t)$ related to each other, $0 < t < 1/\epsilon$?

Except for certain special cases when the system (3) has an asymptotically stable motion, very little is known:

Theorem 1. *For $k = 1$, $\omega \neq 0$ we have $|I(t) - J(t)| < C\epsilon$* [1]. *For $k = 2$, $d(\omega_1/\omega_2)/dt \neq 0$ we have $|I(t) - J(t)| < C\sqrt{\epsilon}\,\log 1/\epsilon$* [2].

The case $k > 2$ has been very little studied. We know only that for systems with an invariant measure $dI\,d\phi$ the quantity $\epsilon\ |I(t) - J(t)|$ is small for small ϵ under the great majority of initial conditions (cf. [3], [4]).

Now let the system (2) be Hamiltonian: $k = l$, $\omega' = I d\phi$.

2. Perturbations of Hamiltonian systems. The Hamiltonian system (2) is of the form

$$\frac{d\phi}{dt} = \frac{\partial H}{\partial I}, \quad \frac{dT}{dt} = \frac{\partial H}{\partial \phi} \qquad H = H_0(I) + \epsilon H_1(I, \phi). \tag{4}$$

Thus the averaging process produces zero:

$$\overline{F}(I) = (2\pi)^{-k} \oint_{T^k} \frac{\partial H_1}{\partial \phi}\, d\phi = 0.$$

Consequently the evolutionary system is a null system, and to the first approximation there is no evolution. Moreover, we have the following theorem.

Theorem 2. (cf. [5]). *Let H_0, H_1 be analytic for $I \in G$, $|\mathrm{Im}\,\phi| < \rho$ and on G*

$$\det\left|\frac{\partial\omega}{\partial I}\right| = \det\left|\frac{\partial^2 H_0}{\partial I^2}\right| \neq 0.$$

Then the greater part of $G \times T^k$ is filled out, for sufficiently small ϵ, with invariant k-dimensional toruses of the system (4).

Theorem 2 can be generalized to certain cases where $\partial^2 H_0/\partial I^2 = 0$, which are important in celestial mechanics. For example, it was in this way that the quasiperiodic motions were discovered in the many-body problem [6]. The assumption of analyticity can also be weakened; it is enough to assume the existence of several hundred derivatives (cf. [7]).

The set of invariant toruses in Theorem 2 has an everywhere dense open complement.

Problem 2. How do the orbits from this complementary set behave? In particular, is it true that for them there is no evolution in the sth approximation; i.e. $|I(t) - I(\theta)| \ll 1$, $0 < t < (1/\epsilon)^s$?

A recent article on "formal stability" [8] apparently leaves this important question open. In the older astronomical literature it was regarded as solved in the affirmative sense. I do not know of any mathematically rigorous proof, except for the case of two degrees of freedom, when we may decide about stability in accordance with the presence of invariant toruses.

3. Invariant toruses and stability. In the case $k = 2$ the invariant toruses of Theorem 2 partition the three-dimensional set of energy level $H = \text{const}$ in system (4). Moreover, if the ratio of the frequencies changes along $H = \text{const}$, that is, if

$$\det\begin{vmatrix} \partial^2 H_0/\partial I^2 & \partial H_0/\partial I \\ \partial H_0/\partial I & 0 \end{vmatrix} \neq 0,$$

then there are many such toruses on each energy level. These toruses enclose each orbit in a narrow layer bounded by them, so that the motion is stable for all initial conditions in the sense that

$$|I(t) - I(0)| < C(\epsilon), \quad -\infty < t < +\infty; \qquad C(\epsilon) \longrightarrow 0 \text{ for } \epsilon \longrightarrow 0.$$

For $k \geq 3$ the toruses T^k do not partition the $(2k-1)$-dimensional level $H = \text{const}$.

Conjecture. The "general case" for a Hamiltonian system (4) with $k \geq 3$ is represented by the situation that for an arbitrary pair of neighborhoods of toruses $I = I'$, $I = I''$, in one component of the level set $H_0(I) = H_0(I'')$ there exists, for sufficiently small ϵ, an orbit intersecting both neighborhoods. That such orbits, joining neighborhoods of distant toruses, are in general possible, is shown by an example [9] of a system which satisfies all the conditions of the theorem on invariant toruses and is therefore stable for the great majority of initial conditions but is not stable for "resonance" initial conditions. The mechanism of "transitional sequences" which occurs in this example is probably of very general application.

For the construction of a nonstable orbit [9] use is made of a family of invariant toruses with simple resonance (of dimension $k - 1$). The first general theorems on existence of such toruses were obtained very recently by V. K. Mel'nikov, J. Moser and A. M. Leontovič. In order to take the final step in the proof of the above conjecture, it is necessary to examine the transition from simple to double resonance (resonance of order 3 and higher is unimportant in topological arguments). As the model problem here we may take the construction of the transitional sequence uniting two periodic orbits and the position of equilibrium in the system on $T^2 \times R^2$:

$$H = \frac{I_1^2 + I_2^2}{2} + u(\phi_1, \phi_2).$$

We may make use of the following elementary theorem proved by E. Gaidukov.

Theorem. *Through each point of a Riemannian T^2 there passes a geodesic which is asymptotic to a closed geodesic homotopic to a given one.*

4. The geometric theorem of Poincaré. Another chapter that is far from solved in the theory of multidimensional Hamiltonian systems is the theory of periodic orbits. For example, the following generalizations of Poincare's "last theorem" appear very probable.

A. Let A: $q \longrightarrow q + f(q)$ be a diffeomorphism of the torus $T^2 = \{q_1, q_2, \bmod d2\pi\}$ preserving the measure $dq_1 dq_2$ and the center of gravity ($\iint f(q) dq_1 dq_2 = 0$). Then A has at least four fixed points, if account is taken of multiplicity, and at least three geometrically distinct fixed points.

B. Let $\Omega = T^k \times B^k$, $T^k = \{q \bmod 2\pi\}$, $B^k = \{p \in R^k, |p| \leq 1\}$ be a toroidal ring with canonical structure $\omega^1 = pdq$, and let A: $\Omega \longrightarrow \Omega$ be a canonical diffeomorphism homotopic to the identity and such that every sphere $q \times \partial B^k$ is linked with its image on the covering edge $T^k \times \partial B^k$. Then A has at least 2^k fixed points, if account is taken of multiplicity, among which $k + 1$ are geometrically distinct.

The proofs, which are based on the Morse-Ljusternik-Šnirel'man theory, can be carried out only under additional restrictions (see [10]).

5. Ergodic properties.

Problem. To determine the ergodic properties of motions in the region complementary to invariant toruses of system (4). In particular, is the entropy of this system positive?

Up to now "typicality" has been proved for only two types of behavior of dynamical systems with invariant measure: the quasiperiodic ones discussed above, and the unstable ones dealt with in a report by Anosov at this Congress. But a number of models of intermediate type are well known, the oricyclic flows [12], systems with quasidiscrete spectrum [13], overlaying of segments [14], etc. For the study of these "systems with slow intermingling" a special technical apparatus has been constructed, of rational approximations, 2^n-entropy [15], etc. But it remains unknown how such systems behave under perturbations, and whether or not they split into components with discrete spectrum and into other components with positive entropy.

It would be very interesting to investigate, for example, perturbations of oricyclic flows which in the algebraic case separate systems with the discrete spectrum from systems with exponential distribution. This will enable us to follow the transition from systems with invariant toruses to a K-system.

To the same general range of ideas belongs the question of the continuity of the integral of a classical dynamical function as a function of the system; here all that has been proved is the boundedness of the entropy of a diffeomorphism of a compact manifold [16].

Particularly interesting is the study of the ergodic properties of systems (4) when ϵ is no longer small. In this case, it would be useful to make a numerical experiment, but such experiments show that the measure of the invariant toruses decreases rapidly [17] with increasing ϵ.

BIBLIOGRAPHY

[1] M. Kruskal, *Adiabatic invariants.*

[2] V. Arnol'd, *Conditions for the applicability, and estimate of the error, of an averaging method for systems which pass through states of resonance in the course of their evolution*, Dokl. Akad. Nauk SSSR **161** (1965), 9–12 = Soviet Math. Dokl. 6 (1965), 331–334. MR 31 #3661.

[3] D. Anosov, *Averaging in systems of ordinary differential equations with rapidly oscillating solutions,* Izv. Akad. Nauk SSSR Ser. Mat. 24 (1960), 721–742. (Russian) MR 23 #A3888.

[4] T. Kasuga, *On the adiabatic theorem for the Hamiltonian system of differential equations in the classical mechanics.* III, Proc. Japan Acad. 37 (1961), 377–382. MR 27 #4404c.

[5] A. N. Kolmogorov, *On conservation of conditionally periodic motions for a small change in Hamilton's function,* Dokl. Akad. Nauk SSSR 98 (1954), 527–530. (Russian) MR 16, 924.

[6] V. Arnol'd, a. *The classical theory of perturbations and the problem of stability of planetary systems,* Dokl. Akad. Nauk SSSR 145 (1962), 487–490 = Soviet Math Dokl. 3 (1962), 1008–1012. MR 25 #5781.

b. *Proof of a theorem of A. N. Kolmogorov on the preservation of conditionally periodic motions under a small perturbation of the Hamiltonian,* Uspehi Mat. Nauk 18 (1963), no. 5 (113), 13–40. (Russian) MR 29 #328.

[7] J. Moser, *On invariant curves of area-preserving mappings of an annulus,* Nachr. Akad. Wiss. Göttingen Math. Phys. Kl. II 1962, 1–20. MR 26 #5255.

[8] J. Glimm, Comm. Pure Appl. Math. 17 (1964), no. 4.

[9] V. Arnol'd, *Instability of dynamical systems with several degrees of freedom,* Dokl. Akad. Nauk SSSR 156 (1964), 9–12 = Soviet Math. Dokl. 5 (1964), 581–585. MR 29 #329.

[10] ———, *Sur une propriété topologique des applications globalement canoniques de la mécanique classique,* C. R. Acad. Sci. Paris 261 (1965), 3719–3722. MR 33 #1861.

[11] L. W. Green, *The Pythagorean group and ergodic flows,* Bull. Amer. Math. Soc. 72 (1966), 44–49. MR 32 #1652.

[12] L. Auslander, F. Mahn and L. Green, *Flows on homogeneous spaces,* Ann. of Math. Studies, no. 53, Princeton Univ. Press, Princeton, N.J., 1963. MR 29 #4841.

[13] L. M. Abramov, *Metric automorphisms with quasi-discrete spectrum,* Izv. Akad. Nauk SSSR Ser. Mat. 26 (1962), 513–530; English transl., Amer. Math. Soc. Transl. (2) 39 (1964), 37–56. MR 26 #606.

[14] A. Katok and A. Stenik, Report at this Congress.

[15] A. Kušnirenko, Report at this Congress.

[16] ———, *An estimate from above for the entropy of a classical dynamical system*, Dokl. Akad. Nauk SSSR 161 (1965), 37–38 = Soviet Math. Dokl. 6 (1965), 360–362. MR 31 #1668.

[17] M. Hénon and C. Heiles, *The applicability of the third integral of motion: Some numerical experiments*, Astronom. J. 69 (1964), 73–79. MR 28 #1969.

Translated by:
S. H. Gould

A remark on the ramification of hyperelliptic integrals as functions of parameters

V.I. Arnold
Translated by Victor Vassiliev

Let $p : E \to B$ be a fiber bundle with fiber E. The fundamental group $\pi_1(B)$ of the base acts on the homology groups of the fiber. The representation A_*, $A_*(\gamma) : H_k(F,\mathbb{Z}) \to H_k(F,\mathbb{Z})$, $\gamma \in \pi_1(B)$, of the fundamental group of the base into the automorphism group of the k-dimensional homology group of the fiber can be called the *k-dimensional monodromy group* (the usual monodromy group occurs for $k = 0$).

In this notice a special case is considered: F is a hyperelliptic curve, and the base B is the space of all non-degenerate hyperelliptic curves of degree n. In this case the one-dimensional monodromy group is a representation of the Artin braid group into the group of symplectic integer matrices. We will show that *the image of this representation coincides with the entire symplectic group when $n = 3,4,6$ and only in these cases.*

A non-degenerate hyperelliptic curve F_n is specified by an equation

$$y = \sqrt{z^n + a_1 z^{n-1} + \ldots + a_n}, \tag{1}$$

containing the radical of a polynomial with complex coefficients, having no multiple roots. F is compactified by adding the ramification point ∞ if n is odd, and two simple points ∞_1, ∞_2 if n is even.

It is well-known that the one-dimensional homology group $H_1(F,\mathbb{Z})$ has $2g$ free generators a_i, b_i ($i = 1,\ldots,2g$, $n = 2g+1$ or $2g+2$), whose intersection indices are (see the figure) $(a_i,a_j) = (b_i,b_j) = (a_i,b_j) = 0$, $(a_i,b_i) = 1$.

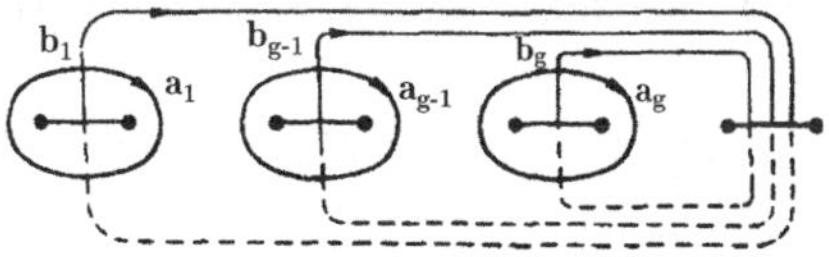

The intersection index is an integer bilinear non-degenerate skew-symmetric 2-form on $H_1(F,\mathbb{Z})$. Automorphisms $A_*(\gamma)$, $\gamma \in \pi_1(B)$, preserve this form. Therefore the matrices of $A_*(\gamma)$ in the basis $a_1, b_1, \ldots, a_g, b_g$ belong to the group of symplectic

Originally publ. in: Funkts. Anal. Prilozh. 2:3, 1-3, © Russ. Acad. Sciences 1968
Original English transl.: Funct. Anal. Appl. 2, 187-189, © Consultants Bureau, New York 1968
Translated by V.Vassiliev.Springer-Verlag Berlin Heidelberg 2014 with kind permission, DOI 10.1007/978-3-642-31031-7_12

(i.e. preserving the form $x_1 \wedge y_1 + \ldots + x_g \wedge y_g$) integer matrices or order $2g$, $A_*(\gamma) \in \mathrm{Sp}(g,\mathbb{Z})$.

The space of curves (1) is naturally identified with the space G_n of complex degree n polynomials with leading coefficient 1, having no multiple roots. Therefore the fundamental group of our fiber bundle is the Artin braid group on n strings, $B(n)$ (see [1]). So, we obtain a representation of the braid group in the symplectic group:

$$A_* : B(n) \to \mathrm{Sp}(g,\mathbb{Z}), \quad n = 2g+1 \text{ or } 2g+2. \tag{2}$$

Theorem 1. *A) For $n = 3,4,6$, and B) for only these values of n we have*

$$A_*B(n) = Sp(g,\mathbb{Z}).$$

Let S be the Riemann sphere of the variable z in (1). Let us fix a hyperelliptic curve F; let $z_1,\ldots,z_{2g+2} \in S$ be its ramification points. Let S' be the sphere with holes obtained by removing sufficiently small discs from S centered at these ramification points. The ramified covering $\varphi : F \to S$ induces the two-fold covering $\varphi' : F' \to S'$ of the surface with holes $F' \subset F$ to the sphere with holes $S' \subset S$.

Any homology class from $H_1(F,\mathbb{Z})$ is represented by a curve γ lying entirely in F'; its projection $\varphi'\gamma$ lies in S'. The homology class of the latter curve in $H_1(S',\mathbb{Z})$ depends on the representative γ. However, the following statement holds.

Lemma 1. *The map φ' induces a homomorphism*

$$\Phi : H_1(F,\mathbb{Z}) \to H_1(S',\mathbb{Z}_2).$$

Proof. Consider the commutative diagram of exact sequences of pairs (F,F') and (S,S') with coefficients in $\mathbb{Z}_2$:

$$\begin{array}{ccccccc}
\to H_2(F/F') & \xrightarrow{\partial} & H_1(F') & \xrightarrow{i_*} & H_1(F) & \longrightarrow & H_1(F/F') = 0 \\
0\downarrow & & \varphi'_*\downarrow & \swarrow & \downarrow & & \\
\to H_2(S/S') & \xrightarrow{\partial} & H_1(S') & \longrightarrow & H_1(S) = 0. & &
\end{array}$$

(We have used the fact that $F/F' \sim S/S'$ is the wedge of $2g+2$ two-dimensional spheres, and the endomorphism in the left-hand column is induced by a map of degree 2 on any sphere of the wedge.) Let $\gamma \in H_1(F)$. $H_1(F/F') = 0$, therefore $\gamma = i_*\beta$, $\beta \in H_1(F')$. We need to prove that $\varphi'_*\beta$ does not depend on the choice of β from $\mathrm{B}_*^{-1}\gamma$. Let $k \in \ker i_*$. Then $k = \partial l$, $l \in H_2(F/F')$. Therefore $\varphi'_*k = \varphi'_*\partial l = \partial(0l) = 0$.

The map Φ takes the action A_* of the braid group on $H_1(F,\mathbb{Z})$ into the action A'_* of the braid group on $H_1(S',\mathbb{Z}_2)$:

$$\begin{array}{ccc}
H_1(F,\mathbb{Z}) & \xrightarrow{A_*(\gamma)} & H_1(F,\mathbb{Z}) \\
\Phi\downarrow & & \Phi\downarrow \\
H_1(S',\mathbb{Z}_2) & \xrightarrow{A'_*(\gamma)} & H_1(S',\mathbb{Z}_2),
\end{array} \qquad \gamma \in B(n).$$

Indeed, S' is a deformation retract of the space S'' obtained from the sphere S by removing $2g+2$ ramification points (for odd n one of them is the point ∞, and for even n none of them is ∞). Consider the fiber bundle with fiber S'', whose base is the space of all spheres with removed $2g+2$ points, obtained from the given sphere S (with the above-mentioned distinction in the cases when n is even or odd). The base of this fiber bundle is naturally identified with the space G_n of complex polynomials without multiple roots with leading coefficient 1 (the removed points are the roots of the polynomial and ∞ if n is odd). The fundamental group of the base is the braid group $B(n)$ and acts on the homology of the fiber S'', and hence acts on the homology of its deformation retract S'.

Lemma 2. *If $n = 2g+1$, then the set $H_1(S',\mathbb{Z}_2)$ can be realized as that of all 2^n subsets of the n-element set $\{e_1,\dots,e_n\}$, and the action A'_* of the braid group $B(n)$ corresponds to permutations of elements $e_1,\dots,e_n$.*

If $n = 2g+2$, then $H_1(S',\mathbb{Z}_2)$ can be realized as the set of all 2^{n-1} partitions of the n-elements set $\{e_1,\dots,e_n\}$ into two parts, and the action A'_* of the braid group $B(n)$ is reduced to permutations of elements $e_1,\dots,e_n$.

Proof of Lemma 2. The group $H_1(S',\mathbb{Z}_2)$ is generated by $2g+2$ circles, forming the boundary of S'; the sum of all of them is homologous to 0. If $n = 2g+1$, then the circles $e_1,\dots,e_n$ centered at finite removed points form the basis of $H_1(S',\mathbb{Z}_2)$. Standard generators of the braid group $B(n)$ permute e_i and e_{i+1}.

If $n = 2g+2$ then the circles $e_1,\dots,e_n$ satisfy the relation $e_1+\dots+e_n \sim 0$. Hence the homology class $\sum \varepsilon_i e_i$ $(\varepsilon_i \in \mathbb{Z}_2)$ is defined by the splitting of the set $\{e_1,\dots,e_n\}$ into two parts: $\{e_i : \varepsilon_i = 0\}$ and $\{e_i : \varepsilon_i = 1\}$. Standard generators of the braid group $B(n)$ permute E_i and E_{i+1}.

Proof of the Theorem, statement B. Let $a_i, b_i \in H_1(F,\mathbb{Z})$ be the standard generators (see the picture). If $g > 1$ (i.e. $n \neq 3,4$) then the symplectic group $\mathrm{Sp}(g,\mathbb{Z})$ contains a transformation T moving a_{g-1} to b_{g-1}. Let us show that if $n \neq 6$ then T does not belong to the monodromy group: $T \neq A_*(\gamma)$. The reason consists in the fact that the path a_{g-1} goes round two ramification points, and the path b_{g-1} goes round four of them.

Let n be odd. By Lemma 2, Φa_{g-1} and $\Phi b_{g-1} \in H_1(S',\mathbb{Z}_2)$ are realized as subsets consisting of two and four elements respectively. No permutation $A'_*(\gamma)$ can move two elements into four.

Let n be even. Then Φa_{g-1} and Φb_{g-1} are realized by partitions of the form $n = 2+(n-2)$ and $n = 4+(n-4)$. If $n \neq 6$ then the parts of both partitions have different numbers of elements. Therefore they cannot be transformed one into the other by permutations $A'_*(\gamma)$. So, for any n $A_*(\gamma)a_{g-1} \neq b_{g-1}$.

Proof of the Theorem, statement A. Let $g = 1$, i.e. $n = 3$ or 4. Applying Picard-Lefschetz formulas to vanishing cycles a and b, we check that transformations $\{a,b\} \mapsto \{a+b,b\}$ and $\{a,b\} \mapsto \{a,a+b\}$ belong to the monodromy group $A_*B(n)$. But these two transformations generate the entire modular group $\mathrm{Sp}(1,\mathbb{Z})$. Hence in the cases $n = 3,4$ we have $A_*B(n) = \mathrm{Sp}(1,\mathbb{Z})$.

Now suppose that $g = 2$, $n = 6$[1]. Let us apply Picard-Lefschetz formulas to six vanishing cycles $a_1, a_2, a_1 + a_2, b_1, b_2, b_1 - b_2$. Denote by $e_{i,j}$ the 4×4 matrix, whose single non-zero element is 1 in the ith row and jth column. Automorphisms $A_*(\gamma)$ are expressed in the basis $a_1, a_2, b_1, b_2 \in H_1(F, \mathbb{Z})$ by (4×4) matrices. The four vanishing cycles indicated above correspond to matrices $A_k = E + V_k$, where

$$V_1 = -e_{1,2}; \quad V_2 = -e_{3,4}; \quad V_3 = V_1 + V_2 - e_{1,4} - e_{3,2};$$
$$V_4 = e_{2,1}; \quad V_5 = e_{4,3}; \quad V_6 = V_4 + V_5 - e_{2,3} - e_{4,1}.$$

Therefore, the matrices $A_* B(6)$ also include

$$A_7 = A_1 A_2 A_3^{-1} = E + e_{2,3} + e_{4,1}; \quad A_8 = A_4 A_5 A_6^{-1} = E + e_{1,4} + e_{3,2}.$$

But the four matrices A_2, A_5, A_7, A_8 form the standard collection of generators of the group $\mathrm{Sp}(2, \mathbb{Z})$ (related with simple roots of the Lie algebra).

References

[1] V.I.Arnold, Braids of algebraic functions and the cohomology of swallowtails, Uspekhi Mat. Nauk (Russian Math. Surveys), 23:4 (1968), 253–254

[2] C. Jordan, Traite des Substitutions, Paris, 1870

Moscow State University Original paper submitted April 2, 1968

[1] As I.R. Shafarevich has indicated to the author, similar calculations for the case $n = 6$ are contained in the C. Jordan's book [2].

SINGULARITIES OF SMOOTH MAPPINGS

V. I. ARNOL'D

The paper is based on a course of lectures on the local theory of singularities delivered in 1966 at a Summer School in Katsiveli and at the Moscow State University.

Contents

Chapter I

THE STRUCTURE OF SINGULARITIES

§1. Examples

The theory of singularities of smooth[2] mappings is concerned with local properties of differentiable mappings of differentiable manifolds,

---

[1] The author is grateful to B. Malgrange, Yu. I. Manin, B. Morin, V.P. Palamodov and R. Thom for fruitful discussions and to S.M. Vishik, A.G. Kushnirenko and A.M. Leontovich for their assistance in preparing these lectures for the printer.

[2] Here and in the sequel "smooth" and "differentiable" mean "infinitely differentiable". The tangent space at a point x in a manifold M is denoted by TM_x. The differential at x of a mapping $f{:}M\to N$ is denoted by f_x: $TM_x \to TN_{f(x)}$.

Originally publ. in: Usp. Mat. Nauk 23:1, 3-44, © Russ. Acad. Sciences 1968
English transl.: Russ. Math. Surv. 23, 1-43, © London Mathematical Society, 1968
Reprint by Springer-Verlag Berlin Heidelberg 2014 with kind permission, DOI 10.1007/978-3-642-31031-7_13

$$f\colon M^m \longrightarrow N^n,$$

invariant under diffeomorphisms

$$h\colon M \longrightarrow M, \qquad k\colon N \longrightarrow N.$$

EXAMPLE 1. Consider the mapping $f\colon \mathbf{R}^1 \to \mathbf{R}^1$ given by the formula $y = f(x) = x^2$ (fig. 1). In the neighbourhood of every $x \neq 0$, f is a diffeomorphism.

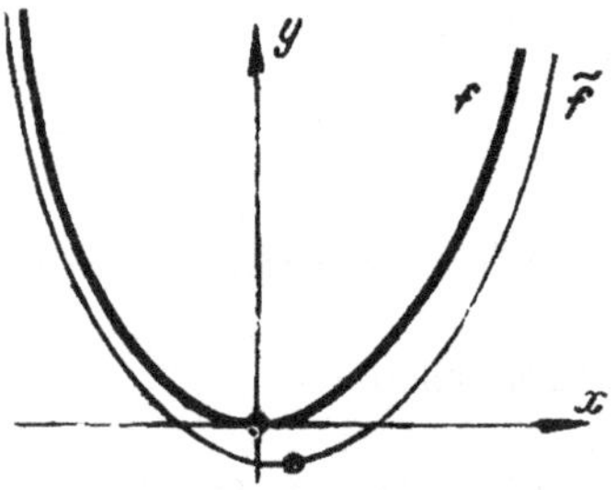

Fig. 1.

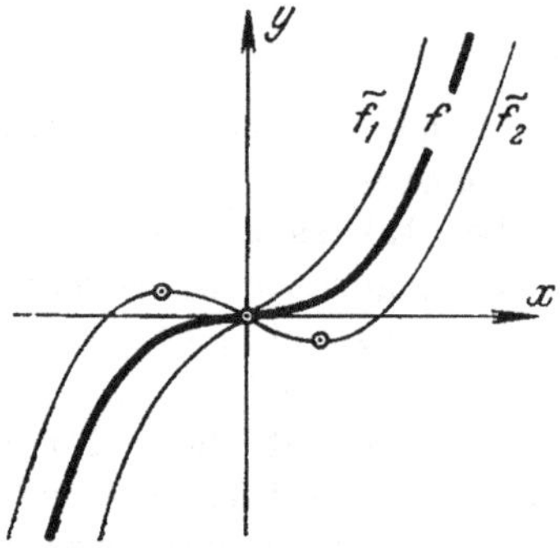

Fig. 2.

This is not the case in the neighbourhood of $x = 0$; f has a *singularity*, the differential of f degenerates at 0. This singularity is *stable*: every mapping $\tilde{f}$ near f whose derivatives are near to those of f has a similar singularity.

EXAMPLE 2. The mapping $y = x^3$ (fig. 2) also has a singularity, but it is *unstable*: for small deformations the singularity can vanish ($\tilde{f}_1$) or split into two ($\tilde{f}_2$).

We give now a general

DEFINITION 1. A differentiable mapping $f\colon M \to N$ is said to be *stable* if for any differentiable mapping $\tilde{f}\colon M \to N$ sufficiently close[1] to f there are diffeomorphisms $h\colon M \to M$, $k\colon N \to N$ close to the identity "converting" f to $\tilde{f}$, that is, such that the diagram

$$\begin{array}{ccc} M & \xrightarrow{f} & N \\ {\scriptstyle h}\downarrow & & \downarrow{\scriptstyle k} \\ M & \xrightarrow{\tilde{f}} & N \end{array}$$

commutes. If there are *homeomorphisms* h and k that are near to 1_M and 1_N and make the diagram commute, then the mapping is said to be topologically stable.

1 The topology on the space of differentiable mappings is that defined by these neighbourhoods of zero in the space of differentiable functions of local coordinates:

$$U(k, \varepsilon) = \left\{ \varphi(x)\colon \max_{|\alpha| \leq k} \left| \frac{\partial^{|\alpha|}\varphi}{\partial x^\alpha} \right| < \varepsilon \right\}$$

Slightly more complicated is the definition of stability at a point.

DEFINITION 2. A smooth mapping $f\colon U \to N^n$ defined on a neighbourhood U of a point x_0 in M^m is said to be *stable at* x_0 if for every mapping $\tilde{f}\colon U \to N^n$ sufficiently near to f there exist neighbourhoods $V \subset U \subset M^m$ of x_0 and $W \subset N^n$ of $y_0 = f(x_0)$, and diffeomorphic embeddings $h\colon V \to U$, $k\colon W \to N^n$ close to the identity and such that the diagram

$$\begin{array}{ccc} M^m \supset U \supset V & \xrightarrow{f} & W \subset N^n \\ \downarrow h & & \downarrow k \\ M^m \supset U & \xrightarrow{\tilde{f}} & N^n \end{array}$$

commutes. If, in addition, the h and k can be taken as homeomorphisms, we get the definition of *topological stability* at x_0; if f, $\tilde{f}$, h and k can be taken as real (complex) analytic functions, we get the definition of real (complex) analytic stability at x_0.

EXAMPLE 3. The implicit function theorem asserts that a mapping whose rank at x_0 is maximal is stable at x_0.

EXAMPLE 4. "Morse's Lemma" (see [1], p. 14) states that a mapping $\mathbf{R}^n \to \mathbf{R}^1$ given by a non-degenerate quadratic form

$$f(x) = x_1^2 + \ldots + x_k^2 - x_{k+1}^2 - \ldots - x_n^2,$$

is stable at 0.

The ideal to which the theory of singularities strives is achieved in the special case of mappings $\mathbf{R}^n \to \mathbf{R}^1$ (Morse theory). The results of this theory that interest us can be stated as follows:

THEOREM 1. 1) *The stable mappings $f\colon M^m \to \mathbf{R}^1$ of a compact manifold M^m to the line form an everywhere dense set in the space of all smooth mappings.*

2) *A mapping f is stable if and only if the following two conditions are satisfied:*

M_1. *f is stable at every point (that is, every critical point of the function f is non-degenerate).*

M_2. *All critical values of f are distinct.*

3) *The mapping $f\colon M^m \to \mathbf{R}^1$ is stable at x_0 if and only if coordinates $x_1, x_2, \ldots, x_m, y$ can be introduced in neighbourhoods of x_0 in M^m and $y_0 = f(x_0)$ in $\mathbf{R}^1$ in such a way that f can be written in one of the $m + 2$ forms:*

MI. $y = x_1$,

MII$_k$. $y = x_1^2 + \ldots + x_k^2 - x_{k+1}^2 - \ldots - x_m^2$ $(k = 0, 1, \ldots, m)$.

The following important case – the theory of two-dimensional manifolds – has received exhaustive study (Whitney [2]).

THEOREM 2. *The mapping $f\colon M^2 \to N^2$ is stable at the point x_0 if and only if it is equivalent in some neighbourhood of x_0 to one of three mappings* (fig. 3):

WI. $y_1 = x_1$, $y_2 = x_2$ *(regular point)*,

WII. $y_1 = x_1$, $y_2 = x_2^2$ *(fold)*,

WIII. $y_1 = x_1$, $y_2 = x_1 x_2 - \frac{1}{3} x_2^3$ *(cusp) of a neighbourhood of 0 in the*

(x_1, x_2)-plane into a neighbourhood of 0 in the (y_1, y_2)-plane.

The stable mappings $f: M^2 \to \mathbf{R}^2$ of a compact surface into the plane form an everywhere dense set in the space of all smooth mappings.

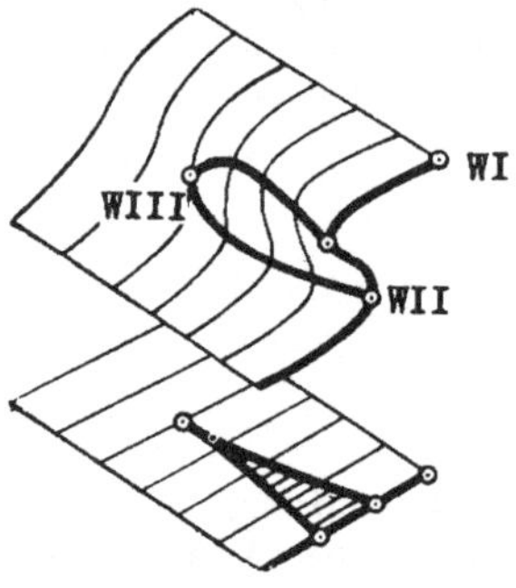

Fig. 3.

The smooth mapping $f: M^2 \to \mathbf{R}^2$ is stable if and only if the following two conditions are satisfied:

WI. *The mapping is stable at every point in M^2.*

WII. *The images of folds intersect only pairwise and at non-zero angles, whereas images of folds and cusps do not intersect.*

EXAMPLE 5. The Whitney mapping W. III. is stable (see fig. 4). Let us examine the structure more closely. Each line $x_1 = C$ goes into the line $y_1 = C$. But if $C < 0$, the image is monotone, while if $C > 0$ it is like $\tilde{f}_2$ in Example 2.

The differential f_* has the matrix

$$\begin{Vmatrix} 1 & 0 \\ x_2 & x_1 - x_2^2 \end{Vmatrix}.$$

The rank of f_* is 2 everywhere except on the parabolic fold $x_1 = x_2^2$. On the fold f_* has a kernel parallel to the x_2-axis. The image of the fold is the semi-cubic parabola $y_1 = x_2^2$, $y_2 = \frac{2}{3}x_2^3$. Every point inside the angle has three pre-images,

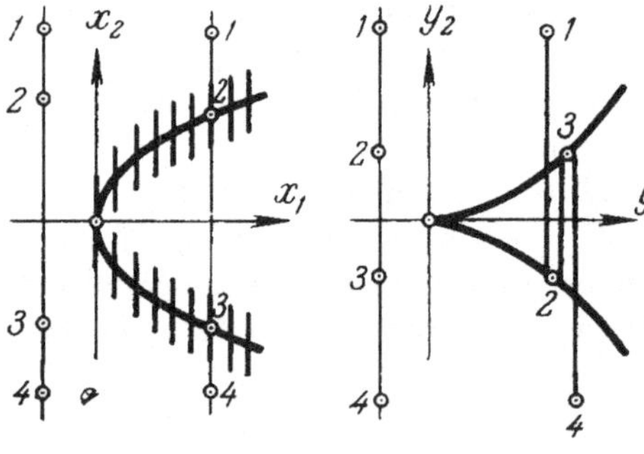

Fig. 4.

each point outside it has one. Note that the fold in the x-plane has no singularity, whereas its image in the y-plane has a singularity at 0. This is explained by the fact that the restriction of f to the parabola $x_1 = x_2^2$ has rank 0 at 0 (the kernel of f_* is tangent to the parabola at 0).

EXAMPLE 6. Consider the mapping $z \to z^2$ of the complex plane onto itself. The singularity at 0 is clearly not one of the types I, II, III. By Whitney's theorem, the mapping is unstable and small deformations can make it into a stable mapping having no singularities apart from folds and cusps. It turns out that it is enough to consider the nearby mappings $z \to z^2 + 2\varepsilon\bar{z}$.

The branch-point splits into three cusps, the circumference into a fold (Fig. 5). The image of the fold is a hypocycloid with three angles;

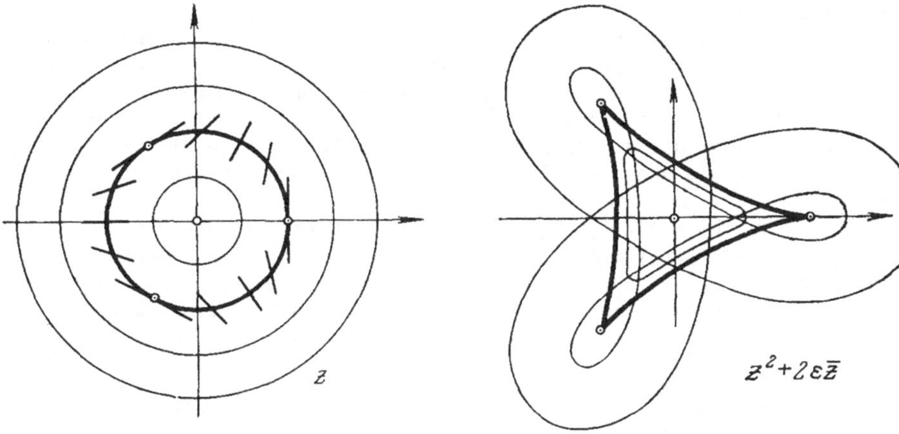

Fig. 5.

the points of the inside have four pre-images, those outside have two. The images of circles of large radius make two turns around the origin.

These examples might give rise to the hope that in higher dimensions every singularity is almost stable, and that stable singularities are easily classifiable. Indeed such a classification exists in dimensions less than six (Whitney [26]). However, the position is completely different in higher dimensions.

THEOREM 3 (Thom [3]). *For $n \geqslant 9$ there exist mappings $\mathbf{R}^n \to \mathbf{R}^n$ unstable at 0 but such that every sufficiently close mapping is unstable.*

Thus, for $n \geqslant 9$ the stable mappings do not form a dense set in the space of all mappings $M^n \to N^n$; there exist *continuous invariants* of smooth mappings f with respect to diffeomorphisms h and k. We give the proof of this theorem in §3.

Naturally the topological classification is coarser.

THEOREM 4 (Thom [4]). *There exists a topological classification of germs of smooth mappings f: $\mathbf{R}^m \to \mathbf{R}^n$ if a set of infinite codimension in the space of all germs is neglected.*[1]

More accurately this means the following. Denote by S the space of all germs of smooth mappings f: $\mathbf{R}^m \to \mathbf{R}^n$ at 0. Then

$$S = \bigcup_{\gamma \in \Gamma} S_\gamma + S_\infty,$$

where Γ is a countable set, each S_γ has finite codimension in S and

[1] The proof has not yet been published.

consists of topologically equivalent[1] germs; and S_∞ has infinite codimension.

We give here the definitions of some of the terms arising. The *germ* of the mapping $f: M^m \to N^n$ at x_0 is the equivalence class of mappings of neighbourhoods of x_0, $\varphi: U \to N^n$, $x_0 \in U \subset M^m$, where we say that two mappings $\varphi_1: U_1 \to N^n$, $\varphi_2: U_2 \to N^n$ are *equivalent* if they coincide on some neighbourhood of x_0 contained in $U_1 \cap U_2$. The *codimension* of a submanifold $A^k \subset B^n$ is $n-k$; that is, "the number of equations giving A locally". This definition carries over to the infinite-dimensional manifold S of germs when one uses the finite-dimensional approximations of S by the spaces of jets.

Two mappings $\varphi_1: U_1 \to N^n$, $\varphi_2: U_2 \to N^n$ are said to have *tangency of order k* at a point in $U_1 \cap U_2$ if $|\varphi_1(x) - \varphi_2(x)| = o(|x|^k)$ in some (and therefore in all) local coordinates. The *jet of order k* of the germ of a mapping $f: M^m \to N^n$ at x in M^m is the set of germs of mappings at x having tangency of order k with f at x. We denote this jet by $j_x^k(f)$. Clearly, $j_x^0(f) = f(x)$, $j_x^1(f)$ is defined by the differential $f_x: TM_x \to TN_{f(x)}$ of f at x.

Let $x_1, \ldots, x_m$ and $y_1, \ldots, y_n$ be choices of local coordinates in M and N, respectively. Then the jet $j_0^k(f)$ is defined by a segment of the Taylor series for f,

$$j_0^k(f) \sim f|_0 + \left.\frac{\partial f}{\partial x}\right|_0 x + \frac{1}{2}\left(\left.\frac{\partial^2 f}{\partial x^2}\right|_0 x, x\right) + \ldots + \frac{1}{k!}\left(\left.\frac{\partial^k f}{\partial x^k}\right|_0 x, \ldots, x\right).$$

Let $J_x^k(M^m, N^n)$ stand for the space of jets of order k of germs $f: M^m \to N^n$ at x. The preceding formula introduces into this space the structure of a finite-dimensional manifold of dimension

$$\dim J_x^k(M^m, N^n) = n + mn + \frac{m(m+1)}{2} n + \ldots + \binom{k+m-1}{m-1} n.$$

There is a natural mapping

$$\pi^k: S \to J_x^k,$$

associating with each germ at x its jet $j_x^k(f)$.

A set $S' \subset S$ is of *finite codimension l*, if for some k,

$$S' = (\pi^k)^{-1} J',$$

where J' is a submanifold of codimension l in J_x^k. In other words, *the set S' of germs has finite codimension l if it is given by l conditions on the Taylor coefficients of fixed order.* Further, S' has *infinite codimension if it lies in the intersection of a sequence of sets of increasing codimensions.*

The set of all jets of order k of germs of mappings $f: M^m \to N^n$ at different points forms a fibre bundle $J^k(M^m, N^n)$ with base $J^0(M, N) = M \times N$. It can be regarded as a vector bundle with the same base, and then we call it the *bundle of k-jets of mappings of M into N*. Finally, proceeding to the limit, as $k \to \infty$, in the sequence of projections $J^{k+1} \to J^k \to \ldots \to J^0$, we get the bundle of jets $J^\infty = J(M, N)$. The notation $J_x^\infty(M, N) = J_x(M, N)$ and $j_x^\infty(f) = j_x(f)$ has a similar meaning.

1 Two germs $f_1, f_2: M^m \to N^n$ at x in M^m are topologically equivalent if there is a germ of a homeomorphism $h: M^m \to M^m$ fixing x and a germ of a homeomorphism $k: N^n \to N^n$ "taking $f_1(x)$ to $f_2(x)$", that is, such that the diagram

$$\begin{array}{ccc} M^m & \xrightarrow{f_1} & N^n \\ h\downarrow & & \downarrow k \\ M^m & \xrightarrow{f_2} & N^n \end{array}$$

commutes.

§2. The classes Σ^I

Let f: $M^m \to N^n$ be a smooth mapping and i an integer $\geqslant 0$.

DEFINITION 1. A point x of M^m lies in the set $\Sigma^i(f) \subset M$ if the kernel of the differential f_*: $TM^m_x \to TN^n_{f(x)}$ has dimension i. We say that f has singularity Σ^i at x or a singularity of class Σ^i.

EXAMPLE 1. For the Whitney singularity (fig. 6)

$$y_1 = x_1, \quad y_2 = x_1 x_2 - \frac{1}{3} x_2^3$$

we have

$$f_x = \begin{Vmatrix} 1 & 0 \\ x_2 & x_1 - x_2^2 \end{Vmatrix}, \quad \dim \operatorname{Ker} f_x = \begin{cases} 0 & \text{if} \quad x_1 \neq x_2^2, \\ 1 & \text{if} \quad x_1 = x_2^2. \end{cases}$$

Thus, Σ^0 is the whole plane excluding the parabola $x_1 = x_2^2$, Σ^1 is the parabola, the kernel is parallel to the x_2-axis; and for $i \geqslant 2$, Σ^i is empty.

We remark that the parabola Σ^1 is a smooth manifold which includes the cusp point 0. This point is distinguished by the fact that the kernel of f_0 is tangent to Σ^1 at it. In other words, f *restricted to* Σ^1, has rank 1 at all points other than 0. Thus, 0 lies in $\Sigma^1(f|\Sigma^1(f))$, and all other points on the parabola lie in $\Sigma^0(f\Sigma^1(f))$. We write $\Sigma^{i_2}(f|\Sigma^{i_1}(f))$ as $\Sigma^{i_1 i_2}(f)$. With this notation the Whitney cusp 0 lies in $\Sigma^{11}(f) \subset \Sigma^1(f)$.

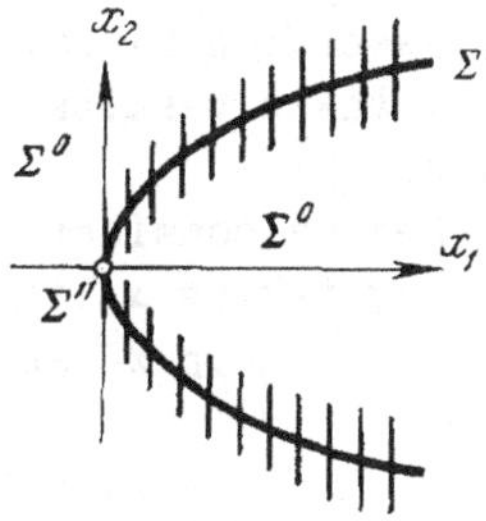

Fig. 6.

For any set $I = i_1, i_2, \ldots, i_n$ of integers the set $\Sigma^I(f)$ is defined by induction as follows.

DEFINITION 2. Let $\Sigma^I(f) = \Sigma^{i_1, \ldots, i_k}(f) \subset M$, M a smooth manifold. Then

$$\Sigma^{i_1, i_2, \ldots, i_k, i_{k+1}}(f) = \Sigma^{i_{k+1}}(f \mid \Sigma^I(f))$$

is the set of all points where the kernel of the differential of the restriction of f to $\Sigma^I(f)$ has dimension i_{k+1}.

REMARK. By the definition, the manifolds

$$M \supset \Sigma^{i_1} \supset \Sigma^{i_1 i_2} \supset \Sigma^{i_1 i_2 i_3} \supset \ldots$$

are embedded in one another. Thus, the kernels of the restrictions of f to these embedded submanifolds are also embedded in one another. So the *sequence of numbers* $i_1, i_2, \ldots$ comprising I must be non-decreasing, $m \geqslant i_1 \geqslant i_2 \geqslant i_3 \geqslant \ldots \geqslant 0$. If one of these inequalities is violated, then Σ^I is empty.

The set $\Sigma^I(f)$ is not necessarily a manifold, therefore the definition given above (due to Thom) does not give a definition of $\Sigma^I(f)$ for all f.

Boardman [5] has proposed a definition of $\Sigma^I(f)$ in terms of the space of jets. For any set $I = i_1, \ldots i_k$ of integers he defines a subset Σ^I of the space of k-jets $J^k(M^m, N^n)$, not depending on any mapping f (see below, p. 11). He has proved:

THEOREM 1. *For any* $I = i_1, \ldots, i_k$ *the set* Σ^I *is a (not necessarily closed) submanifold of codimension* $\nu_I(m, n)$ *in* $J^k(M^m, N^n)$. *(The formula for* ν_I *is given below,* p. 9).

The significance of Σ^I is that a "good" mapping f has singularity $\Sigma^I(f)$ at x in the sense of the preceding definition if and only if the jet of f at x lies in Σ^I.

DEFINITION 3. Let $f\colon M^m \to N^n$ be a smooth mapping. The *induced mapping* $\bar{f}\colon M^m \to J^k(M^m, N^n)$ associates with each point x of M^m the jet of f at x:

$$\bar{f}(x) = j_x^k(f).$$

A mapping is called "good" if its induced mapping $\bar{f}$ is transversal[1] on Σ^I.

Boardman has proved

THEOREM 2. 1) *If f is good, then* $\Sigma^I(f) = \bar{f}^{-1}(\Sigma^I)$; *that is,* $\Sigma^I(f)$ *is a manifold of codimension* $\nu_I(m, n)$ *in* M^m, *and* $x \in \Sigma^I(f)$ *if and only if the jet of f at x lies in* Σ^I.

2) *Every smooth mapping can be approximated, together with an arbitrary number of its derivatives, as closely as desired by a good mapping.*

Assertion 2) follows from 1) and Thom's transversality lemma. For $k = 1$ these results were established by Thom [3] and for $k = 2$ by Levine [3].

Let us consider the case $k = 1$ in more detail. A 1-jet of a smooth mapping taking x_0 in M^m to y_0 in N^n is given in local coordinates $x_1, \ldots, x_m$ in M^m and $y_1, \ldots, y_n$ in N^n by the matrix of the differential

$$f_{x_0} = \left\| \left. \frac{\partial y_i}{\partial x_j} \right|_{x_0} \right\| \qquad (i = 1, \ldots, n;\ j = 1, \ldots, m).$$

The $m \times n$ matrices form an mn-dimensional linear space L. Consider the set L_r of matrices of rank r. The numbers $k = m - r$ and $l = n - r$ can be called the coranks.

LEMMA 1. *The matrices of rank r form a smooth (non-closed) submanifold L_r in the space L of all $m \times n$ matrices, and the codimension of L_r is the product of the coranks:*

$$\dim L_r = mn - kl = mn - (m - r)(n - r).$$

PROOF. Since $GL(n, \mathbf{R}) \times GL(m, \mathbf{R})$ acts transitively on L_r, it is sufficient to consider the neighbourhood of the following matrix in L_r:

---

[1] Let A, B, C be smooth manifolds, $f\colon A \to B$ and $g\colon C \to B$ smooth mappings. Then f and g are said to be transversal if for every pair $a \in A$, $c \in C$ of points for which $f(a) = g(c) = b$ we have $f_*(TA_a) + f_*(TC_c) = TB_b$. If g is an *embedding*, then we speak of the transversality of f on the manifold C. Here $f^{-1}(C)$ is a submanifold of A and its codimension in A is the codimension of C in B (implicit function theorem).

Thom's "Transversality Lemma" asserts that *the set of mappings $f\colon A \to B$ transversal to a given mapping $g\colon C \to B$ is everywhere dense in the space of all differentiable mappings.* Further, *the set of mappings f such that f is transversal on an arbitrary submanifold of the space of jets is everywhere dense (and clearly open).*

$$A_0=\begin{array}{r|c|c|l} \multicolumn{1}{c}{} & \multicolumn{1}{c}{r} & \multicolumn{1}{c}{l} & \\ \cline{2-3} r & \begin{matrix}1_1 & & \\ & \ddots & \\ & & 1\end{matrix} & 0 & \\ \cline{2-3} k & 0 & 0 & \\ \cline{2-3} \multicolumn{1}{c}{} & \multicolumn{2}{c}{\underbrace{\qquad\qquad\qquad}_{n}} & \end{array}\Bigg\}\, m$$

For a nearby matrix A to lie in L_r it is necessary and sufficient that the kl bordering minors are 0. The resulting equations have the form

$$a_{ij}+O(a^2)=0,\ \ r<i\leqslant n,\ \ r<j\leqslant m,$$

and are therefore independent for $a = 0$, as required.

EXAMPLE 2. The manifold of square matrices of corank 1 has codimension 1; the manifold of square matrices of corank 2 has codimension 4, and in general, the codimension is k^2 for corank k.

COROLLARY 1. *Suppose that the mapping f: $M^m \to \mathbf{R}^n$ is "good" in the sense that the induced mapping $x \to$ (matrix of the differential at x) is transversal on the manifold L_r of matrices of rank r. Then the set $\Sigma^i(f)$, $i = m - r$, defined at the beginning of this section, is a submanifold of M^m of codimension $(m - r)(n - r) = i(n - m + i)$.*

EXAMPLE 3. Let us consider mappings of manifolds of the same dimension ($m = n$). The codimension of $\Sigma^k(f)$, where the rank of a "good" mapping splits into k units, is k^2.

On the other hand, a mapping "of general type" is good, because every mapping can be approximated by a good one (Thom's Lemma [3]).

It follows from this, for example, that the set Σ^2 of singularities has codimension 4 and so cannot arise for mappings of the plane to the plane (see Example 5 in §1.). But it can even be *non-removable* for mappings $\mathbf{R}^n \to \mathbf{R}^n$ when n is greater than 3.

EXAMPLE 4. The two good mappings $f_\pm$ given by the formulae

$$\begin{aligned} y_1&=x_1,\\ y_2&=x_2,\\ y_3&=x_3^2\pm x_4^2+x_1x_3+x_2x_4,\\ y_4&=x_3x_4, \end{aligned}$$

both have a non-removable point Σ^{20} at zero.

In the next section it will be shown that the germs of $f_\pm$ at 0 are *not equivalent*. This means that the classification of singularities into the classes Σ^I is not complete.

Boardman's formula for the codimension of Σ^I, $I = i_1, i_2, \ldots, i_n$ is

$$\nu_I(m, n)=(m-n+i_1)\,\mu(i_1, i_2, \ldots, i_n)-(i_1-i_2)\,\mu(i_2, i_3, \ldots, i_n)-\ldots \\ \ldots-(i_{n-1}-i_n)\,\mu(i_n),$$

where $\mu(i_1, i_2, \ldots, i_n)$ is the number of sequences $j_1, j_2, \ldots, j_n$ of integers satisfying the conditions:

a) $j_1 \geqslant j_2 \geqslant \ldots \geqslant j_n$,

b) $i_r \geqslant j_r \geqslant 0$ for all r $(1 \leqslant r \leqslant n)$ when $j_1 > 0$.

EXAMPLE 5. For $n = 1$ we have $I = i$, $\mu(i) = i$, so that $\nu_I(m, n) = (m - n + i)i$, that is, formula (*) for the product of coranks.

EXAMPLE 6. For $I = \underbrace{1, 1, \ldots, 1}_{k}$ we have $\mu(1, 1, \ldots, 1) = k$, $\nu_I(m, n) = (m - n + 1)k$.

In particular, the Whitney singularity of class Σ^{11} has codimension 2 when $m = n$, and is therefore non-removable for mappings of the plane to the plane at discrete points, for mappings $\mathbf{R}^3 \to \mathbf{R}^3$ on curves, and so on.

B. Morin [6] has considered good mappings of class $\Sigma^{1,1,\ldots,1,0}$ for all m, n. It turns out that they are always stable if $m > \nu_I = (m - n + 1)k$, and completely characterized by their own class. For instance, the following assertions are equivalent for the mapping $f\colon \mathbf{R}^n \to \mathbf{R}^n$:

a) $x \in \Sigma^{\overbrace{1,1,\ldots,1,0}^{n}}$

b) The germ of f at x is equivalent to that of the "generalized Whitney mapping"

$$\begin{aligned} y_1 &= x_1, \\ &\ldots\ldots\ldots \\ y_{n-1} &= x_{n-1}, \\ y_n &= x_1 x_n + x_2 x_n^2 + \ldots + x_{n-1} x_n^{n-1} + x_n^{n+1}. \end{aligned}$$

EXAMPLE 7. For $I = i, j$ we have $\mu(i, j) = i(1 + j) - \dfrac{j(j-1)}{2}$, from which we get the formula found earlier by Levine,

$$\nu_{i,j}(m, n) = (m - n + i)\,i + \frac{j}{2}\,[(m - n + i)(2i - j + 1) - i + j].$$

In particular, for $m = n$,

$$\nu_{i,j}(n, n) = i^2 + \frac{j}{2}\,[2i^2 - ij + j] = i^2 + j\left[i^2 - \frac{j\,(i-1)}{2}\right].$$

Hence it follows that *a singularity of class* $\Sigma^{i,j}$ *first appears as non-removable for* $m = n$ *if*

i,j	1, 0	1, 1	2, 0	2, 1	2, 2	3, 0	3, 1	3, 2	3, 3	4, 0	4, 1	4, 2	4, 3	4, 4
$\nu_{i,j} = n$	1	2	4	7	10	9	16	22	27	16	29	40	49	56

EXAMPLE 8. For $m = n \leqslant 16$ the following classes are realized as points (for good mappings):

$n=\nu_I$	n	4	7	9	10	13	15	16
I	1_n	2	2, 1	3	2, 2 2, 1_2	2, 1_3	2, 2, 1	3, 1 4 2, 1_4

where 1_n denotes $\underbrace{1, 1, \ldots, 1}_{n}$.

BOARDMAN'S DEFINITION OF Σ^I. We first give the definition in non-invariant terms, using local coordinates $x_1, \ldots, x_m$ in M^m and $y_1, \ldots, y_n$ in N^n.

DEFINITION 4. Let B be an ideal in the ring A of germs of infinitely differentiable functions $\varphi(x_1, \ldots, x_m)$. The ideal $\Delta_k(B)$ of A generated by B and the Jacobians $\det\left|\frac{\partial\varphi_i}{\partial x_j}\right|$, consisting of partial derivatives of functions in B, is called a *Jacobian extension of B*.

REMARK. The Jacobian extension is invariant, that is, it does not depend on the coordinate system $x_1, \ldots, x_m$. This is because $\frac{\partial\varphi}{\partial x'} = \frac{\partial\varphi}{\partial x}\frac{\partial x}{\partial x'}$, and the determinant is multilinear.

LEMMA 2. *$\Delta_{k+1}(B) \subset \Delta_k(B)$ for $k = 1, 2, \ldots$.*

This follows on expanding determinants by rows.

DEFINITION 5. A Jacobian extension $\Delta_k(B)$ is called *critical* if $\Delta_k(B) \neq A$, $\Delta_{k-1}(B) = A$.

In other words, for a critical extension the order of the adjoint minors is the least order for which the extension does not coincide with the whole ring.

EXAMPLE 9. Take $m = 4$, and let B be the ideal generated by x_1, x_2, x_3^2, x_4^2. Then the critical Jacobian extension $\Delta_3(B)$ is generated by x_1, x_2, x_3, x_4. For this ideal $\Delta_3(B)$ the critical extension is the fifth: $\Delta_5\Delta_3(B) = \Delta_3(B)$.

EXAMPLE 10. Take $m = 1$ and let B be the ideal generated by x^3. We have the sequence of critical extensions: $\Delta_1(B) = Ax^2$, $\Delta_1\Delta_1(B) = Ax$, $\Delta_2\Delta_1\Delta_1(B) = Ax$.

For convenience we use a different numbering for extensions in what follows.

Notation: $\Delta^k = \Delta_{m-k+1}$. Thus, in Example 9 the critical extensions are Δ^2 and $\Delta^0\Delta^2$; in Example 10 they are Δ^1, $\Delta^1\Delta^1$ and $\Delta^0\Delta^1\Delta^1$.

Let I be a set of integers $i_1 \geqslant i_2 \geqslant \ldots \geqslant i_k$.

DEFINITION 6. Suppose that the germ of the mapping $f: M^m \to N^n$ is given in coordinates x, y by the formulae $y_i = f_i(x)$, $f(0) = 0$. *We say that f has a singularity of class Σ^I at 0 if the successive critical Jacobian extensions of the ideal generated by the functions $f_i(x)$, $i = 1, \ldots, n$, are the $\Delta^{i_k}\Delta^{i_{k-1}} \ldots \Delta^{i_1}$.*

EXAMPLE 11. The mapping $y = x^3$ has a singularity of class Σ^{110} at 0; $y = x^{k+1}$ has a singularity of class $\Sigma^{1_k,0}$. The mapping $y_1 = x_1x_2$, $y_2 = x_1^2 - x_2^2$ and the mapping of Example 4 have singularity of class $\Sigma^{2,0}$.

REMARK. It is clear that the definition given above imposes a restriction only on the coefficients of the Taylor expansion of orders up to k inclusive. It is easy to check that, in fact, *the conditions do not depend on the coordinate system* and are imposed only on the k-jet $j_0^k(f)$. The set of all k-jets satisfying these conditions also defines the intersection of $\Sigma^I \subset J^k(M, N)$ with the fibre bundle $J^k(M, N) \to M \times N$.

§3. The quadratic differential

The rank of the first differential gives rise to the singularity classes Σ^I. An investigation of the quadratic part of the mapping yields a finer classification: with each singularity we associate, in an invariant manner, its family of quadratic forms. The second differential is defined uniquely only on the kernel of the first differential, and only to within the image of the first. Therefore we define the quadratic differential of the mapping $f\colon M^m \to N^n$ at x in M^m to be a quadratic[1] mapping

$$f_x\colon \operatorname{Ker}(f_x) \longrightarrow \operatorname{Coker}(f_x),$$

where $\operatorname{Ker} f_x \subset TM_x$ is the kernel of the first differential $f_x\colon TM_x \to TN_{f(x)}$, and $\operatorname{Coker} f_x = TN_{f(x)}/f_x TM_x$ is its cokernel.

Firstly we define f_{xx} by means of local coordinates

$$X\colon TM_x^m \longrightarrow M^m, \qquad Y\colon TN_{f(x)}^n \longrightarrow N^n,$$

where

$$X(0) = x,\ Y(0) = f(x),\ \frac{d}{dt}\Big|_{t=0} X(\xi t) = \xi, \qquad \frac{d}{dt}\Big|_{t=0} Y(\xi t) = \xi.$$

In these coordinates f takes the form

$$\varphi\colon TM_x \longrightarrow TN_{f(x)}, \quad \text{where} \quad \varphi = Y^{-1} \circ f \circ X.$$

DEFINITION 1. The value of f_{xx} at ξ in $\operatorname{Ker} f_x$ is

$$f_{xx}(\xi) = \lim_{t\to 0} \frac{\varphi(t\xi)}{t^2} \Big/ f_* TM_x \in \operatorname{Coker} f_x.$$

LEMMA 1. *The quadratic differential f_{xx} is independent of the choice of local coordinates X, Y.*

The proof is clear from Taylor's formula (Fig. 7),

$$\varphi(\eta) = \frac{\partial\varphi}{\partial\xi}\Big|_0 \eta + \frac{1}{2}\left(\frac{\partial^2\varphi}{\partial\xi^2}\Big|_0 \eta, \eta\right) + O(\eta^3).$$

If $\eta \in \operatorname{Ker} f_x$, $\eta = O(t)$, $\zeta = O(t^2)$, then

$$\varphi(\eta+\zeta) = f_x(\eta+\zeta) + \frac{1}{2}\left(\frac{\partial^2\varphi}{\partial\xi^2}\Big|_0 \eta, \eta\right) + \left(\frac{\partial^2\varphi}{\partial\xi^2}\Big|_0 \eta, \zeta\right) + \frac{1}{2}\left(\frac{\partial^2\varphi}{\partial\xi^2}\Big|_0 \zeta, \zeta\right) + \ldots =$$
$$= f_x\zeta + \frac{1}{2}\left(\frac{\partial^2\varphi}{\partial\xi^2}\Big|_0 \eta, \eta\right) + O(t^3)$$

---

1 A mapping $a\colon A \to B$ of linear space is said to be *quadratic* if there is symmetric bilinear mapping $a'\colon A + A \to B$ such that $a = a' \circ \Delta$, where Δ is the diagonal mapping $\Delta\colon A \to A + A$, $\Delta(x) = (x, x)$.

and so on. At the same time we have proved that, for a local system of coordinates in which $\eta_1, \ldots, \eta_k$ are the coordinates of $\eta \in \operatorname{Ker} f_x$ and $\varphi_1, \ldots, \varphi_l$ the coordinates in Coker f_x, the quadratic differential is given by the formula

$$(f_{xx}(\eta))_i = \frac{1}{2} \sum_{j,k} \frac{\partial^2 \varphi_i}{\partial \xi_j \partial \xi_k} \eta_j \eta_k.$$

EXAMPLE 1. For the mapping in Example 4 of §2, the quadratic differential is given by f_{xx}: $x_3, x_4 \to x_3^2 \pm x_4^2, x_3 x_4$.

REMARK. The *cubic* differential cannot be defined in a similar way. To give a definition of invariant differentials of higher orders we have to iterate a construction of Porteous, which associates with each mapping g: $F_1 \to F_2$ of vector bundles over M an invariantly defined "inner derivative" at x in M, (dg): $TM_x \to \operatorname{Hom}(\operatorname{Ker} g|_x, \operatorname{Coker} g|_x)$.

In particular, with the differential f_x we associate the mapping $TM \to f^*TN$ of the tangent bundle of M to the inverse image of the tangent bundle of N. The derivative (dg) defines a bilinear mapping

$$TM_x \times \operatorname{Ker} g|_x \to \operatorname{Coker} g|_x,$$
$$(\xi, \eta) \to (dg)(\xi)\eta,$$

which reduces to the quadratic differential when $\xi = \eta$:

$$dg(\eta, \eta) = f^* 2 f_{xx}(\eta).$$

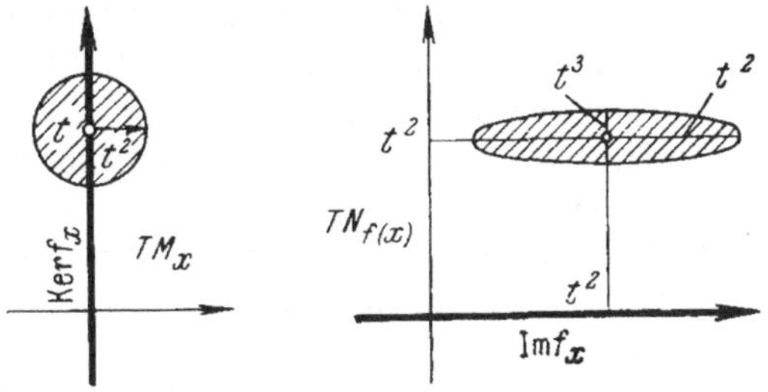

Fig. 7.

The quadratic differential f_{xx} is associated with a bundle L of quadratic forms, in an invariant fashion. Denote by F the linear space of all real quadratic forms on Ker f_x, and by C' the dual space of Coker f_x. To every form

$$\alpha: \operatorname{Coker} f_x \to \mathbf{R}^1, \quad \alpha \in C'$$

there corresponds a quadratic form $\alpha \circ f_{xx} \in F$.

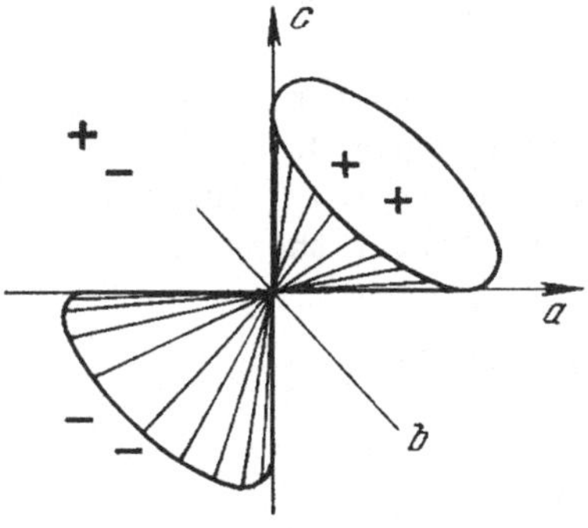

Fig. 8.

DEFINITION 2. *The linear mapping of the cokernel of f_x into the space of quadratic forms on the kernel of f_x defined by $L(\alpha) = \alpha \circ f_{xx}$, L: $C' \to F$, is called the bundle of quadratic forms corresponding to f_{xx}.*

EXAMPLE 2. Suppose that Ker f_x and Coker f_x both have dimension 2. The space F of quadratic forms in two variables is 3-dimensional, and L is a mapping of the two-dimensional plane C' into the 3-dimensional space F. This space F has an additional structure defined by classifying forms according to their indices of inertia. Let x_1, x_2 be coordinates in Ker f_x. We take coordinates in F so that the point with coordinates a, b, c means the form

$$ax_1^2 + 2bx_1x_2 + cx_2^2.$$

Forms of rank 1 (parabolic) form the cone $b^2 = ac$ (Fig. 8). The vertex is the null form. The insides of the two halves of the cone contain elliptic forms of type + + and − −, and the outside contains hyperbolic forms of type + −.

The bundle L is represented by the subspace $L(C')$ of F, and this subspace can be situated in one of seven ways:

1. A plane lying entirely outside the cone (all forms hyperbolic). Example: $\alpha_1(x_1^2 - x_2^2) + \alpha_2 x_1 x_2$.
2. A plane intersecting the cone (two parabolic forms in the bundle). Example: $\alpha_1 x_1^2 + \alpha_2 x_2^2$.
3. A plane tangent to the cone (one parabolic form). Example: $\alpha_1 x_1^2 + \alpha_2 x_1 x_2$.
4. A line inside the cone. Example: $\alpha_1(x_1^2 + x_2^2) + \alpha_2 \cdot 0$.
5. A line outside the cone. Example: $\alpha_1 x_1 x_2 + \alpha_2 \cdot 0$.
6. A line tangent to the cone. Example: $\alpha_1 x_1^2 + \alpha_2 \cdot 0$.
7. The point 0 ($\alpha_1 \cdot 0 + \alpha_2 \cdot 0$).

Thus, *the disposition of the subspace $L(C') \subset F$ is an invariant of the singularity.*

DEFINITION 3. Let $F(\mathbf{R}^k)$ stand for the space of all quadratic forms in k indeterminates, and let

$$H(c, k) = \mathrm{Hom}(\mathbf{R}^c, F(\mathbf{R}^k))$$

be the linear space of all bundles of c quadratic forms in k variables.

The group $GL(c, \mathbf{R}) \times GL(k, \mathbf{R})$ acts on $H(c, k)$ in a natural way, according to the formula

$$g_c \times g_k L(\xi_c)\xi_k = L(g_c^{-1}\xi_c)\, g_k^{-1}\xi_k,$$
$$\xi_c \in \mathbf{R}^c, \quad \xi_k \in \mathbf{R}^k, \quad g_c \in GL(c, \mathbf{R}), \quad g_k \in GL(k, \mathbf{R}).$$

EXAMPLE 3. $H(2, 2)$ is 6-dimensional and splits into 7 orbits under the action of $GL(2, \mathbf{R}) \times GL(2, \mathbf{R})$, as listed in Example 2.

We recall now that with every germ of a mapping $f\colon M \to N$ there is associated a quadratic differential, and with the latter there is associated a bundle of quadratic forms

$$L_f \in \mathrm{Hom}(C', F(K)), \quad C' = \mathrm{Coker}\, f_x, \quad K = \mathrm{Ker}\, f_x.$$

If we *choose* the identifications

$$C' \approx \mathbf{R}^c, \quad c = \dim \mathrm{Coker}\, f_x, \quad K \approx \mathbf{R}^k, \quad k = \dim \mathrm{Ker}\, f_x,$$

then L_f corresponds to an element of $H(c, k)$. A change in these identifications of linear spaces gives rise to an operation of $GL(c, \mathbf{R}) \times GL(k, \mathbf{R})$ on $H(c, k)$, as described above. So we have proved:

LEMMA 2. *The construction described associates with each germ of a mapping $f\colon M \to N$ at x an orbit of the action of $GL(c, \mathbf{R}) \times GL(k, \mathbf{R})$ on the space $H(c, k)$, where $c = \dim \mathrm{Coker}\, f_x$, $k = \dim \mathrm{Ker}\, f_x$; and the correspondence is invariant under diffeomorphisms of M and N.*

COROLLARY 1. *Consider the two mappings $f_\pm\colon \mathbf{R}^4 \to \mathbf{R}^4$ given by the* *formulae of Example 4 in §2:*

$$\begin{pmatrix} x_1 \\ x_2 \\ x_3 \\ x_4 \end{pmatrix} \to \begin{pmatrix} x_1 \\ x_2 \\ x_3^2 \pm x_4^2 + x_1x_3 + x_2x_4 \\ x_3x_4 \end{pmatrix}.$$

Their germs at 0 *are not equivalent, that is, there do not exist germs of diffeomorphisms* h, k *with* $h(0) = 0$, $k(0) = 0$ *for which the diagram*

$$\begin{array}{ccc} \mathbf{R}^n & \xrightarrow{f_+} & \mathbf{R}^n \\ h\downarrow & & \downarrow k \\ \mathbf{R}^n & \xrightarrow{f_-} & \mathbf{R}^n \end{array}$$

is commutative.

PROOF. In this case $c = k = 2$, and we have the conditions of Example 2. The bundles $L_{\pm}$ are given by the formulae

$$\alpha_1(x_3^2 \pm x_4^2) + \alpha_2 x_3 x_4$$

and define a plane in the (a, b, c) space. In the case of f_- this plane lies outside the cone, hence f_- corresponds to the first of the orbits in Example 2. For f_+ the plane intersects the cone, so that f_+ corresponds to the second orbit in that Example. These orbits are different, hence the germs of f_+ and f_- are not equivalent, as we wanted to show.

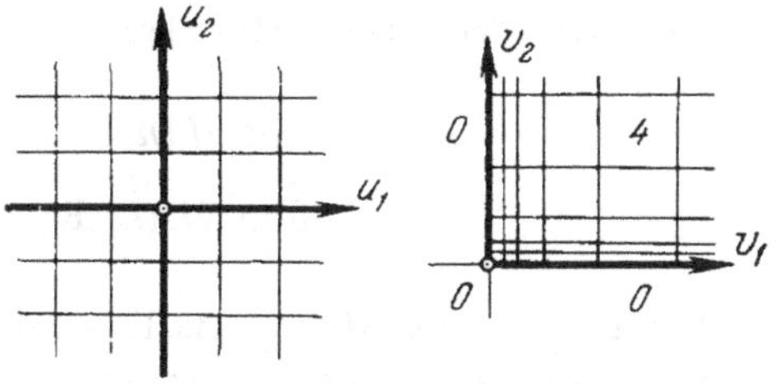

Fig. 9.

REMARK 1. It is easily checked that f_+ and f_- have an unstable singularity of type Σ^{20} at 0. Therefore the classification by the Σ^I is incomplete.

REMARK 2. It is easy to check that f_+ and f_- are distinct in the topological as well as the differentiable sense. To understand their structure better we may regard them as mappings of the plane

$$\begin{matrix} x_3 \\ x_4 \end{matrix} \to \begin{matrix} x_3^2 \pm x_4^2 + x_1x_2 + x_2x_4, \\ x_3x_4, \end{matrix}$$

depending on parameters x_1, x_2. For $x_1 = x_2 = 0$ we get either the complex mapping $z \to z^2$ (for f_-) or a "corner" (for f_+) equivalent to the mapping

$$\begin{matrix} u_1 \\ u_2 \end{matrix} \to \begin{matrix} u_1^2 \\ u_2^2 \end{matrix} = \begin{matrix} v_1, \\ v_2 \end{matrix} \qquad \text{(see fig. 9).}$$

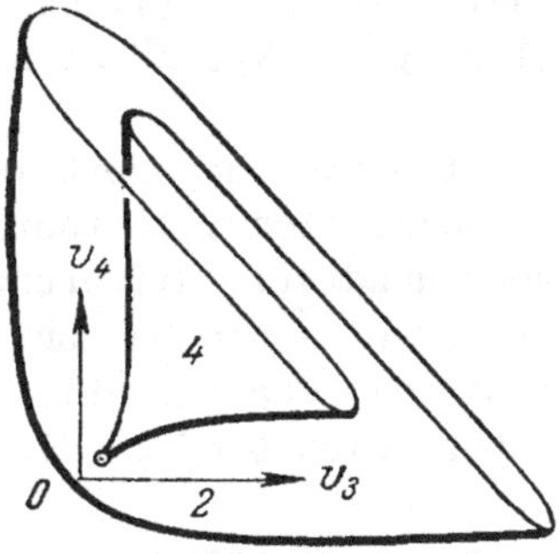

Fig. 10.

Even from this it is clear that f_+ and f_- are topologically inequivalent; the image of f_- covers $\mathbf{R}^4$, whereas that of f_+ does not. For small x_1, x_2 we get nearby mappings (i) in the case of f_- as in Example 6 of §1, (ii) in the case of f_+ as described in Fig. 10.

REMARK 3. Later (§6) we shall prove that *the germs of* f_+ *and* f_- *are analytically stable; it can be shown that every mapping* $M^4 \to N^4$ *may be approximated by a mapping whose germ at each point is equivalent to one of the 7 stable germs given by the formulae:*

$$\begin{aligned}
\Sigma^0:&\quad y_i = x_i,\quad i=1,\ldots,4,\\
\Sigma^{10}:&\quad y_i = x_i,\quad i=1,\ldots,3,\quad y_4 = x_4^2,\\
\Sigma^{110}:&\quad y_i = x_i,\quad i=1,\ldots,3,\quad y_4 = x_1x_4 + x_4^3,\\
\Sigma^{1110}:&\quad y_i = x_i,\quad i=1,\ldots,3,\ y_4 = x_1x_4 + x_2x_4^2 + x_4^4,\\
\Sigma^{11110}:&\quad y_i = x_i,\quad i=1,\ldots,3,\quad y_4 = x_1x_4 + x_2x_4^2 + x_3x_4^3 + x_4^5,\\
f_+ = \Sigma^{20}_+:&\quad y_i = x_i,\quad i=1,2,\quad y_3 = x_3^2 + x_4^2 + x_1x_3 + x_2x_4,\quad y_4 = x_3x_4,\\
f_- = \Sigma^{20}_-:&\quad y_i = x_i,\quad i=1,2,\quad y_3 = x_3^2 - x_4^2 + x_1x_3 + x_2x_4,\quad y_4 = x_3x_4.
\end{aligned}$$

We note another consequence of the lemma:

THEOREM 1 (Thom [3]). *The set of stable mappings* $M^{n^2} \to N^{n^2}$ *is not everywhere dense in the space of all smooth mappings for* $n \geqslant 3$.

The proof is based on the following remark.

LEMMA 3. *The codimension of every orbit of* $GL(n, \mathbf{R}) \times GL(n, \mathbf{R})$ *in the space* $H(n, n)$ *of bundles of n quadratic forms in n variables is positive for* $n \geqslant 3$.

PROOF. We have, clearly,

$$\dim H(n, n) = \frac{n^2(n+1)}{2},$$
$$\dim GL(n, \mathbf{R}) \times GL(n, \mathbf{R}) = 2n^2.$$

There is a one-dimensional subgroup (the scalars) leaving all points of H fixed. Therefore the codimension of each orbit is not less than

$$\frac{n^2(n+1)}{2} - (2n^2 - 1) \geqslant 1 \quad \text{for} \quad n \geqslant 3.$$

This proves the lemma.

Consider now a mapping $f\colon M^{n^2} \to N^{n^2}$ having a transversal singularity at 0 of type Σ^n. By the formula for the product of codimensions, $\Sigma^n(f)$ has codimension n^2 and every nearby mapping has a singularity of type Σ^n at some nearby point.

Next we examine the quadratic differential f_{xx} at 0 and the orbit in $H(n, n)$ corresponding to it. Since this orbit has codimension $\geqslant 1$, there is a mapping $\tilde{f}$ in an arbitrary neighbourhood of f whose quadratic differential at a point of $\Sigma^n(\tilde{f})$ corresponds to *another* orbit (such an $\tilde{f}$ can be constructed easily, by changing f only on a jet of order 2). Consequently, the germ of every mapping $f\colon M^{n^2} \to N^{n^2}$ is unstable at the point 0 of $\Sigma^n(f)$, and this proves the theorem.

REMARK 4. Let us consider the "plane" of pairs of natural numbers (m, n). Then (m, n) lies in the "region of stability" if every mapping $\mathbf{R}^m \to \mathbf{R}^n$ can be approximated by a stable mapping.

Thus all points $(m, 1)$ lie in the region of stability (Morse, Theorem 1 of §1) as do the point (2, 2) (Whitney, Example 5) and all points (m, n) with $n > 2m$ (Whitney's embedding theorem). On the other hand, as we have seen, (9, 9) lies in the region of instability. Mather announced in the autumn of 1965 that the boundary of the region of stability is as shown in Fig. 11.

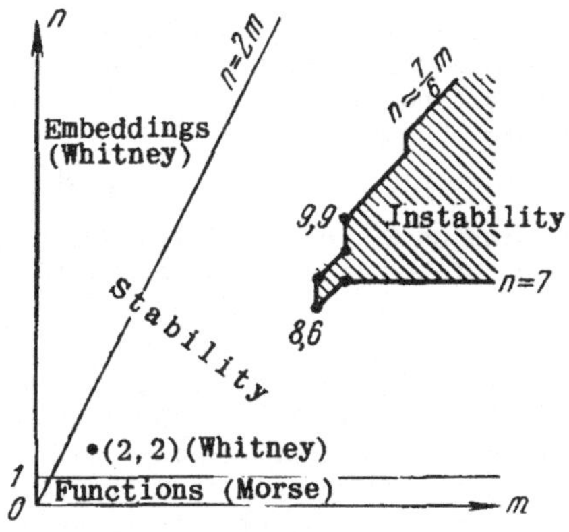

Fig. 11.

REMARK 5. We can restate the theorem mentioned above as the assertion that the differentiable singularities of mappings $M^{n^2} \to N^{n^2}$ for large n have "moduli" (that is, invariants changing continuously with the mapping). For example, it can be seen from our proof that there is at least one modulus for $n \geqslant 3$.

Thom has shown that for sufficiently large n the number of moduli is infinite, that is, the space of non-equivalent differentiable singularities is infinite-dimensional.

§4. The local ring of a singularity and the Weierstrass preparation theorem

The local ring of a singularity is a very powerful invariant. This ring is "a ring of functions on infinitesimal pre-images of a point" and is defined as follows.

Let f be the germ at 0 of a smooth map

$$f\colon \mathbf{R}^m \to \mathbf{R}^n, \quad y = f(x), \quad x \in \mathbf{R}^m, \quad y \in \mathbf{R}^n, \quad f(0) = 0.$$

DEFINITION 1. *The local ring* Q of the germ of f at 0 is the ring

$$Q = \mathcal{E}(x)/f^*[\mathfrak{M}\mathcal{E}(y)]\mathcal{E}(x),$$

where

$\mathcal{E}(x)$ is the ring of germs of smooth functions $\varphi(x)$ at 0,

$\mathcal{E}(y)$ is the ring of germs of smooth functions $\varphi(y)$ at 0,

$\mathfrak{M}\mathcal{E}(y) \subset \mathcal{E}(y)$ is the maximal ideal of $\mathcal{E}(y)$, consisting of germs $\varphi(y)$ vanishing at 0, $\varphi(0) = 0$,

$f^*[\mathfrak{M}\mathcal{E}(y)] \subset \mathcal{E}(x)$ is the image of the maximal ideal in the ring of functions of x, that is, the set of germs of the form $\varphi(f(x))$, $\varphi \in \mathfrak{M}\mathcal{E}(y)$,

$f^*[\mathfrak{M}\mathcal{E}(y)]\mathcal{E}(x)$ is the ideal generated by this last set in $\mathcal{E}(x)$.

EXAMPLE 1. Consider the mapping $\mathbf{R}^1 \to \mathbf{R}^1$ given by $y = x^2$. In this case

$$\mathfrak{M}\mathcal{E}(y) = y\mathcal{E}(y), \quad f^*[\mathfrak{M}\mathcal{E}(y)] = x^2\mathcal{E}(x^2), \quad f^*[\mathfrak{M}\mathcal{E}(y)]\mathcal{E}(x) = x^2\mathcal{E}(x).$$

Thus, Q is the two-dimensional local ring

$$Q = \mathcal{E}(x)/x^2\mathcal{E}(x) = \mathbf{R}[x]/x^2\mathbf{R}[x]$$

of linear functions $a_0 + a_1x$ (with multiplication according to the rule $x^2 = 0$).

MOTIVATION. The ring Q can be regarded as the ring of functions on infinitesimal pre-images of $y = 0$ (in our example, the ring of functions on two coincident points). To give these words an exact meaning we consider the family of mappings

$$f_\varepsilon: \mathbf{R}^1 \longrightarrow \mathbf{R}^1, \quad y = x^2 - \varepsilon^2.$$

depending on ε. For $\varepsilon \neq 0$ the pre-images of $y = 0$ consist of two points $x = \pm \varepsilon$. Instead of the local ring we consider the analogous ring

$$Q_\varepsilon = F(x)/f_\varepsilon^*[\mathfrak{M}F(y)]\,F(x),$$

where $F(x)$, $F(y)$ are the rings of infinitely differentiable functions of x and y.

PROPOSITION 1. *The ring Q_ε is a ring of functions on the pre-images of the point $y = 0$, that is, on ε and $-\varepsilon$.*

This follows easily from the fact that the ring $f_\varepsilon^*[\mathfrak{M}F(y)]\,F(x)$ is the ring of all smooth functions of x that vanish at the points $x = \pm\varepsilon$. It follows from Proposition 1 that every element of Q_ε can be regarded as a linear function $a_0 + a_1x$. As we have seen, the local ring of Example 1 also consists of linear functions of x.

PROPOSITION 2. *As $\varepsilon \to 0$, the multiplication in Q_ε tends to that in the local ring Q.*

This follows from the fact that both pre-images $x = \pm\varepsilon$ tend to 0 as $\varepsilon \to 0$. Thus, *the local ring Q is the limit of the rings Q_ε of functions on the inverse image of $f_\varepsilon^{-1}(0)$, where the pre-images merge as $\varepsilon \to 0$.* In particular, it is clear from this that it is no accident that the dimension of Q is the number of pre-images in the more general situation.

EXAMPLE 2. Consider the mapping $\mathbf{R}^1 \to \mathbf{R}^1$, $y = x^n$. Here

$$\mathfrak{M}\mathscr{E}(y) = y\mathscr{E}(y), \quad f^*[\mathfrak{M}\mathscr{E}(y)] = x^n\mathscr{E}(x^n), \quad f^*[\mathfrak{M}\mathscr{E}(y)]\,\mathscr{E}(x) = x^n\mathscr{E}(x).$$

Therefore Q is the n-dimensional ring

$$Q = \mathscr{E}(x)/x^n\mathscr{E}(x) = \mathbf{R}[x]/x^n\mathbf{R}[x]$$

of polynomials $a_0 + a_1x + \ldots + a_{n-1}x^{n-1}$, with multiplication determined by $x^n = 0$. (This is called the ring of *truncated* polynomials). The dimension n is the multiplicity of the pre-image of 0.

EXAMPLE 3. This ring has a generalized Whitney singularity:

$$f: \mathbf{R}^{n-1} \longrightarrow \mathbf{R}^{n-1}, \quad y_i = x_i \ (i = 1, 2, \ldots, n-2),$$
$$y_n = x_1x_{n-1} + x_2x_{n-1}^2 + \ldots + x_{n-2}x_{n-1}^{n-2} + x_{n-1}^n.$$

EXAMPLE 4. The two mappings $f_\pm$: $\mathbf{R}^4 \to \mathbf{R}^4$ of class Σ^{20} (Example 4 of §2),

$$\begin{matrix} x_1 & & x_1, \\ x_2 & & x_2, \\ x_3 & \longrightarrow & x_3^2 \pm x_4^2 + x_1x_3 + x_2x_4, \\ x_4 & & x_3x_4 \end{matrix}$$

have *non-isomorphic* 4-dimensional rings

$$Q = \mathbf{R}[x_3, x_4]/\{x_3, x_4, x_3^2 \pm x_4^2\}\,\mathbf{R}[x_3, x_4].$$

The equivalence classes of the monomials 1, x_3, x_4, x_3^2, for instance, can be taken as four linear generators for Q. The multiplication in Q then takes the form $x_3x_4 = 0$, $x_4^2 = \mp x_3^2$, $x_3^3 = 0$, $x_4^3 = 0$.

REMARK. If the germs of f_1 and f_2 at 0 are differentiably equivalent, then the rings $Q_1=\mathscr{E}_1/I_1$ and $Q_2=\mathscr{E}_2/I_2$ are equivalent in the sense that there is a commutative diagram

$$\begin{array}{ccccccccc} 0 & \to & I_1 & \to & \mathscr{E}_1 & \to & Q_1 & \to & 0 \\ & & \downarrow\uparrow & & \downarrow\uparrow & & \downarrow\uparrow & & \\ 0 & \to & I_2 & \to & \mathscr{E}_2 & \to & Q_2 & \to & 0. \end{array}$$

Up to equivalence the ring Q represents the strongest invariant of a singularity.

Mather has announced the following results:

1. *A stable singularity is uniquely determined by its ring; that is, if the germs f_1 and f_2 are stable and the rings Q_1 and Q_2 are equivalent, then f_1 and f_2 are differentiably equivalent:*

$$\begin{array}{ccc} M & \xrightarrow{f_1} & N \\ h\downarrow & & \downarrow k \\ M & \xrightarrow{f_2} & N \end{array}$$

where h and k are germs of diffeomorphisms.

2. *If the germ $f\colon \mathbf{R}^m \to \mathbf{R}^n$, $n \geqslant m$, lies in Σ^{20} at 0, then the local ring Q is equivalent to one of the following:*

$\mathrm{I}_{a,b}$ $\mathbf{R}[[x, y]]/\{xy, x^a+y^b\}$, $b \geqslant a \geqslant 2$,
$\mathrm{II}_{a,b}$ $\mathbf{R}[[x, y]]/\{xy, x^a-y^b\}$, $b \geqslant a \geqslant 2$ even,
$\mathrm{III}_{a,b}$ $\mathbf{R}[[x, y]]/\{x^a, y^b, xy\}$, $n > m$,
IV_a $\mathbf{R}[[x, y]]/\{x^2+y^2, x^a\}$, $a \geqslant 3$,
V_a $\mathbf{R}[[x, y]]/\{x^a, y^b, x_y\}$, $a \geqslant 3$, $n > m$.

The codimensions of the corresponding singularities can also be computed:

$$\operatorname{codim} \mathrm{I}_{a,b} = \operatorname{codim} \mathrm{II}_{a,b} = (a+b-1)(m-n+1)+1,$$
$$\operatorname{codim} \mathrm{III}_{a,b} = (a+b-2)(n-m+2)+2,$$
$$\operatorname{codim} \mathrm{IV}_a = (2a+1)(n-m+1)+1,$$
$$\operatorname{codim} \mathrm{V}_a = (2a+2)(n-m+2)+2.$$

EXAMPLES 5. A stable singularity of the mapping $f\colon \mathbf{R}^n \to \mathbf{R}^n$, $y = f(x)$, with local ring of type I_{ab} (II_{ab}), $n = a + b$, is given by the formulae

$$y_i = x_i \ (i=1, \ldots, a-1),\ y_i' = x_i' \ (i=1, \ldots, b-1),\ y_a = x_a x_b',$$
$$y_b' = x_1 x_a + x_2 x_a^2 + \ldots + x_{a-1} x_a^{a-1} + x_a^a + x_1' x_b' + x_2' x_b'^2 + \ldots + x_{b-1}' x_b'^{b-1} \pm x_b'^b$$

for a suitable choice of coordinates

$$x = x_1, \ldots, x_{a-1}, x_a, x_1', \ldots, x_{b-1}', x_b';\ y = y_1, \ldots, y_{a-1}, y_a, y_1', \ldots, y_{b-1}', y_b'.$$

The stability of this singularity will be proved later, in §6.

The fundamental technical tool for working with singularities of smooth mappings is the Weierstrass "preparation theorem" and its generalizations. As before, let $f\colon \mathbf{R}^m \to \mathbf{R}^n$, $y = f(x)$, $f(0) = 0$, be the germ of a smooth mapping.

THEOREM 1. *Suppose that the local ring Q is finite-dimensional as a real linear space, and let $e_1(x), \ldots, e_r(x)$ in $\mathscr{E}(x)$ be representatives of generators of Q. Then every germ $\varphi(x)$ in $\mathscr{E}(x)$ has a representation of the form*

$$\varphi(x)=\varphi_1(y)e_1(x)+\ldots+\varphi_r(y)e_r(x),\quad y=f(x),\quad \varphi_k\in\mathscr{E}(y),$$

that is, $\mathscr{E}(x)$ *is a finite-dimensional* $\mathscr{E}(y)$ *-module.*

This is the form in which Malgrange [7], [8] proved the theorem. This proof, which is very complicated, was recently simplified by Mather [9]. An analytic version of the theorem (see [10]) preceded Malgrange's paper.

Let $f\colon \mathbf{C}^m\to\mathbf{C}^n$, $y=f(x)$, $f(0)=0$, be the germ of an analytic mapping at 0, $\mathscr{H}(x)$ the ring of germs of functions of x analytic at 0, $\mathscr{H}(y)$ the same thing for y.

THEOREM 2. *If the local ring*

$$Q=\mathscr{H}(x)/(f^*[\mathfrak{M}\mathscr{H}(y)])\mathscr{H}(x)$$

is finite-dimensional over $\mathbf{C}$ *and* $e_1(x),\ \ldots,\ e_r(x)$ *denote representatives of generators of* Q *in* $\mathscr{H}(x)$, *then every germ* $\varphi(x)$ *in* $\mathscr{H}(x)$ *has a representation of the form*

$$\varphi(x)=\varphi_1(y)e_1(x)+\ldots+\varphi_r(y)e_r(x),\quad y=f(x),\quad \varphi_k\in\mathscr{H}(y),$$

that is, $\mathscr{H}(x)$ *is a finite-dimensional* $\mathscr{H}(y)$*-module.*

This theorem will be useful in the sequel; we give a proof below.

APPENDIX

A PROOF OF THE PREPARATION THEOREM

1. First of all we make some preliminary remarks. Let Q be an r-dimensional local ring of mappings $y=f(x)$, $f(0)=0$.

LEMMA 1. *The* r*-th power of every element of the maximal ideal of* Q *is* 0: $\mathfrak{M}^r=0$.

PROOF. Take a in $\mathfrak{M}$; for $\lambda\neq0$, $(a-\lambda e)^{-1}$ *exists*. We interpret multiplication by a as an operator $A\colon Q\to Q$, and show that all eigenvalues of A are zero. Let ξ be an eigenvector, $a\xi=\lambda\xi$, $\lambda\neq0$. Then $(a-\lambda e)\xi=0$. Multiplying by $(a-\lambda e)^{-1}$ we get $\xi=0$, contrary to assumption. Therefore $\lambda=0$, hence $A^r=0$, so that $a^r=0$ and the lemma is proved.

COROLLARY 1. *Consider the germs* $x_1,\ \ldots,\ x_m$ *of local coordinates in* $\mathbf{C}^m$ *at* 0. *Then there exist* m *decompositions*

$$x_i^r=\sum_{j=1}^{n}y_j\lambda_{ji}(x),\quad \lambda_{ji}(x)\in\mathscr{H}(x). \tag{1}$$

PROOF. The class of x_i in Q lies in the maximal ideal. By the lemma, the class of x_i^r is zero, as required.

To (1) we add the obvious decompositions

$$x^{\mathbf{k}}=\sum_{i=1}^{r}a_{\mathbf{k}i}e_i(x)+\sum_{j=1}^{n}y_j\lambda_{j\mathbf{k}}(x),\quad \lambda_{j\mathbf{k}}\in\mathscr{H}(x),\quad a_{\mathbf{k}i}\in\mathbf{C}, \tag{2}$$

where $x^{\mathbf{k}}$ denotes the monomial $x_1^{k_1}x_2^{k_2}\ \ldots\ x_m^{k_m}$ and all $k_i<r$. There are only finitely many such monomials.

2. Formal preparation theorem. Every germ $\varphi \in \mathscr{H}(x)$ is given by a power series

$$\varphi(x) = \sum_{\mathbf{k} \geqslant 0} \varphi_{\mathbf{k}} x^{\mathbf{k}}, \quad x^{\mathbf{k}} = x_1^{k_1} \ldots x_m^{k_m}. \tag{3}$$

Our immediate aim is to construct power series for the $\varphi_i(y)$ such that, formally[1]

$$\varphi(x) = \varphi_1(y) e_1(x) + \ldots + \varphi_r(y) e_r(x), \text{ where } y = f(x). \tag{4}$$

With each power series (3) we associate the polynomial

$$\rho[\varphi] = \sum_{\substack{k_i < r \\ i=1, \ldots, m}} \varphi_{\mathbf{k}} x^{\mathbf{k}}. \tag{5}$$

We divide the remainder of the series by x_i^r. To this end we use the notation

$$\tau_i[\varphi] = \sum_{k_i \geqslant r} \varphi_{\mathbf{k}} \frac{x^{\mathbf{k}}}{x_i^r} \quad (i = 1, \ldots, m), \tag{6}$$

and divide φ by x_1^r, the remainder by x_2^r, and so on:

$$\sigma_i[\varphi] = \tau_i\left[\varphi - \sum_{j=1}^{i-1} x_j^r \sigma_j[\varphi]\right], \quad \sigma_1 = \tau_1 \quad (i = 1, \ldots, m). \tag{7}$$

Then

$$\varphi = \rho[\varphi] + \sum_{i=1}^{m} \sigma_i[\varphi] x_i^r. \tag{8}$$

Every monomial $\varphi_{\mathbf{k}} x^{\mathbf{k}}$ for which $k_1, \ldots, k_{i-1} < r$, $k_i \geqslant r$, gives a contribution $\varphi_{\mathbf{k}} \frac{x^{\mathbf{k}}}{x_i^r}$ to $\sigma_i[\varphi]$; if all $k_i < r$, then $\varphi_{\mathbf{k}} x^{\mathbf{k}}$ occurs in $\rho[\varphi]$.

We transform $\rho[\varphi]$ and x_i^r by means of (1) and (2). We get

$$\varphi = p[\varphi] + \sum_{j=1}^{n} y_j s_j[\varphi], \tag{9}$$

where

$$p[\varphi] = \sum_{i=1}^{r} p_i[\varphi] e_i, \quad p_i[\varphi] = \sum_{\substack{k_l < r \\ l=1, \ldots, m}} \varphi_{\mathbf{k}} a_{\mathbf{k}i} \tag{10}$$

---

1 Here $f(x)$ can be understood as a formal power series and the $e_i(x)$ as formal power series in x representing the generators of the ring

$$\hat{Q} = \Phi(x)/f^*[\mathfrak{M}\Phi(y)]\,\Phi(x),$$

where $\Phi(x)$ ($\Phi(y)$) is the ring of formal power series in x (in y). The formal Weierstrass theorem will then be proved; *if a formal mapping f has a finitely generated ring $\hat{Q}$ and $e_i(x)$ are representatives of the generators, then* (4) *is valid for any $\varphi \in \Phi(x)$.*

and

$$s_j[\varphi] = \sum_{\substack{k_l<r \\ l=1,\ldots,m}} \varphi_{\mathbf{k}}\lambda_{j\mathbf{k}}(x) + \sum_{i=1}^{m} \sigma_i[\varphi]\lambda_{ji}(x) \quad (j=1,\ldots,n). \tag{11}$$

Iterating (9) we arrive

$$\varphi = p[\varphi] + \sum_{j=1}^{n} y_j p s_j[\varphi] + \sum_{j_1,\,j_2=1}^{n} y_{j_1}y_{j_2}s_{j_2}s_{j_1}[\varphi]$$

and generally for each l,

$$\varphi = p[\varphi] + \sum_{j_1=1}^{n} y_{j_1}ps_{j_1}[\varphi] + \sum_{j_1,\,j_2=1}^{n} y_{j_1}y_{j_2}ps_{j_2}s_{j_1}[\varphi] + \ldots$$
$$\ldots + \sum_{j_1,\ldots,j_l=1}^{n} y_{j_1}\ldots y_{j_l}ps_{j_l}\ldots s_{j_1}[\varphi] +$$
$$+ \sum_{j_1,\ldots,j_l,\,j_{l+1}=1}^{n} y_{j_1}\ldots y_{j_l}y_{l+1}s_{j_{l+1}}s_{j_l}\ldots s_{j_1}[\varphi]. \tag{12}$$

Collecting the coefficients of e_i we get finally the required expression (4) in the form $\varphi = \sum_{i=1}^{r} \varphi_i(y)e_i$, where the power series

$$\varphi_i(y) = \sum_{l=0}^{\infty} \sum_{j_1,\ldots,j_l=1}^{n} \varphi_{ij_1\ldots j_l}y_{j_1}\ldots y_{j_l} \tag{13}$$

has coefficients

$$\varphi_{ij_1\ldots j_l} = p_i s_{j_l}\ldots s_{j_1}[\varphi]. \tag{14}$$

This proves the formal theorem.

3. Investigation of convergence. If φ is the germ of an analytic function, then by Cauchy's inequality there exist $C > 0$, $N > 0$ such that

$$|\varphi_{\mathbf{k}}| < CN^{|\mathbf{k}|}, \quad |\mathbf{k}| = k_1 + \ldots + k_m. \tag{15}$$

This is expressed by the symbol $\varphi \prec (C, N)$. The inequality (15) is satisfied by some $C = C(N)$ if $N > 1/R$, where R is the radius of convergence. The expressions (1) and (2) are fixed and finite in number.

Suppose that

$$|a_{\mathbf{k}i}| < A, \quad A > 0. \tag{16}$$

Further we may assume that

$$\lambda_{j\mathbf{k}} \prec \left(1, \frac{1}{2}\right), \quad \lambda_{ji}(x) \prec \left(1, \frac{1}{2}\right) \tag{17}$$

(this can be achieved by a suitable choice of the scales of x and y). From From (15), (16), (7) we find that

$$\varphi \prec (C, N) \Rightarrow \sigma_i[\varphi] \prec (CN^r, N), \tag{18}$$

while from (15), (16), (5), (10),

$$\varphi \prec (C, N) \Rightarrow |p_i[\varphi]| < A_1 \cdot ACN^r, \tag{19}$$

where $A_1 = A_1(r, m) > 0$. Furthermore it is easy to prove:

LEMMA 2. *Suppose that* $\varphi \prec (1, \theta)$, $\psi \prec (C, N)$, $0<\theta<1<N$. *Then*

$$\varphi\psi < (A_2 C, N), \text{ where } A_2 = A_2(\theta, m) > 0. \tag{20}$$

PROOF OF LEMMA. If

$$\varphi = \sum_{\mathbf{k}\geqslant 0} \varphi_{\mathbf{k}} x^{\mathbf{k}}, \ |\varphi_{\mathbf{k}}| < \theta^{|\mathbf{k}|}, \quad \psi = \sum_{\mathbf{k}\geqslant 0} \psi_{\mathbf{k}} x^{\mathbf{k}}, \ |\psi_{\mathbf{k}}| < CN^{|\mathbf{k}|},$$

then

$$\varphi\psi = \sum_{\mathbf{k}\geqslant 0} (\varphi\psi)_{\mathbf{k}} x^{\mathbf{k}},$$

where

$$|(\varphi\psi)_{\mathbf{k}}| = \Big|\sum_{0\leqslant \varkappa\leqslant \mathbf{k}} \varphi_{\varkappa}\psi_{\mathbf{k}-\varkappa}\Big| \leqslant \sum_{0\leqslant \varkappa\leqslant \mathbf{k}} \theta^{|\varkappa|} CN^{|\mathbf{k}-\varkappa|} \leqslant$$

$$\leqslant \sum_{0\leqslant l\leqslant |\mathbf{k}|} \theta^{l} CN^{|\mathbf{k}|}(l+1)^m \leqslant CN^{|\mathbf{k}|} \sum_{0\leqslant l\leqslant |\mathbf{k}|} \theta^{l}(l+1)^m \leqslant CN^{|\mathbf{k}|} A_2(\theta, m),$$

where this time

$$A_2 = \sum_{0\leqslant l\leqslant \infty} \theta^{l}(l+1)^m < \infty,$$

as we wanted to show.

Next, we find from (11), (17), (18), (20) that

$$|(s_j[\varphi])_{\mathbf{k}}| \leqslant r^m CN^r + mA_2 CN^r N^{|\mathbf{k}|},$$

that is,

$$\varphi \prec (C, N) \Rightarrow s_j[\varphi] \prec (A_3 CN^r, N), \text{ where } A_3(r, m) = mA_2\left(\tfrac{1}{2}, m\right) + r^m. \tag{21}$$

Repeating (21) we get

$$s_{j_l} \ldots s_{j_1}[\varphi] \prec (C(A_3 N^r)^l, N). \tag{22}$$

It follows from (14), (22) and (19) that

$$|\varphi_{ij_1\ldots j_l}| \leqslant A_1 \cdot A \cdot N^r (A_3 N^r)^l C. \tag{23}$$

Rewrite (13) in the form

$$\varphi_i(y) = \sum_{\boldsymbol{\lambda}} \varphi_{i\boldsymbol{\lambda}} y^{\boldsymbol{\lambda}}, \quad y^{\boldsymbol{\lambda}} = y_1^{\lambda_1} \ldots y_n^{\lambda_n}, \quad |\boldsymbol{\lambda}| = l. \tag{24}$$

Every monomial $y^{\boldsymbol{\lambda}}$ can be written in the form $y_{j_1} \ldots y_{j_l}$ in not more than n^l ways. It is then clear from (13) and (23) that, for any $i = 1, \ldots, r$,

$$|\varphi_{i\boldsymbol{\lambda}}| \leqslant n^l A_1 A N^r (A_3 N^r)^l C < DM,$$

where $D = A_1 A N^r C$, $M = nA_3 N^r$. Thus the series (24) represents the germ of an analytic function $\varphi_i(y) \in \mathscr{H}(y)$, as required.

Chapter 2

DEFORMATIONS OF SINGULARITIES

§5. "Infinite-dimensional Lie groups" acting on "infinite-dimensional manifolds"

We want to discuss the infinite-dimensional analogue of the following situation. Let G be a Lie group, M a smooth manifold, and let $G \times M \to M$ be a given operation of G on M. The point m_* of M is called *stable* if every point m of M sufficiently near it is the image of m_* under some element of G, that is, if the orbit Gm_* of m_* contains a neighbourhood of m_* in M. Further, m_* is said to be *infinitesimally stable* if every point m of M sufficiently near m_* is the image of m_* under an infinitely small displacement in G. In other words, the mapping L of the tangent space of G at the identity into the tangent space of M at m_* given by

$$L\colon TG_e \longrightarrow TM_{m_*}, \qquad L = \left.\frac{\partial gm}{\partial g}\right|_{e,\, m_*} \tag{1}$$

must be a mapping *onto*. We could also say that m_* is infinitesimally stable if the operator L has a (one-sided) inverse

$$\overline{L}\colon TM_{m_*} \longrightarrow TG_e, \qquad L\overline{L} = E. \tag{2}$$

It follows quickly from the implicit function theorem that

THEOREM 1 (on stability). *Every infinitesimally stable point is stable.*

Here I have in mind a proof not depending on the implicit function theorem that lends itself better to a generalization to the infinite-dimensional case.

Suppose that coordinate systems have been introduced in a neighbourhood of the identity of G and in a neighbourhood of m_* in M. Then we can identity the points g in G and m in M in our neighbourhoods with elements of the linear spaces TG_e and TM_{m_*}. The action of G on M is then given by the formulae

$$gm = m + Lg + R(g, m), \tag{3}$$

where $L = L(m_*)$ is the differential of the action given by (1) and R is a remainder term of second order of smallness:

$$|R(g, m)| \leqslant C_1 |g| (|g| + |m|). \tag{4}$$

In these terms stability of m_* means the solubility of the equation $gm = 0$ in g for sufficiently small m.

We solve this equation by successive approximations

$$g_s = -\overline{L}m_{s-1}, \quad m_s = g_s m_{s-1} = R(g_s, m_{s-1}) \quad (s = 1, 2, \ldots), \ m_0 = m. \tag{5}$$

The solution is given by the infinite product

$$g = \lim_{s\to\infty} g_s \ldots g_2 g_1. \tag{6}$$

The convergence of the approximations follows for sufficiently small m_0 from the estimates

$$|g_s| < C_2 |m_{s-1}|, \quad |m_{s+1}| < C_3 |m_s|^2 \quad (C_2 = \|\overline{L}\|, \quad C_3 = C_1C_2(1 + C_2)), \tag{7}$$

which come from (4) and (5). It is clear from (7) that the sequence $|m_s|$ decreases faster than any geometric progression for sufficiently small $|m_0|$, so that the product (6) converges to a solution of the equation $gm = 0$.

The advantage of applying these arguments (over applying the implicit function theorem) to the solution of $m + Lg + R(g, m) = 0$ lies in the fact that *we use the value of the inverse operator L only at the single point* m_* (whereas the value of L is needed in a neighbourhood of m_* for the implicit function theorem). Of course, in the finite-dimensional case the difference is inessential; on the other hand, in infinite-dimensional cases it turns out to be impossible, or very difficult, to evaluate L in a neighbourhood of m_*.

The stability theorem admits of a generalization to the infinite-dimensional case when the g and m in (3) are elements of function spaces. A similar generalization has essentially been used in the paper [11] of H. Cartan, and particularly in Kolmogorov [12]. An abstract formulation of the conditions of this generalization would be too cumbersome[1] because algebraic and topological notions turn up which have not yet been given names. For this reason I have restricted myself to an indication of the character of the restrictions imposed, and to a detailed discussion of an example in §6 and §7.

The most essential condition is that the operator L must have *finite order*. Suppose, for example, that we are considering spaces of analytic functions. Here L will be of finite order ν if it takes functions m analytic in a neighbourhood of a domain U to functions analytic in a domain V such that for any $\delta > 0$,

$$|Lm|_V \leqslant C|m|_{U+\delta}\delta^{-\nu}, \tag{8}$$

where $U + \delta$ denotes the δ-neighbourhood[2] of U.

It turns out that *if the operator L in* (3) *has finite order* (and satisfies certain other conditions whose formulation would take up too much space, but which are always satisfied in natural questions of analysis) *then the successive approximations converge for sufficiently small* $m = m_0$ *to a solution of the equation* $gm = 0$.

We mention now some analytic questions whose solutions arise as special cases of the infinite-dimensional stability theorem not stated above.

EXAMPLE 1. *The stability of the germ of an analytic mapping from one complex space to another.*

Let

$$y = f(x), \quad f(0) = 0, \quad x \in X, \quad y \in Y, \tag{9}$$

---

1 And would certainly not lie on the Procrustean bed of the theory of infinite-dimensional manifolds with Banach charts.

2 For instance, the operator $\frac{d^\nu}{dz^\nu}$ has order ν, as can be seen from Cauchy's formula

$$\frac{d^\nu}{dz^\nu} m(z) = \frac{\nu!}{2\pi i} \oint_{|\zeta - z| = \delta} \frac{m(\zeta)\, d\zeta}{(\zeta - z)^{\nu+1}}.$$

be an analytic mapping of a domain in an n-dimensional complex affine space X into another such space Y. The germ of f at 0 can be regarded as a point in the "infinite-dimensional manifold" M of germs of mappings. An "infinite-dimensional local Lie group G" acts on M, namely the direct product of local Lie groups of analytic mappings of domains of X into X, and of domains of Y into Y. The elements $h \times k$ of G operate on M as change of coordinates in image and in pre-image:

$$(h \times k) f = k \circ f \circ h^{-1}.$$

The question of the stability of (9) is that of the stability of f in M under the action of the "group" G.

In connection with the general stability theorem it is sufficient to have infinitesimal stability and the estimate (2) for the inverse operator. In this case it turns out that it is enough for infinitesimal stability to know that the following n linear equations in $2n$ germs at 0 of the analytic $n \times n$ matrices $H_i(x)$, $K_i(y)$ have a solution:

$$x_i E = \frac{\partial f}{\partial x} H_i(x) + K_i(f(x)) \quad (E \text{ the } n \times n \text{ unit matrix}. \tag{10}$$

Here the inverse operator $\overline{L}$ has finite order and we obtain

COROLLARY 1. *If the n equations* (10) *are soluble, then f is stable at* 0.

A similar proof is given in §§6 and 7. Mather proved an analogous theorem in [13] for infinitely differentiable functions.

EXAMPLE 2. *Every analytic function is equivalent to a polynomial in the neighbourhood of an isolated critical point.*

Let

$$f: \ \mathbf{C}^n \to \mathbf{C}^1, \quad f(0) = 0, \quad df|_0 = 0,$$

be the germ of an analytic function having 0 as an isolated critical point. The group of germs at 0 of bi-analytic diffeomorphisms g: $\mathbf{C}^n \to \mathbf{C}^n$, $g(0) = 0$, operates on the space of functions f: $\mathbf{C}^n \to \mathbf{C}^1$ as a group of change of variables:

$$gf = f \circ g.$$

Let M be the "manifold" consisting of functions f having a fixed jet of order k at 0:

$$M = \{f\colon j_0^k(f) = j\}.$$

Consider the subgroup of the group of germs of diffeomorphisms leaving fixed a k-jet of f at 0:

$$j_0^k(gf) = j_0^k(f).$$

This subgroup G contains, for example, all germs that coincide with the identity up to order k: $g(x) - x = o(|x|^k)$. The "infinite-dimensional Lie group G" acts on the "infinite-dimensional manifold M".

Suppose now that 0 is an *isolated* critical point. Then, for sufficiently large k, the point f in M is infinitesimally stable. Furthermore, the inverse operator $\overline{L}$ turns out to have finite order, and we get from the general theorem:

COROLLARY 2. *If 0 is an isolated critical point of a complex analytic function f, then an analytic system of coordinates can be chosen in a neighbourhood of 0 such that f is a polynomial in those coordinates.*

See §8 for the proof (and the dissertation [14] of Tougeron for another proof).

EXAMPLE 3. *A normal form for the germ of an analytic mapping of $\mathbf{C}^n$ into itself at a fixed point.*

Let

$$m:\ z \to \mu z + \ldots,\ \mu = dm|_{z=0}, \tag{11}$$

be the germ of an analytic mapping of complex n-dimensional space $\mathbf{C}^n$ into itself. Clearly m leaves 0 fixed. We ask whether it is possible to "change coordinates"

$$g:\ z \to w = z + \ldots \tag{12}$$

in the neighbourhood of 0 in such a way that m is brought to linear normal form

$$g \circ m \circ g^{-1}:\ w \to \mu w.$$

This is a question of the stability of the point m of the "infinite-dimensional manifold" M of germs of type (11); the "infinite-dimensional group" G of germs of the form (12) acts on M as a group of change of variables:

$$gm = g \circ m \circ g^{-1}.$$

Straightforward calculation shows that the point μ is infinitesimally stable and the inverse operator $\bar{L}$ has finite order if the eigenvalues μ_i of the endomorphism μ satisfy the condition

$$\Big|\mu_i - \prod_{j=1}^{n} \mu_j^{k_j}\Big| > C|k|^{-\nu} \text{ for all integers } k_j \geqslant 0,\ \ |k| = \sum_{j=1}^{n} k_j \geqslant 2. \tag{13}$$

From the general theorem we get:

COROLLARY 3. *If the eigenvalues of μ satisfy* (13), *then* (11) *reduces to linear normal form under a bi-analytic change of coordinates.*

This result is contained in C.L. Siegel's book [15] and is very non-trivial even for $n = 1$. The same method can be used to give his results on normal linear form for systems of analytic ordinary differential equations in the neighbourhood of a fixed point, or of periodic motion (see [16]); and also some of the results of [17] on normal forms for analytic differential equations on the torus.

EXAMPLE 4. *Decomposition of analytic matrix functions.*

Let $D = D_1 \cap D_2$ be the intersection of two sufficiently good bounded domains in the complex z-plane. We consider the "infinite-dimensional manifold" M of non-degenerate matrix functions analytic in D, m: $D \to GL(n, \mathbf{C})$. The "infinite-dimensional Lie groups" of analytic non-degenerate matrix functions in D_1 and in D_2 act on M by left and right multiplication:

$$(g_1 \times g_2)\, m = g_1 m g_2^{-1}, \qquad g_1:\ D_1 \longrightarrow GL(n, \mathbf{C}), \qquad g_2:\ D_2 \longrightarrow GL(n, \mathbf{C}).$$

It turns out that the unit matrix $m = E$ is infinitesimally stable with respect to such an operation; from the general theorem we get

COROLLARY 4. *For any two closed domains $F_1 \subset D_1$, $F_2 \subset D_2$, every matrix function $m(z)$ sufficiently near a constant in D can be represented on the intersection $F_1 \cap F_2$ in the form of a product $m(z) = g_1(z) g_2^{-1}(z)$ of matrices analytic in both domains: g_1: $F_1 \to GL(n, \mathbf{C})$, g_2: $F_2 \to GL(n, \mathbf{C})$.*

This result is due to H. Cartan [11].

§6. The stability theorem

We consider an analytic mapping f of a neighbourhood of 0 in an n-dimensional complex affine space X into another such space Y:

$$y=f(x), \quad x=x_1, \ldots, x_n; \quad y=y_1, \ldots, y_n, \quad f=f_1, \ldots, f_n; \quad f(0)=0. \tag{1}$$

We recall that f is said to be *(analytically) stable* at 0 if every nearby mapping f_0 is equivalent to it. To be precise, if for every neighbourhood U of 0 in X and every mapping f_0: $U \to Y$ sufficiently near f there exist neighbourhoods U', W of 0, $0 \in U' \subset U \subset X$, $0 \in W \subset f(U) \subset Y$ and analytic diffeomorphic embeddings h: $U' \to U$, k: $W \to Y$ near to the identity and carrying f_0 to f so that in U',

$$k^{-1} \circ f_0 \circ h = f, \qquad \begin{array}{ccc} X \supset U \supset U' & \xrightarrow{f} & W \subset Y \\ h \downarrow & & \downarrow k \\ U & \xrightarrow{f_0} & Y \end{array} \tag{2}$$

Here nearness is to be understood relative to the topology given by these neighbourhoods of zero

$$U(\varepsilon, r) = \{\varphi\colon \max_{|x| \leq r} |\varphi(x)| < \varepsilon\}$$

in the space of analytic functions.

The main result of this chapter is the following.

THEOREM 1 (on stability[1]). *The mapping* (1) *is stable at* 0 *if in some neighbourhood of* $x = 0$ *there exist n decompositions*

$$x_i E = f' H_i + K_i \qquad (i = 1, \ldots, n), \tag{3}$$

where E is the $n \times n$ unit matrix; $f'(x) = \dfrac{\partial f}{\partial x}$ is the Jacobian matrix of f; $H_i(x)$, $K_i(y)$ are $n \times n$ matrices analytic in neighbourhoods of $x = 0$ (of $y = 0$); and $y = f(x)$.

EXAMPLE 1. The mapping $y = x^2$ of the line is stable at 0, because for any $\varphi(x)$ there exists a decomposition of the form

$$\varphi(x) = 2xh(x) + k(x^2).$$

On the other hand, the mapping $y = x^3$ is unstable at 0, because there is no decomposition $x = 3x^2h(x) + k(x^3)$ of the type (3).

EXAMPLE 2. Consider the Whitney cusp Σ^{110}:

$$y_1 = x_1 x_2 - \frac{1}{3} x_1^3, \quad y_2 = x_2, \quad n = 2, \tag{4}$$

where to get decompositions (3) we may take

$$H_1 = \begin{vmatrix} 0 & 1 \\ 1 & x_1 \end{vmatrix}, \quad -K_1 = \begin{vmatrix} 0 & y_2 \\ 1 & 0 \end{vmatrix}, \quad H_2 = 0, \quad K_2 = y_2 E.$$

[1] The infinite-dimensional analogue of this theorem has been proved recently by Mather [13].

COROLLARY 1. *The mapping* (4) *is stable at* 0 (Whitney's Theorem [2]).

EXAMPLE 3. Consider the more general singularity of class $\Sigma^{1_n,0}$: $C^n \to C^n$:

$$\left.\begin{aligned} y_1 &= x_1x_2 - x_1^2x_3 + x_1^3x_4 - \ldots + (-1)^n x_1^{n-1}x_n + \frac{1}{n+1}(-x_1)^{n+1}, \\ y_2 &= x_2, \ \ldots, \ y_n = x_n. \end{aligned}\right\} \qquad (5)$$

To obtain decompositions (3) in this case it is sufficient to take $H_i = 0$,

$$H_i = 0, \quad K_i = y_iE \quad \text{for} \quad i > 1,$$

$$H_1 = \begin{vmatrix} 0 & & & & 1 \\ 1 & x_1 & & & 2x_3 \\ & 1 & x_1 & & 3x_4 \\ \cdots & \cdots & \cdots & \cdots & \cdots \\ & & 1 & x_1 & (n-1)\,x_n \\ & & & 1 & x_1 \end{vmatrix} \quad -K_1 = \begin{vmatrix} 0 & & & & y_2 \\ 1 & 0 & & & 2y_3 \\ & 1 & 0 & & 3y_4 \\ \cdots & \cdots & \cdots & \cdots & \cdots \\ & & 1 & 0 & (n-1)\,y_n \\ & & & 1 & 0 \end{vmatrix}.$$

COROLLARY 2. *The mapping* (5) *is stable at* 0 (theorem of B. Morin [6] and Jo Ging-tzung [18]).

EXAMPLE 4. Consider a non-parabolic mapping of class $\Sigma^{2,0}$, $C^4 \to C^4$:

$$y_1 = x_1, \quad y_2 = x_2, \quad y_3 = x_3x_4, \quad y_4 = \frac{1}{2}(x_3^2 \pm x_4^2) + x_1x_3 + x_2x_4. \qquad (6)$$

For the matrices in (3) we can take

$$H_i = 0, \quad K_i = y_iF \qquad (i = 1, 2),$$

$$H_3 = \begin{vmatrix} x_1 + x_3 & 0 & 0 & 1 \\ 0 & x_3 & \mp 1 & 0 \\ -x_3 & 0 & 0 & 0 \\ 0 & 0 & 1 & 0 \end{vmatrix}, \quad K_3 = \begin{vmatrix} -y_1 & 0 & 0 & -1 \\ 0 & 0 & \pm 1 & 0 \\ y_3 & 0 & 0 & 0 \\ 0 & -y_3 & -y_2 & 0 \end{vmatrix},$$

and similar expressions for H_4, K_4.

COROLLARY 3. *The singularity* (6) *is stable at* 0 (see Whitney [26]).

REMARK 1. If f and the decompositions (3) are real, then f is real analytically stable. In particular, this applies to the mappings (4), (5), (6) of the preceding examples.

REMARK 2. In fact, we prove a stronger assertion than that of the stability theorem. For example, every mapping f_0 having tangency of sufficiently high order with f at 0 is already sufficiently near to f for the existence of a commutative diagram (2). Thus, in particular, such a mapping is equivalent to a polynomial, namely to a segment of its Taylor series.

The proof of the stability theorem is given below. Condition (3) is nothing more than the condition of infinitesimal stability of f at 0 (see §5).

INFINITESIMAL STABILITY. We recall that a mapping f is said to be infinitesimally stable if every mapping f_0 "infinitely near" to f is the image of f under "infinitely small" diffeomorphisms h, k. To write down the corresponding formulae, we consider three arbitrary mappings

$$f_0:\quad x \longrightarrow f(x) + \boldsymbol{\varphi}(x), \quad h:\quad x \longrightarrow x + \mathbf{h}(x), \quad k:\quad y \longrightarrow y + \mathbf{k}(y).$$

Here $\boldsymbol{\varphi}(x)$, $\mathbf{h}(x)$, $\mathbf{k}(y)$ are sets of germs of n analytic functions, which we shall (illegally) call "vector fields".

LEMMA 1. *The mapping $k \circ f \circ h$ is given by the formula*

$$\left.\begin{aligned} &k \circ f_0 \circ h:\ x \longrightarrow f(x) + \overline{\boldsymbol{\varphi}}(x), \text{ где } \overline{\boldsymbol{\varphi}}(x) = \Sigma_0 + \Sigma_1 + \Sigma_2 + \Sigma_3, \\ &\quad \Sigma_0 = \boldsymbol{\varphi}(x) + f'(x)\,\mathbf{h}(x) + \mathbf{k}(f(x)), \\ &\quad \Sigma_1 = f(x + \mathbf{h}(x)) - f(x) - f'(x)\,\mathbf{h}(x), \\ &\quad \Sigma_2 = \boldsymbol{\varphi}(x + \mathbf{h}(x)) - \boldsymbol{\varphi}(x), \\ &\quad \Sigma_3 = \mathbf{k}(f(x + \mathbf{h}(x)) + \boldsymbol{\varphi}(x + \mathbf{h}(x))) - \mathbf{k}(f(x)). \end{aligned}\right\} \tag{7}$$

The proof is by an obvious substitution.

If we take $\boldsymbol{\varphi}$, $\mathbf{h}$, $\mathbf{k}$ to have values of the first order of smallness then, as is clear from (7), Σ_0 is also of the first order of smallness, and Σ_1, Σ_2, Σ_3 of the second order. Thus for the infinitesimal stability of f the equation $\Sigma_0 = 0$ must have a solution in $\mathbf{h}$, $\mathbf{k}$.

DEFINITION 1. *A mapping f is said to be infinitesimally stable at 0 if for every vector field $\boldsymbol{\varphi}(x)$ analytic at 0 there exist analytic vector fields $\mathbf{h}(x)$ and $\mathbf{k}(y)$ at 0 such that $\boldsymbol{\varphi}$ is expressed by the formula*

$$-\boldsymbol{\varphi}(x) = f'(x)\,\mathbf{h}(x) + \mathbf{k}(f(x)) \tag{8}$$

in some neighbourhood of $x = 0$.

We denote by V a sufficiently small neighbourhood of 0 in Y and by U the component of 0 in the pre-image $f^{-1}V$. Let $\mathcal{L}$ be an operator from the space of functions of x analytic in U into the space of functions analytic in x (or in y) in the closure of U (or of V).

DEFINITION 2. *The operator $\mathcal{L}$ has degree p and order ν if for any function $\boldsymbol{\varphi}$,*

$$\| \mathcal{L}\boldsymbol{\varphi} \|_{U \text{ (or } V)} < C\left(\| \varphi \|_{U+\delta} \right)^{p} \delta^{-\nu} \tag{9}$$

for arbitrary δ, $0 < \delta < \delta_0$.

Here $\| \ \|_D = \sup\limits_{z \in D}$, and $U + \delta$ is the δ-neighbourhood of U.

Now we state a theorem concerning the solubility of equation (8).

THEOREM 2 (on infinitesimal stability). *If there exist n decompositions (3), then f is infinitesimally stable. Furthermore, there exist linear operators $\mathbf{H}$ and $\mathbf{K}$ of degree 1 and of finite order ν with vector values and vector arguments,*

$$\mathbf{H}[\boldsymbol{\varphi}(x)] = \mathbf{h}(x), \quad \mathbf{K}[\boldsymbol{\varphi}(x)] = \mathbf{k}(y),$$

satisfying (8):

$$-\boldsymbol{\varphi} = f'\mathbf{H}[\boldsymbol{\varphi}] + K[\boldsymbol{\varphi}].$$

Here the order ν does not exceed $r - 1$, where r is the multiplicity of f at 0.

The proof of this theorem is given on pages 31 – 33. Using the theorem on infinitesimal stability it is easy to prove the theorem of stability by the successive approximations that are described in abstract form in §5. In our case these approximations come to the following.

PROOF OF THEOREM 1 (on stability). Suppose that $y = f_0(x)$, $f_0(x) = f(+) + \varphi(x)$, is a mapping near $y = f(x)$. We want to establish the equivalence of f and f_0 and introduce the following notation for the remainder term in (7) for $\Sigma_0 = 0$:

$$R[\varphi, \mathbf{h}, \mathbf{k}] = \Sigma_1 + \Sigma_2 + \Sigma_3, \tag{10}$$

where the expressions for Σ_1, Σ_2, Σ_3 are indicated in (7). We set

$$\mathbf{R}[\varphi] = R[\varphi, H[\varphi], K[\varphi]], \tag{11}$$

where **H** and **K** are the operators in the statement of the theorem on infinitesimal stability.

Starting with φ_0, we define inductively vector fields

$$\varphi_s = \mathbf{R}[\varphi_{s-1}], \qquad \mathbf{h}_s = \mathbf{H}[\varphi_{s-1}], \qquad \mathbf{k}_s = \mathbf{K}[\varphi_{s-1}], \tag{12}$$

and mappings

$$h_s\colon x \longrightarrow x + \mathbf{h}_s(x), \qquad k_s\colon y \longrightarrow y + \mathbf{k}_s(y), \qquad f_s\colon x \longrightarrow f(x) + \varphi_s(x), \tag{13}$$

and further mappings

$$h^s = h_1 \circ h_2 \circ \ldots \circ h_s, \qquad k^s = k_s \circ \ldots \circ k_1. \tag{14}$$

Finally we consider mappings whose convergence will be proved in §7:

$$h = \lim_{s\to\infty} h^s, \quad k = \lim_{s\to\infty} k^s. \tag{15}$$

Note that by Lemma 1

$$k_s \circ f_{s-1} \circ h_s = f_s, \tag{16}$$

(because $\Sigma_0 = 0$ by definition of **H** and **K**). And so, by (14), (16),

$$k^s \circ f_0 \circ h^s = f_s. \tag{17}$$

Since $\lim_{s\to\infty} f_s = f$, we get from (15), (17) that

$$k \circ f_0 \circ h = f, \tag{18}$$

which proves the stability theorem.

PROOF OF THE THEOREM ON INFINITESIMAL STABILITY. We shall prove that the existence of n decompositions as in (3) is equivalent to infinitesimal stability.

LEMMA 1. *The mapping $y = f(x)$ is infinitesimally stable at 0 if and only if for every function $\varphi(x)$ analytic at 0 there exist $n \times n$ matrices $H(x)$ and $K(y)$ analytic at 0 such that*

$$\varphi(x)E = f'H + K(f(x)), \ E \text{ the unit } n\times n \text{ matrix.} \tag{1}$$

PROOF. Suppose that the decompositions (1) exist, and that

$$\varphi(x) = \sum_{i=1}^{n} \varphi_i(x)\mathbf{e}_i.$$

is the coordinate representation of the field φ. Then

$$-\varphi=f'\mathbf{h}+\mathbf{k}, \quad \text{where} \quad \mathbf{h}=-\sum_{i=1}^{n}\mathbf{h}_i, \quad \mathbf{k}=-\sum_{i=1}^{n}\mathbf{k}_i, \tag{2}$$

where $\mathbf{h}_i$ and $\mathbf{k}_i$ are the i-th columns of the matrices H, K, respectively, that arise in $\varphi_i E$. Conversely, suppose that a decomposition (8) exists. Then

$$-\varphi(x)\mathbf{e}_i=f'\mathbf{h}_i+\mathbf{k}_i \qquad (i=1, \ldots, n).$$

Therefore the matrices H, K with the columns $-\mathbf{h}_i$, $-\mathbf{k}_i$ satisfy (1), as we wanted to show.

In particular, if f is infinitesimally stable at 0, then there exist n decompositions (3). The converse is also true: having n such decompositions, one can construct an expression (1) for any function $\varphi(x)$ and so, by Lemma 1, all the decompositions (8).

LEMMA 2. *If the functions φ_1 and φ_2 have decompositions like* (1),

$$\varphi_1 E=f'H_1+K_1, \quad \varphi_2 E=f'H_2+K_2,$$

then their product also has such a decomposition:

$$\varphi_1\varphi_2 E=f'H_{12}+K_{12},$$

where

$$H_{12}=\varphi_1 H_2+H_1K_2, \quad K_{12}=K_1K_2.$$

The proof of Lemma 2 is straightforward. According to the lemma we derive from (3) that there is a decomposition (1) for any polynomial $\varphi(x)$. On passage to the limit it then follows that we can construct a decomposition (1) for any analytic function $\varphi(x)$; however, insofar as we need estimates for H and K in terms of φ, it is convenient to derive (1) by other means, namely by using the Weierstrass preparation theorem.

LEMMA 3. *If decompositions* (3) *exist, then f has finite multiplicity at* 0.

LEMMA 4. *If f is an analytic function of finite multiplicity r at* 0, *then every function $\varphi(x)$ analytic at* 0 *can be written as a sum of r terms,*

$$\varphi(x)=\sum_{k=1}^{r}\psi_k(y)e_k(x), \quad y=f(x), \tag{3}$$

where $e_1(x), \ldots, e_r(x)$ are monomials in $x_1, \ldots, x_n$ not depending on φ, and $\psi_1(y), \ldots, \psi_r(y)$ are functions of y analytic at 0. *Further, there exist (linear) operators $\mathbf{W}_k$ of degree* 1 *and finite order $r-1$ for which*

$$\psi_k=\mathbf{W}_k[\varphi] \quad (k=1, \ldots, r). \tag{4}$$

The proof of Lemma 3 is given below. Lemma 4 represents an analytic refinement of the Weierstrass preparation theorem and was proved[1] in [19].

Theorem 2 of §6 on infinitesimal stability follows easily from Lemmas 1 - 4. By Lemma 3, the multiplicity r is finite. By Lemma 4 there exist monomials e_k and operators $\mathbf{W}_k$. Using Lemma 2 and starting from the decompositions (3), we decompose the monomials e_k like this:

$$e_k E = f' \mathscr{H}_k(x) + \mathscr{K}_k(y) \qquad (k = 1, \ldots, r). \tag{5}$$

Next we define operators $\mathscr{H}$ and $\mathscr{K}$, acting on functions $\varphi(x)$ and taking values that are $n \times n$ matrices analytic in x (or y), by the following relations:

$$\mathscr{H}[\varphi] = \sum_{k=1}^{r} \mathbf{W}_k[\varphi] \mathscr{H}_k, \quad \mathscr{K}[\varphi] = \sum_{k=1}^{r} \mathbf{W}_k[\varphi] \mathscr{K}_k.$$

Then in view of (3) - (5),

$$\varphi(x) E = f' \mathscr{H}[\varphi](x) + \mathscr{K}[\varphi](y). \tag{6}$$

Relation (6) reduces to (1) with $\mathscr{H}[\varphi] = H$, $\mathscr{K}[\varphi] = K$. By Lemma 1 the mapping f is infinitesimally stable at 0.

Furthermore, $\mathscr{H}$ and $\mathscr{K}$ are linear operators of order ν, because the $\mathbf{W}_k$ are. The operators $\mathbf{H}$: $\varphi \to \mathbf{h}$, $\mathbf{K}$: $\varphi \to \mathbf{k}$ are linear of order ν, because $\mathbf{h}$ and $\mathbf{k}$ are obtained from $\mathscr{H}[\varphi]$, and $\mathscr{K}[\varphi]$ by using (2). This proves the theorem on infinitesimal stability.

PROOF OF LEMMA 3. Suppose that f has infinite multiplicity at 0, that is, for each r every neighbourhood contains r different points with the same value $f(x)$. Then 0 is not an isolated point of $f^{-1}(0)$, and $f^{-1}(0)$ contains a curve γ passing through 0 (see for instance Hervé [21], Chapter IV, Theorem 11). Clearly, $\det |f'| = 0$ around this curve. Take a sequence of points $\xi_i \to 0$ on γ, and consider the $n + 1$ linear subspaces

$$f'|_\xi (TX_\xi) = T_\xi \subseteq TY_0 \qquad (\xi = \xi_i, \xi_{i+1}, \ldots, \xi_{i+n}).$$

Since $\det |f'|_\xi = 0$, we must have $\dim T_\xi < n$. Now let $\varphi(x)$ be a "vector field". If the $n + 1$ planes $\varphi(\xi) + T_\xi$, $\xi = \xi_i, \ldots, \xi_{i+n}$ in the n-dimensional space TY_0 do not intersect in a single point, then the field $\varphi(x)$ cannot be represented in the form

$$-\varphi = f'\mathbf{h} + \mathbf{k}(f(x))$$

for any $\xi_i, \ldots, \xi_{i+n}$ contained in a neighbourhood of 0.

But for fixed i and almost all φ, the $n + 1$ planes $\varphi(\xi) + T_\xi$, $\xi = \xi_i, \ldots, \xi_{i+n}$, do not intersect in a single point in the space of vector polynomials of sufficiently high degree. Thus, for almost all vector polynomials φ, the same must be true for all i. So for almost all φ a decomposition $-\varphi = f'\mathbf{h} + \mathbf{k}$ is impossible in every region containing $\xi_i, \ldots, \xi_{i+n}$ for any i, that is, in any neighbourhood of 0. On the other hand, it follows from Lemmas 1 and 2 that φ admits a decomposition $-\varphi = f'\mathbf{h} + \mathbf{k}$. This contradiction proves Lemma 3.

---

1 We remark that this lemma is the only difficult part of our proof of the stability theorem, and the only place where analyticity is used in an essential way. If an estimate similar to Lemma 4 could be found in the infinitely differential case, then the successive approximations in §5 combined with the Nash-Moser smoothing [20] would lead immediately to an infinite-dimensional stability theorem (and such a theorem has been proved by Mather, using a completely different method).

§7. Proof of convergence

Notation. To ease the discussion we introduce some abbreviated notation. Let V be a sufficiently small closed neighbourhood of $y = 0$, $V \subset V_0$. Then we denote the component of 0 in $f^{-1}(V)$ by U. If φ is a func-function[1] of y analytic in V, then we write

$$|\varphi|_V = \max_{y \in V} |\varphi(y)|.$$

If φ is a function of x analytic in U, then

$$|\varphi|_V = \max_{x \in U} |\varphi(x)|.$$

For $\delta > 0$, the symbol $V - \delta$ denotes the set of points having a δ-neighbourhood lying entirely within V. In what follows the following obvious fact will often be used:

LEMMA 1. *If* $C < \left\| \frac{\partial f}{\partial x} \right\|^{-1}$, *then* $f^{-1}(V - \delta) \subset U - C\delta$.

PROOF. For any x_0 in $f^{-1}(V - \delta)$, $f(x_0) \in V - \delta$. Let us consider the segment $f(x_0 + \theta h)$, $0 \leqslant \theta \leqslant 1$, $|h| \leqslant C\delta$. We have

$|f(x_0 + \theta h) - f(x_0)| \leqslant \left\| \frac{\partial f}{\partial x} \right\|_V |h| \leqslant \delta$, so that $x_0 + \theta h \in U$, as required.

Let L be an operator taking functions φ analytic in one domain to functions $L[\varphi]$ analytic in another, and let $p \geqslant 0$, $\nu \geqslant 0$, $\nu_0 \geqslant 0$. We write

$$L[\varphi] \prec \varphi^p (\nu \mid \nu_0)$$

and say that L is an operator of degree p if, for some $\delta_0 > 0$, we have

$$|L[\varphi]|_{V-\delta} \leqslant \frac{|\varphi|_V^p}{\delta^\nu}, \qquad \text{whenever} \qquad |\varphi|_V < \delta^{\nu_0},\ 0 < \delta < \delta_0,\ 0 \in V \subset V_0.$$

A similar definition holds for operators $L[\varphi_1, \ldots, \varphi_k]$ of several arguments. The relation

$$L[\varphi_1, \ldots, \varphi_k] \prec \Phi(\varphi_1, \ldots, \varphi_k) \qquad (\nu \mid \nu_{01}, \ldots, \nu_{0k})$$

means that

$$|L[\varphi]_{V-\delta}| \leqslant \frac{\Phi(|\varphi_1|_V, \ldots, |\varphi_k|_V)}{\delta^\nu}, \quad \text{whenever} \quad |\varphi_i|_V < \delta^{\nu_{0i}},$$

$$i = 1, \ldots, k, \quad 0 < \delta < \delta_0.$$

Examples. EXAMPLE 1. *Put* $L = \frac{\partial^\alpha}{\partial x^\alpha}$. *Then* $L[\varphi] \prec \varphi(\alpha + 1 \mid 0)$.

This is because by Cauchy's formula and Lemma 1,

$$\left| \frac{\partial^\alpha \varphi}{\partial x^\alpha} \right|_{V-\delta} \leqslant \frac{\alpha! \, |\varphi|_V}{(C\delta)^\alpha} \leqslant \frac{|\varphi|_V}{\delta^{\alpha+1}} \text{ when } \delta_0 < \left(\alpha! \left\| \frac{\partial f}{\partial x} \right\|_V^\alpha \right)^{-1}.$$

EXAMPLE 2. *Put* $L[\varphi] = f(x + \varphi) - f(x)$. *Then* $L[\varphi] \prec \varphi(1 \mid 2)$.

For if $x \in f^{-1}(V - \delta)$, $|\varphi| < \delta^2$, $\delta < \delta_0 < \left\| \frac{\partial f}{\partial x} \right\|_V^{-1}$, then

[1] With values in a normed space of numbers, vectors or matrices.

$$|L\varphi|_{V-\delta} \leqslant \left\| \frac{\partial f}{\partial x} \right\|_V |\varphi| \leqslant |\varphi|\,\delta^{-1}.$$

Similarly, *if* $L[\varphi] = f(x + \varphi) - f(x) - f(x)\varphi$, *then* $L[\varphi] \prec \varphi^2(1\,|\,2)$.

EXAMPLE 3. *If* $L[\varphi, h] = \varphi(x + h) - \varphi(x)$, *then* $L[\varphi, h] \prec \varphi h\,(1\,|\,0{,}2)$. The proof is similar.

EXAMPLE 4. *If* $L[\varphi] \prec \varphi^l(\lambda\,|\,\lambda_0)$, $M[\varphi] \prec \varphi^m(\mu\,|\,\mu_0)$, *then*

$(L+M)[\varphi] \prec \varphi^r(\nu\,|\,\nu_0)$, *where* $r = \min(l, m)$, $\nu = \max(\lambda, \mu) + 1$, $\nu_0 = \max(\lambda_0, \mu_0)$,

$L[\varphi]M[\varphi] \prec \varphi^r(\nu\,|\,\nu_0)$, *where* $r = l + m$, $\nu = \lambda + \mu$, $\nu_0 = \max(\lambda_0, \mu_0)$,

$M[L[\varphi]] \prec \varphi^r(\nu\,|\,\nu_0)$, *where* $r = lm$, $\nu = m\lambda + \mu + 1$, $\nu_0 = \max\left(\lambda_0 + 1, \dfrac{\lambda + \mu_0 + 1}{l}\right)$.

The proof is obvious.

Estimates for the operators Σ, RH, K, R. We turn now to an estimate of the operators Σ_1, Σ_2, Σ_3 in formula (7) of §6. We find that

$$\Sigma_1[\mathbf{h}] \prec \mathbf{h}^2(1\,|\,2),$$
$$\Sigma_2[\boldsymbol{\varphi}, \mathbf{h}] \prec \boldsymbol{\varphi}\mathbf{h}\,(1\,|\,0, 2),$$
$$\Sigma_3[\varphi, \mathbf{h}, \mathbf{k}] \prec \mathbf{k}(\boldsymbol{\varphi} + \mathbf{h})\,(2\,|\,3, 3, 0).$$

The estimates of Σ_1 and Σ_2 are contained in Examples 2 and 3 above. To deal with Σ_3 we use the representation $\Sigma_3 = S_3[\mathbf{k}, \Delta[\mathbf{h}, \boldsymbol{\varphi}]]$, where

$$S_3[\mathbf{k}, \Delta] = \mathbf{k}(f(x) + \Delta(x)) - \mathbf{k}(f(x)) \prec \mathbf{k}\Delta\,(1\,|\,0, 2),$$
$$\Delta[\mathbf{h}, \boldsymbol{\varphi}] = f(x + \mathbf{h}) - f(x) + \boldsymbol{\varphi}(x + \mathbf{h}) \prec \mathbf{h} + \boldsymbol{\varphi}\,(1\,|\,2, 0).$$

Finally, for $R = \Sigma_1 + \Sigma_2 + \Sigma_3$ we get

$$R[\boldsymbol{\varphi}, \mathbf{h}, \mathbf{k}] \prec \mathbf{h}^2 + \mathbf{h}\boldsymbol{\varphi} + \mathbf{k}(\boldsymbol{\varphi} + \mathbf{h})\,(3\,|\,3, 3, 0).$$

Now let H and K be the operators in the theorem on infinitesimal stability. These operators have order not more than $r - 1$, where r is the multiplicity of the mapping at 0. By Lemma 1 we can write in our new notation

$$\mathbf{H}[\boldsymbol{\varphi}] \prec \boldsymbol{\varphi}(r\,|\,0), \quad \mathbf{K}[\boldsymbol{\varphi}] \prec \boldsymbol{\varphi}(r\,|\,0). \tag{1}$$

Combining these estimates with those for $R[\boldsymbol{\varphi}, \mathbf{h}, \mathbf{k}]$ we get

$$\mathbf{R}[\boldsymbol{\varphi}] \prec \boldsymbol{\varphi}^2(2r + 1\,|\,r + 4). \tag{2}$$

for the operator $\mathbf{R}[\boldsymbol{\varphi}] = R[\boldsymbol{\varphi}, \mathbf{H}[\boldsymbol{\varphi}], \mathbf{K}[\boldsymbol{\varphi}]]$. This shows that the remainder term has degree higher than one, and this ensures the rapid convergence of the successive approximations.

A convergence lemma. For the proof of convergence we use some obvious assumptions.

LEMMA 2. *Suppose that we are given a sequence* $\delta_1 > \delta_2 > \ldots > 0$ *of numbers, a sequence* $U_0 \supset U_1 \supset \ldots \supset 0$ *of domains and a sequence*

$$h_s\colon\ U_s \to U_{s-1}, \quad h_s(x) = x + \mathbf{h}_s(x),$$

of analytic mappings for which the following conditions hold:

A) $U_s \subset U_{s-1} - C\delta_s$, $\| \mathbf{h}_s \| \leqslant C\delta_s$,

B) $\left\| \frac{\partial \mathbf{h}_s}{\partial x} \right\|_{U_s} < \delta_s$, $\sum_{s=1}^{\infty} \delta_s < \infty$.

Then the sequence of mappings $h^s\colon U_s \to U_0$, $h^s = h_1 \circ h_2 \circ \ldots \circ h_s$ *on the intersection* $U_\infty = \bigcap_{s \to \infty} U_s$ *converges to a differentiable mapping* $h\colon U_\infty \to U_0$, $x \to x + \mathbf{h}(x)$.

PROOF. If $x \in U_{s+1}$, then by A) the segment $x + \theta \mathbf{h}_{s+1}(x)$, $0 < \theta \leqslant 1$, lies in U_s. Therefore

$$| h^{s+1}x - h^s x | = | h^s(x + \mathbf{h}_{s+1}(x)) - h^s(x) |$$
$$\leqslant \left\| \frac{dh^s}{dx} \right\|_{U_s} \cdot \| \mathbf{h}_{s+1} \|_{U_{s+1}}.$$

But, by B),

$$\left\| \frac{dh^s}{dx} \right\|_{U_s} \leqslant \prod_{i=1}^{s} \left\| \frac{dh_s}{dx} \right\|_{U_s} \leqslant \prod_{i=1}^{s} \left(1 + \left\| \frac{dh_i}{dx} \right\|_{U_i}\right) \leqslant \prod_{i=1}^{s} (1+\delta_i) < C_1.$$

Thus, $| h^{s+1}x - h^s x |_{U_{s+1}} < C_1 C \delta_{s+1}$, so that the sequence h^s is uniformly convergent in U_∞. Further, $\frac{dh^{s+1}}{dx} = \frac{dh^s}{dx} \frac{dh_{s+1}}{dx}$ so that

$$\left\| \frac{dh^{s+1}}{dx} - \frac{dh^s}{dx} \right\| \leqslant \left\| \frac{dh^s}{dx} \right\| \left\| \frac{dh_{s+1}}{dx} \right\| \leqslant C_1 \delta_{s+1}.$$

Therefore the product converges uniformly in U_∞, and this concludes the proof of the lemma.

LEMMA 3. *Let* A_s *be linear operators such that* $\| A_s \| \leqslant \delta_s \leqslant \frac{1}{2}$, $\sum_{s=1}^{\infty} \delta_s = C < \infty$.

Then $\lim_{s \to \infty} \prod_{i=1}^{s} (E + A_i) = \Pi$ *exists and* $\| \Pi \| \leqslant e^C$, $\| \Pi^{-1} \| \leqslant e^{2C}$.

PROOF.

$$\| \Pi \| \leqslant \prod_{s=1}^{\infty} (1+\delta_s) \leqslant e^C, \quad \| \Pi^{-1} \| \leqslant \prod_{s=1}^{\infty} (1-\delta_s)^{-1} \leqslant \prod_{s=1}^{\infty} (1+2\delta_s) \leqslant e^{2C}.$$

LEMMA 4. *Set* $\mathbf{k}'[\mathbf{k}, \varphi, \mathbf{h}] = \mathbf{k}(f(x+\mathbf{h}(x)) + \varphi(x+\mathbf{h}(x)))$, $\mathbf{k}''[\varphi] = \mathbf{k}'[\mathbf{K}[\varphi], \varphi, \mathbf{H}[\varphi]]$. *Then*

$$\mathbf{k}''[\varphi] \prec \varphi(r+1 \mid r+3). \tag{3}$$

The proof of (3) is similar to the argument above; first we introduce the estimate

$$\mathbf{k}'[\mathbf{k}, \varphi, \mathbf{h}] \prec \mathbf{k}(0 \mid 0, 2, 2),$$

and then use the estimates (1).

Convergence of the sequences of mappings h^s, k^s. Set $V_0 = \{y\colon |y| \leqslant \rho_0\}$, and assume that $|\varphi_0|_{V_0} \leqslant M_0$. We have to show that for sufficiently small M_0, the sequences $h^s = h_1 \circ \ldots \circ h_s$, $k^s = k_s \circ \ldots \circ k_1$ defined by the relations

$$h_s\colon x \longrightarrow x + \mathbf{h}_s(x), \qquad k_s\colon y \longrightarrow y + \mathbf{k}_s(y),$$
$$\mathbf{h}_s = \mathbf{H}[\varphi_{s-1}], \quad \mathbf{k}_s = \mathbf{K}[\varphi_{s-1}], \quad \varphi_s = \mathbf{R}[\varphi_{s-1}],$$

$(s = 1, 2, \ldots)$ converge in some neighbourhood of 0.

We use the estimates (1), (2), (3) already established:

$$\mathbf{H}[\varphi] \prec \varphi(r\,|\,0),\ \ \mathbf{K}[\varphi] \prec \varphi(r\,|\,0),\ \ \mathbf{R}[\varphi] \prec \varphi^2(2r+1\,|\,r+4),$$
$$\mathbf{k}''[\varphi] \prec \varphi(r+1\,|\,r+3),$$

and the following immediate consequence of (1):

$$\frac{\partial}{\partial x}\mathbf{H}[\varphi] \prec \varphi(r+2\,|\,0),\quad \frac{\partial}{\partial y}\mathbf{K}[\varphi] \prec \varphi(r+2\,|\,0). \tag{4}$$

Let δ_0 be the constant in (1) – (4), C the constant of Lemma 1, and let δ_1 be a sufficiently small positive number:

$$0 < \delta_1 \leqslant \min(0.1;\ 0{,}1\rho;\ \delta_0;\ 0.5C).$$

We show that, *for the convergence of the sequences h^s, k^s, it is sufficient that*

$$|\varphi_0|_{V_0} \leqslant M_0 = \delta_1^T,\quad T = 4r+2. \tag{5}$$

Define a sequence $\delta_1 > \delta_2 > \ldots > 0$ of numbers and a sequence $V_0 \supset V_1 \supset \ldots \supset 0$, $U_0 \supset U_1 \supset \ldots \supset 0$, of domains by the rules

$$\delta_{s+1} = \delta_s^{1\frac{1}{2}},\quad V_s = V_{s-1} - \delta_s,\quad U_s = f^{-1}V_s \qquad (s = 1, 2, \ldots). \tag{6}$$

By the choice of δ_1, we have $\sum_{s=1}^{\infty} \delta_s < 0.2\rho$. Therefore $V_\infty = \bigcap_{s\geqslant 1} V_s$ contains a neighbourhood of 0. We use the notation $|\cdot|_s = |\cdot|_{V_s}$.

LEMMA 5. *The functions in U_s and in V_s listed above are well-defined, analytic, and satisfy the following inequalities*:

$$|\varphi_s|_s \leqslant M_s = \delta_{s+1}^T \qquad (s = 1, 2, \ldots), \tag{7}$$

$$\max\left(|\mathbf{h}_s|_s,\ |\mathbf{k}''[\varphi_s]|_s,\ \left|\frac{d}{dx}\mathbf{h}_s\right|_s,\ \left|\frac{d}{dy}\mathbf{k}_s\right|_s\right) \leqslant \delta_s^2. \tag{8}$$

PROOF. For $s = 0$ we have (7) = (5). Suppose that $|\varphi_{s-1}|_{s-1} \leqslant \delta_s^T$. Then $|\varphi_{s-1}|_{s-1} \leqslant \delta_{s+1}^{r+4}$, because $\frac{2}{3}T > r + 4$. So we apply the operator $\mathbf{R}$ and by (2), (6) we see that $|\varphi_s|_s = |\mathbf{R}[\varphi_{s-1}]|_s \leqslant \delta_s^{2T}\delta_s^{-(2r+1)} = \delta_s^{\frac{3}{2}T} = \delta_{s+1}^T$, which proves (7) for all $s = 1, 2, \ldots$. The inequalities (8) follow from (7) by means of (1), (3), (4), (6). This proves the lemma.

CONVERGENCE OF h^s. It follows, in particular, from (8) that the mapping h^s: $x \to x + \mathbf{h}_s(x)$ is defined on U_s. The image h_sU_s lies in U_{s-1}; for $|\mathbf{h}_s|_s \leqslant \delta_s^2 < C\delta_s$ by (8), and so $U_s \subset U_{s+1} - C\delta_s$ by Lemma 1.

Thus, the mappings h_s, $s = 1, 2, \ldots$ satisfy all the requirements of Lemma 2 (inequalities A) and B) follow easily from (8)). By Lemma 2, the sequence of mappings $h^s = h_1 \circ \ldots \circ h_s$ converges in U_∞ to an analytic mapping h: $U_\infty \to U_0$. By Lemma 3, h has a non-degenerate differential; therefore h defines a diffeomorphism in some neighbourhood of $x = 0$.

CONVERGENCE OF k^s. Consider next the mappings k_s: $y \to y + \mathbf{k}_s(y)$. It follows from the estimate for $\mathbf{k}''$ in (8) that k_s is defined and analytic in $f_{s-1} \circ h_s(U_s)$, where f_{s-1}: $x \to f(x) + \varphi_{s-1}(x)$ in U_s. And in U_s we have from (16) of §6 that $f_s = k_s \circ f_{s-1} \circ h_s$. We show that *the mapping* $k^s = k_s \circ \ldots \circ k_1$ *is defined and analytic in* $f_0 \circ h^s(U_s)$. For $s = 1$ this has already been done. Suppose that k^{s-1} is defined and analytic in $f_0 \circ h^{s-1}(U_{s-1})$, and that $f_{s-1} = k^{s-1} \circ f_0 \circ h^{s-1}$ in U_{s-1}. Then we get $f_s = k_s \circ (k^{s-1} \circ f_0 \circ h^{s-1}) \circ h_s$ in U_s, since $h_s\colon U_s \to U_{s-1}$; and it follows that $k^s = k_s \circ k^{s-1}$ is defined and analytic in $f_0 \circ h^s(U_s)$, and also that $f_s = k^s \circ f_0 \circ h^s$.

Furthermore, using the estimate (8) for $\|\mathbf{k}''\|$, we see that k^s converges uniformly in $V' = f_0 \circ h(U_\infty)$ to the mapping $\mathbf{k}\colon V' \to V_\infty$.

The fact that $\mathbf{k}$ is differentiable and non-degenerate follows from Lemma 3 and the estimate (8) for $\left|\frac{d}{dy}\mathbf{k}\right|$. Consequently, $\mathbf{k}$ defines a diffeomorphism from some neighbourhood of $f_0 \circ h(0)$ to a neighbourhood of $y = 0$. It is easy to derive[1] the limit relation $f = k \circ f_0 \circ h$ from (7), (8) and the fact that $f_s = k^s \circ f_0 \circ h^s$.

Finally, the estimate (8) shows that for sufficiently small M_0 the mappings h and k are arbitrarily near to the identity. This proves the stability theorem.

REMARK. *Every mapping* f_0 *having tangency of sufficiently high order* m *with* f *at* 0 *is equivalent to* f *at* 0.

For suppose that $f_0(x) = f(x) + \varphi_0(x)$, $|\varphi_0(x)| \leqslant C_1|x|^{m+1}$. Since f has finite multiplicity r, we get

$$x^r \in (\mathfrak{M}\mathcal{H}_y)\mathcal{H}_x, \quad |x|^r \leqslant C_2|f(x)|.$$

(see the Appendix to §4). Therefore $|\varphi_0(x)|_{V_0} \leqslant C_3\rho^{\frac{m+1}{r}}$. If $\frac{m+1}{r} > T = 4r+2$ and ρ is sufficiently small, then (5) is satisfied for $\delta_1 = 0.1\rho$. So, if[2] $m \geqslant 4r^2 + 2r$, then f_0 is equivalent to f in a sufficiently small neighbourhood of $x = 0$. At the same time we have proved:

COROLLARY 1. *The mapping* f *is equivalent to a polynomial at* 0, *namely to its Taylor polynomial of degree* m.

§8. In the neighbourhood of an isolated critical point every analytic function is equivalent to a polynomial

The germs of functions $f\colon \mathbf{C}^n \to \mathbf{C}$, $g\colon \mathbf{C}^n \to \mathbf{C}$ at 0 are *equivalent* if there exists a bi-analytic diffeomorphism h of neighbourhoods U, V of 0 in $\mathbf{C}^n$ such that $h(0) = 0$ and the diagram

1 Take x in U_∞. Then $h(x)$ and $h^s(x)$ can be connected by a curve in $h^s(U_s)$ whose length is small, as $s \to \infty$. Further, f_0 has derivative in $h^s(U_s)$ bounded by a constant independent of s. Thus, $f_0 \circ h(x)$ and $f_0 \circ h^s(x)$ can also be joined by a short curve in $f_0 \circ h^s(U_s)$. The derivative of k^s is also bounded in $f_0 \circ h^s(U_s)$ independently of s; thus, for large s the points $k^s \circ f_0 \circ h^s(x)$ and $k^s \circ f_0 \circ h(x)$ are near each other. Finally, $k^s \circ f_0 \circ h(x)$ is near to $k \circ f_0 \circ h(x)$ for large s. Thus, $k \circ f_0 \circ h = \lim_{s\to\infty} k^s \circ f_0 \circ h^s = \lim_{s\to\infty} f_s = f$.

2 Actually this is more than we need for our argument.

$$\begin{array}{ccc} U & \xrightarrow{f} & \mathbf{C} \\ {\scriptstyle h}\uparrow & \nearrow_{g} & \\ V & & \end{array}$$

is commutative.

It is easy to see that every analytic function of a single variable is equivalent to a polynomial in the neighbourhood of any critical point. N. Levison [22] has proved the analogous result for functions of two variables. Whitney [23] has given examples of analytic functions of three variables that are not equivalent to a polynomial in the neighbourhood of a critical point. An example of such a function is

$$f(x, y, z) = xy(x-y)(x-yz)(x-ye^z).$$

The proof of the non-equivalence is based on the following remark. A surface transversal to the z-axis at the point z_0 intersects the level set $f = 0$ in five curves passing through the point $x = y = 0$, $z = z_0$. Consider the cross-ratios formed by the tangents at z_0; they are independent of the choice of surface. The cross-ratio α corresponding to the first four factors of f define z_0. Therefore the cross-ratio β corresponding to the last four factors is a function $\beta(\alpha)$. If f were equivalent to a polynomial at 0, then $\beta(\alpha)$ would be algebraic, whereas e^z is transcendental.

Note that the critical point $x = y = z = 0$ in Whitney's example is not isolated.

THEOREM 1. *The germ of an analytic function at an isolated critical point is equivalent to the germ of its Taylor polynomial of sufficiently high degree.*

PROOF[1]. Let $f(x) = f(x_1, \dots, x_n)$ be the given function and suppose that $x = 0$ is an isolated critical point. Let $\mathbf{h}(x) = (h_1(x), \dots, h_n(x))$ be the germ of a "vector field" analytic at zero and $\varphi(x)$ the germ of a function analytic at zero.

LEMMA 1. *The following identity holds:*

$$f(x+\mathbf{h}(x)) + \varphi(x+\mathbf{h}(x)) = f(x) + \Sigma_0 + \Sigma_1 + \Sigma_2,$$

where

$$\Sigma_0 = f'(x)\,\mathbf{h}(x) + \varphi(x),$$
$$\Sigma_1 = f(x+\mathbf{h}(x)) - f(x) - f'(x)\,\mathbf{h}(x),$$
$$\Sigma_2 = \varphi(x+\mathbf{h}(x)) - \varphi(x).$$

Corresponding to the general theme of §5 we must investigate the "infinitesimal" equation $\Sigma_0 = 0$.

LEMMA 2. *Suppose that $\varphi(x)$ has sufficiently high order N at zero. Then there exists a field $\mathbf{h}(x)$ such that*

$$\varphi(x) + f'(x)\,\mathbf{h}(x) = 0.$$

The construction of $\mathbf{h}$ is performed as follows. We consider the mapping

[1] Another proof can be found in Tougeron's dissertation [14].

$f'\colon x \longrightarrow \left(\frac{\partial f}{\partial x_1}, \ldots, \frac{\partial f}{\partial x_n}\right)$. The critical points of f are the inverse images of zero under f'. Since $x = 0$ is an isolated critical point, the analytic function f' has finite multiplicity at zero (the "theorem on zeros"). Consequently the local ring of f' at zero has finite dimension r. It follows that there exist n decompositions

$$x_i^r = f'(x)\,\mathbf{h}_i(x) \qquad (i = 1, \ldots, n),$$

where the $\mathbf{h}_i(x)$ are the germs of vector fields analytic at zero (see the Appendix to §4). We assume that $\mathbf{h}_i(x)$ is analytic in the unit disc with respect to the first coordinate (if not, change the scale).

If N is sufficiently large ($N \geqslant nr$), then the germ of the function $\varphi(x)$, which is analytic at zero and of order N, can be written as

$$\varphi(x) = \sum_{i=1}^{n} x_i^r\,(\varphi_i(x)), \text{ where the } \varphi_i = \sigma_i[\varphi] \qquad (i = 1, \ldots, n)$$

are obtained by the successive division of $\varphi(x)$ by x_i^r (see the Appendix to §4). We introduce the operator

$$\mathbf{H} = -\sum_{i=1}^{n} \mathbf{h}_i \sigma_i. \tag{1}$$

The field $\mathbf{h} = \mathbf{H}[\varphi]$ satisfies the conditions of the lemma, so that the proof of Lemma 2 is complete.

Suppose now that $f_0(x) = f(x) + \varphi_0(x)$ is a mapping close to f. We define by recursion functions φ_s and fields $\mathbf{h}_s$ starting with $\varphi_0(x)$ and continuing with

$$\varphi_s = \mathbf{R}[\varphi_{s-1}], \quad \mathbf{h}_s = \mathbf{H}[\varphi_{s-1}] \qquad (s = 1, 2, \ldots),$$
$$\mathbf{R}[\varphi] = R[\varphi, \mathbf{H}[\varphi]], \qquad R[\varphi, \mathbf{h}] = \Sigma_1 + \Sigma_2.$$

We define further mappings:

$$h_s\colon x \longrightarrow x + \mathbf{h}_s(x), \qquad f_s\colon x \longrightarrow f(x) + \varphi_s(x), \qquad h^s = h_1 \circ \ldots \circ h_s.$$

By Lemma 1, $f_{s-1} \circ h_s = f_s$, so that $f_0 \circ h^s = f_s$. Thus, our theorem follows from the following proposition.

LEMMA 3. *If φ_0 has sufficiently high order at 0, then $h = \lim_{s\to\infty} h^s$ is a diffeomorphism in a sufficiently small neighbourhood of $x = 0$, and $f_0 \circ h = f$.*

For the convergence proof we introduce the following notation. Let φ be a function analytic in $|x| \leqslant \rho < 1$ and having order not less than a at 0. Set

$$|\varphi|_{a,\rho} = \max_{|x|\leq\rho} \frac{|\varphi(x)|}{|x|^a}.$$

Further, let L be an operator taking analytic functions to analytic functions, and suppose that a, b, ν, $\nu_0 \geqslant 0$. The notation

$$|L[\varphi]|_a \prec \Phi(|\varphi|_b)(\nu|\nu_0)$$

means that there exists a constant $\delta_0 > 0$ such that

$|L[\varphi]|_{a,\rho-\delta} \leqslant \frac{\Phi(|\varphi|_{b,\rho})}{\delta^\nu}$, if $0<\delta<\left(\delta_0, \frac{\rho}{2}\right)$ and $|\varphi|_{b,\rho} < \delta^{\nu_0}$. The meaning of the symbol $\prec$ for operators of several arguments is defined in a similar way.

EXAMPLE 1. *The following estimate holds*:

$$\left|\frac{d\varphi}{dx}\right|_{a-1} \prec |\varphi|_a \ (2|0).$$

For by using Cauchy's formula for $x \neq 0$, $d=\delta\frac{|x|}{\rho-\delta}$, $0<\delta<\frac{\rho}{2}$, we find $\left|\frac{d\varphi}{dx}\right| \leqslant \frac{|\varphi|_a(|x|+d)^a}{d} \leqslant \frac{2^a}{\delta}|x|^{a-1}$, as required.

In the same way one proves

$$|\Sigma_1[\mathbf{h}]|_a \prec |\mathbf{h}|_a^2(1|1), \quad |\Sigma_2[\varphi, \mathbf{h}]|_{a+b-1} \prec |\varphi|_a|\mathbf{h}|_b(2|0, 2). \tag{2}$$

LEMMA 4. *The operator* $\mathbf{H}$ *satisfies*

$$|\mathbf{H}[\varphi]|_{N-r} \prec |\varphi|_N(n+1|0), \ N \geqslant nr. \tag{3}$$

PROOF. As is clear from (1), it is enough to establish that

$$|\sigma_i[\varphi]|_{N-r} \prec |\varphi|_N(n+1|0).$$

This is done by a simple calculation of the Taylor coefficients. If $|\varphi(x)| \leqslant M|x|^N$ for $|x| \leqslant \rho$, then $\varphi(x) = \sum_{\mathbf{k}} \varphi_{\mathbf{k}} x^{\mathbf{k}}$, where $\mathbf{k} = (k_1, \ldots, k_n)$, and $|\varphi_{\mathbf{k}}| \leqslant M\rho^{N-k}$ $(k = k_1 + \ldots + k_n, \ k \geqslant N)$.

By definition of $\sigma_i[\varphi]$, the coefficients $\varphi_{i,\mathbf{k}}$ of the Taylor expansion

$$\sigma_i[\varphi] = \sum_{\mathbf{k}} \varphi_{i,\mathbf{k}} x^{\mathbf{k}} \qquad (i=1, \ldots, n)$$

are obtained from $\varphi_{\mathbf{k}}$ by displacing the index r positions. Thus,

$$|\varphi_{i,\mathbf{k}}| \leqslant M\rho^{N-k-r} \qquad (k \geqslant N-r).$$

Suppose now that $0 < \delta < 0.5\rho$, so that $\theta = 1 - \delta/r > 0.5$. Then

$$|\sigma_i[\varphi]|_{N-r,\rho-\delta} \leqslant M\sum_{\mathbf{k}}\left|\frac{x}{\rho}\right|^{k-N+r} \leqslant M\theta^{r-N}\sum_{\mathbf{k}}\theta^{\mathbf{k}} \leqslant \frac{M2^N}{(1-\theta)^h} \leqslant \frac{M2^N\rho^n}{\delta^n},$$

as required.

Comparing (2) and (3) we get for $N \geqslant nr$, $n \geqslant 2$,

$$|\mathbf{R}[\varphi]|_N \prec |\varphi|_N^2(2n+4|n+3). \tag{4}$$

PROOF OF LEMMA 3. We prove convergence for

$$|\varphi_0(x)| \leqslant C|x|^m, \ m \geqslant 4n+7+N, \ N=nr, \ n \geqslant 2.$$

Let δ_0 be the constant in (3) and (4). Define sequences δ_s, ρ_s $(s=1,2,\ldots)$ of numbers by the rules $\delta_{s+1} = \delta_s^{1.5}$, $\rho_s = \rho_{s-1} - \delta_s$, $\delta_1 = 0.1\rho_0$.

If ρ_0 is sufficiently small, then $\delta_s < 0.5\rho_{s-1}$, $\delta_s < \delta_0$, $\rho_s > 0.5\rho_0$ and

$$|\varphi_0|_{N,\rho_0} \leqslant \delta_1^T,\quad T = 4n + 6.$$

Here we use the notation $|\cdot|_s = |\cdot|_{N,\rho_s}$.

LEMMA 5. *The following estimates hold*:

$$|\varphi_s|_s \leqslant \delta_{s+1}^T \quad (s = 1, 2, \ldots). \tag{5}$$

For $s = 0$, (5) is clearly true. If $|\varphi_{s-1}|_{s-1} \leqslant \delta_s^T$, then by (4),

$$|\varphi_s|_s = |\mathbf{R}[\varphi_{s-1}]|_s \leqslant \delta_s^{2T}\delta_s^{-(2n+4)} = \delta_s^{\frac{3}{2}T} = \delta_{s+1}^T,$$

as required.

It follows from (3) and (5) that

$$|\mathbf{h}_s(x)| < \delta_s^2,\quad \left|\frac{d}{dx}\mathbf{h}_s(x)\right| \leqslant \delta_s^2. \tag{6}$$

for $|x| < \rho_s$. The convergence of the sequence h^s is a consequence of (6) by Lemma 2 of §7 (p.). This proves Lemma 3 and with it the theorem.

REMARK 1. We have shown that the germ of a function is equivalent to its Taylor polynomial of degree $m = nr + 4n + 6$ at zero. It is easy to see that in fact $m = 2r$ will do.

REMARK 2. A.M. Samoilenko has carried the proof over to the case of infinitely differentiable functions, using the Nash-Moser smoothing [20]. Of course, instead of assuming that the critical points are isolated we have to assume that the local ring of mappings $x \to \frac{\partial f}{\partial x}$ is finite-dimensional.

References

[1] J. Milnor, Morse Theory, Princeton 1963.
Translation: *Teoriya Morsa*, Izdat. "Mir", Moscow 1965.

[2] H. Whitney, On singularities of mappings of Euclidean spaces. I. Mappings of the plane into the plane, Ann. of Math. 62 (1955), 374-410.

[3] R. Thom and H. Levine, Singularities of differentiable mappings I, Bonn 1959.

[4] R. Thom, Local topological properties of differentiable mappings, Bombay Colloquium on Differentiable Analysis, Oxford University Press 1964.

[5] J.M. Boardman, Singularities of differentiable maps, University of Warwick (1965), 1-105.

[6] B. Morin, Formes canoniques des singularités d'une application différentiable, Compt. Rendus Acad. Sci. Paris 260 (1965), 5662-5665, 6503-6506.

[7] B. Malgrange, Le théorème de préparation en géométrie différentiable, Séminaire H. Cartan (1962/63), Nos. 11, 12, 13, 22.

[8] B. Malgrange, Ideals of differentiable functions, Oxford University Press, 1966.

[9] J. Mather, On the preparation theorem of Malgrange, Princeton 1966.

[10] C. Houzel, Géométrie analytique locale I, Séminaire H. Cartan (1960/1961), No. 18.

[11] H. Cartan, Sur les matrices holomorphes de n variables complexes, J. Math. Pures. Appl. 19 (1940), 1-26.

[12] A.N. Kolmogorov, On the preservation of conditionally periodic motions, Dokl. Akad. Nauk SSSR 98 (1954), 527-530. MR 16, 294.

[13] J. Mather, Structural stability of mappings, Princeton 1966.
[14] J.C. Tougeron, Idéaux des fonctions différentiables, Université de Rennes, 1967.
[15] C.L. Siegel, Vorlesungen über Himmelsmechanik, Springer-Verlag, Berlin 1965. Translation: *Lektsii po necebnoi mekhanik*, Izdat.-Lit, Moscow 1959.
[16] C.L. Siegel. Über die Normal form analytischer Differentialgleichungen in der Nähe einer Gleichgeirichtslösung, Nachr. Akad. Wiss. Göttingen, Math.-Phys. Kl. 1952, 21-30. MR 15, 222.
[17] V.I. Arnol'd, Small denominators. I. Mappings of a circle onto itself, Izv. Akad. Nauk Ser. Mat. 25 (1961), 21-86. MR 25, # 4113.
[18] Jo Ging-tzung, Singularities S_{ro}, Sci. Sinica 14 (1965), 816-830.
[19] V.I. Arnol'd, A remark on the Weierstrass preparation theorem, Funktsional. Anal i prilozhen. 1 (1967), 1-8.
[20] J. Moser, A new technique for the construction of solutions of non-linear differential equations, Proc. Acad. Nat. Sci. U.S.A., 47 (1961), 1824-1831.
[21] M. Hervé, Functions of several complex variables, Oxford University Press 1963.
Translation: *Funktsii mnogikh kompleksnykh peremennykh*, Izdat. "Mir", Moscow 1965.
[22] N. Levinson, A polynomial canonical form for certain analytic functions of two variables at a critical point, Bull. Amer. Math. Soc. 66 (1960), 366.
[23] H. Whitney, Local properties of analytic varieties; in the book: Differential and combinatorial topology, Princeton 1965.
[24] H. Whitney, The general type of singularity of a set of $2n$-1 smooth functions of n variables, Duke. Math. J. 10 (1943), 161-172.
[25] H. Whitney, The self-intersections of a smooth n-manifold in $2n$-space, and ($2n$-1)-space, Ann. of Math. Ser. 2, 45 (1944), 220-293.
[26] H. Whitney, On singularities of mappings of euclidean spaces, Symposium Internacional de Topologia Algebraica, Mexico 1958.
[27] R. Thom, Les singularités des applications différentiables, Ann. Inst. Fourier 6 (1956), 43-87.
[28] R. Thom, Remarques sur les problèmes comportant des inéquations différentielles globales, Bull. Soc. Math. France 87 (1959), 455-461.
[29] R. Thom, L'équivalence d'une fonction différentiable et d'un polynome, Topology 3 (1965), 297-307.
[30] A. Haefliger, Sur les self-intersections des applications différentiables, Bull. Soc. Math. France 87 (1959), 351-359.
[31] A. Haefliger, Quelques remarques sur les applications différentiables d'une surface dans le plan, Ann. Inst. Fourier 10 (1960), 47-60.
[32] A. Haefliger, Points multiples d'une application et produit cyclique reduit, Amer. J. Math. 83 (1961), 57-70.
[33] A.W. Tucker, Branched and folded coverings, Bull. Amer. Math. Soc. 42 (1936), 859-862.
[34] Z. Wolfsohn, On differentiable maps of Euclidean n-space into Euclidean m-space, Bull. Amer. Math. Soc. 61 (1955), 171.
[35] N. Levinson, Transformation of an analytic function of several variables to a canonical form, Duke. Math. J. 28 (1961), 345-353.
[36] H.I. Levine, The singularities S_1^q, Illinois, K. Math. 8 (1964), 152-168.
[37] H.I. Levine, Elimination of cusps, Topology 3 (1965), 263.
[38] H.I. Levine, Mappings of manifolds into the plane, Amer. J. Math. 88 (1966), 357-365.
[39] L.S. Pontryagin, Vector fields on manifolds, Mat. Sb. 24 (66) (1949), 155-162. = Amer. Math Soc. Translations, no. 13 (1949).

Received by the Editors September 4, 1967.

Translated by J. Wiegold.

Remarks on singularities of finite codimension in complex dynamical systems

V.I. Arnold
Translated by Victor Vassiliev

It is natural to study singular points of dynamical systems from the viewpoint of the singularity theory of maps (see [1], [2]).

Singularities "in general position" are explored by Poincaré [3] and Siegel [4]. They have shown that in a neighborhood of a generic singular (i.e. fixed) point an analytic dynamical system becomes linear in appropriate coordinate system.

At first glance, more complicated singular points should not occur in the nature, because any system can be approximated by one, all whose singular points are generic. This is, however, not completely true, if (as it often happens) we are interested not in an individual system, but in a family of systems depending, say, on k parameters. In such a family, non-generic singular points can occur in such a way that they cannot be removed by small deformations of the family: namely, the points of those classes whose codimensions in the functional space of all systems do not exceed k.

Therefore we arrive at the problem of studying singular points of codimension k. Despite many investigations of different exotic singular points, the simplest case of $k = 1$ is still poorly explored.

We shall consider the topological and differentiable structure of complex solutions in a neighborhood of a singular point of a dynamical system depending on parameters. The change of parameters can change the topological type of the singular point; we are interested in the transformation experienced when the parameters are passing through critical or "resonance" values corresponding to singular points of complex codimension $k = 1$, i.e. to resonances of multiplicity 1. It turns out that under transversal crossing of such a critical value an invariant manifold of codimension $k = 1$ splits off from the separatrices of the singular point.[1]

Moscow State University. Translated from Funkts. Anal. Prilozh. 3:1, 3–44.
Original paper submitted September 18, 1968

[1] A similar phenomenon occurs not only in the complex case, but also in real analytic and even finitely differentiable ones. However sometimes the surgery of the topology in the complex domain does not affect the topological arrangement of real solutions (examples: node and saddlepoint).

Originally publ. in: Funkts. Anal. Prilozh. 3:1, 1-6, © Russ. Acad. Sciences 1969
Original English transl.: Funct. Anal. Appl. 3, 1-5, © Consultants Bureau, New York 1969
Translated by V. Vassiliev. Springer-Verlag Berlin Heidelberg 2014 with kind permission, DOI 10.1007/978-3-642-31031-7_14

For the study of this phenomenon, a mapping of our dynamical system onto a one-dimensional one is useful. Such a mapping can be called a one-dimensional cocycle (since the embeddings of a one-dimensional system into ours one are called cycles). With any resonance we relate (at least in the class of formal series) some one-dimensional cocycle; singular points of the one-dimensional system correspond to (codimension 1) invariant manifolds of the initial one.

A different method of investigation of a complex neighborhood of a singular point consists in the study of the intersection of solutions with a small sphere centered at the singular point. In this case a real foliation on the sphere is related in an invariant way with the complex system.

1. Resonances

Let $\lambda_1, \dots, \lambda_n$ be the eigenvalues of the linear part

$$\dot{\mathbf{z}} = A\mathbf{z}, \quad \mathbf{z} \in \mathbb{C}^n, \tag{1}$$

of a system of ordinary differential equations

$$\dot{\mathbf{z}} = \mathbf{a}(\mathbf{z}), \quad \mathbf{a}(\mathbf{z}) = A\mathbf{z} + \dots, \tag{2}$$

where $\mathbf{a}(\mathbf{z})$ is a vector field, analytic in a neighborhood of $\mathbf{z} = 0$.

A *resonance* is the relation

$$\lambda_i = (\mathbf{m}, \lambda), \tag{3}$$

where $\mathbf{m}$ is a vector with integer non-negative components m_j, $|\mathbf{m}| \equiv \sum m_j \geq 2, \lambda = \lambda_1, \dots, \lambda_n$, $(\cdot, \cdot)$ is the scalar product.

Relation (3) can be rewritten in symmetric form:

$$(\mu, \lambda) = 0, \quad \text{where } \mu = (m_j - \delta_{ij}). \tag{4}$$

In the space $\mathbb{C}^n = \{\lambda\}$ of collections of eigenvalues the resonance equation (4) specifies the *resonance plane* Γ_μ.

The entire space of collections of eigenvalues splits into two parts.

The *Poincaré domain* $\mathscr{P}$ is the set of vectors λ such that the convex hull of $\lambda_1, \dots, \lambda_n$ does not contain 0.

The *Siegel domain* $\mathscr{S}$ is its complement, $\mathscr{S} = \mathbb{C}^n \setminus \mathscr{P}$.

The Poincare domain is open, and the Siegel domain is closed. It is easy to check that in the Poincaré domain the resonance planes are placed discretely (i.e. any point of Poincaré domain has a neighborhood, intersecting only a finite number of resonance planes), and in Siegel domain everywhere dense. Suppose that all n eigenvalues $\lambda_1, \dots, \lambda_n$ are distinct.

Poincaré theorem. *If $\lambda \in \mathscr{P} \setminus \cup \Gamma_\mu$ then the system (2) can be reduced to linear form (1) by a biholomorphic change of coordinates in a neighborhood of* 0.

Siegel theorem. *For almost all (over the Lebesque measure) λ the system (2) can be reduced to the linear form (1) by a holomorphic change of coordinates in a neighborhood of* 0.

A point λ is called *k-fold resonance point* if among relations (4) satisfied by it there are k but no $k+1$ linearly independent ones. The codimension of the set of points of k-fold resonance is equal to k. 1-fold resonance points are those points λ of a resonance plane Γ_μ which do not belong to the intersection of Γ_μ with other resonance planes.

We are interested in the mode of the change of the behavior of solutions of a non-linear system (2) when λ passes through a one-fold resonance (4). Here the following analogy with the system of differential equations on a torus is useful. Let $z_1, \dots, z_n$ be a coordinate system in which the linear part A of system (2) is diagonal. The substitution $\mathbf{z} = e^{i\varphi}$ reduces (2) to the form

$$\dot{\varphi}_j = \omega_j + \sum a_{\mathbf{m}} e^{i\mathbf{m}\varphi}, \qquad \text{where } \omega_j = -i\lambda_j.$$

On the torus, a linear integer relation $(\mu, \omega) = 0$ between the frequencies ω_j defines an invariant manifold of codimension 1, which is close to $(\mu, \varphi) = c$. Therefore it is natural to seek an invariant manifold of the system (2), which is close to $\mathbf{z}^\mu = c$.

2. Invariant manifold M_ε

Consider an analytic family of dynamical systems

$$\dot{\mathbf{z}} = \mathbf{a}(\mathbf{z}, \varepsilon), \quad \mathbf{a}(\mathbf{z}, 0) = \mathbf{a}(\mathbf{z}) = A\mathbf{z} + \dots, \qquad (2_\varepsilon)$$

depending analytically on the complex parameter ε. Suppose that all eigenvalues $\lambda_1, \dots, \lambda_n$ of the linear part A are not equal to 0, distinct, and form a vector λ of one-fold resonance (3).

Conjecture. *For almost all families (2_ε), in a neighborhood of $\mathbf{z} = 0$ there is a manifold M_ε of codimension 1, invariant under (2_ε), which for $\varepsilon = 0$ becomes a separatrix manifold of the system (2_0). More precisely, there is a change of variables $\mathbf{w} = w(\mathbf{z}.\varepsilon)$, bianalytic in a neighborhood of 0 and depending analytically on ε, such that in the new coordinates the invariant manifold M_ε is defined by equation*

$$F(\varepsilon, \mathbf{w}) = 0,$$

where F is a function, analytic in a neighborhood of $\varepsilon = 0, \mathbf{w} = 0$, whose Taylor series at 0 begins as

$$F = c_0 \varepsilon w_i + c_1 \mathbf{w}^m + \dots, \qquad (5)$$

where c_0, c_1 are constants, $\mathbf{w}^m = w_1^{m_1} \dots w_n^{m_n}$.

Therefore the invariant manifold M_ε is defined in the first approximation by an equation of the form $\mathbf{w}^\mu = c\varepsilon$.

Example. If $\lambda_1 = i\omega, \lambda_2 = -i\omega$, then the approximate equation of the invariant manifold has the form $w_1 w_2 = c\varepsilon$. If the initial system is a real one, written in a complex self-conjugate basis of eigenvectors, then $w_2 = \bar{w}_1$. So, on distance ε from the resonance (in the parameter space) there is a limit cycle, whose radius $|w|$ is proportional to $\sqrt{\varepsilon}$. This situation is well studied (see [5]–[8], [10]).

Remark. The parameter ε is introduced solely for the convenience of writing down the formula (5).

3. Calculations

Since all λ_i are not equal to 0, by the implicit function theorem we can choose coordinates $\mathbf{z}'(\mathbf{z},\varepsilon)$, depending analytically on ε, in such a way that the fixed point of system (2_ε) for all ε will be the point $\mathbf{z}' = 0$.

Since all λ_i are distinct, we can turn the matrix of the linear part to diagonal form by a new change of coordinates $\mathbf{z}''(\mathbf{z},\varepsilon)$. After that we can, following Poincaré, eliminate in (2_ε) arbitrarily many "non-resonant terms", so that our system in new coordinates $\mathbf{w}(\mathbf{z}'',\varepsilon)$ will have the form

$$\dot{w}_j = w_j(\lambda_j(\varepsilon) + b_j(\varepsilon)\mathbf{w}^\mu + \dots), \quad j = 1,\dots,n. \tag{6}$$

Suppose that the terms of degrees bigger than $|\mu|+1$, denoted by dots, are equal to 0. Consider the function $\varphi = \mathbf{w}^\mu$. By (6) we have

$$\dot{\varphi} = \varphi[c_0(\varepsilon) + c_1(\varepsilon)\varphi], \tag{7}$$

where $c_0(\varepsilon) = (\mu,\lambda(\varepsilon))$, $c_1(\varepsilon) = (\mu, b)$. The manifold M_ε, defined by the equation

$$M_\varepsilon = \{\mathbf{w} : c_0(\varepsilon) + c_1(\varepsilon)\varphi(\mathbf{w}) = 0\},$$

is invariant with respect to (6). Notice that $c_0(0) = 0$ (since $(\mu,\lambda(0)) = 0$ by (4)). Therefore $c_0(\varepsilon) = c_0\varepsilon + \dots$; it remains to set $c_1 = c_0(0)$ and multiply the equation of M_ε by w_i, in order to reduce it to the form (5).

The terms $(\dots)$, removed above, can indeed be eliminated by an analytic choice of coordinates, if the vector λ belongs to the Poincaré domain. So, in the Poincaré case the above arguments can be turned into a proof. This was done by N.N. Brushlinskaya.

In the Siegel case, generally speaking, one cannot remove the terms $(\dots)$ in (6); we can only push them to arbitrarily high degrees, replacing $\lambda + b\mathbf{w}^\mu$ by a polynomial in degrees of $\mathbf{w}^\mu$. However, one can hope that these perturbations of high degrees will not destroy the invariant manifold M_0.

4. Cocycles

A *dynamical system* (M, g^t) is a one-parametric group g^t of diffeomorphisms of the manifold M.

A *cycle* or *invariant manifold* of the system (M, g^t) is a dynamical system (N, h^t) together with an embedding $i : N \to M$, for which $ih^t = g^t i$. The dimension of N is called the dimension of the cycle.

A *cocycle* or *factor* of the system (M, g^t) is a dynamical system (K, f^t) together with a map φ of the manifold M onto manifold K, such that $\varphi g^t = f^t \varphi$. The dimension of K is called the dimension of the cocycle.

In these terms we can say that the map $\varphi : \mathbf{w} \mapsto \mathbf{w}^\mu$ defines the one-dimensional cocycle (7), and the invariant manifold M_ε is a cycle of codimension 1, corresponding to the fixed point of the cocycle (7) (more precisely, one should speak on the "germ of a meromorphic cocycle".

Thus, in the Poincaré domain traversing a one-fold resonance we obtain the bifurcation of the fixed point $\varphi = 0$ of the one-dimensional cocycle (7).

In the Siegel domain an one-dimensional cocycle, similar to (7),

$$\dot{\varphi} = \varphi(b_0 + b_1\varphi + b_2\varphi^2 + \dots) \tag{8}$$

also exists, but only in the class of formal power series. It follows from [9] that the corresponding series can be divergent for a majority of $\lambda \in \mathscr{S}$.

5. Intersection with the sphere

Complex trajectories of the system (2) form a two-dimensional (in the real sense) foliation with singularity at 0 in a neighborhood of $\mathbf{z} = 0$. To understand the structure of this singularity, let us intersect the foliation by a sphere of small radius centered at the origin.

Suppose that the linear part of the system (2) is diagonal in coordinates $z_1, \dots, z_n$, i.e. $\dot{z}_j = \lambda_j z_j + \dots$, $j = 1, \dots, n$.

Theorem. *If the collection λ of eigenvalues belongs to the Poincaré domain, then any sphere $|z_1|^2 + \dots + |z_n|^2 = r^2$ of sufficiently small radius intersects the foliation (2) transversally.*

Proof. Consider first a linear system. We have

$$dr^2 = \sum(\bar{z}_j dz_j + z_j d\bar{z}_j) = A dt + \bar{A} d\bar{t}, \quad A = \sum |z_j|^2 \lambda_j.$$

The transversality condition of a fiber with the sphere means that the 1-form dr^2 should not vanish on the tangent plane to the fiber. But the form $Adt + \bar{A}d\bar{t}$ vanishes only if $A = 0$. The condition $A = 0$ is not satisfied in Poincaré case (and only in Poincaré case) for any $\mathbf{z} \neq 0$. Thus in the linear case theorem is proved: the fibers intersect the sphere with a non-zero angle $\alpha(\mathbf{z})$.

Let α_0 be the minimal value of angles $\alpha(\mathbf{z})$ on the sphere $|\mathbf{z}| = r$. The value α_0 does not depend on r (since $\alpha(c\mathbf{z}) = \alpha(\mathbf{z})$). So, $\alpha(\mathbf{z}) \geq \alpha_0 > 0$ for all $\mathbf{z} \neq 0$.

Consider now a non-linear system. The angle between directions of the non-linear system and its linear part is small together with $|\mathbf{z}|$. Therefore in a sufficiently small neighborhood of $\mathbf{z} = 0$ it is smaller than α_0, and trajectories of the non-linear system (2) intersect the sphere transversally. Theorem is proved.

Corollary 1. *Intersections of trajectories of system* (2) *with a sphere of sufficiently small radius* $|\mathbf{z}| = r$ *form a one-dimensional foliation without singular points on this sphere* S^{2n-1}.

Corollary 2. *The foliations obtained on all spheres of sufficiently small radii* $|\mathbf{z}| = r$ *are diffeomorphic to one another. Moreover, the differentiable type of the foliation does not change at any deformation of the secant sphere, if it remains smooth and transversal to the fibers.*

Thus, the two-dimensional foliation (2) is homeomorphic to the cone over the one-dimensional foliation on the sphere in a neighborhood of the singular point $\mathbf{z} = 0$.

Remark 1. The resulting one-dimensional foliation on the sphere S^{2n-1} decomposes the sphere into trajectories of a real dynamical system without fixed points. This follows from the orientability of both the sphere and the foliation (2).

Remark 2. In the non-resonant case, accordingly to the Poincaré theorem, the system (2) is linear in appropriate coordinate system in a sufficiently small neighborhood of the point $\mathbf{z} = 0$. Therefore the differential type of the foliation on S^{2n-1} in the non-resonant case is the same as that for the linear part of system (2).

We conclude that the differentiable type of the foliation on S^{2n-1} remains the same as for the linear part of the system not only in a neighborhood of the origin, where the Poincaré series do converge, but also far beyond its boundary. Indeed, close to the resonance the domain of convergence of Poincaré series shrinks to 0, while the size of the domain of transversality in the above theorem remains bounded from below. Therefore we can watch the passage of the resonance by the change of the one-dimensional foliation, obtained in the intersection with the sphere of a fixed radius, $|\mathbf{z}| = r_0$.

6. Example

Let us describe the change of the topological type of the foliation in S^3 at the passage through the resonance $\lambda_1 = 2\lambda_2$ in the system

$$\dot{z}_1 = \lambda_1 z_1 + \dots, \quad \dot{z}_2 = \lambda_2 z_2 + \dots \tag{9}$$

We are in the Poincaré domain if the ratio of eigenvalues $\lambda = \lambda_1/\lambda_2$ is not a negative number. Consider first the foliation in S^3 corresponding to the linear part of system (9).

The separatrices $z_1 = 0$ and $z_2 = 0$ intersect the sphere along great circles, which are cycles of a system on S^3. Their linking number is equal to 1.

If λ is not a real number (the case of a "focus") then all other trajectories unwind from one cycle and wind onto the other. The recurrence functions, corresponding to our two cycles, are complex analytic (for an appropriate choice of of the complex structure in the two-dimensional slice). Indeed, these recurrence functions are differentiably equivalent to complex recurrence functions of separatrices (both in the linear and non-linear cases). Therefore the eigenvalues of our (real two-dimensional) functions are equal to $e^{\pm 2\pi i\lambda}$ and $e^{\pm 2\pi i\lambda^{-1}}$.

The foliations on S^3, corresponding to all foci, are homeomorphic to one another, but not all of them are diffeomorphic: the number $\lambda^2 + \lambda^{-2}$ is an invariant of diffeomorphisms.

If λ is real positive ("node"), then we also are in the Poincaré domain. In this case the space of S^3 between two linked cycles is fibered into two-dimensional tori, filled in by windings with winding number equal to λ (for all tori).

Consider now the non-linear system (9).

In the case of a focus a resonance is impossible, therefore the foliation in S^3, corresponding to the non-linear system, is diffeomorphic to the above-described foliation of the linear system. The same is true for a non-resonant node, i.e. for all $\lambda > 0$, except for the case when λ or $1/\lambda$ is an integer number.

Consider, for instance, the resonance $\lambda = 2$. In this case (if the coefficient c_1 in (5) is not equal to 0) the system (9) has only one separatrice, and system on S^3 only one cycle. Replace $\lambda = 2$ by a close non-real value. The obtained system in S^3 should be close to a resonant system with one cycle. On the other hand, it should be diffeomorphic to the above-considered system corresponding to the linear focus, and therefore should have two cycles with linking number 1.

One can show that one of these cycles, C_1, is close to the unique cycle of the resonant system. The other cycle, C_2, lies in a thin torus with the axial loop C_1 and closes after two rotations along C_1 and one rotation along the meridian (so that the linking number of C_2 and C_1 is equal to 1).

So, the transformation of the system in S^3 at the passage through the resonance $\lambda = 2$ consists in the bifurcation of a two-fold periodic trajectory from a periodic trajectory with eigenvalues $-1, -1$.

The author is thankful to N.N. Brushlinskaya, A.D.Bruno and A.G. Koushnirenko for useful discussion, and also to the referee for valuable remarks.

References

[1] Singularities of differentiable mappings. Collection of translations to Russian[2]. Moscow, Mir Publ., 1968, 268 p

[2] Arnold V.I. Singularities of differentiable mappings. Russian Math. Surveys, 1968, **23**(1), 3–44

[3] Poincaré H. Thèse, 1879. In: Oeuvres de Henri Poincaré. Vol. 1. Paris, 1928, 132+382 p

[4] Siegel C.L., Ueber die Normalform analytischer Differentialgleichungen in der Nahe einer Gleichgewichtslosung; Nachr. Acad. Wiss. Goettingen, Math. Phys. Kl, IIA, 21–30 (1952)

[5] Landau L.D., Lifshitz E.M. Mechanics of Continuous Media: Fluid Mechanics and Theory of Elasticity, Moscow, OGIZ, GITTL, 1944

[6] Naimark Ju.I., On some cases of dependence of periodic motions on parameters, Dokl. Akad. Nauk SSSR 129(4), 1959, 736–739

[7] Brushlinskaya N.N., Qualitative integration of a system of n differential equations in a region containing a singular point and a limit cycle, Dokl. Akad. Nauk SSSR, 1961, 139(1), 9–12

[8] Sacker R.J. On Invariant Surfaces and Bifurcation of Periodic Solutions of Ordinary Differential Equations. New York University, 1964

[9] Bruno A.D. On divergence of transformations of differential equations to normal forms. Dokl. Akad. Nauk SSSR, 1967, 174(5), 1003–1006

[10] Hopf E., Abzweigung einer periodishen Lösung von einer stationären Lösung. Berich. Sächs. Acad. Wiss., Leipzig, Math.-Phys. Kl., 1942, 94(19), 3–22

---

[2] Containing articles by J.Mather, R.Thom. B. Malgrange, B. Morin, and others

Braids of algebraic functions and the cohomology of swallowtails

V.I. Arnold
Translated by Gerald Gould

There is an interesting connection between the theory of algebraic functions and Artin's braid theory: the space G_n of nth-degree polynomials not having multiple roots is the space $K(\pi, 1)$ for the group $B(n)$ of braids on n strands:

$$\pi_1(G_n) = B(n), \qquad \pi_i(G_n) = 0 \quad \text{for} \quad i > 1. \tag{1}$$

This connection can be used in both directions: both for the study of braid groups and for the study of algebraic functions. Here are some examples.

A) Along with the monodromy group, which describes the rearrangements of the leaves of a function when going round its ramification locus, there is a finer invariant of an algebraic function, namely, the *braid group of the function*. This group takes into account not only the rearrangement of the function values after going round the ramification locus, but also how they go round each other in the plane of function values. The monodromy group is a representation of the fundamental group of the complement of the ramification manifold in the permutation group. The braid group of an algebraic function is a representation of the same fundamental group in the Artin braid group.

B) The space G_n can be regarded as the *space of hyperelliptic curves of degree n*.

On the one hand, one can derive from this remark the representation of the braid group in the group of symplectic integer-valued matrices (namely, matrices of automorphisms of the homology of a curve induced by contours in the coefficient space). It can be shown that this representation is a representation on the entire symplectic group in the cases $n = 3, 4, 6$ and only in those cases.

On the other hand, we obtain information on the branching of hyperelliptic integrals as functions of the paramters: relations between the Picard-Lefschetz matrices follow from the relations between the generators of the braid group.

C) The space G_n can be regarded as the *set of regular values of the map* Σ^{1n}. Thus, the relation (1) and the theorems stated below provide us with information on the topology of the simplest singularities of complex analytic maps.

Translation of Uspekhi Mat. Nauk **23**, No. 4 (1968), 247–248

Originally publ. in: Usp. Mat. Nauk 23:4, 247-248, © Russ. Acad. Sciences 1968
Translated by G. Gould. Springer-Verlag Berlin Heidelberg 2014 with kind permission, DOI 10.1007/978-3-642-31031-7_15

D) The space G_n can be regarded as the *complement of the ramification manifold of the universal n-valued entire algebraic function* $z(a)$:

$$z^n + a_1 z^{n-1} + \cdots + a_n = 0 \tag{2}$$

(root of a general nth-degree equation as a function of its coefficients).

It is clear from this remark what the significance of the cohomology of G_n is. Indeed, an n-valued entire algebraic function (respectively, algebroid, pseudo-algebraic function) $z(x)$ is induced by a polynomial (respectively, analytic, continuous) map $a(x)$ of the space of arguments x into the space of arguments a of the universal function (2). Under this map the cohomology classes of G_n induce special cohomology classes in the complement of the ramification manifold of the function $z(x)$. On the other hand, it follows from (1) that the cohomology of G_n coincides with the cohomology of the braid group on n strands (the action of $\mathbb{Z}$ is trivial):

$$H^i(G_n, \mathbb{Z}) = H^i(B_n, \mathbb{Z}).$$

E) *Cohomology of the braid group.* The complex and algebraic structure of G_n turn out to be very useful for the study of the cohomology of braid groups.

First of all we point out that G_n is an n-dimensional Stein manifold (because G_n is given in the space $\mathbf{C}^n$ of all nth-degree polynomials of the form (2) by the polynomial condition $\Delta(a) \neq 0$, where Δ is the discriminant). Consequently, $H_i(G_n) = H^i(B(n)) = 0$ for $i > n$. Further results are obtained in a more detailed study of the geometry of stratified manifolds $\Delta(a) = 0$ (these manifolds can be called multidimensional swallowtails, since the case $n = 4$ corrresponds to the surface "queue d'aronde").

Table of braid cohomology groups $H^i(B(n), \mathbb{Z})$, $n < 12$

n \ i	0	1	2	3	4	5	6	7	8	9
2, 3	$\mathbb{Z}$	$\mathbb{Z}$	**0**	**0**	**0**	**0**	**0**	**0**	**0**	**0**
4, 5	$\mathbb{Z}$	$\mathbb{Z}$	**0**	$\boxed{\mathbb{Z}_2}$	**0**	**0**	**0**	**0**	**0**	**0**
6, 7	$\mathbb{Z}$	$\mathbb{Z}$	**0**	$\mathbb{Z}_2$	$\boxed{\mathbb{Z}_2}$	**0**	**0**	**0**	**0**	**0**
8, 9	$\mathbb{Z}$	$\mathbb{Z}$	**0**	$\mathbb{Z}_2$	$\mathbb{Z}_2$	$\boxed{\mathbb{Z}_6}$	$\mathbb{Z}_3$	$\mathbb{Z}_2$	**0**	**0**
10, 11	$\mathbb{Z}$	$\mathbb{Z}$	**0**	$\mathbb{Z}_2$	$\mathbb{Z}_2$	$\mathbb{Z}_6$	$\boxed{\mathbb{Z}_{6?3}}$	$\mathbb{Z}_{2?1}$	$\mathbb{Z}_2$	$\mathbb{Z}_5$

FINITENESS THEOREM. *The cohomology groups are finite apart from* H^0 *and* H^1. *Here* $H^i(B(n)) = 0$ *for* $i > n$.

REPETITION THEOREM. *All the braid cohomology groups of odd number of strands are the same as for the preceding even number of strands:*

$$H^i(B(2n+1)) \cong H^i(B(2n)).$$

STABILIZATION THEOREM. *As n increases, the ith cohomology group of the braid group of n strands stabilizes: $H^i(B(n)) \cong H^i(B(2i-2))$ for $n \geq 2^i - 2$.*

Thus, the first stable cohomology groups (they are singled out in the above table) are $\mathbb{Z}, \mathbb{Z}, \mathbf{0}, \mathbb{Z}_2, \mathbb{Z}_2, \mathbb{Z}_6, \ldots$.

Similar theorems hold for the cohomology groups of spaces of polynomials having at most k roots of multiplicity q.[1]

References

[1] V. I. Arnol'd, *Cohomology ring of coloured braids*, Mat. Zametki **4** (1968), no. 6.

[2] V. I. Arnol'd, *A remark on the ramification of hyperelliptic integrals as functions of the parameters*, Functional. Anal. i Prilozhen. **2** (1968), no. 3.

Submitted on April 11, 1968

[1] A more detailed exposition of this report will be published in "Trudy Moskov. Mat. Obshch." **21**(1969).

Hamiltonian nature of the Euler equations in the dynamics of a rigid body and of an ideal fluid

V.I. Arnold
Translated by Gerald Gould

The following simple arguments in the theory of Lie groups (given below) imply new conservation laws in the hydrodynamics of an ideal fluid.

1 Euler equation

Let G be a Lie group, $\mathfrak{A}$ its Lie algebra, $\mathfrak{A}^*$ its dual space, and $[\ ,\ ]$ the Lie commutator in $\mathfrak{A}$. Each element $a \in \mathfrak{A}$ acts on $\mathfrak{A}^*$ as a linear operator Ad_a^*. We denote $Ad_a^*\xi$ by $\{a,\xi\}$ (where $a \in \mathfrak{A}$, while ξ and $\{a,\xi\}$ belong to $\mathfrak{A}^*$). Then, by definition,

$$(b,\{a,b\}) = ([a,b],\xi).$$

Let H be a function on $\mathfrak{A}^*$. The cotangent space to $\mathfrak{A}^*$ at each point is naturally identified with $\mathfrak{A}$. The equation

$$\dot{M} = \{\omega,M\}, \quad \text{where } M \in \mathfrak{A}^*,\ \omega(M) = dH|_M \in \mathfrak{A}. \tag{1}$$

is called the *Euler equation for a rigid body with group G and energy H* (see [1]).

In the case of an ordinary rigid body $G = SO(3)$, and M and ω are the angular momentum and the angular velocity respectively. In the case of an ideal fluid $G = S\mathrm{Diff}\, D$ is the group of diffeomorphisms of the Riemannian domain D that preserve the volume element. In this case ω is the velocity field and M is the circulation.

Translation of Uspekhi Mat. Nauk **24**, No. 3, (1969), 225–226.

Originally publ. in: Usp. Mat. Nauk 24:3, 225-226, © Russ. Acad. Sciences 1969
Translated by G. Gould. Springer-Verlag Berlin Heidelberg 2014 with kind permission, DOI 10.1007/978-3-642-31031-7_16

2 Hamilton equation

The orbits of the group G action in $\mathfrak{A}^*$ are invariant manifolds of equation (1) (law of conservation of momentum). On each of these orbits there is a natural symplectic structure (see [2]). Namely, let ξ and η be vectors tangent to the orbit V at the point $M \in \mathfrak{A}^*$. Then in the algebra there are f and g in $\mathfrak{A}$ for which

$$\xi = \{f,M\}, \quad \eta = \{g,M\}.$$

We set

$$\Omega(\xi,\eta) = ([f,g],M).$$

The value of $\Omega(\xi,\eta)$ does not depend on the choice of the representatives f and g. The differential 2-form Ω on the orbit V is non-degenerate and closed and therefore defines the structure of a symplectic manifold on V.

On the symplectic manifold (V,Ω) any function F defines a Hamiltonian vector field X_F in accordance with the formula "$\Omega(\xi,X_F) = (dF,\xi)$ for each vector ξ tangent to V". The equation $\dot{x} = X_F$, $x \in V$ is called the *Hamilton equation with Hamilton function F*. The following result is straightforward.[1]

Theorem. The Euler equation (1) on each orbit V is Hamiltonian with Hamilton function H.

Proof. Each vector ξ tangent to V at the point M has the form

$$\xi = \{f,M\}.$$

In particular, the Euler field (1) has the form $X = \{dH,M\}$. It follows from the definition of Ω and $\{\ ,\ \}$ that

$$\Omega(\xi,X) - ([f,dH],M) = (dH,\{f,M\}) = (dH,\xi),$$

as required.

3 Integral invariants and first integrals

Applying general theorems of Hamiltonian dynamics to equation (1) we get the following result.

Corollary 1. On each orbit V Euler's equation (1) has an invariant 2-form Ω and also an exterior power of Ω.

Let M be an equilibrium point of equation (1) on V. Applying standard methods [3] we get the next result.

[1] This fact was also noted by A. M. Vershik.

Corollary 2. An elliptic non-degenerate equilibrium of Euler's equation (1) on an orbit V is a density point of conditionally-periodic trajectories. In a neighbourhood of such a point M equation (1) has n formal first integrals on V (where $2n = \dim V$).

If the equilibrium M is not isolated on V, so that there exists a vector ξ at M tangent to V and to the manifold of equilibria of equation (1), then we have the following corollary.

Corollary 3. Linearized Euler's equation (1) in a neighbourhood of M on V has a linear first integral.

Namely, such an integral is

$$\Phi(\eta) = \Omega(\xi, \eta).$$

4 Hydrodynamical case

Applying the above to the equations of an ideal fluid one obtains the conservation for quantities of an unusual nature, Ω and I. The quantity Ω is a functional of three infinitesimally close isovorticed fields. The corresponding integral

$$I = \iint \Omega$$

is taken over a two-dimensional chain in a functional space. The form Ω is preserved in the linearized problem, and the integral I is preserved exactly.

Example. Consider flows of an ideal fluid in a bounded simply connected domain D of the Euclidean (x, y)-plane. The Lie algebra $\mathfrak{A}$ of the group $G = S\mathrm{Diff}\, D$ can be identified with the algebra of stream functions that vanish on the boundary; here the Lie commutator becomes the Jacobian of the stream functions

$$J(\phi, \psi) = \frac{\partial(\phi, \psi)}{\partial(x, y)}.$$

The dual space $\mathfrak{A}^*$ can be identified with curl functions r. The operation $\{\ ,\ \}$ then reduces to the Jacobian of the stream function and the curl function.

Consider some initial velocity field $v \in \mathfrak{A}$ and two infinitesimally close fields $v_1 = v + \delta v_1$ and $v_2 = v + \delta v_2$ in $\mathfrak{A}$. These fields are isovorticed with v if their curls can be represented in the form $r + \delta r_i$, where $\delta r_i = J(\delta\phi_i, r)$, $\delta\phi_i \in \mathfrak{A}$. Under the fluid motion with initial conditions v, v_1 and v_2 the velocity fields will remain isovorticed all the time, so that for all t we shall have $\delta r_i = J(\delta\phi_i, r)$ (with the fields $\delta\phi_i$ dependent on t). The following quantity will be preserved (in the linearized problem):

$$\Omega(\delta v_1, \delta v_2) = \iint rJ(\delta\phi_1, \delta\phi_2)\,dxdy.$$

The author thanks V. I. Yudovich for useful discussions.

References

[1] V. I. Arnol'd, *Sur la géométrie différentielle des groupes de Lie de dimension infinie et ses applications à l'hydrodynamique des fluides parfaits*, Annales Inst. Fourier **16**, No. 1, 319–361 (1966).

[2] A. A. Kirillov, *Characters of unitary representations of Lie grouops*, Functional. Anal. i Prilozhen. **2** No. 1, 40–55 (1968).

[3] V. I. Arnol'd, *Small denominators and problems of stability of motion in classical and celestial mechanics*, Uspekhi Mat. Nauk **18**, No. 6, 91–196 (1963).

Submitted on February 17, 1969

On the one-dimensional cohomology of the Lie algebra of divergence-free vector fields and rotation numbers of dynamical systems

V.I. Arnold
Translated by Victor Vassiliev

1. Definitions

Let $\mathfrak{A}$ be a Lie algebra. The *commutant* $[\mathfrak{A},\mathfrak{A}]$ is the set of finite sums of commutators of elements of $\mathfrak{A}$. The one-dimensional cohomology group $H^1(\mathfrak{A},\mathbb{R})$ is the linear space dual to $\mathfrak{A}/[\mathfrak{A},\mathfrak{A}]$.

Let G be the group of diffeomorphisms, preserving the volume element τ of an infinitely differentiable compact connected n-dimensional manifold M. The corresponding Lie algebra $\mathfrak{A}$ consists of divergence-free C^∞-vector fields on (M,τ).

A divergence-free vector field X on (M,τ) is also called a *dynamical system* (X,M,τ). The *rotation class* [1], [2] of a dynamical system X is the one-dimensional homology class $\lambda(X) \in H_1(M,\mathbb{R})$ defined in the following way.

The natural isomorphism between dynamical systems on (M,τ) and closed $(n-1)$-forms on M allows us to associate a cohomology class $h \in H^{n-1}(M,\mathbb{R})$ with any dynamical system X. The Poincaré isomorphism $D: H^{n-1} \to H_1$ takes h into the rotation number $\lambda(X)$. The map $\lambda: \mathfrak{A} \to H_1$ is a homomorphism of Lie algebras (where H_1 is a trivial algebra).

Theorem 1. *The commutant of* $\mathfrak{A}$ *is the Lie algebra of all dynamical systems with zero rotation number:* $[\mathfrak{A},\mathfrak{A}] = \ker\lambda$, $H^1(\mathfrak{A},\mathbb{R}) = H^1(M,\mathbb{R})$.

2. Proof

Consider the corresponding local problem. Denote by I^n the unit closed cube in $\mathbb{R}^n = \{x = (x_1,\dots,x_n)\}$. Let us call a function (or a form, or a vector field, ...)

Moscow State University. Translated from "Ob odnomernyh kogomologijah algebry Li bezdivergentnyh vektornyh polej i o chislah vrashhenija dinamicheskih sistem," Funktsional'nyi Analiz i ego Prilozheniya, 1969, Vol. 3, No. 4, pp. 77–78. Original article submitted May 15, 1969.

Originally publ. in: Funkts. Anal. Prilozh. 3:4, 77-78, © Russ. Acad. Sciences 1969
Original English transl.: Funct. Anal. Appl. 3, 319-321. © Consultants Bureau, New York 1969
Translated by V. Vassiliev. Springer-Verlag Berlin Heidelberg 2014 with kind permission, DOI 10.1007/978-3-642-31031-7_17

finite, if it vanishes in a neighborhood of the boundary of the cube. Denote by $\mathfrak{A}_0$ the Lie algebra of finite divergence-free C^∞-vector fields in I^n.

Lemma 1. *If $n \neq 2$ then the Lie algebra $\mathfrak{A}_0$ coincides with its commutant.*

Let $n > 2$. The Theorem follows from the Lemma, since by a partition of unity any divergence-free field X in M can be represented as a finite sum of divergence-free fields with small supports.

The proof of the Lemma is based on an explicit representation of any vector field as a sum of Poisson brackets $\{P_i, Q_i\}$, where the fields P_i *are fixed*. The fields Q_i are then determined by linear equations, which can be solved. Here we can take for P_i even non-finite divergence-free vector fields, because any divergence-free field in I^n can be made finite by changing it in an arbitrarily small neighborhood of the boundary, where anyway $Q_i \equiv 0$.

Let $X \in \mathfrak{A}_0$. The field X is associated with a closed $(n-1)$-form. This form can be represented as $d\Omega$, where the $(n-2)$-form Ω is finite (by the Poincaré lemma and the homology of the sphere). The form Ω can be split into basic ones: $\Omega = a_{12}dx_3 \wedge \ldots \wedge dx_n + \ldots$. The coefficients a_{ij} are finite. If the integral of a_{12} is equal to 0, then the field corresponding to the summand with a_{12} can de represented as a sum of Poisson brackets of basic fields $(P_i = e_i)$ with fields $Q_i \in \mathfrak{A}_0$, since this problem can be reduced[1] to the representation of a finite function, whose integral is equal to 0, in the form of the divergence of a finite field, which can be done with the help of Poincaré lemma. So, *the commutant* of $\mathfrak{A}_0$ *contains all fields X, such that the integrals of all their coefficients a_{ij} are equal to zero.*

The condition $n \geq 3$ is exploited in the following way in the fight with integrals of a_{ij}.

Consider vector fields $Y = x_3e_2 - x_2e_3$, $Z = v_1e_3 - v_3e_1$, where $v_i = \partial v/\partial x_i$, v a finite function. The fields Y and Z are divergence-free, and Z is finite. Poisson bracket $\{Y, Z\}$ is the vector field corresponding to the $(n-2)$-form Ω, all whose coefficient are equal to zero, except for $a_{12} = x_3v_3$, $a_{23} = x_3v_1$, and $a_{13} = x_2v_3$. All integrals of a_{ij} vanish except for $\int a_{12}\tau = -\int v\tau$.

Hence, *the commutant of $\mathfrak{A}_0$ contains the fields X with any prescribed values of integrals of a_{ij}.* Lemma is proved.

By the way, we see that the second commutant of $\mathfrak{A}$ (or $\mathfrak{A}_0$) coincides with the first one. In a similar way one can verify that the commutant of the Lie algebra of *all* vector fields on a manifold coincides with this algebra (cf. [3]).

3. Symplectic case

In the case $n = 2$ Lemma 1 is not valid. However, it is better to consider this case within a different series of algebras.

[1] here half-page of calculations is omitted

Let (M^{2n}, ω^2) be a symplectic manifold, i.e. ω^2 is a closed non-degenerate 2-form on M^{2n}. The corresponding Lie algebra $\mathfrak{A}$ is called the algebra of hamiltonian vector fields.

Theorem 2[2]**.** *The commutant of $\mathfrak{A}$ is the Lie algebra of hamiltonian fields with single-valued Hamiltonians*: $\mathfrak{A}/[\mathfrak{A},\mathfrak{A}] = H^1(M,\mathbb{R})$; $H^1(\mathfrak{A},\mathbb{R}) = H_1(M,\mathbb{R})$.

Lemma 2. *The commutant of algebra $\mathfrak{A}_0$ of finite hamiltonian fields (in the cube I^{2n} of the space $\mathbb{R}^{2n}$, $\omega^2 = dp_1 \wedge dq_1 + \ldots + dp_n \wedge dq_n$) is the Lie algebra of hamiltonian fields defined by finite Hamilton functions with zero mean value.*

Lemma 3. *Any function on a connected manifold M with volume element τ, whose mean value is equal to zero, can be represented as a sum of functions, also with zero mean values, with arbitrarily small supports.*

Lemma 3 follows immediately from the connectedness of the nerve of any cover of M (see also [4]). Theorem 2 follows from Lemmas 2 and 3 and Darboux theorem [5] in the same way how theorem 1 follows from Lemma 1.

Simultaneously we notice that *the commutant of the Lie algebra of Hamilton functions on (M,ω) coincides with the Lie algebra of functions on M with zero mean value.*

4. Two-dimensional case

On a two-dimensional manifold the sets of hamiltonian and divergence-free fields coincide. Therefore Theorem 1 for $n = 2$ follows from Theorem 2. Notice that in the two-dimensional case Theorems 1 and 2 look as follows[3].

Theorem 3. *Any 2-form α on a connected compact two-dimensional manifold M, whose integral is equal to 0, is induced from the standard form $\omega = dp_1 \wedge dq_1 + \ldots + dp_n \wedge dq_n$ in some Euclidean space $\mathbb{R}^{2n}$ by a map $f: M \to \mathbb{R}^{2n}$, so that $\alpha = f^*\omega$.*

For instance, *any function on the two-dimensional sphere with zero mean value can be represented as a sum of three Jacobians of maps of the sphere to the plane.*

5. Remark

Any hamiltonian field is divergence-free, and therefore has a rotation class. The problem arises: (A) *does any hamiltonian field with zero rotation class have a single-valued Hamiltonian?* Equivalent questions: (B) *does any homology class in*

[2] This theorem was also proved by Palais and Calabi

[3] This formulation is inspired by a V.A. Rokhlin's talk concerning the works by Nash on embeddings of Riemannian metrics

$H_1(M,\mathbb{R})$ *contain a hamiltonian field?* (C) *Is the operator of multiplication by* ω^{n-1} *an isomorphism* $H^1 \to H^{2n-1}(M^{2n},\mathbb{R})$*?* The answer is affirmative, if (M^{2n},ω^2) admits a Kaehler structure.

References

[1] Swartzman S., *Ann. Math.*, 1957, **66**(2), 270–289
[2] Gelfand I.M., Piatetsky-Shapiro I.I. On a theorem by Poincaré. *Dokl. Ac. Sci. USSR*, 1959, **127**(3), 490–493 (in Russian)
[3] Gelfand I.M., Fuchs D.B. Cohomology of the Lie algebra of vector fields on a circle. *Functional Analysis and Its Applications*, **2**(4), 342–343
[4] Moser J.K. On the volume elements on a manifold. *Trans. Amer. Math. Soc.*, 1965, **120**(2), 286–294
[5] Darboux G. Sur le problème de Pfaff. *Bull. Sci. Math. Astron.*, 1882, **6**, 14–30, 49–68

The cohomology ring of the colored braids group

V.I. Arnold
Translated by Victor Vassiliev

The cohomology ring of the space of ordered subsets of n different points of the plane is calculated.

The group $I(n)$ of colored Artin braids is defined as the fundamental group of the space M_n of ordered collections of n pairwise distinct points of the plane[1] It is easy to show that M_n is the $K(\pi,1)$ space for the group $I(n)$:

$$\pi_1(M_n) = I(n), \qquad \pi_i(M_n) = 0 \quad \text{for } i > 1.$$

Therefore the cohomology ring of the group $I(n)$ coincides with that of the space M_n (one assumes the trivial action on $\mathbb{Z}$):

$$H^*(I(n)) \simeq H^*(M_n, \mathbb{Z}).$$

In this article that cohomology ring is described. We use the realization of M_n as the complex affine space $\mathbb{C}^n = \{z = (z_1, \dots, z_n)\}$ with "removed diagonals":

$$M_n = \{z \in \mathbb{C}^n : z_k \neq z_l \ \forall k \neq l\}.$$

Denote by $A(n)$ the exterior graded ring generated by $\binom{n}{2}$ one-dimensional elements $\omega_{k,l} = \omega_{l,k}$, $1 \leq k \neq l \leq n$ which satisfy the $\binom{n}{3}$ relations

$$\omega_{k,l}\omega_{l,m} + \omega_{l,m}\omega_{m,k} + \omega_{m,k}\omega_{k,l} = 0. \tag{1}$$

Moscow State University. Translated from Mat. Zametki 5:2, 227-231,
Original article submitted April 29, 1968.

[1] The name is motivated by another definition: $I(n)$ is the kernel of the natural homomorphism $B(n) \to S(n)$ of the braid group of n strings to the symmetric group of permutations of endpoints of these strings. In other words, $I(n)$ consists of braids whose strings are individualized (each colored its own color) and end where they begin.

Originally publ. in: Mat. Zametki 5:2, 227-231, © Russ. Acad. Sciences 1969
Original English transl.: Math. Notes 5, 138-140, © Consultants Bureau, New York 1969
Translated by V. Vassiliev. Springer-Verlag Berlin Heidelberg 2014 with kind permission, DOI 10.1007/978-3-642-31031-7_18

Theorem 1. *The cohomology ring of the colored braid group is isomorphic to $A(n)$. The isomorphism $H^*(M_n, \mathbb{Z}) \simeq A(n)$ is defined by formulas*

$$\omega_{k,l} = \frac{1}{2\pi i}\frac{dz_k - dz_l}{z_k - z_l}. \tag{2}$$

In other words, the one-dimensional generators $\omega_{k,l}$ correspond to rotations around diagonals $z_k = z_l$.

Corollary 1. *The cohomology groups of the colored braid group are torsion-free.*

Corollary 2. *The Poincaré polynomial of the manifold M_n is equal to*

$$p(t) = (1+t)(1+2t)\cdots(1+(n-1)t).$$

In other words, the cohomology groups of the manifold M_n (or group $I(n)$) are the same as of the direct product of the circle, the wedge of two circles, ..., wedge of $n-1$ circles.

Corollary 3. *An additive basis of the ring $A(n)$ consists of all products of the form*

$$\omega_{k_1,l_1}\omega_{k_2,l_2}\cdots\omega_{k_p,l_p}, \qquad \text{where } k_s < l_s, \quad l_1 < l_2 < \cdots < l_p. \tag{3}$$

Corollary 4. *The subring of the ring of exterior differential forms generated by C_n^2 forms (2) is isomorphic to $A(n)$.*

Corollary 5. *An exterior polynomial on differential forms (2) is cohomologous to 0 in M_n if and only if it is equal to 0.*

Corollary 6. *The symmetrization of any exterior polynomial of degree ≥ 2 in differential forms (2) is equal to 0.*

Example. There is a non-evident identity

$$\sum_{120} \omega_{1,2} \wedge \omega_{2,3} \wedge \omega_{3,4} \wedge \omega_{4,5} = 0,$$

where the sum is taken over all 120 permutations of indices 1, ..., 5.

It is easy to prove

Lemma 1. *There is a fiber bundle $M_n \xrightarrow{p} M_{n-1}$; its fiber is the plane without $n-1$ points. The action of the fundamental group of the base M_{n-1} in the cohomology of the fiber is trivial. The fiber bundle p has a cross-section.*

Indeed, set $p(z_1,\ldots,z_n) = (z_1,\ldots,z_{n-1})$. Then the fiber F_{n-1} is equal to $\{z \in \mathbb{C} : z \neq z_1,\ldots,z_{n-1}\}$. The fiber F_{n-1} is homotopy equivalent to the wedge of $n-1$ circles. The one-dimensional (co)homology group of the fiber is isomorphic to $\mathbb{Z} + \cdot \ldots \cdot + \mathbb{Z}$ ($n-1$ times). The fundamental group of the base is the group $I(n-1)$ of colored braids with $n-1$ strings. Its action in the fiber is the usual action of the braid group in the plane with removed points. But the braids of $I(n-1)$ are colored, therefore they do not permute the removed points. Therefore $I(n-1)$ acts trivially in the (co)homology of the fiber. The cross-section can be defined by the formula

$$z_n = \frac{z_1 + \cdots + z_{n-1}}{n-1} + 2 \max_{1 \le i,j \le n-1} |z_i - z_j| + 1.$$

The simple proof of Theorem 1 given below is due to D.B. Fuchs.

Consider the cohomology spectral sequence of the fiber bundle $M_n \to M_{n-1}$. The group $\pi_1(M_{n-1})$ acts trivially in the cohomology of the fiber F_{n-1}, therefore the term $E_2^* = H^*(M_{n-1}, H^*(F_{n-1}))$ is the same as if the fiber bundle were a direct product. The only possible differential d_2 is equal to 0. (This fact follows easily from the existence of the cross-section.) Therefore $E_2 = E_\infty$. So, the (co)homology groups of M_n are the same as if it were a direct product of M_{n-1} and F_{n-1}. Setting consecutively $n = 2, 3, \dots$ ($M_1 = \mathbb{C}$) we obtain that the (co)homology groups of M_n are the same as those of the direct product of the circle, lemniscate, ..., wedge of $n-1$ circles. Corollaries 1 and 2 are proved.

Let us construct an additive basis of $H^*(M_n, \mathbb{Z})$. Our spectral sequence implies that such a basis can be obtained from the image of an additive basis of $H^*(M_{n-1}, \mathbb{Z})$ under the map p^* by adding to this image the products of elements from this image with $n-1$ one-dimensional cohomology classes which are mapped into generators of $H^1(F_{n-1}, \mathbb{Z})$ by the homomorphism i^* (where $i : F_{n-1} \hookrightarrow M_n$). Notice that for such one-dimensional classes one can take the cohomology classes of differential forms $\omega_{1,n}, \omega_{2,n}, \dots, \omega_{n-1,n}$ from (2). Setting consecutively $n = 2, 3, \dots$ we obtain that the products (3) form an additive basis of $H^*(M_n, \mathbb{Z})$.

The differential forms (2) satisfy relations (1) as can be verified by substitution. The cohomology classes of these differential forms (2) a fortiori satisfy these relations. We can therefore construct a ring homomorphism $\varphi : A(n) \to H^*(M_n, \mathbb{Z})$ by sending the generators $\omega_{k,l} \in A(n)$ to differential forms in accordance with formula (2). Above we have proved that φ is surjective. Let us prove that it has no kernel. It is easy to prove

Lemma 2. *The ring $A(n)$ is additively generated by products (3).*

Indeed, the skew commutativity implies that $A(n)$ is generated by products $\omega_{k_1,l_1} \cdot \ldots \cdot \omega_{k_p,l_p}$, where $k_s < l_s, l_s \le l_{s+1}$. The relation (1) allows us to remove products having equal indices l. For instance,

$$\omega_{k_1,l}\omega_{k_2,l} = \omega_{k_1,k_2}\omega_{k_2,l} - \omega_{k_1,k_2}\omega_{k_1,l}.$$

In all summands the indices of the first factor are less than l. Therefore all products of $\omega_{k,l}$ can be expressed as sums of products where $k_s < l_s$, $l_s < l_{s+1}$. (See (3).) The lemma is proved.

This implies that the kernel of the ring homomorphism $\varphi : A(n) \to H^*$ is trivial. Indeed, the products (3), generating $A(n)$ additively, go into independent elements of H^* (above we have proved that they form an additive basis in H^*). Therefore the kernel of φ is trivial, and φ is a ring isomorphism. Theorem 1 is proved.

Simultaneously Corollary 3 is proved, since we already know that products (3) form an additive basis in the ring H^*. Corollaries 4 and 5 follow from the fact that, on one hand, cohomology classes of forms, generated by (2), form the ring $H^*(M_n, \mathbb{Z})$ isomorphic to $A(n)$, and on the other hand, the differential forms (2) themselves satisfy the relations (1).

Corollary 6 follows from Corollary 5 and the finiteness of cohomology groups $H^i(B(n))$, $i > 1$ ($B(n)$ is the braid group with n strings).

Remark. Let M be a manifold obtained from $\mathbb{C}^n$ by removing an arbitrary finite collection of hyperplanes:

$$M = \{z \in \mathbb{C}^n : a_k(z) \neq 0, k = 1, \ldots, N\}.$$

Probably the ring $H^*(M, \mathbb{Z})$ has no torsion and is generated by one-dimensional classes $\omega_k = \frac{1}{2\pi i} \frac{da_k}{a_k}$ and an exterior polynomial in the ω_k is cohomologous to 0 in H^* only if that polynomial is equal to zero.

The author is thankful V.P. Palamodov and D.B. Fuchs for useful discussions.

References

[1] V.I.Arnold, On braids of algebraic functions and cohomology of swallowtails. Russian Math. Surveys 23(4), 1968, 247–248

On cohomology classes of algebraic functions invariant under Tschirnhausen transformations

V.I. Arnold
Translated by Victor Vassiliev

There is an analogy between algebraic functions and vector bundles, where the role of the Grassmann manifold is played by the space $K(\pi,1)$ of the braid group. Among the cohomology classes of the ramification variety of an algebraic function, one can distinguish the classes induced from the cohomology classes of the braid group. Some of these classes (including the one-dimensional mod 2 classes and the three-dimensional classes) are invariant under Tschirnhausen transformation. This implies, in particular, that the function $w(a,b,c)$ given by the equation $w^4+aw^2+bw+c=0$ cannot be represented as the Tschirnhausen transformation with polynomial coefficients of any algebraic function $z(u,v)$ of two variables u,v that are polynomials in a,b,c.

1. Cohomology classes of an algebraic function

An *algebraic* (entire n-valued) *function* $z(x)$ of k variables x is induced by a polynomial map $p:\mathbb{C}^k\to\mathbb{C}^n$ of the space of its arguments to the space $\mathbb{C}^n=\{a\}$ of arguments of the universal function $z(a)$ given by the equation

$$z^n+a_1z^{n-1}+\cdots+a_n=0.$$

The *ramification variety* of the universal function $z(a)$ is the variety in $\mathbb{C}^n$ specified by the equation $\Delta(a)=0$, where Δ is the discriminant. The ramification variety $\Sigma(p)$ of the function $z(x)$ induced by the map p is the variety $p^{-1}(\Sigma)$. Their complements will be denoted $G=\mathbb{C}^n\setminus\Sigma$ and $G(p)=\mathbb{C}^k\setminus\Sigma(p)$. The function $z(x)$ is *properly n-valued* if $G(p)$ is not empty. The restrictions of p to $\Sigma(p)$ and $G(p)$ will be denoted simply by p. The map p also induces a map of cohomology groups,

Moscow State University Translated from Funkts. Anal. Prilozh. 4:1, 84–85.
Original article submitted September 29, 1969.

Originally publ. in: Funkts. Anal. Prilozh. 4:1, 84-85, © Russ. Acad. Sciences 1970
Original English transl.: Funct. Anal. Appl. 4, 74-75, © Consultants Bureau, New York 1970
Translated by V. Vassiliev. Springer-Verlag Berlin Heidelberg 2014 with kind permission, DOI 10.1007/978-3-642-31031-7_19

$p^* : H^i(G,\mathbb{Z}) \to H^i(G(p),\mathbb{Z})$. Let $\alpha \in H^i(G)$. The *cohomology class* $\alpha(p)$ *of a* properly n-valued *algebraic function* induced by the map p is defined as the class $p^*\alpha \in H^i(G(p))$.

Some information on cohomology classes of the universal function $H^i(G)$ is given in [1]. In particular, for $n \geq 4$ we have $H^0 = H^1 = \mathbb{Z}$, $H^2 = 0$, $H^3 = \mathbb{Z}_2$. Denote by $\alpha_1 = \mathrm{cl}(2\pi i)^{-1} d \ln \Delta$ the generator of H^1, and by α_3 the generator of H^3. The classes α_1 and α_3 have the following geometrical sense. Let X be an algebraic subvariety in $\mathbb{C}^N$. Denote by ${}_FH$ the homology groups with closed supports, and by $I : H^i(\mathbb{C}^N \setminus X) \to {}_FH_{2N-i-1}(X)$ the Alexander isomorphism. Let Σ^k be the algebraic subvariety in $\mathbb{C}^n$ consisting of all polynomials a having at least k double roots, i.e. having the form $a = a_k^2 a_{n-2k}$. Then $I\alpha_1 = \mathrm{cl}\ \Sigma^1$, $I\alpha_3 = \mathrm{cl}\ \Sigma^2$.

2. Tschirnhausen transformation

Let $\mathbb{C}_a^n$ be the space of polynomials of the form $z^n + a_1 z^{n-1} + \cdots + a_n$ with roots $z_1, \ldots, z_n$. Let $\mathbb{C}_b^n$ be the space of polynomials b of the form $b_1 z^{n-1} + \cdots + b_n$. The *Tschirnhausen map* T sends the pair (a,b) to the polynomial c of the form $w^n + c_1 w^{n-1} + \cdots + c_n$ whose roots are $w_i = b(z_i)$. It is easy to check that $T : \mathbb{C}_a^n \times C_b^n \to \mathbb{C}_b^n$ is a polynomial map.

Let $p : \mathbb{C}^k \to \mathbb{C}_a^n$ be a polynomial map, defining an algebraic function $z(x)$. Let $q : \mathbb{C}^k \to \mathbb{C}_b^n$ be another polynomial map. The *Tschirnhausen transformation of the function* $z(x)$ *with coefficients* q sends $z(x)$ to the algebraic function $w(x)$ defined by the map $r : \mathbb{C}^k \to \mathbb{C}^n$, where

$$r(x) = T(p(x), q(x)).$$

The Tschirnhausen transformation can only enlarge the ramification variety. Therefore there is the natural embedding of complements, $i : G(r) \to G(p)$, inducing a cohomology map $i^* : H^k(G(p)) \to H^k(G(r))$. We will assume that the function $w(x)$ is properly n-valued.

3. Theorem

The cohomology classes $\alpha_1 \bmod 2$ *and* α_3 *are invariant under the Tschirnhausen transformation*:

$$\alpha_1(r) = i^*\alpha_1(p) + 2\beta, \alpha_3(r) = i^*\alpha_3(p).$$

Here $I\beta$ is the fundamental cycle of some hypersurface in $\mathbb{C}^k$.

4. Calculations

Let $f : \mathbb{C}^M \to \mathbb{C}^N$ be a polynomial map, $X \subset \mathbb{C}^N$ an algebraic submanifold, $Y = f^{-1}X$. The Alexander isomorphism transforms the cohomology map $f^* : H^i(\mathbb{C}^N \setminus X) \to H^i(\mathbb{C}^M \setminus Y)$ to the homology map $f^! = If^*I^{-1} : {}_FH_{2N-i-1}(X) \to {}_FH_{2M-i-1}(Y)$. If some quasitransversality condition is satisfied, then $f^!$ can be interpreted as the taking of pre-images accounting for multiplicities.

In particular, take for (f,X) the Tschirnhausen transformation $T : \mathbb{C}^{2n} \to \mathbb{C}^n_c$ and the variety $\Sigma^1 \subset \mathbb{C}^n_c$. Denote by $\Psi^k \subset \mathbb{C}^{2n} = \mathbb{C}^n_a \times \mathbb{C}^n_b$ the set of pairs (a,b) such that the polynomial b takes equal values $b(z_i) = b(z_j)$ at k non-intersecting pairs of roots of the polynomial a. Let $\Sigma^k_1 = \Sigma^k \times \mathbb{C}^n_b \subset \mathbb{C}^{2n}$ be the set of pairs (a,b) such that the polynomial a has at least k double roots. Denote by $[v]$ the fundamental homology class of the closure of v. Counting the multiplicities of components of $T^{-1}(\Sigma^k)$ shows that

$$T^![\Sigma^k] = [\Sigma^k_1] + 2[\Sigma^{k-1}_1 \cap \Psi^1] + 4[\Sigma^{k-2}_1 \cap \Psi^2] + \cdots + 2^k[\Psi^k].$$

The theorem of §3 follows from this formula with $k = 1$ and 2. The invariance of α_1 mod 2 can be obtained also by taking the logarithm of the identity $\Delta(T(a,b)) = \Delta(a)Q^2(a,b)$ (where Q is a polynomial) relating the discriminants of the function and its Tschirnhausen transformation.

5. Corollary

If a properly n-valued algebraic function of an arbitrary number of variables x can be obtained by a Tschirnhausen transformation with polynomial coefficients $b = q(x)$ from an algebraic function $z(u,v)$ in two variables u,v that depend polynomially on x, then the cohomology class α_3 for this function is equal to 0.

Indeed, $H^3(\mathbb{C}^2 \setminus \Sigma) = 0$ for any algebraic curve $\Sigma \subset \mathbb{C}^2$.

The class α_3 for the function $w(a,b,c)$, where $w^4 + aw^2 + bw + c = 0$, is non-trivial. Thus, this function cannot be obtained in the above-described way from algebraic functions in two variables.

6. Remarks

1. The polynomiality of maps p, q, r is not essential for our considerations, therefore the result can be easily generalized to non-algebraic functions (e.g. to algebroid germs).

2. D.B. Fuchs has indicated that the comparison of cohomology of braid groups and symmetric groups allows one to extend the theorem of §3 to some other cohomology classes.

Note added in proof. In a similar way, one can prove that the function in $k = n - 1$ variables, specified by the equation

$$z^n + a_1 z^{n-2} + \cdots + a_k = 0, \qquad n = 2^r, r \geq 2,$$

cannot be represented as a superposition of algebraic functions in a lower number of variables, that has exactly n values. In particular the formula for the solution in radicals of equations of degree 4 defines a 72-valued function.

References

[1] V.I.Arnold, On braids of algebraic functions and cohomology of swallowtails. Russian Math. Surveys 23(4), 1968, 247–248

Trivial problems

V.I. Arnold
Translated by Gerald Gould

Define a problem to be a decomposition of a functional space into subsets. For instance, in the problem of stability of an equilibrium for systems of ordinary differential equations one deals with the decomposition of the space of vector fields with zero at O into two classes: the class of fields for which the point O is stable, and the class of fields for which the point O is unstable.

The problem is called trivial if the decomposition of the k-jets corresponding to this problem is finite and semi-algebraic for each k. In this case the question to which class a given function belongs can be answered with the help of a finite number of arithmetical manipulations with Taylor coefficients, except for the "special case" of infinite codimension.

For instance, the problem of whether the point O is a point of strict local minimum of an infinitely differentiable function is trivial.

In this talk the following hypotheses are discussed:

1. The problem of the asymptotic (or Lyapunov) stability of equilibrium is trivial.
2. The problem of the topological classification of equilibria for the system of real differential equations is trivial. It should be noted that the similar problem considered in the complex domain is not trivial.

Translation of V.I. Arnold – Trivial Problems. Proceedings 5th International Conference on Nonlinear Oscillations (Kiev 1969), Vol. 1. Mathematics Institute of the Ukrainian Academy of Sciences, Kiev. pp. 630–631 (1970)

Originally publ. in: Proc. 5th Int. Conf. on Nonlinear Oscillations, Kiev 1969. Vol. 1, 630-631, © Ukrain. Acad. Sciences, Kiev 1970
Translated by G. Gould. Springer-Verlag Berlin Heidelberg 2014 with kind permission, DOI 10.1007/978-3-642-31031-7_20

LOCAL PROBLEMS OF ANALYSIS

V. I. Arnol'd

Vestnik Moskovskogo Universiteta. Matematika, Vol. 25, No. 2, pp. 52-56, 1970

UDC 517

The aim of this paper is to indicate a number of natural problems which arise in investigating the local behavior of various analytic objects. Let us begin with some examples.

Examples of analytic objects:

1. A real function f which is analytic in a neighborhood of the point 0 of the space R^n.

2. The system of differential equations

$$\dot{x} = v(x), \tag{1}$$

specified by a vector field which is analytic in the neighborhood of point 0.

3. An analytic mapping F of a neighborhood of the point 0 into a neighborhood of 0.

Examples of local problems:

1. Does the function f have a local minimum at point 0?

2. Is the equilibrium position 0 of system (1) (or a stationary point 0 of mapping F) stable (either Lyapunov-stable or asymptotically stable)?

3. Are the following two systems topologically equivalent in a neighborhood of the equilibrium position 0: (1) and

$$\dot{x} = u(x)? \tag{2}$$

This includes, for example, the problem of distinguishing center and focus.

The list of examples could easily be continued; any student of analysis can find many such problems.

A problem will be understood to be a further breakdown of the space of shoots* of functions which are analytic in 0 into a certain number of classes. Thus in Example 1 we are dealing with a partition into two classes: the class of shoots of functions which have a minimum in 0 and of functions which do not have a minimum.

Certain problems may be solved by a finite number of arithmetic operations on coefficients of the Taylor expansion. An example is provided by the problem of the minimum of functions of one variable. These problems are encountered so frequently and have so many common features that they deserve a special name. We will call them trivial problems.

Their formal definition consists in the following. Let S be a space of analytic objects of interest to us (a space of analytic shoots of some mappings at point 0). We consider the corresponding spaces of streams** J and natural projections of the infinite-dimensional space S onto the finite-dimensional spaces of the streams

$$p_k: S \to J_k, \quad S \to \ldots \to J_k \to \ldots \to J_1 \to J_0.$$

The problem is the partition of S into nonintersecting subsets $S = \cup S_i$, where $i \in I$ and I is some set of indices. A problem is called trivial if for each k there exists a partition of the space of streams into nonintersecting semialgebraic*** subsets

$$J_k = J'_k + \sum_{i \in I} J_{ki},$$

such that

*A shoot of a mapping at point 0 is an equivalent class of the mapping of the neighborhoods of 0; two mappings are equivalent if they coincide in some (generally speaking smaller) neighborhood 0. A shoot of an analytic function in 0 may be identified with its Taylor series in 0.

**A stream of a mapping at points 0 is an equivalent class of the mapping of the neighborhoods of 0; two mapping are equivalent if they are tangential of order k, i.e., $|f(x) - g(x)| = O(|x|^k)$. The set J_k of all k-streams in 0 may be identified with the space of segments Taylor series of length k ([1]).

***A semialgebraic set in the neighborhood of each of its points if a finite union of sets each of which is specified by a finite number of polynomial equations ($P_i = 0$) and inequalities ($Q_i \neq 0$) (see, e.g., [1]).

Originally publ. in: Vestn. Mosk. Univ. Ser. I, Mat. Mekh. 25:2, 52-56, © Izdat. Moscow Univ. 1970
English transl.: Moscow Univ. Math. Bull. 25 (1970), 77-80 © Allerton Press, Inc., New York, 1970
Reprint by Springer-Verlag Berlin Heidelberg 2014 with kind permission, DOI 10.1007/978-3-642-31031-7_21

1) $p_k^{-1}J_{ki} \subseteq S_i$;

2) $\lim_{k\to\infty}(\dim J_k - \dim J_k') = \infty$.

In other words, for each k we can determine, by arithmetic operations on the Taylor coefficients up to order k, either to which of the classes S_i a given shoot s belongs (and then all shoots which have a tangent to it of order k), or whether we should include derivatives of order greater than k, i.e., that $s \in p_k^{-1}J_k'$.

Condition 2 means that the set of shoots for which it is impossible to determine by any finite number of derivatives whether they belong to S_i $(\bigcap_{k=1}^{\infty} p_k^{-1}J_k')$, is of codimension ∞ in S; this "neutral" case is thus completely exclusive.

Example. In the problem of the minimum of the function $f(x) = a_1x + a_2x^2 + \ldots$ we have $J_k = J_k' + J_{k1} + J_{k2}$, where the set J_k is specified by the condition $a_1 = \ldots = a_k = 0$ and the set J_{k1} by the condition that the first nonzero coefficient appear with an even power of x not greater than k and positive; the set J_{k2} contains all the remaining cases.

Thus our problem is trivial.

Unlike this simple example, the enumeration of all possible cases is virtually impossible in many trivial problems. Fortunately, the basic issue, i.e., whether or not a given problem is really trivial, may sometimes be solved without writing out all the special cases explicitly.

Example. The Sturm-Tarskii-Seidenberg theorems [2] (the image of a semialgebraic set is semialgebraic under a polynomial mapping) imply that the problem of the local minimum of a function of n variables is trivial.

Moreover, algebraic equations and inequalities for Taylor coefficients which ensure a minimum become rapidly intractable as k and n increase.

In exactly the same way it may be proved, without writing out the conditions for a center explicitly, that the center-focus problem is trivial.

Let us consider some other problems from this point of view.

Problem 1. Is the problem of the Lyapunov ability of the equilibrium position of system (1) trivial? Is the problem of asymptotic stability trivial?

Roughly speaking, we are asking whether the boundary between regions of (asymptotic) stability and instability (for fixed derivatives on the right side up to order k in O) is specified by algebraic equations and inequalities or whether this boundary has a more complex structure.

Problem 2. Is the problem of the existence of an analytic Lyapunov function for (1) trivial?

In view of what was stated above regarding the minimum problem, it would be sufficient to investigate the following problem: is it possible to represent as the derivative of some analytic function in the direction of a given vector field v at least one analytic function, which is positively defined in the neighborhood of O,

$$\varphi = \sum \frac{\partial f}{\partial x_i} v_i > 0 \quad \text{for} \quad x \neq 0?$$

Problem 3. Is the problem of the topological classification of the equilibrium positions of system (1) in n-dimensional space trivial?

It is a consequence of Bendixon's result [3] that this problem is trivial for n = 2.

That the problem of topological classification is trivial means that it is possible to partition the space S of shoots of analytic vector fields in O into a denumerable number of pairwise nonintersecting parts,

$$S = S' \bigcup S_i, \qquad i = 1, 2, \ldots,$$

such that 1) one of the parts, S', is of codimension ∞; 2) each of the remaining parts, S_i, is specified by algebraic conditions on a finite number of Taylor coefficients $S_i = p_k^{-1}J_{ki}$, where J_{ki} is a semialgebraic subset of the space of k-streams J_k; k = k(i); 3) any two systems (1), (2) belonging to one set S_i are topologically equivalent in the neighborhood of O, i.e., there exists a homeomorphism of the neighborhood of O which carries the phase curves of system (1) into the phase curves of system (2).

We should note that we are thus dealing with topological classification to within a set S' of codimension ∞. It is natural to expect that S' will include, e.g., linear systems with more than one pair of purely imaginary eigenvalues; the topological type of this system may be varied by restructuring only the Taylor coefficients of arbitrarily high order.

Example of a nontrivial problem. The problem of topological classification of the phase curves of system (1) in the neighborhood of the equilibrium position 0 in complex space C^n, $n \geqslant 2$, is nontrivial.

This is a consequence of the following facts.

1. An analytic system of type (1), n = 2, is analytically equivalent to its linear part:

$$\dot{z}_1 = \lambda_1 z_1, \quad \dot{z}_2 = \lambda_2 z_2 \tag{3}$$

for almost all (in the sense of the Lebesgue measure) negative values of the number $\lambda = \frac{\lambda_1}{\lambda_2}$ (Siegel's theorem [4]).

2. The topological types of the linear system (3) varies continuously with a change in λ. This is a consequence of the fact that rotations of the neighborhood through various angles are not topologically conjugate [5].

3. The set of systems (1) with fixed linear parts (3) is of finite codimension in the space of all systems.

Classification of special cases. In formulating the problem of topological classification we neglected special cases of codimension ∞. This is due to the following considerations.

Nongeneral cases, i.e., those that belong to a subset of nonzero codimension in the space of all possible cases, disappear under small deformations. For example, at the critical point of a function of "general form" the second differential must not degenerate, while the eigenvalues of the linear part of "general" system (1) must be arithmetically independent, and so forth. Indeed, if such is not the case for our system, it may always be approximated by another one for which all the singular points will be nonsingular (of codimension 0).

The need for considering cases which belong to a subset of nonzero codimension arises in studying families of systems which depend on parameters. For example, in a one-parameter system we may encounter a singularity of codimension 1 which cannot be eliminated by small deformations of the family; in a k-parametric system we may encounter one of codimension k. Here a small deformation of a k-parametric family may be used in such a way that the singularities of codimension l will appear only "transversely" on submanifolds of dimension k - l in the space of the parameters (transversality theorem of Thom [1]). In particular, cases whose codimension is greater than k may be avoided entirely. This implies that the appearance of cases which belong to a set of infinite codimension may always be avoided by a small deformation of a family with an arbitrary finite number of parameters.*

On the basis of what has been said we may conclude that we should always pay attention to the codimensions in classifying special cases. It is, of course, first necessary to study the most important (and usually the simplest) general case (codimension 0).

In studying cases of high codimension we should bear in mind that the greater the codimension of the case, the less frequently it is encountered (and, as a rule, the more difficult it is to study). Therefore investigation of special cases of codimension 2 is meaningless as long as cases of codimension 1 remain undealt with, etc.

Although the principles we have enumerated here are obvious and have been known for a long time,** students of analysis rarely pay attention to them. In the forest of explicitly and implicitly written conditions it is difficult to see what the codimension of the case in question is, or even whether or not it is finite. For example, in the theory of singular points of differential equations (1), cases of codimensions 1 have been quite unsatisfactorily investigated, and this of course largely invalidates a considerable number of investigations of more complex singular points.

We encounter a similar situation in other problems of analysis. For example, in the oblique-derivative problem we deal with a vector field which is specified on the edge of a manifold and is tangent to that manifold. Here the local behavior of the field in the neighborhood of an edge point is important. At a "general" point the field is transversal to the edge; at certain points of the edge, however, the field may be tangential. In the "general case" these points form an edge manifold of codimension 1. At certain points of this submanifold the field may be tangential to it, etc. It may be shown that the general case in this problem will be as follows: in the neighborhood of each edge point there exists a coordinate system $x_1, \ldots, x_n$ in which the vector field has the components (1,0,...,0)

*In addition, the case of codimension ∞ may appear as a result of a particular symmetry of the problem due to conservation laws, etc. Examples are Hamiltonian systems, invertible systems, etc.

**For example, Andronov and his school in Gor'kii systematically studied the "degree of refinements" (codimension) of the special cases.

and the edge is given by the equation

$$x_1^k + x_2x_1^{k-2} + x_3x_1^{k-3} + \ldots + x_k = 0, \qquad 1 \leqslant k \leqslant n.$$

All the more complex cases are of higher codimension; they may be avoided by an arbitrarily small deformation of the field.

There has been a series of investigations of such complex special cases of large codimension in the oblique-derivative problem; the general case (codimension 0), however, has been fully accounted for only in the two-dimensional problem.

Note added in proof. Recently the author has shown that the problem of Lyapunov stability of a neighborhood of the equilibrium position in three-dimensional space is nontrivial. Hence no algebraic algorithm exists either for determining Lyapunov stability or for determining the topological types of a singular point, even if we neglect special cases of infinite codimension.

REFERENCES

1. Collection: Singularities of Differential Mappings [Russian translation], Mir, Moscow, 1968.

2. E. A. Gorin, "Asymptotic properties of polynomials and algebraic functions of several variables," UMN, 16, no. 1, (97), 91-117, 1961.

3. I. Bendixon, "Curves defined by differential equations," UMN, no. 9, 191-253, 1941.

4. K. L. Siegel, "Normal form of analytic differential equations in the neighborhood of an equilibrium position," collection: Matematika [Russian translation], 5:2, 1-19, 1961

5. V. I. Arnol'd, "Singular points of finite codimension in complex dynamic systems," Funkts. analiz i ego prilozheniya, 3, no. 1, 1-6, 1969.

Editor's remark on translation. In this paper, Russian terms translated as "shoot" and "stream" were supposed to be translated as "germ" and "jet" respectively. Furthermore, the first two sentences on the second page of this paper should read:

"In other words, for each k we can determine, by arithmetic operations on the Taylor coefficients up to order k, either to which of the classes S_i a given germ s (and then every germ tangent to it to order k) belongs, or whether we should include derivatives of order greater than k, i.e., that $s \in p_k^{-1}J_k'$.

Condition 2 means that the set $\bigcap_{k=1}^{\infty} p_k^{-1}J_k'$ of germs, for which it is impossible to determine whether they belong to S_i by any number of derivatives, is of codimension ∞ in S; this 'undetermined' case is thus completely exceptional."

1970 г. март—апрель *т. XXV, вып. 2 (152)*
УСПЕХИ МАТЕМАТИЧЕСКИХ НАУК

СООБЩЕНИЯ МОСКОВСКОГО МАТЕМАТИЧЕСКОГО ОБЩЕСТВА

АЛГЕБРАИЧЕСКАЯ НЕРАЗРЕШИМОСТЬ ПРОБЛЕМЫ УСТОЙЧИВОСТИ И ПРОБЛЕМЫ ТОПОЛОГИЧЕСКОЙ КЛАССИФИКАЦИИ ОСОБЫХ ТОЧЕК АНАЛИТИЧЕСКИХ СИСТЕМ ДИФФЕРЕНЦИАЛЬНЫХ УРАВНЕНИЙ

В. И. Арнольд

Решение обеих проблем хорошо известно в случае общего положения, когда среди собственных чисел линейной части нет чисто мнимых. В этом случае топологический тип особой точки определяется количествами собственных чисел в левой и в правой полуплоскости. Поэтому существуют алгебраические критерии, позволяющие определить топологический тип особой точки общего положения конечным числом арифметических действий над коэффициентами Тейлора правой части дифференциального уравнения.

Подобные критерии для определения топологического типа и, в частности, для распознавания устойчивости по Ляпунову известны и в некоторых случаях не общего положения.

Оказывается, *критерии такого вида не могут существовать в достаточно сложны особых случаях, даже если мы согласимся пренебрегать особыми случаями бесконечной коразмерности* (или даже коразмерности 103).

Рассмотрим пространство r-струй векторных полей в точке 0 пространства $\mathbf{R}^n$. Особая точка векторного поля с заданной r-струей (т. е. с заданным разложением Тейлора до членов степени r, если фиксировать систему координат) может быть: (I) устойчивой (по Ляпунову) независимо от продолжения ряда Тейлора; (II) неустойчивой независимо от продолжения; (III) устойчивой или неустойчивой, в зависимости от продолжения. В соответствии с этим мы называем струю *устойчивой* (I), *неустойчивой* (II) или *нейтральной* (III).

Теорема 1. *Разбиение пространства струй на части* I, II, III *не является полуалгебраическим* [1]), *если размерность фазового пространства* n *больше* 2, *а порядок струй* r *больше* 4.

Это и означает, что алгебраический критерий устойчивости по Ляпунову (а значит, и критерий для определения топологического типа) в особых случаях достаточно большой коразмерности не существует.

Я не пытался точно определить наименьшую коразмерность, где это явление наступает впервые. При $n \geqslant 3$, $r \geqslant 5$ можно указать такую плоскость коразмерности 102 в пространстве струй, что в этой плоскости часть границы между струями классов I и II является трансцендентной аналитической гиперповерхностью.

---

[1]) Полуалгебраическим множеством называется конечное объединение разностей алгебраических, т. е. множество, определенное при помощи конечного числа полиномиальных уравнений и неравенств.

Originally publ. in: Usp. Mat. Nauk 25:2, 265-266 (Russian), © Russ. Acad. Sciences 1970
Reprint by Springer-Verlag Berlin Heidelberg 2014 with kind permission, DOI 10.1007/978-3-642-31031-7_22

Пример. Рассмотрим четырехпараметрическое семейство систем в $\mathbf{R}^3$

$$\left.\begin{aligned} \dot{x}_1 &= x_2x_3^4 + x_1(F+G), \\ \dot{x}_2 &= -x_1x_3^4 + x_2(F+G) - \mu x_1^3x_3^2, \\ \dot{x}_3 &= x_3G, \end{aligned}\right\} \tag{1}$$

где

$$F = ax_3^4 + b(x_1^2+x_2^2)x_3^2 + c(x_1^2+x_2^2)^2,$$
$$G = x_3^4 - 4(x_1^2+x_2^2-x_3^2)^2$$

и где a, b, c, μ — параметры. Рассмотрим точку M пространства параметров:

$$a = -1, \quad b = 2, \ c = -1, \ \mu = 0.$$

Теорема 2. *Граница, разделяющая устойчивые и неустойчивые системы* (1) *в четырехмерном пространстве параметров* (a, b, c, μ), *имеет трансцендентную аналитическую компоненту, проходящую через* M.

Утверждение останется верным, если добавить к правым частям уравнений (1) любые добавки выше 5-й степени. Поэтому теорема 1 следует из теоремы 2.

Доказательство теоремы 2 основано на рассмотрении соответствующего системе (1) слоения на проективной плоскости. Это слоение описывается в аффинной карте дифференциальными уравнениями вида

$$x' = P(x, y), \qquad y' = Q(x, y), \tag{2}$$

где P и Q — многочлены. Уравнения (1) подобраны так, что смена устойчивости при изменении параметров в системе (1) происходит в момент рождения пары предельных циклов системы (2): одного устойчивого и одного неустойчивого.

Условие рождения циклов не может быть алгебраическим. Ибо иначе существовал бы эффективный способ находить число предельных циклов для систем (2).

Можно и непосредственно доказать, что условие рождения пары предельных циклов (вдали от особых точек) — трансцендентное. Например, для системы (2), соответствующей (1), полезно рассмотреть поведение аналитического продолжения границы устойчивости вблизи точки

$$a = b = c = 0$$

при фиксированном $\mu \neq 0$. Уравнение касательного конуса в этой точке выражается через эллиптические интегралы, и легко проверить, что конус трансцендентен.

Приведенные примеры заставляют предположить, что «патологии», обнаруженные в последние годы в теории дифференциальных уравнений на многомерных многообразиях (и заключающиеся в том, что задача топологической классификации, видимо, не допускает разумного решения, даже если ограничиться случаями «общего положения»), могут проявиться также и в локальных задачах при исследовании окрестности особой точки, если только размерность фазового пространства и коразмерность рассматриваемого случая достаточно велики.

В частности, можно ожидать, что граница устойчивости, утратив полуалгебраичность и ничем более не сдерживаемая, будет представлять патологии уже на теоретико-множественном уровне. Например, множество устойчивых струй в конечномерном алгебраическом подмногообразии пространства струй фиксированного порядка может, вероятно, иметь бесконечное число компонент связности или быть всюду плотным вместе со своим дополнением. Точно так же можно ожидать, что в задаче о топологической классификации особых точек встретится (при изучении случаев достаточно большой, но конечной коразмерности) бесконечное число различных топологических типов, образующих столь же скверное разбиение пространства струй, как топологические типы векторных полей на многообразии в аналогичной глобальной задаче.

Поступило в Правление общества 25 ноября 1969 г.

Trudy Moskov. Mat. Obšč.
Tom 21 (1970)

Trans. Moscow Math. Soc.
Vol. 21 (1970)

ON SOME TOPOLOGICAL INVARIANTS OF ALGEBRAIC FUNCTIONS[1)]

UDC 513.83

V. I. ARNOL'D

CONTENTS

There are some interesting connections between the theory of algebraic functions and Artin's theory of braids. For instance, the space G_n of polynomials of degree n not having multiple roots is the space $K(\pi, 1)$ for the group $B(n)$ of braids with n strings:[2)]

$$\pi_1(G_n) = B(n), \quad \pi_i(G_n) = 0 \quad \text{for} \quad i > 1. \tag{1}$$

It is possible to make use of this relation in both directions, both for the study of the braid group and for the study of algebraic functions. Here are a few examples.

A) **The braid group of an algebraic function.** Apart from the monodromy group, which describes the permutation of the sheets of a function as we go round the branching manifold, there exists a more delicate invariant of the algebraic function (but not of the corresponding covering): the *braid group* of the function. This

1) The results in this article were announced in [26]. The author thanks S. P. Novikov, V. P. Palamodov and D. B. Fuks for some useful discussions about the results obtained.

2) The group $B(n)$ of braids with n strings has $n-1$ generators $\sigma_1, \cdots, \sigma_{n-1}$ and relations $\sigma_i\sigma_j = \sigma_j\sigma_i$ ($|i-j| > 1$) and $\sigma_i\sigma_{i+1}\sigma_i = \sigma_{i+1}\sigma_i\sigma_{i+1}$ ($i = 1, 2, \cdots, n-2$). It can be defined as the fundamental group of the space of unordered sets of n distinct points in a plane. A path representing an element of this fundamental group yields an n-valued mapping of the segment $0 \le t \le 1$ into the (x, y)-plane. The graph of this mapping in three-dimensional (x, y, t)-space has the form of a braid with n strings. For the theory of braids see [5]–[25].

Originally publ. in: Tr. Mosk. Mat. Obsc., 27-46, © Izdat. Moscow Univ. 1970
English transl.: Transact. Math. Moscow Soc. 21, 30-52, © American Math. Society, Providence, RI, 1970
Reprint by Springer-Verlag Berlin Heidelberg 2014 with kind permission, DOI 10.1007/978-3-642-31031-7_23

group takes into account not only the permutation of the values of the function after the circuit, but also how they circulate around one another in the plane of values of the function. For example, the monodromy groups of the functions $\sqrt{z}$ and $\sqrt{z^3}$ are identical, while the braid groups differ. The monodromy group is a representation of the fundamental group of the complement of the branching manifold in the symmetric group. The braid group of an algebraic function is a representation of the same fundamental group in Artin's braid group. Certain noteworthy subgroups of the braid group are connected with the behavior of the function in the neighborhood of the branching manifold and on the manifold itself.

B) The space G_n of polynomials without multiple roots. Some other applications of the connection mentioned above are based on the following three interpretations of G_n.

1) G_n can be thought of as *the space of hyperelliptic curves of degree n.*

From this remark one can extract, on the one hand, a representation of the braid group in the group of symplectic integral matrices (namely, the matrices of automorphisms of the homologies of the curves, induced by circuits in the coefficient plane). It can be shown that this yields a representation on the whole of the symplectic group in the cases $n = 3, 4, 6$, and only in these cases.

On the other hand, we obtain information about the branching of hyperelliptic integrals as functions of the parameters: the relations between the generators of the braid group give rise to relations between the Picard-Lefschetz matrices (see [27] and [28]).

2) G_n can be regarded as *the set of regular values of the mapping* Σ^{1n} (see [1], [2]). Thus the relation (1) and other results of the present article yield facts about the topology of the simplest singularities of complex analytic mappings.

3) G_n can be considered as *the complement to the branching manifold of a universal n-valued entire algebraic function* $z(a_1, \cdots, a_n)$:

$$z^n + a_1 z^{n-1} + \ldots + a_n = 0 \tag{2}$$

(a root of an equation of degree n is regarded as a function of the coefficients). The manifold G_n is the base of the n-sheeted covering corresponding to this function.

One can see from this remark what significance the cohomology classes of $H^i(G_n)$ have. In fact, an n-valued entire algebraic (respectively, algebroid or quasi-algebraic) function $z(x)$ gives rise to a polynomial (respectively, analytic or continuous) mapping $a(x)$ of the space of arguments $\{x\}$ into the space of arguments $\{a\}$ of the universal function (2). Under this mapping, the cohomology classes of $H^i(G_n)$ induces special cohomology classes in the complement to the branching manifold of $z(x)$.

On the other hand, it follows from (1) that the cohomology groups of G_n coincide with those of the braid group with n strings (with trivial action on $\mathbf{Z}$):

$$H^i(B(n)) = H^i(G_n, \mathbf{Z}).$$

C) The cohomology of the braid group. The complex and algebraic structures of G_n have proved very useful in investigating the cohomology of braid groups.

First of all we remark that G_n is an n-dimensional Stein manifold (since G_n is given in the space $\mathbf{C}^n$ of all polynomials of degree n of the form (2) with the condition that $\Delta(a) \neq 0$, where $\Delta(a)$ is the discriminant of a). Hence, for $i > n$

$$H^i(B(n)) = H^i(G_n) = 0. \tag{3}$$

We show (see §4) that for $n < 12$ the cohomology groups of the braid group with n strings are given by the following table (the empty spaces in the table all stand for 0):

Cohomology table of the braid groups $H^i(B(n))=H^i(G(n))$, $n<12$

n \ i	0	1	2	3	4	5	6	7	8	9
2,3	$\mathbf{Z}$	$\mathbf{Z}$								
4,5	$\mathbf{Z}$	$\mathbf{Z}$	0	$\boxed{\mathbf{Z}_2}$						
6,7	$\mathbf{Z}$	$\mathbf{Z}$	0	$\mathbf{Z}_2$	$\boxed{\mathbf{Z}_2}$	$\mathbf{Z}_3$				
8,9	$\mathbf{Z}$	$\mathbf{Z}$	0	$\mathbf{Z}_2$	$\mathbf{Z}_2$	$\boxed{\mathbf{Z}_6}$	$\mathbf{Z}_3$	$\mathbf{Z}_2$		
10,11	$\mathbf{Z}$	$\mathbf{Z}$	0	$\mathbf{Z}_2$	$\mathbf{Z}_2$	$\mathbf{Z}_6$	$\boxed{\mathbf{Z}_6? \, 3}$	$\mathbf{Z}_2? \, 1$	$\mathbf{Z}_2$	$\mathbf{Z}_5$

We shall in addition prove (see §3) the following three theorems:

Finiteness Theorem. *With the exception of $H^0 \cong H^1 \cong \mathbf{Z}$, the cohomology groups of the braid groups are all finite. Further,*

$$H^i(B(n)) = 0 \text{ for } i \geqslant n.$$

Repetition Theorem. *All the cohomology groups of a braid group with an odd number of strings are the same as those of the braid group with the preceding even number of strings:*

$$H^i(B(2n+1)) = H^i(B(2n)).$$

Stability Theorem. *As n increases, the ith cohomology group of the braid group with n strings becomes stable.:*

$$H^i(B(n)) = H^i(B(2i-2)) \text{ for } n \geqslant 2i-2.$$

Thus the first stable cohomology groups (in the boxes in the above table) are

i	0	1	2	3	4	5	6
$H^i(B)$	$\mathbf{Z}$	$\mathbf{Z}$	0	$\mathbf{Z}_2$	$\mathbf{Z}_2$	$\mathbf{Z}_6$	$\mathbf{Z}_6$? 3

The proofs of all these results are based on the relation (1) and on a study of the manifold G_n. Beginning with Artin, many authors have given proofs of the first part of the relation (1) (in other terms, of course).[1)]

The second part of (1) is almost obvious (for a detailed proof see [4], where of course the assertion is stated in different terms).

Thus to study the cohomology of braid groups it is sufficient to study the cohomology of the manifold G_n. To this end we consider the natural stratification of the algebraic manifold of polynomials with multiple roots by a Young diagram. In fact, we obtain more information about this stratification than indicated above. To state these results (see §§3 and 4) we need first to introduce some notation.

§0. Notation

A) **The transition to homology.** We adopt the unusual notation $H_{(i)}$ for the integral homology groups relative to a point, counted "downwards", so that $H_{(0)}$ is the highest group, $H_{(1)}$ has codimension 1 and so on:

$$H_{(i)}(X) = H_{N-i}(X, x; \mathbf{Z}), \tag{4}$$

where X is an N-dimensional complex with base point x. By a *polynomial* we mean a complex polynomial of the type (2) in the single variable z with leading coefficient 1.

Let S^{2n} be the one-point compactification of the space $\mathbf{C}^n$ of all polynomials of degree n. We are interested in the open subset $G_n \rightarrowtail S_n$ consisting of

1) We mention that yet another proof can be derived from Zariski's theorems on the fundamental group of the complement to an algebraic manifold. We further remark that, conversely, the theory of braids throws some light on the nature of Zariski's arguments. It seems a reasonable conjecture that the fundamental group of the complement to an affine manifold M does not have elements of finite order. The conjecture would be proved if it were known that every element of this fundamental group could be represented by a nontrivial braid of some algebraic function with branching manifold M.

Another proof of the trivality of the higher homotopy groups of G_n can be derived from the contraction of a component of I^n in the group of homeomorphisms of the plane with points removed. (It is necessary to consider the action of the group of homeomorphisms of G_n.)

polynomials without multiple roots.[1] The complement $\Sigma = S^{2n}\setminus G_n$ is the one-point compactification of the space of polynomials having at least one multiple root. Σ and S^{2n} have base point ∞. The dimension of Σ is equal to $2n-2$. By the duality theorems of Lefschetz and Alexander we have

$$H^i(G_n) = H_{2n-i}(S^{2n}, \Sigma) = H_{(i-1)}(\Sigma), \quad i > 0. \tag{5}$$

Therefore the formulation of the above-stated results on the cohomology of G_n follows from the corresponding theorems on the homology groups $H_{(i)}(\Sigma)$. To prove them it will be useful to consider certain algebraic submanifolds of Σ.

B) The manifolds Σ_n^k. Our constructions will be dependent on an integer $q > 0$. This number is fixed and will usually be omitted from the notation. The case $q = 2$ is especially important. This case alone will be used for the calculation of the cohomology of the braid groups. If desired, it may be assumed henceforth that $q = 2$.

Denote by ${}_q^\circ\Sigma_n^k$ the set of polynomials of degree $kq+n$ having at least k q-fold roots (the other n roots being arbitrary). Obviously, ${}_q^\circ\Sigma_n^k$ is a closed irreducible algebraic $(n+k)$-dimensional affine submanifold of the space of all polynomials of degree $kq+n$:

$${}_q^\circ\Sigma_n^k = \{p_k^q p_n\} \rightarrowtail \mathbf{C}^{kq+n}, \quad \dim_{\mathbf{C}} {}_q^\circ\Sigma_n^k = n + k$$

We mention the natural homeomorphisms

$${}_q^\circ\Sigma_0^k \sim \mathbf{C}^k, \quad {}_q^\circ\Sigma_n^0 \sim \mathbf{C}^n.$$

Witt's Theorem and the fundamental theorem of algebra establish the homeomorphism between $\mathbf{C}^n$ and the symmetric nth power of $\mathbf{C}^1$: each polynomial corresponds to the set of its roots. This homeomorphism is compatible with one-point compactification: at least one root tends to ∞ precisely when at least one coefficient tends to ∞.

By ${}_q\Sigma_n^k = \Sigma_n^k$ we denote the one-point compactification of the manifold ${}_q^\circ\Sigma_n^k$ with the distinguished point ∞. In this notation, the manifold Σ of subsection A takes the form $\Sigma = \Sigma_{n-2}^1 = {}_2\Sigma_{n-2}^1$ (at least one double root). We mention the natural embedding ${}_q\Sigma_n^k \leftarrowtail \Sigma_{n-q}^{k+1}$ $(n \geq q)$ as an algebraic submanifold of complex codimension $q-1$ completed by the point ∞.

C) The mappings ϕ. The multiplication of polynomials is defined by a system of natural mappings of tensor products of the spaces Σ, which we describe directly.

---

1) The symbol $\rightarrowtail$ denotes an embedding, while $\twoheadrightarrow$ is an onto mapping (of sets, manifolds, groups, and so on); the symbol $\mapsto$ stands for the action of a mapping on an element (for example z^2: $\mathbf{C}^1 \twoheadrightarrow \mathbf{C}^i$, $i \mapsto -1$).

Let (X, x) and (Y, y) be spaces with base point. The *tensor product* $X \# Y$ is defined to be the space with base point

$$X \# Y = X \times Y/(X \times y) \vee (Y \times x)$$

(the direct product with contraction of the pairs involving x or y). The operation $\#$ is associative. If X is the k-dimensional sphere, then $X \# Y$ is called the *k-fold suspension* of Y. In our notation (4), suspension does not alter homology:

$$H_{(i)}(S^k \# Y) \cong H_{(i)}(Y). \tag{6}$$

Multiplication of polynomials is defined by the continuous "onto" mapping

$$\varphi : (\mathbf{C}^m + \infty) \# (\mathbf{C}^n + \infty) \twoheadrightarrow (\mathbf{C}^{m+n} + \infty), \quad (p_m, p_n) \mapsto p_m p_n.$$

Passing to the manifolds Σ, we obtain mappings (also onto)

$${}_q\varphi^{l,k}_{m,n} : \Sigma^l_m \# \Sigma^k_n \twoheadrightarrow \Sigma^{k+l}_{m+n}, \quad (p_m p_l^q, p_n p_k^q) \mapsto p_m p_n (p_l p_k)^q.$$

As a rule, we omit the indices from the mappings ϕ, so that these mappings are natural; that is, they yield commutative diagrams when combined with the inclusions mentioned in B). The letter ϕ is also used to denote other analogous natural mappings, for instance

$$\varphi : \Sigma^l_m \# \Sigma^k_n \# \Sigma^t_s \twoheadrightarrow \Sigma^{l+k+t}_{m+n+s}.$$

We note that $\phi(\infty) = \phi^{-1}(\infty) = \infty$, this being the base point of all our spaces.

D) The degrees of the mappings ϕ. We remark that Σ^k_n is a complex algebraic irreducible affine manifold, compactified at the point ∞. Therefore $H_{(0)}(\Sigma^k_n) = \mathbf{Z}$. The complex structure provides Σ^k_n with an orientation (at the nonsingular points) and likewise fixes a generator of $H_{(0)}$. ϕ induces a mapping of $H_{(0)}$ given by the integer to which the generator passes. This integer is called the *degree* of ϕ. Insofar as the mapping ϕ is consistent with the complex structure at the nonsingular points, the degree of ϕ is simply equal to the number of preimages of some (and thus of any) regular point.

Lemma 1. *The mapping*

$$\varphi : \Sigma^l_m \# \Sigma^k_n \twoheadrightarrow \Sigma^{l+k}_{m+n}$$

has degree $C^n_{m+n} C^k_{l+k}$.

Proof. A general point of Σ^{l+k}_{m+n} is a polynomial having $m+n+k+l$ distinct roots, $m+n$ of these being simple and $k+l$ q-fold. The number of preimages of such a point is equal to the number of ways of choosing n simple roots from $m+n$ and k q-fold roots from $k+l$, that is, $C^n_{m+n} C^k_{k+l}$, as required.

We indicate the degree of ϕ by a number above the arrow:

$$\Sigma^l_m \# \Sigma^k_n \xrightarrow{C^n_{m+n} C^k_{k+l}} \Sigma^{l+k}_{m+n}.$$

We mention two special cases: the case $m = 0$ and the case $l = 0$. In these cases Σ^l_m is a sphere, and tensor multiplication by Σ^l_m is a multiple suspension. We introduce two notations for the $2m$-fold suspension of Σ^k_n:

$$E_m \Sigma^k_n = \Sigma^0_m \# \Sigma^k_n, \quad E^m \Sigma^k_n = \Sigma^m_0 \# \Sigma^k_n.$$

In these notations there exist mappings ϕ:

$$E_m \Sigma^k_n \xrightarrow{C^n_{m+n}} \Sigma^k_{m+n}, \quad E^m \Sigma^k_n \xrightarrow{C^k_{m+k}} \Sigma^{m+k}_n.$$

In particular, for $k = 0$

$$S^{2(m+n)} = E^m \Sigma^0_n \xrightarrow{1} \Sigma^m_n.$$

Thus each of the manifolds Σ^m_n is obtained from a sphere under a certain mapping of degree 1. The investigation of the homology of Σ^m_n will be based upon a study of the singularities of this mapping.

§1. Squares and cubes

Our calculation of the homology of Σ^k_n will be in essence the calculation of the spectral sequence of the filtration given by the embeddings

$$\Sigma^k_n \leftarrowtail \Sigma^{k+1}_{n-q} \leftarrowtail \Sigma^{k+2}_{n-2q} \leftarrowtail \cdots.$$

To find the differentials we use geometrical information in the form of commutative diagrams of two types: squares and cubes.

A) The square and the sequence of a square. Consider the manifolds Σ^k_n, $n \geq q$. We have the commutative diagram

$$\begin{array}{ccc} E^k \Sigma^0_n & \leftarrowtail & E^k \Sigma^1_{n-q} \\ {\scriptstyle 1}\downarrow & & \downarrow{\scriptstyle k+1} \\ \Sigma^k_n & \leftarrowtail & \Sigma^{k+1}_{n-q} \end{array} \qquad \begin{array}{ccc} S & \leftarrowtail & A \\ \downarrow & & \downarrow \\ \Sigma & \leftarrowtail & B \end{array},$$

where the horizontal arrows stand for the natural embeddings and the vertical for the mappings ϕ; the numbers beside the arrows denote the degrees. We call (7) the *square* for Σ^k_n.

We mention that 1) in the upper left-hand corner there is a sphere $E^k \Sigma^0_n = S^{2(n+k)}$; 2) in each column there stand complexes of the same dimension: 3) the right-hand column is embedded in the left with complex codimension $q-1$.

Lemma 2. *The mapping ϕ establishes a homeomorphism of complements* $E^k\Sigma_n^0\setminus E^k\Sigma_{n-q}^1 \sim \Sigma_n^k\setminus\Sigma_{n-q}^{k+1}$.

Proof. If a polynomial has the form $p_n p_k^q$ but not the form $p_{n-q}p_{k+1}^q$, then the factor p_k^q is uniquely determined.

Remark. In a surprising way the difference $\Sigma_n^k\setminus\Sigma_{n-q}^{k+1}$ turns out to be homeomorphic to a domain of Euclidean space, although the algebraic manifold Σ_n^k has complicated singularities outside Σ_{n-q}^{k+1}.

Corollary. *For $n<q$, Σ_n^k is homeomorphic to a sphere*

$$ {}_q\Sigma_n^k \sim S^{2(n+k)},\quad n<q. \tag{8}$$

For Σ_{n-q}^{k+1} is empty for $n<q$.

Lemma 3. *If in the diagram of complexes with base point*

$$\begin{array}{ccc} S & \leftarrow\!\!< & A \\ {\scriptstyle\varphi}\downarrow & & \downarrow{\scriptstyle\varphi} \\ \Sigma & \overset{j}{\leftarrow\!\!<} & B \end{array} \qquad \begin{array}{l} \dim S=\dim\Sigma=N \\ \\ \dim A=\dim B=N-r \end{array}$$

$S=S^N$ is a sphere, the horizontal arrows are embeddings of closed subcomplexes, and the vertical arrows are continuous mappings onto, establishing the homeomorphism $S\setminus A\sim\Sigma\setminus B$, then $H_{(0)}(\Sigma)=\mathbf{Z}$ and for $i>1-r$ there is an exact sequence

$$\ldots\to H_{(i-1)}(B)\xrightarrow{j_*}H_{(i+r-1)}(\Sigma)\xrightarrow{\partial}H_{(i)}(A)\xrightarrow{\varphi_*}H_{(i)}(B)\to\ldots. \tag{9}$$

Proof. We form a commutative diagram from the exact sequences of the pairs (S,A) and (Σ,B):

$$\begin{array}{ccccccccc} \ldots\to & H_{(i+r-1)}(S) & \to & H_{(i+r-1)}(S/A) & \to & H_{(i)}(A) & \to & H_{(i+r)}(S) & \to\ldots \\ & \downarrow & & \downarrow & \ddots & \downarrow & & \downarrow & \\ \ldots\to & H_{(i+r-1)}(\Sigma) & \to & H_{(i+r-1)}(\Sigma/B) & \to & H_{(i)}(B) & \to & H_{(i+r)}(\Sigma) & \to\ldots. \end{array}$$

Insofar as $H_{(i+r-1)}(S)=0$, the mapping $H_{(i+r-1)}(S/A)\to H_{(i)}(A)$ is an isomorphism. By the hypothesis of the lemma, $H_{(i+r-1)}(S/A)\to H_{(i+r-1)}(\Sigma/B)$ is also an isomorphism. This establishes an isomorphism $H_{(i+r-1)}(\Sigma/B)\cong H_{(i)}(A)$ which transforms the sequence of the pair (Σ,B) into the sequence (9). For $i=1-r$ we discover from the same diagram that $H_{(0)}(\Sigma)=\mathbf{Z}$. This proves the lemma.

We remark that, by Lemma 2, the square (7) satisfies the conditions of Lemma 3 (with $N=2(n+k)$ and $r=2(q-1)$). We call the sequence (9), constructed from the square (7), *the sequence of the square* for Σ_n^k. Since suspension does not alter homology (see (6)), this sequence has the form $(i > 1 - r,\ r = 2(q-1))$

$$\ldots \to H_{(i-1)}(\Sigma_{n-q}^{k+1}) \to H_{(i+r-1)}(\Sigma_n^k) \to H_{(i)}(\Sigma_{n-q}^1) \overset{\varphi_*}{\to} H_{(i)}(\Sigma_{n-q}^{k+1}) \to \ldots . \qquad (10)$$

B) The cube and the diagram of sequences of the cube. The sequence (10) links the homology of Σ_n^k with the homologies of Σ_{n-q}^1 and of Σ_{n-q}^{k+1}. In both spaces the lower index is less than n. To compute $H_{(i)}(\Sigma_n^k)$ by induction on n it will be useful to know ϕ_*. Information about ϕ_* is derived from the following commutative cube. Consider the mapping ϕ: $E^k\Sigma_n^l \to\to \Sigma_n^{k+l}$. It can be included in the mapping of the suspended square for Σ_n^l onto the square for Σ_n^{k+l}; we have a commutative diagram

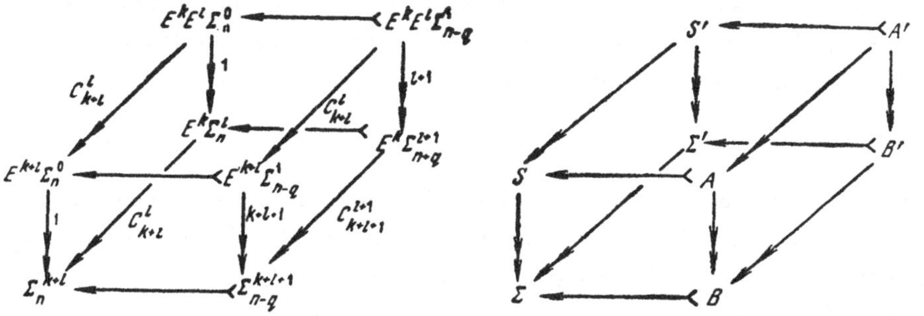

(11)

Here the horizontal arrows are embeddings, while the vertical and oblique arrows are the mappings ϕ; the numbers beside the arrows are the degrees. The diagram (11) is called the *cube* for the mapping ϕ: $E^k\Sigma_n^l \to\to \Sigma_n^{k+l}$. This mapping corresponds to the left-hand lower edge.

Lemma 4. *Let* $(S, A; \Sigma, B)$ *and* $(S', A'; \Sigma', B')$ *be two commutative squares satisfying the conditions of Lemma 3. Then the commutative cube mapping the second square onto the first induces a commutative diagram of exact sequences of squares* (*for* $i > 1 - r$, $r = 2(q-1)$):

$$\begin{array}{ccccccccc} \ldots \to & H_{(i-1)}(B') & \to & H_{(i+r-1)}(\Sigma') & \to & H_{(i)}(A') & \to & H_{(i)}(B') & \to \ldots \\ & \downarrow & & \downarrow & & \downarrow & & \downarrow & \\ \ldots \to & H_{(i-1)}(B) & \to & H_{(i+r-1)}(\Sigma) & \to & H_{(i)}(A) & \to & H_{(i)}(B) & \to \ldots . \end{array} \qquad (12)$$

This follows from the natural exact sequences of pairs and the compatibility of the identifications $H_{(i+r-1)}(\Sigma'/B') \cong H_{(i)}(A')$ and $H_{(i+r-1)}(\Sigma/B) \cong H_{(i)}(A)$, constructed in the proof of Lemma 3.

We call (12) the *diagram of sequences of the cube*. Note that the cube (11) satisfies the hypothesis of Lemma 4, the suspension E^k of the square for Σ_n^l representing the square involving the primed symbols. Since suspension does not alter homology (see (6)), the diagram of sequences of the cube (11) has the form

$$\begin{array}{ccccccccc}
\ldots \to H_{(i-1)}(\Sigma^{l+1}_{n-q}) & \to & H_{(i+r-1)}(\Sigma^{l}_{n}) & \to & H_{(i)}(\Sigma^{1}_{n-q}) & \xrightarrow{l+1} & H_{(i)}(\Sigma^{l+1}_{n-q}) & \to \ldots \\
\downarrow c^{l+1}_{k+l+1} & & \downarrow c^{l}_{k+l} & & \downarrow c^{l}_{k+l} & & \downarrow c^{l+1}_{k+l+1} & \\
\ldots \to H_{(i-1)}(\Sigma^{k+l+1}_{n-q}) & \to & H_{(i+r-1)}(\Sigma^{k+l}_{n}) & \to & H_{(i)}(\Sigma^{1}_{n-q}) & \xrightarrow{k+l+1} & H_{(i)}(\Sigma^{k+l+1}_{n-q}) & \to \ldots
\end{array} \tag{13}$$

The vertical arrows are induced by the oblique edges of the cube; that is, by the mappings ϕ. We remark that the mapping $\phi\colon E^kE^l\Sigma^1_{n-q} \twoheadrightarrow E^{k+1}\Sigma^1_{n-q}$ simply induces multiplication by C^l_{k+l} in the cohomology of Σ^1_{n-q} in all dimensions. In the remaining cases, the number beside the arrow in the diagram (13) indicates the degree; that is, the action of ϕ_* on $H_{(0)}$.

Diagram (13) allows us to obtain, recursively with respect to n, information about the group $H_{(i)}(\Sigma^k_n)$ and the mappings ϕ_*, using (8) for the initial conditions. This information is sufficient to yield the table on page 32 and the Finiteness Theorem (page 32). Additional facts about $H(\Sigma)$ are obtained by comparing Σ^k_n with Σ^k_{n-1}, and this we carry out below.

§2. Punctures

The set Σ^k_n is not a manifold in the topological sense. We now consider the structure of neighborhoods of certain special points of Σ^k_n, and also the spaces obtained from Σ^k_n when punctured at isolated points.

A) Neighborhoods of the point O. First consider the "most singular" point $O \in \Sigma^k_n$, the polynomial all of whose $N = qk + n$ roots are equal to zero. The collection of all polynomials all of whose roots have modulus less than ϵ forms a neighborhood $V(\epsilon)$ of O.

Lemma 5. *The neighborhood $V(\epsilon)$ is homeomorphic to the finite part, ${}^\circ\Sigma^k_n = \Sigma^k_n\backslash\infty$, of Σ^k_n. The point O is a deformation retract of $V(\epsilon)$ and of ${}^\circ\Sigma^k_n$.*

Proof. In accordance with B) of §0 the set $V(\epsilon)$ (or ${}^\circ\Sigma^k_n$) can be regarded as the collection of unordered sets of N points of $\mathbf{C}^1$. The corresponding collection of ordered sets forms the polycylinder $D = \{z \in \mathbf{C}^N\colon |z| < \epsilon\}$ ($|z| = \max|z_i|$) (or the whole of $\mathbf{C}^N$). The homeomorphism $z \longmapsto z(\epsilon - |z|)^{-1}$, $D \twoheadrightarrow \mathbf{C}^N$ gives rise to a homeomorphism $V(\epsilon) \twoheadrightarrow {}^\circ\Sigma^k_n$. The retraction $z \longmapsto tz$, $z \in D$, $0 \le t \le 1$, (or $z \in \mathbf{C}^N$) gives rise to the retraction of $V(\epsilon)$ (or ${}^\circ\Sigma^k_n$) to O. This is because they are permuted under the action of the symmetric group.

We note that the homeomorphism and the retraction just constructed are natural in the sense that they do not affect the multiplicities of roots, and, in particular, the associated manifold Σ^l_m is preserved. This property of naturality is also possessed by the homeomorphisms and the retractions constructed in the proofs of the following two lemmas.

Lemma 6. *The complement $\Sigma^k_n\backslash V(\epsilon)$ is a deformation retract of the complement of a point $\Sigma^k_n\backslash O$.*

Proof. In the notation of the proof of Lemma 5 the retraction is given by the formulas

$$z \mapsto \varepsilon t|z|^{-1}z \quad (0<|z|\leqslant \varepsilon t), \ z \mapsto z\,(\varepsilon t \leqslant |z| < \infty), \quad 0 \leqslant t \leqslant 1.$$

B) **Neigborhoods of the point $*$.** Now consider the "next most singular" point, $* \in \Sigma_n^k$, the polynomial that has one root equal to 1 and the remaining $kq + n - 1$ roots all equal to 0.

Lemma 7. *In Σ_n^k, the point $*$ has a neighborhood U homeomorphic to the direct product $\mathbf{C}^1 \times {}^\circ\Sigma_{n-1}^k$; furthermore, $\Sigma_n^k \setminus U$ is a deformation retract of $\Sigma_n^k \setminus *$.*

Proof. Consider the collection $U(\epsilon)$ of all polynomials having one root that differs from 1 by less than ϵ and the remaining $kq + n - 1$ roots less in modulus then ϵ. If $\epsilon < 1/2$, then $U = U(\epsilon)$ is naturally represented in the form of the direct product of the circle $|z-1| < \epsilon$ and the neighborhood $V_{n-1}^k(\epsilon)$ of the point $O \in \Sigma_{n-1}^k$. Lemma 5 guarantees the homeomorphism $U \sim \mathbf{C}^1 \times {}^\circ\Sigma_{n-1}^k$.

Let $z = (z_0, z_1, \cdots, z_{N-1}) \in \mathbf{C}^N$. We give a deformation retract of $\{z: 0 < |z| \le \epsilon\}$ onto $\{z: |z| = \epsilon\}$ by the formulas in the proof of Lemma 6. We compare the point z and the polynomial with the roots $1 + z_0, z_1, z_2, \cdots, z_{N-1}$. The retraction we have constructed yields a retraction of $\overline{U}\setminus *$ into the boundary $\partial U = \overline{U}\setminus U$ and thus a retraction of $\Sigma_n^k \setminus *$ onto $\Sigma_n^k \setminus U$.

C) **The space $*\Sigma_n^k$.** We denote by $*\Sigma_n^k$ *the space* $\Sigma_n^k \setminus *$ *obtained by puncturing Σ_n^k at the "next most singular" point $*$. When $n = 0$ it coincides with Σ_n^k.*

The homology groups of Σ_n^k and $*\Sigma_n^k$ are connected by the following exact sequence.

Lemma 8. *There exists an exact sequence* $(i \geq 0)$

$$\ldots \to H_{(i)}(*\Sigma_n^k) \to H_{(i)}(\Sigma_n^k) \to H_{(i)}(\Sigma_{n-1}^k) \xrightarrow{\partial} H_{(i+1)}(*\Sigma_n^k) \to \ldots. \tag{14}$$

Proof. Consider the exact sequence of the pair $(\Sigma_n^k, \Sigma_n^k \setminus U)$:

$$\to H_{(i)}(\Sigma_n^k \setminus U) \to H_{(i)}(\Sigma_n^k) \to H_{(i)}(\Sigma_n^k, \Sigma_n^k \setminus U) \to H_{(i+1)}(\Sigma_n^k \setminus U) \to \ldots.$$

By Lemma 7 the embedding $\Sigma_n^k \setminus U \rightarrowtail *\Sigma_n^k$ induces an isomorphism of homology groups. The same lemma yields a homeomorphism $(\Sigma_n^k/(\Sigma_n^k \setminus U)) \sim (\overline{U}/\partial U) \sim E_1\Sigma_{n-1}^k$. By (6) we obtain (14). We note that this sequence is natural in the same sense as the homeomorphisms and retractions of Lemmas 5–7, by means of which it was constructed.

D) **The space $*E^m\Sigma_n^k$.** We denote by $*E^m\Sigma_n^k$ the set $(E^m\Sigma_n^k)\setminus *$ obtained by puncturing $E^m\Sigma_n^k$ at the single point $* = (z^{qm}, z^{qk+n-1}(z-1))$. This notation is somewhat ambiguous, as the symbol $*$ is being used for points of different spaces. However, this ambiguity is harmless, because the choice of the point $*$ is compatible with the natural embeddings and with the mappings ϕ. We also write $* = (z^m, z^l, z^{kq+n-1}(z-1)) \in E^m E^l \Sigma_n^k$.

Lemma 9. *Under the mappings* ϕ:

$$E^m \Sigma_n^k \twoheadrightarrow \Sigma_n^{m+k}, \quad E^m E^l \Sigma_n^k \twoheadrightarrow E^{m+l} \Sigma_n^k, \quad E^m E^l \Sigma_n^k \twoheadrightarrow E^m \Sigma_n^{l+k} \tag{15}$$

the image and the complete inverse image of the point $*$ *is once again the point* $*$:

$$\varphi(*) = \varphi^{-1}(*) = *.$$

Proof. Since $z = 1$ is a simple root of the polynomial $*$, the factors z^{mq} and z^{lq} will necessarily be distinguished in E^m and E^l.

Corollary. *The mappings* ϕ *indicated in* (15) *yield mappings of spaces with base points* $*$. *If stars are placed on all six spaces in* (15), *then we obtain mappings that are natural with respect to the embeddings* $\Sigma_n^k \leftarrow \Sigma_{n-q}^{k+1}$, *etc. We denote these mappings, as before, by* ϕ.

We mention a standard property of suspensions: embeddings of the form $E^m \Sigma_n^k \rightarrowtail {}^*E^m \Sigma_n^k$ give rise to isomorphisms of homology groups.

$$H_{(i)}({}^*E^m \Sigma_n^k) \cong H_{(i)}(E^{m*}\Sigma_n^k) \cong H_{(i)}({}^*\Sigma_n^k),$$
$$H_{(i)}({}^*E^m E^l \Sigma_n^k) \cong H_{(i)}({}^*\Sigma_n^k). \tag{16}$$

Corollary. *The diagrams of the square* (7) *and the cube* (11), *as well as Lemma* 2, *remain valid if stars are placed at all the vertices.*

The resulting diagrams will be called the diagram of the square for ${}^*\Sigma_n^k$ and the diagram of the cube for the mapping ${}^*E^k\Sigma_n^l \twoheadrightarrow {}^*\Sigma_n^{k+l}$, respectively.

Lemma 10. *The space* ${}^*\Sigma_n^k$ *is homeomorphic to the Euclidean space* $\mathbf{R}^{2(n+k)}$ *for* $0 < n < q$, *and to the sphere* S^{2k} *for* $n = 0$.

This is because, for $0 < n < q$, ${}^*\Sigma_n^k$ is obtained from the sphere (8) by throwing away the single point $*$, and for $n = 0$ it coincides with this sphere.

Lemma 11. *For the space* ${}^*\Sigma_n^k$, $n \geq q$, *we have the exact sequence of the square* (*10) *of the form* (10) *and the diagram of sequences of the cube* (*13) *of the form* (13) *for* $i \geq 1 - r$, $r = 2(q-1)$. *These are obtained from* (10) *and* (13) *by placing stars on all the spaces.*

This is proved as in §1, with (15) taken into account. We note that instead of the sphere $S = E^k\Sigma_n^0$ in the upper left-hand corners of the squares, there stands the Euclidean space ${}^*E^k\Sigma_n^0$, and therefore $H_{(i+r-1)}(\Sigma/B) \cong H_{(i)}(A)$ even for $i = 1 - r$.

§3. Finiteness, repetition and stability theorems

Keeping in view the sequence (14), we begin our study of the homology of Σ_n^k with an examination of the homology groups of the spaces ${}^*\Sigma_n^k$.

A) **The homology of** $^*\Sigma_n^k$. We will prove the following statement.

Theorem. *All the homology groups* $H_{(i)}(^*\Sigma_n^k)$, $n > 0$, *are finite. They can differ from zero only for* $n = \lambda q$ *and only for* λ *values of* i, *namely for* $\lambda(r-1) < i \leq \lambda r$, $r = 2(q-1)$:

$$H_{(i)}(^*\Sigma_n^k) = 0 \quad \text{for} \quad n \neq \lambda q \text{ and for } n = \lambda q, \ i > \lambda r \quad \text{or} \quad i < \lambda(r-1), \ \lambda > 0. \tag{17}$$

The values of the finite groups $H_{(i)}(^*E_{\lambda q}^k)$, $\lambda(r-1) < i \leq \lambda r$, are calculated for $\lambda \leq 5$ in §4.

Proof. Let $n = \lambda q + s$, $0 \leq s < q$. We proceed by induction on λ, using Lemma 11. Consider the following three assertions:

$$A_\lambda : H_{(i)}(^*\Sigma_{\lambda q+s}^k) = 0 \ \forall i, \ k, \quad \text{if} \quad s \neq 0;$$

$$B_\lambda : H_{(i)}(^*\Sigma_{\lambda q}^k) \quad \text{is a finite group} \quad \forall i, \ k;$$

$$C_\lambda : H_{(i)}(^*\Sigma_{\lambda q}^k) = 0 \ \forall k, \ \forall i > \lambda r, \ \forall i < \lambda(r-1).$$

From the exact sequence (*10) of the square for $^*\Sigma_{n+q}^k$ it is clear that

$$A_\lambda \Rightarrow A_{\lambda+1}, \quad B_\lambda \Rightarrow B_{\lambda+1}, \quad C_\lambda \Rightarrow C_{\lambda+1}.$$

The assertion A_0 follows from Lemma 10. We prove assertions B_1 and C_1. According to the exact sequence (*10) of the square for $^*\Sigma_q^k$.

$$H_{(i)}(^*\Sigma_q^k) = 0 \quad \text{for} \quad i \neq r, \ H_{(r)}(^*\Sigma_q^k) = \mathbf{Z}_{k+1}, \tag{18}$$

where $r = 2(q-1)$. In fact, by Lemma 10, $A = {}^*E^k\Sigma_0^1 = S^{2(k+1)}$ and $B = {}^*\Sigma_0^{k+1} = S^{2(k+1)}$ in the square in question, and by Lemma 1 the degree of ϕ: $A \twoheadrightarrow B$ is equal to $k+1$. Thus the statements B_1 and C_1 are proved. Hence assertion A_λ is true for all $\lambda \geq 0$, and B_λ and C_λ are true for all $\lambda > 0$, as required.

B) **The homology of** Σ_n^k. Combining the sequence (14) with the theorem just proved, we obtain the following results.

Repetition Theorem. *The homology groups of the space* Σ_n^k, *where* n *is not a multiple of* q, *are isomorphic to the homology groups of the preceding space* Σ_{n-1}^k:

$$H_{(i)}(\Sigma_{\lambda q+s}^k) \cong H_{(i)}(\Sigma_{\lambda q}^k) \quad \text{for} \quad 0 < s < q. \tag{19}$$

Finiteness Theorem. *The homology groups of the space* $\Sigma_{\lambda q}^k$ *are finite, except for the first one, and equal to* 0 *except for the first* $\lambda r + 1$ *groups* $(r = 2(q-1))$:

$$H_{(i)}(\Sigma_{\lambda q}^k, \mathbf{R}) = 0 \quad \text{for} \quad i > 0; \qquad H_{(i)}(\Sigma_{\lambda q}^k) = 0 \quad \text{for} \quad i > \lambda r. \tag{20}$$

Note that, by Lemma 3, $H_{(0)}(\Sigma^k_n) = \mathbf{Z}$ for all n, k.

Stability Theorem. *As* $n \to \infty$ *the homology groups of the space* Σ^k_n *become stable; namely,*

$$H_{(i)}(\Sigma^k_{\lambda q}) \cong H_{(i)}(\Sigma^k_{\lambda_0 q}) \quad for \quad \lambda > \lambda_0 \geqslant \frac{i}{r-1} - 1, \tag{21}$$

where $r = 2(q-1)$.

Proof. Consider the exact sequence

$$\ldots \to H_{(i)}({}^*\Sigma^k_n) \to H_{(i)}(\Sigma^k_n) \to H_{(i)}(\Sigma^k_{n-1}) \xrightarrow{\partial} H_{(i+1)}({}^*\Sigma^k_n) \to \ldots . \tag{14}$$

For $n \neq \lambda q$, the end terms are equal to zero by the previous theorem (17), and so the middle terms are equal, which proves the repetition theorem (19).

By the preceding theorem, all the homology groups $H_{(i)}({}^*\Sigma^k_n)$ are finite. From the exact sequence (14), we conclude by induction on n that for $i > 0$ all the groups $H_{(i)}(\Sigma^k_n)$ are finite (we begin the induction with the sphere Σ^k_0). Let $n = \lambda q$. For $i > \lambda r$, by (17), $H_{(i)}({}^*\Sigma^k_n) = 0$. From the exact sequence (14) we conclude by induction on λ that $H_{(i)}(\Sigma^k_n) = 0$ (by the repetition theorem, $H_{(i)}(\Sigma^k_{n-1}) = H_{(i)}(\Sigma^\lambda_{(\lambda-1)q})$). The finiteness theorem (20) is proved.

Let $n = \lambda q$ and $\lambda > \lambda_0 > i/(r-1) - 1$. Then $(\lambda_0 + 1)(r-1) \geq i$. By (17), $H_{(i)}({}^*\Sigma^k_n) = H_{(i-1)}({}^*\Sigma^k_n) = 0$.

From the exact sequence (14), in view of (19), we obtain the stability theorem (21).

Corollary. *The homology groups of the space* ${}_qG^k_n = S^{2(qk+n)} \setminus {}_q\Sigma^k_n$ *of polynomials of degree* $qk+n$ *having less than* k q*-fold foots satisfy the theorems of repetition, finiteness and stability:*

$$H^i({}_qG^k_{\lambda q+s}) \cong H^i({}_qG^k_{\lambda q}) \quad for \quad 0 \leqslant s < q; \tag{22}$$

$$H^i({}_qG^k_n, \mathbf{R}) = 0 \quad for \quad i > 1; \quad H^i({}_qG^k_{\lambda q}) = 0 \quad for \quad i > \lambda r + 1; \tag{23}$$

$$H^i({}_qG^k_{\lambda q}) \cong H^i({}_qG^k_{\lambda_0 q}) \quad for \quad \lambda > \lambda_0 > \frac{i-1}{r-1} - 1, \tag{24}$$

where $r = 2(q-1)$.

Proof. Combining (5) with (19), (20) and (21), we obtain (22), (23) and (24).

Remark 1. Setting $q = 2$ and $k = 1$ in (22), (23) and (24), we obtain the theorems of repetition, finiteness and stability for the cohomology of the braid groups, as stated on page 32. The finiteness theorem (23) leads to interesting identities between the generators $(dz_i - dz_j)/(z_i - z_j)$ of the cohomology ring of the space of ordered sets of n distinct points $\{(z_1, \cdots, z_n) \in \mathbf{C}^n, z_i \neq z_j\}$ (see [29]).

Remark 2. There exists a natural embedding of the braid group with $n-1$ strings into the braid group with n strings. It is compatible with the natural embedding $j\colon G_{n-1} \rightarrowtail G_n$ which extends to the embedding $j\colon \mathbf{C}^{n-1} \rightarrowtail \mathbf{C}^n$ (and even to $j\colon S^{2(n-1)} \rightarrowtail S^{2n}$). This embedding is constructed in the following way. Let $p \in \mathbf{C}^{n-1}$ be a polynomial of degree $n-1$ with roots $z_1, \cdots, z_{n-1}$ (with any multiplicities). Then $jp \in \mathbf{C}^n$ is the polynomial whose roots are

$$z_1, z_2, \ldots, z_{n-1} \text{ and } z_n = z_0 + \max_{1 \leqslant i < n} |z_i - z_0| + 1, \; z_0 = \frac{z_1 + \ldots + z_{n-1}}{n-1}.$$

It is easily seen that j preserves multiplicities of all roots, and therefore preserves the associated G, Σ^k, etc.

From the constructions of §2 it is not difficult to discern that $j^*\colon H^i(G_n^k) \to H^i(G_{n-1}^k)$ is dual to the mapping $H_{(i)}(\Sigma_n^k) \to H_{(i)}(\Sigma_{n-1}^k)$ in (14). Therefore the stability and repetition isomorphisms (22) and (24) are induced by the embeddings j.

Remark 3. There also exist natural operations of "addition" of braids $B(x) \times B(n) \to B(m+n)$ and of "multiplication" $B(m) \times B(n) \to B(mn)$, connected with the mappings $G_m \times G_n \to G_{m+n}$ and $G_m \times G_n \to G_{mn}$. It would be interesting to study their effect on cohomology.

§4. Computations for small values of n

In the computation of the groups $H_{(i)}(\Sigma_n^k)$ we come across certain special exact sequences; it is convenient to write these out beforehand.

A) **Arithmetical lemmas.** We denote by $\mathbf{Z}_k$ the cyclic group of order k, and by $\mathbf{Z}$ the free cyclic group, each group having a *fixed generator*. We denote by

$$\mathbf{Z}_k \xrightarrow{m} \mathbf{Z}_l$$

the homomorphism carrying the generator of $\mathbf{Z}_k$ into the mth power of the generator of $\mathbf{Z}_l$ (not excepting the cases when k or l is equal to 1 or ∞). Of course, mk must be divisible by l, and m defined modulo l. Consider the exact sequence

$$\mathrm{Ker} \rightarrowtail \mathbf{Z}_k \xrightarrow{m} \mathbf{Z}_l \twoheadrightarrow \mathrm{Coker}.$$

Lemma 12. *In the kernel and in the cokernel the generator can be chosen in such a way that the above sequence takes the form*

$$\mathbf{Z}_\varkappa \overset{u}{\rightarrowtail} \mathbf{Z}_k \xrightarrow{m} \mathbf{Z}_l \overset{1}{\twoheadrightarrow} \mathbf{Z}_\lambda, \tag{25}$$

where $k = u\kappa$, $l = u\lambda$, $m = \mu\lambda$ and μ is relatively prime to u. This choice of generators is unique and natural in the sense that the commutative diagram

$$\begin{array}{ccc} \mathbf{Z}_k & \xrightarrow{m} & \mathbf{Z}_l \\ {\scriptstyle p}\downarrow & & \downarrow{\scriptstyle q} \\ \mathbf{Z}_{k'} & \xrightarrow{m'} & \mathbf{Z}_{l'} \end{array}$$

can be extended to the diagram

$$\begin{array}{ccccccc} \mathbf{Z}_\varkappa & \overset{u}{\rightarrowtail} & \mathbf{Z}_k & \xrightarrow{m} & \mathbf{Z}_l & \overset{1}{\twoheadrightarrow} & \mathbf{Z}_\lambda \\ \downarrow{\scriptstyle v} & & \downarrow{\scriptstyle p} & & \downarrow{\scriptstyle q} & & \downarrow{\scriptstyle q} \\ \mathbf{Z}_{\varkappa'} & \overset{u'}{\rightarrowtail} & \mathbf{Z}_{k'} & \xrightarrow{m'} & \mathbf{Z}_{l'} & \overset{1}{\twoheadrightarrow} & \mathbf{Z}_{\lambda'} \end{array},$$

where $u'v = up$ *and* $m'p = mq$.

Proof. For the generator of the kernel we take the least power of the generator of $\mathbf{Z}_k$ that lies in the kernel, and for the generator of the cokernel we take the image of the generator of $\mathbf{Z}_l$.

We require several special cases of the sequence (25). We denote by (a, b) the greatest common divisor of the integers a and b. We introduce the notation

$$a/b = \frac{a}{(a, b)}, \quad a/b/c = a/(b/c).$$

By means of the identity $C_b^a = (b/a)\, C_{b-1}^{a-1}$, one can readily check

Lemma 13. *We have the following exact sequences:*

$$\mathbf{Z}_{2/b} \rightarrowtail \mathbf{Z}_2 \xrightarrow{C_b^2} \mathbf{Z}_b \overset{1}{\twoheadrightarrow} \mathbf{Z}_{b/2}, \qquad (26)_b$$

$$\mathbf{Z}_{3/b} \overset{1}{\rightarrowtail} \mathbf{Z}_3 \xrightarrow{C_b^3} \mathbf{Z}_{b/2} \overset{1}{\twoheadrightarrow} \mathbf{Z}_{b/6}, \qquad (27)_b$$

$$\mathbf{Z}_{2/b/2} \overset{1}{\rightarrowtail} \mathbf{Z}_2 \xrightarrow{C_b^4} \mathbf{Z}_{b/6} \overset{1}{\twoheadrightarrow} \mathbf{Z}_{b/12}, \qquad (28)_b$$

$$\mathbf{Z}_{5/b} \overset{1}{\rightarrowtail} \mathbf{Z}_5 \xrightarrow{C_b^5} \mathbf{Z}_{b/12} \overset{1}{\twoheadrightarrow} \mathbf{Z}_{b/60}, \qquad (29)_b$$

$$\mathbf{Z}_{(p,b)} \overset{1}{\rightarrowtail} \mathbf{Z}_p \xrightarrow{b} \mathbf{Z}_{p/b} \overset{1}{\twoheadrightarrow} 0, \text{ where } p \text{ is simple.} \qquad (30)_{p,b}$$

We remark that the group Ker(on the left) has the following values:

$$(26) \begin{cases} 0 & b \text{ even} \\ \mathbf{Z}_2 \end{cases} \quad (27) \begin{cases} 0 & b \text{ a multiple of } 3 \\ \mathbf{Z}_3 \end{cases} \quad (28) \begin{cases} 0 & b \text{ a multiple of } 4 \\ \mathbf{Z}_2 \end{cases}$$

$$(29) \begin{cases} 0 & b \text{ a multiple of } 5 \\ \mathbf{Z}_5 \end{cases} \quad (30) \begin{cases} \mathbf{Z}_p & b \text{ a multiple of } p \\ 0 \end{cases}$$

B) The homology groups $H_{(i)}({}^*\Sigma^k_{\lambda q})$. By (17) the groups can be nontrivial only for those λ for which i lies in the range $\lambda(r-1) < i \le \lambda r$, $r = 2(q-1)$. We list them by the index $j = 1, \cdots, \lambda$, where $i = \lambda(r-1) + j$. We put

$$h^k_{\lambda,j} = H_{(\lambda(r-1)+j)}({}^*\Sigma^k_{q\lambda}) = H_{2k+3\lambda-j}({}^*\Sigma^k_{q\lambda}, \infty).$$

Theorem. *For* $1 \le \lambda \le 5$ *the homology groups* $H_{(i)}({}^*\Sigma^k_{q\lambda})$ *are cyclic,* $h^k_{\lambda,j} \cong \mathbf{Z}_{\alpha^k_{\lambda,j}}$, *and the orders* $\alpha^k_{\lambda,j}$ *are given by the following table:*

The orders $\alpha^k_{\lambda,j}$ *of the group* $h^k_{\lambda,j}$

$\lambda \backslash j$	1	2	3	4	5
1	$k+1$				
2	$2/k$	$(k+2)/2$			
3	$2/k$	$3/k$	$(k+3)/6$		
4	$2/k$	$(3,\ k+1)$	$2/k/2$	$(k+4)/12$	
5	$2/k$	1	$(6,\ k+5)$	$(2,(k+3/2))(5,k)$	$(k+5)/60$

The homology groups $h^1_{\lambda,j}$ *in the case* $k=1$

$\lambda \backslash j$	1	2	3	4	5	6
1	$\mathbf{Z}_2$					
2	$\mathbf{Z}_2$	$\mathbf{Z}_3$				
3	$\mathbf{Z}_2$	$\mathbf{Z}_3$	$\mathbf{Z}_2$			
4	$\mathbf{Z}_2$	0	$\mathbf{Z}_2$	$\mathbf{Z}_5$		
5	$\mathbf{Z}_2$	0	$\mathbf{Z}_6$	$\mathbf{Z}_{10}$	0	
6	$\mathbf{Z}_2$	0	$\mathbf{Z}_6$	$\mathbf{Z}_2$	0	$\mathbf{Z}_7$

Proof. Consider the following assertions A_λ and B_λ. A^k_λ: $h^k_{\lambda,j} \cong \mathbf{Z}_{\alpha^k_{\lambda,j}}$ $\forall q,j$ *as a group with fixed generator.*

${}^kB^l$: *the homomorphism of cyclic groups with generators fixed (according to* A_λ*)* $\phi_*\colon h^l_{\lambda,j} \to h^{k+l}_{\lambda,j}$, *induced by* $\phi\colon {}^*E^k\Sigma^l_{\lambda q} \to\!\!\to \Sigma^{k+l}_{\lambda q}$, *has the form* $\mathbf{Z}_{\alpha^k_{\lambda,j}} \xrightarrow{m} \mathbf{Z}_{\alpha^{k+l}_{\lambda,j}}$, *where* $m = C^l_{k+l}$ *for* $j < \lambda$ *and* C^{l+1}_{k+l+1} *for* $j = \lambda$.

By A^k_0 we mean the choice of generator in $H_{(0)}({}^*\Sigma^k_0) \cong \mathbf{Z}$ (Lemma 10) in conformity with the complex structure. By ${}^kB^l_0$ we understand the group homomorphism with fixed generator given in Lemma 1

$$\varphi_*\colon H_{(0)}({}^*\Sigma^l_0) \to H_{(0)}({}^*\Sigma^{k+l}_0), \quad \mathbf{Z} \xrightarrow{C^l_{k+l}} \mathbf{Z}$$

(the stars are fictitious, as $*$ belongs neither to Σ^l_0 nor to $E^k\Sigma^l_0$).

Thus the assertions A^k_0 and ${}^kB^l_0$ are true for all k, $l \ge 1$.

Denote by ${}^*(10)^k_\lambda$ and ${}^k_*(13)^l_\lambda$ the sequence of the square (10) for ${}^*\Sigma^k_{\lambda q}$ and the diagram of sequences of the cube (13) for $^*E^k\Sigma^l_{\lambda q} \twoheadrightarrow {}^*\Sigma^{k+l}_{\lambda q}$. The vertical endomorphism of $H_{(i)}({}^*\Sigma^1_{(\lambda-1)q})$ in ${}^k_*(13)^l_\lambda$ is known to us: it is simply multiplication by C^l_{k+l}. Furthermore, the assertions $A_{\lambda-1}$ contain a description of the groups $H_{(i)}$ on the right-hand square of $*(13)_\lambda$, and the $B_{\lambda-1}$ contain a description of the homomorphisms appearing in this square. Having such information for all i, we obtain assertions A_λ and B_λ from the diagram ${}_*(13)_\lambda$ by means of Lemmas 12 and 13.

Setting $\lambda = 1, 2, 3, 4, 5$ in turn, we see that

$$^*(10)^k_\lambda + A^1_{\lambda-1} + A^{k+1}_{\lambda-1} + {}^kB^1_{\lambda-1} + X_\lambda = \mathfrak{A}^k_\lambda \Rightarrow A^k_\lambda, \tag{31}$$

$$^k_*(13)^l_\lambda + A^l_\lambda + A^{k+l}_\lambda + {}^kB^{l+1}_{\lambda-1} + Y_\lambda = {}^k\mathfrak{B}^l_\lambda \Rightarrow {}^kB^l_\lambda, \tag{32}$$

where X_λ and Y_λ stand for arithmetical information from Lemmas 13 and 12 respectively. More precisely,

$$X_1 = \{0 \overset{1}{\rightarrowtail} \mathbf{Z} \overset{k+1}{\longrightarrow} \mathbf{Z} \overset{1}{\twoheadrightarrow} \mathbf{Z}_{k+1}\},$$
$$X_2 = (26)_{k+2},\ X_3 = (27)_{k+3} + (30)_{2,k+1},$$
$$X_4 = (28)_{k+4} + (30)_{3,k+1} + (30)_{2,k+1},$$
$$X_5 = (29)_{k+5} + (30)_{2,k+1} + (33)_{k+1},$$

where

$$\mathbf{Z}_{(2,b)} \overset{1}{\rightarrowtail} \mathbf{Z}_2 \overset{b}{\rightarrow} \mathbf{Z}_{2/b/2} \overset{1}{\twoheadrightarrow} \mathbf{Z}_{2,(b+2/2)}. \tag{33$_b$}$$

The assertion A^k_5 is obtained in weakened form, because Lemma 12 does not fix the generator of $h^k_{\lambda,j}$ for $\lambda = 5$, $j = 3$ and 4. The assertions B_5, generally speaking, lose their meaning. Nevertheless from ${}^1\mathfrak{B}^1_5$ we obtain

j	1	2	3	4	5
φ^*	$\mathbf{Z}_2 \to 0$	$0 \to 0$	$\mathbf{Z}_6 \to 0$	$\mathbf{Z}_{10} \to \mathbf{Z}_5$	$0 \to \mathbf{Z}_7$

$\mathfrak{A}^1_6$ now yields the last line of the table of the homology groups $h^1_{\lambda,j}$, and the theorem is proved.

Remark. The groups $h^k_{\lambda,j} = H_{2k+3\lambda-j}({}^*_q\Sigma^k_{\lambda q}, \infty)$ found above do not depend on q.

C) The homology groups $H_{(i)}(\Sigma^k_{\lambda q})$. Combining the results of the preceding computations with the exact sequence (14) we can compute many of the homology groups $H_{(i)}(\Sigma^k_{\lambda q})$, particularly for large values of q. For $\lambda < r = 2(q-1)$ the sequence $H_{(i)}(\Sigma^k_{\lambda q})$, $i = 0, 1, \cdots$, consists of segments of the form $H_{(i)}({}^*\Sigma^k_{\mu q})$, $\mu \leq \lambda$, separated by zeros. For example, for $q = 3$ the homology groups $H_{(i)}({}_3\Sigma^1_6)$

are given by the table

i	0	1	2	3	4	5	6	7	8	$\geqslant 9$
$H_{(i)}$	$\mathbf{Z}$	0	0	0	$\mathbf{Z}_2$	0	0	$\mathbf{Z}_2$	$\mathbf{Z}_3$	0

The first ten groups ($i \leq 9$) are stable, in accordance with (21).

In the case $q = 2$, which is of particular interest to us, we have the

Theorem. *The homology groups $H_{(i)}(\Sigma_n^k)$, $n < 10$, are cyclic. $H_{(0)} \cong \mathbf{Z}$, $H_{(1)} \cong 0$; and the orders of the others are given by the following table:*

The orders $\beta_{n,i}^k$ of the groups $H_{(i)}(\Sigma_n^k)$

$n \backslash i$	2	3	4	5	6	7	8
2,3	$k+1$						
4,5	$k+1$	$2/k$	$(k+2)/2$				
6,7	$k+1$	$2/k$	$((k+2)/2)(2/k)$	$3/k$	$(k+3)/6$		
8,9	$k+1$	$2/k$	$((k+2)/2)(2/k)$	$6/kv$	$(3,k+1)((k+3)/12)u$	$(3/k)(2/k/2)$	$(k+4)/12$

Here u, $v = 1$ if $k \neq 4s + 1$. But if $k = 4s + 1$, then u, $v = 1$ or 2, $u + v = 3$.

The homology groups $H_{(i)}(\Sigma_n^1)$

$n \backslash i$	0	1	2	3	4	5	6	7	8
0,1	$\mathbf{Z}$								
2,3	$\mathbf{Z}$	0	$\mathbf{Z}_2$						
4,5	$\mathbf{Z}$	0	$\mathbf{Z}_2$	$\mathbf{Z}_2$	$\mathbf{Z}_3$				
6,7	$\mathbf{Z}$	0	$\mathbf{Z}_2$	$\mathbf{Z}_2$	$\mathbf{Z}_6$	$\mathbf{Z}_3$	$\mathbf{Z}_2$		
8,9	$\mathbf{Z}$	0	$\mathbf{Z}_2$	$\mathbf{Z}_2$	$\mathbf{Z}_6$	$\mathbf{Z}_6$? 3	$\mathbf{Z}_2$? 1	$\mathbf{Z}_2$	$\mathbf{Z}_5$

The stable groups $H_i(\Sigma_\infty^k)$

i	0	1	2	3	4	5
$H_{(i)}(\Sigma_\infty^k)$	$\mathbf{Z}$	0	$\mathbf{Z}_{k+1}$	$\mathbf{Z}_{2/k}$	$\mathbf{Z}_{(k+2)/2} + \mathbf{Z}_{2/k}$	$\mathbf{Z}_{6/kv}$
$H_{(i)}(\Sigma_\infty^1)$	$\mathbf{Z}$	0	$\mathbf{Z}_2$	$\mathbf{Z}_2$	$\mathbf{Z}_6$	$\mathbf{Z}_6$ for 3

Remark. Passing to cohomology by means of formula (5), we obtain the cohomology table of the complements $2G_n^k = S^{2(2k+n)} \setminus {}_2\Sigma_n^k$; in particular, for $k = 1$ we have the cohomology tables of the braid groups given on pages 32 and 33.

Proof of the theorem. Consider the following assertions:

A^k_λ: $H_{(i)}(\Sigma^k_{2\lambda}) \cong \mathbf{Z}_{\beta^k_{2\lambda,i}}$ *as a group with fixed generator.*

${}^kB^l_\lambda$: *The homomorphism* $\phi_*: H_{(i)}(\Sigma^l_{2\lambda}) \to H_{(i)}(\Sigma^{k+l}_{2\lambda})$ *has the form* $\mathbf{Z}_{\beta^k_{2\lambda,i}} \xrightarrow{m_i} \mathbf{Z}_{\beta^{k+l}_{2\lambda,i}}$, *where* m_i *is given by the table*

i	0	1	2	3	4	5	6
m_i	C^l_{k+l}	C^l_{k+l}	C^{l+1}_{k+l+1}	C^l_{k+l}	C^{l+2}_{k+l+2}	C^l_{k+l}	C^{l+3}_{k+l+3}

By A^k_0 we understand the choice of generator of $H_{(0)}(\Sigma^k_0)$ in conformity with the complex structure, and by ${}^kB^l_0$ we understand the homomorphism ϕ_*: $\mathbf{Z} \xrightarrow{C^l_{k+l}} \mathbf{Z}$ (Lemma 1) of groups with fixed generator.

The assertions A^k_0 and ${}^kB^l_0$ are thus true for all k and l.

The proofs of A_λ and B_λ, $\lambda = 1, 2, 3$, are obtained from (31) and (32) by deleting the stars and replacing X_λ and Y_λ by X'_λ and Y'_λ, where $X'_\lambda = X_\lambda + \mathfrak{A}_{\lambda-1}$, $Y'_\lambda = Y_\lambda + \mathfrak{B}_{\lambda-1}$, $\mathfrak{A}_0 = A_0$ and $\mathfrak{B}_0 = B_0$. The assertion A^k_3 for odd k is obtained in weakened form, because Lemma 12 does not fix the generator for $\lambda = 3$, $i = 4$. Thus for $i = 4$, ${}^kB^l_3$ is not determined if k or l is odd. Nevertheless, from $\mathfrak{B}_3$ the kernel of the mapping $\phi_*: H_{(4)}(\Sigma^1_6) \to H_{(4)}(\Sigma^{k+l}_6)$ is obtained up to a direct factor $\mathbf{Z}_2$:

$$\mathbf{Z}_{3/k} + \mathbf{Z}_u \rightarrowtail \mathbf{Z}_6 \xrightarrow{\varphi_*} \mathbf{Z}_{(k+3)/2} + \mathbf{Z}_{(2,k)},$$

where $u = 1$ for $k \neq 4s + 1$, and for $k = 4s + 1$ either $u = 1$ or $u = 2$.

$\mathfrak{A}_4$ now yields the values of $\beta^k_{8,i}$ indicated in the table, and the theorem is proved.

Remark. All the groups $H_{(i)}(\Sigma^k_{2\lambda})$ found above turn out to be equal to the direct sums $\Sigma^\lambda_{\mu=0} H_{(i)}({}^*\Sigma^k_{2\mu})$.

Added in translation. Recently $H^*(B(n), \mathbf{Z}_2)$ has been completely calculated by D. B. Fuks [34]. He found that the natural mapping

$$H^*(O(n), \mathbf{Z}_2) \to H^*(B(n), \mathbf{Z}_2)$$

is an epimorphism; that is, that all the $\mathbf{Z}_2$ cohomology classes of braid groups are generated by the Stiefel-Whitney classes.

This gives a new proof to our theorem on the Tschirnhausen transformations of algebraic functions [30]. One can deduce from [34] a theorem on the impossibility of reducing the number of variables of an algebraic function by the superposition of entire algebraic functions (see [33]).

Received 10 MAY 68

BIBLIOGRAPHY

[1] B. Morin, *Formes canoniques des singularités d'une application différentiable*, C. R. Acad. Sci. Paris 260 (1965), 5662–5665, 6503–6506. MR 32 #8354.

[2] V. I. Arnol'd, *Singularities of smooth mappings*, Uspehi Mat. Nauk 23 (1968), no. 1 (139), 3–44 = Russian Math. Surveys 23 (1968), no. 1, 1–43.

[3] O. Zariski, *Algebraic surfaces*, Ergebnisse der Mathematik, Bände 3, 5, Springer-Verlag. Berlin, 1935.

[4] E. Fadell and L. Neuwirth, *Configuration spaces*, Math. Scand. 10 (1962), 111–118. MR 25 #4537.

[5] R. H. Fox and L. Neuwirth, *The braid groups*, Math. Scand. 10 (1962), 119–126. MR 27 #742.

[6] E. Artin, *Theorie der Zöpfe*, Abh. Math. Sem. Univ. Hamburg 4 (1925), 47–72.

[7] F. Klein, *Vorlesungen über höhere Geometrie*, 3rd ed., Springer, Berlin, 1926; Russian transl., ONTI, Moscow, 1929.

[8] W. Burau, *Über Zopfinvarianten*, Abh. Math. Sem. Univ. Hamburg 9 (1932), 117–124.

[9] ———, *Über Zopfgruppen und gleichsinnig verdrillte Verkettungen*, Abh. Math. Sem. Univ. Hamburg 11 (1935), 179–186.

[10] A. A. Markov, *Über die freie Äquivalenz der geschlossenen Zöpfe*, Mat. Sb. 1 (43) (1936), 73–78; Trudy Mat. Inst. Steklov. 16 (1945), (Russian) MR 8, 131.

[11] E. Artin, *Theory of braids*, Ann. of Math. (2) 48 (1947), 101–126. MR 8, 367.

[12] ———, *Braids and permutations*, Ann. of Math. (2) 48 (1947), 101–126. MR 9, 6.

[13] F. Bohnenblust, *The algebraical braid group*, Ann. of Math. (2) 48 (1947), 127–136. MR 8, 367.

[14] W. L. Chow, *On the algebraical braid group*, Ann. of Math. (2) 49 (1948), 654–658. MR 10, 98.

[15] E. Artin, *The theory of braids*, Amer. Sci. 38 (1950), 112–119. MR 11, 377.

[16] J. E. Reeve, *A summary of results in the topological classification of plane algebroid singularities*, Rend. Sem. Math. Torino 14 (1954/55), 159–187. MR 20 #3146.

[17] B. J. Gassner, *On braid groups*, Abh. Math. Sem. Univ. Hamburg 25 (1961), 10–22. MR 24 #A174.

[18] S. Lipschutz, *On a finite matrix representation of the braid group*, Arch. Math. 12 (1961), 7–12. MR 23 #A2469.

[19] E. Fadell and J. Van Buskirk, *On the braid groups of E^2 and S^2*, Bull. Amer. Math. Soc. 67 (1961), 211–213. MR 23 #A2877.

[20] J. A. H. Shepperd, *Braids which can plaited with their threads tied together at each end,* Proc. Roy. Soc. Ser. A 265 (1961/62), 229–224. MR 24 A#2959.

[21] G. Burde, *Zur Theorie der Zöpfe,* Math. Ann. 151 (1963), 101–107. MR 27 #1942.

[22] S. Lipschutz, *Note on a paper by Shepperd on the braid group,* Proc. Amer. Math. Soc. 14 (1963), 225–227. MR 26 #4350.

[23] G. Burde, *Über Normalisatoren der Zopfgruppe,* Abh. Math. Sem. Univ. Hamburg 27 (1964), 97–115. MR 30 #1188.

[24] J. Van Buskirk, *Braid groups of compact 2-manifolds with elements of finite order,* Trans. Amer. Math. Soc. 122 (1966), 81–97. MR 32 #6440.

[25] W. Magnus, A. Karrass and D. Solitar, *Combinatorial group theory,* Interscience, New York, 1966. MR 34 #7617.

[26]* V. I. Arnol'd, *Braids of algebraic functions and the cohomologies of swallow tails,* Uspehi Mat. Nauk 23 (1968), no. 4 (142), 247–248. (Russian) MR 38 #156.

[27]* ———, *A remark on the branching of hyperelliptic integrals as functions of the parameters,* Funkcional. Anal. i Priložen. 2 (1968), no. 3, 1–3 = Functional Anal. Appl. 2 (1968), 187–189.

[28]* A. N. Varčenko, *On the bounding of multiple integrals dependent on the parameters,* Funkcional. Anal. i Priložen. 3 (1969), no. 4, 79–80 = Functional Anal. Appl. 3 (1969), 322–324.

[29]* V. I. Arnol'd, *The cohomology ring of the group of dyed braids,* Mat. Zametki 5 (1969), 227–231. (Russian) MR 39 #3529.

[30]* ———, *On cohomology classes of algebraic functions preserved under Tschirnhausen transformations,* Funkcional. Anal. i Priložen. 4 (1970), no. 1 no. 1, 84–85 = Functional Anal. Appl. 4 (1970) (to appear)

[31]* E. A. Gorin and V. Ja. Lin, *The braid group and algebraic equations with continuous coefficients,* Uspehi Mat. Nauk 24 (1969), no. 2 (146), 225–226. (Russian) MR 39 #3479.

[32]* ———, *Algebraic equations with continuous coefficients and some problems of the algebraic theory of braids,* Mat. Sb. 78 (120) (1969), 579–610 = Math. USSR Sb. 7 (1969), 569–596.

* The articles marked with an asterisk appeared while the present article was being printed.

[33]** V. I. Arnol'd, *Topological invariants of algebraic functions*. II, Funkcional. Anal. i Priložen. 4 (1970), no. 2, 1–9 = Functional Anal. Appl. 4 (1970) (to appear).

[34]** D. B. Fuks, Mod 2 *cohomology of braid groups*, Funkcional. Anal. i Priložen. 4 (1970), no. 2, 62–73 = Functional Anal. Appl. 4 (1970) (to appear).

[35]** J. Birman, *Mapping class groups and their relationship to braid groups*, Comm. Pure Appl. Math. 22 (1969), 213–242. MR 39 #4840.

[36]** W. Magnus and A. Peluso, *On a theorem of V. I. Arnol'd*, Comm. Pure Appl. Math. 22 (1969), 683–692.

Translated by:
D. L. Johnson

***Editor's note*. References [33] – [36] were added by the author for the English translation.

TOPOLOGICAL INVARIANTS OF ALGEBRAIC FUNCTIONS. II*

V. I. Arnol'd

In an earlier paper of the same name [2] the connection between algebraic functions and braids was used in order to calculate the cohomologies of braid groups. In this work the cohomologies of braid groups are used to confirm the non-representability of algebraic functions of a certain number of variables as superpositions of algebraic functions of a smaller number of variables.

The reason for this non-representability is topological: a representation of a function as a superposition prevents the classes of cohomologies of the complement to the branch manifold of the function from being induced by the class of cohomologies of the braid groups.

§1. DEFINITION OF THE SUPERPOSITION OF MANY-VALUED FUNCTIONS

The standard designations of many-valued functions are at present ambiguous. For example, how many values are attained by the function $\sqrt{z}+\sqrt{z}$: two, three, or more? We will make use of the following formalisms:†

A. Symmetric Power. Let X be a topological space. The space of unordered collections of n not necessarily distinct points of X is called the n-th symmetric power of X and denoted by $\mathrm{Sym}^n X$. Thus the symmetric power $\mathrm{Sym}^n X$ is a topological space obtained from the Cartesian n-th power X^n factorized by means of the real symmetric group S(n) of coordinate permutations: $\mathrm{Sym}^n X = \dfrac{X^n}{n!}$.

Example. Let C be the complex plane. Then $\mathrm{Sym}^n C = C^n$ by the theorem of Vieta and the "fundamental theorem of algebra."

B. Many-valued Mappings. A mapping $f: X \to \mathrm{Sym}^n Y$ is called an n-valued mapping of X into Y and written in the form $f: X \underset{n}{\to} Y$. For the point $x \in X$ the corresponding point $y_i \in Y$ $(i = 1, \ldots, n)$ is called the value of the mapping f at the point x.

Example. The n-th order equation

$$z^n + a_1 z^{n-1} + \ldots + a_n = 0 \tag{1}$$

defines an n-valued mapping $z: \mathbf{C}^n \underset{n}{\to} \mathbf{C}$, whose value at the point a with coordinates $a_1, \ldots, a_n$ is the set of n roots of Eq. (1). The mapping z is called a universal (entire, n-valued) algebraic function. (In standard notation $z = z(a)$.)

C. Superpositions. Let $f: X \underset{n}{\to} Y$ and $g: Y \underset{m}{\to} Z$ be multiple-valued mappings. The superposition $g \circ f$ of the mappings g and f is a mapping $g \circ f: X \to Z$, whose value at the point $x \in X$ consists of every one of the mn values of g at the n image points in Y of the point x under the mapping f. The operation $\circ$ is associative, whence we will write $f \circ (g \circ h)$ in the form $f \circ g \circ h$.

---

*The author is grateful to B. B. Venkov, D. K. Faddeev, and D. B. Fuks for their helpful discussions. The author is, on the other hand, particularly grateful to D. B. Fuks for making available his recent paper [5].

†We will define the superposition of algebraic functions as their composite product.

Moscow State University, Moscow. Translated from Funktsional'nyi Analiz i Ego Prilozheniya, Vol. 4, No. 2, pp. 1-9, April-June, 1970. Original article submitted December 14, 1969.

Originally publ. in: Funkts. Anal. Prilozh. 4:2, 1-9, © Russ. Acad. Sciences 1970
English transl.: Funct. Anal. Appl. 4, 91-98, © Consultants Bureau, New York 1970
Reprint by Springer-Verlag Berlin Heidelberg 2014 with kind permission, DOI 10.1007/978-3-642-31031-7_24

Example. Let $p : C^k \to C^n$ be a polynomial (single-valued) mapping and let $z : C^n \underset{n}{\to} C$ be the universal function of (1). Then the n-valued mapping $f = z \circ p : C^k \underset{n}{\to} C^1$ is said to be an algebraic (entire, n-valued) function on C^k with coefficients p. We note that an algebraic function f uniquely defines the polynomial mapping p. (In other words we have $f(x) = z(p(x))$.)

D. An Entire Superposition. Together with the superposition ($\circ$) we will examine the concept of an entire superposition (*). An entire superposition $g * f$ is an mn-valued mapping of X into the Cartesian product $Y \times Z$. The value of the mapping $g * f : X \underset{mn}{\to} Y \times Z$ at a point $x \in X$ is the pair (y, z), where y is one of the values of f at the point x, while z is one of the values of g at the point y. The operation * is non-associative, and we will follow the convention that $h * g * f$ denotes $h * (g * f)$.

Example. Let $f : C^k \underset{n}{\to} C^1$ be an algebraic function, $q : C^{k+1} \to C^1$ – a polynomial, and $1 : C^k \to C^k$ the identity operator. Then the n-valued algebraic function $F = q \circ (f * 1) : C^k \underset{n}{\to} C^1$ is called a Tschirnhausen transformation of the function f with coefficients q. (In functional notation $F(x) = q(f(x), x)$.)

E. The Representation of a Function by a Superposition. Let $\varphi_i : C^l \underset{n_i}{\to} C^1$ $(i = 1, \ldots, N)$ be algebraic function of l variables. Let $p_i : C^{k+i-1} \to C^l$ $(i = 1, \ldots, N)$ be polynomial mappings. For the functions φ_i and p_i we construct the composite function f. Let us define algebraic functions $\Phi_i = \varphi_i \circ p_i : C^{k+i-1} \underset{n_i}{\to} C^1$. We now define the sequence

$$F_1 = \Phi_1 * 1 : C^k \underset{n_1}{\to} C^{k+1},\ F_2 = \Phi_2 * F_1 : C^k \underset{n_1 n_2}{\longrightarrow} C^{k+2},\ \ldots,\ F_N = \Phi_N * F_{N-1} : C^k \underset{n_1 \ldots n_N}{\longrightarrow} C^{k+N}$$

(where $1 : C^k \to C^k$ is the identity mapping). The function $f : C^k \underset{m}{\to} C$ is represented in the form of a superposition of algebraic functions of l variables, φ_i, if $f = \Phi_N \circ F_{N-1}$, i.e.,

$$f = \Phi_N \circ \Phi_{N-1} * \ldots * \Phi_1 * 1. \tag{2}$$

(Actually writing this term is awkward; yet for N = 2 we have $f(x) = \varphi_2\{p_2[\varphi_1(p_1(x)), x]\}$.) We note that f is an $m = n_1 \ldots n_N$-valued function.

Example. The formula for the solution of the cubic equation $z^3 + 3x_1 z + 2x_2 = 0$

$$f = \Phi_3 + \Phi_2,\quad \Phi_3 = \sqrt[3]{-x_2 - \Phi_1},\quad \Phi_2 = \sqrt[3]{\Phi_1 - x_2},\quad \Phi_1 = \sqrt{x_2^2 + x_1^3}$$

defining an 18-valued algebraic function of two variables x_1, x_2 is representable in the form of a superposition of algebraic functions of one variable $\varphi_1 = \sqrt{\ }$, $\varphi_2 = \sqrt[3]{\ }$, $\varphi_3 = \sqrt[3]{\ }$. The original three-valued function does not coincide with this 18-valued function, and is only one of its "indecomposable components."

F. Remark. The polynomial mapping $p_i : C^{k+i-1} \to C^l$ in the definition of representability of a function as a superposition can be taken as a linear function (and indeed as a projection onto a coordinate subspace). The class of representable functions for this is not changed if we take $l \geq 2$. In fact every polynomial can be represented in the form of a superposition of single-valued algebraic functions of the two variables $z = x + y$, $z = xy$.

Every function f representable in the form of a superposition of algebraic functions (2) is algebraic.

The proof is obvious. We note that the coefficients of an algebraic function f are polynomials in x. The coefficients of the latter are expressed as polynomials using the coefficients of the polynomials p_i and those of all polynomials of l variables, constituting the coefficients of the algebraic functions φ_i.

§2. THEOREM ON NON-REPRESENTABILITY

The following theorem is proved in part 4 with the aid of techniques of part 3.

A. THEOREM. An algebraic function of k variables $a_1, \ldots, a_n$

$$z^n + a_1 z^{n-2} + a_2 z^{n-3} + \ldots + a_k = 0 \qquad (k = n - 1) \tag{3}$$

cannot be represented as a superposition of algebraic functions of less than k variables, if $n = 2^r$, $r \geq 2$.

In particular the roots of a polynomial of even order

$$z^4 + a_1 z^2 + a_2 z + a_3 = 0,$$

considered as functions of the coefficients, cannot be represented as the superposition of algebraic functions of two variables.

The following lemma reduces the problem to a particular case already considered in [3].

LEMMA 1. If the function (3) can be represented as a superposition (2), then all of the algebraic functions $\varphi_1, \ldots, \varphi_N$, except one, are single-valued.

B. Indecomposable Coverings. The proof of Lemma 1 is based on the following remark.

Let p_1: $Y \to X$ and p_2: $Z \to Y$ be (unbranched) coverings. Then $p_3 = p_1 \circ p_2$: $Z \to X$ is also a covering. We will say that the covering p_3 is indecomposable if for any of its composite representations $p_3 = p_1 \circ p_2$ of two coverings, one of the factors is trivial (reciprocally single-valued). It is easily proved that

LEMMA 2. If the monodromy group of an n-sheeted covering is the symmetric group S(n), then it is indecomposable.

Proof of Lemma 2. The monodromy group M_1 of the covering p_1 is a factor group of the monodromy group S(n) of p_3. Let $n \geq 5$. Then $M_1 = 1$, for either S(n) or Z_2, since S(n) has as normal divisors only S(n), 1 and the group A(n) of even permutations. In the first case p_1 is reciprocally single-valued, while in the second case p_2 is. In the third case $p_1^{-1}(x)$ consists of two points y_1, y_2. The group A(n), operating on n points of the set $p_3^{-1}(x)$, does not affect n/2 points of $p_2^{-1}(y_1)$. But the group A(n) operates on n points of $p_3^{-1}(x)$ transitively. Therefore the third case is impossible.

In the case n = 2 or 3 the proof is obvious, since the numbers 2 and 3 are prime. If n = 4, then the coverings p_1 and p_2 may be two-sheeted, but then the monodromy group is smaller than S(4). Lemma 2 is thus proved.

C. Coverings by Means of Given Functions. Let f: $C^k \underset{m}{\to} C^1$ be an algebraic function all m of whose values at some point $x \in C^k$ are distinct. Then the set G of all such points x is open and is the complement to some algebraic hypersurface Σ in C^k; this set is called the branch manifold. To such an algebraic function f we can assign a non-branching m-sheeted covering p: $E \to G$. Namely, $p^{-1}(x) = (f(x), x) \in C^{k+1}$ consists of m points of the graph of f in $C^k \times C^1$.

Example. To the algebraic function (3) corresponds an n-sheeted covering p: $E \to G$. It is easily seen that its monodromy group is the symmetric group S(n). By Lemma 2 the covering p is indecomposable.

D. Coverings by Means of Given Superpositions. Let us now assume that our algebraic function f: $C^k \underset{m}{\to} C$ can be represented in the form of a superposition (2). Then the corresponding covering p: $E \to G$ can be resolved into the composition of N coverings

$$E = E_N \xrightarrow{p_N} E_{N-1} \xrightarrow{p_{N-1}} \ldots \xrightarrow{p_1} G, \qquad p = p_1 \circ p_2 \circ \ldots \circ p_N. \tag{4}$$

Formally we can define E_i as the subset of C^{k+i} consisting of the values y of the mapping F_i: $C^k \xrightarrow[n_1 \ldots n_i]{} C^{k+i}$ at the points x of the set G; for this $p_{i+1}^{-1}(y) = \Phi_{i+1}(y)$. We note that p_i is an n_i-sheeted covering, since all of the n_i points $\Phi_{i+1}(y)$ are distinct. In the opposite case not every $m = n_i \ldots n_N$ value of the function f at the point $x \in G$ would be distinct.

E. Proof of Lemma 1. Let the covering p of formula (4) corresponding to the superposition (2) be indecomposable. Then we conclude from D that in (2) every number n_i must equal 1, except one value, which must be equal to n.

Consequently, in the representation of such a function by a superposition (2) only one of the functions φ_i is not single-valued. By C the covering corresponding to the function (3) is indecomposable. Lemma 1 is proved.

A single-valued algebraic function is a polynomial. Therefore the resolution (2) in which we find only one n-valued algebraic function φ_i: $C^l \underset{n}{\to} C^1$, while the remaining functions φ_j are single-valued, can be written in the form $f = Q \circ ((\varphi \circ P) * 1)$, i.e.,

$$f(x) = Q(\varphi(P(x)), x), \tag{5}$$

where Q: $C^{k+1} \to C^1$ and P: $C^k \to C^l$ are polynomial mappings.

For the proof of the theorem of paragraph A it is sufficient to prove that a function of k variables (3) cannot be represented in the form (5) for any choice of an n-valued algebraic function φ of less variables.

The impossibility of such a representation for the roots of a polynomial of even order (3) considered as a function of the coefficients a_1, a_2, a_3, is proved in [3] on the basis of information about the cohomology class of the braid group, found in [1] and [2]. A generalization of the results of [3], set forth below, is based on the following considerations of cohomologies of braid groups, due to D. B. Fuks [5].

§3. THE COHOMOLOGIES OF THE COMPLEMENT TO A BRANCH MANIFOLD

The complement to a branch manifold of an algebraic function can be used as a basis for some of the stratifications (with distinct group structures) related to this function. Accordingly we may extract from among the classes of cohomologies of the complement that class of cohomologies induced by the mappings in the classification space.

A. The Braid Group. We will denote by $\mathrm{Sym}_0^n X$ the symmetric part of n-th order of X, formed by the choice of n pairwise distinct points of X. If X is a manifold then $\mathrm{Sym}_0^n X$ has the natural structure of the manifold of the same class.

Example. $\mathrm{Sym}_0^n C$ can be identified with the complement to the branch manifold of the universal k-valued algebraic function (1) (see part 1, section A).

The fundamental group of the space $\mathrm{Sym}_0^n C$ is called the braid group of n threads and denoted by B(n). It is easily seen that $\mathrm{Sym}_0^n C$ is the space K(B(n), 1), i.e., that $\pi_i(\mathrm{Sym}_0^n C) = 0$ for $i > 1$.

Let $f: C^k \underset{n}{\rightarrow} C^1$ be an algebraic n-valued function, having at least one point at which it attains n equal values. We denote by G_f the complement to the branch manifold of the function f. The space G_f admits a natural mapping into the complement of the branch manifold of the universal function (1). Namely, such a mapping is equivalent to a contraction to G_f of the polynomial mapping p: $C^k \rightarrow C^n$, assigned to the function f (see part 1, section C).

The image of the fundamental group under the induced mapping

$$p_* : \pi_1(G_f) \rightarrow \pi_1(\mathrm{Sym}_0^n C) = B(n)$$

is called the braid group of the function f.

B. Cohomologies of an Algebraic Function. Every class of cohomologies α of the braid group (i.e., the space $\mathrm{Sym}_0^n C$) induces a class of cohomologies $\alpha[f] = p^*\alpha$ among the cohomologies of the complement to the branch manifold of the function f under the mapping of the cohomologies p^*: $H^*(\mathrm{Sym}_0^n C) \rightarrow H^*(G_f)$. The induced class will be called the class of cohomologies $\alpha[f]$ of the algebraic function f.

Example. Let f be the algebraic function of (3) (for n not necessarily of the form 2^r). Then the mappings p^* and p_* are isomorphisms.

This follows from the fact that p imbeds G_f into the complement to the branch manifold of the universal function (1) in its capacity as a factor in the direct product resolution $\mathrm{Sym}_0^n C \approx C \times G_f$.

Such a resolution depends on the replacement of the problem of choosing n roots z_i of Eq. (1) having the "center of gravity" $z_0 = -a_1/n$ by that of choosing "relative coordinates" $y_i = z_i - z_0$.

In view of what has been said the function (3) and the complement to its branch manifold in homotopic problems can be replaced by the universal function (1) and the manifold $\mathrm{Sym}_0^n C$.

C. A Covering. The classification space of n-sheeted coverings (of fibers with the structure of the group S(n)) can be constructed in the following manner. Let us examine the manifold $\mathrm{Sym}_0^n C^N$ of unordered choices of n pairwise distinct points in the multi-dimensional space C^N. This manifold serves as a basis for the smooth fibers with the symmetric structure of the group S(n). The fiber space is obtained from the Cartesian power $(C^N)^n$ by discarding the "diagonal" and by means of a transformation is factored into a rearrangement of its co-factors.

It is easily seen that the homotopy groups of the fiber space of dimension less than $2N-1$ are trivial (see i.e., [4]). Consider the natural imbedding $\mathrm{Sym}_0^n C^N \rightarrow \mathrm{Sym}_0^n C^{N+1}$. There exists a limiting principal stratification $E \rightarrow \mathrm{Sym}_0^n C^\infty$, which is universal for the S(n)-stratifications: any n-sheeted covering is induced by means of the mapping of the basis into $\mathrm{Sym}_0^n C^\infty = K(S(n), 1) = BS(n)$; this mapping is unique up to a homotopy.

Example. A covering over $\mathrm{Sym}_0^n C^1$ corresponding to the universal algebraic functions (1) is yielded by a natural imbedding i: $\mathrm{Sym}_0^n C^1 \to \mathrm{Sym}_0^n C^\infty$. This mapping induces a homomorphism of the fundamental groups i_*: B(n) → S(n) and of the cohomologies i^*: H*(S(n)) → H*(B(n)) (i.e., i: H*($\mathrm{Sym}_0^n C^\infty$) → H*($\mathrm{Sym}_0^n C$)).

D. The Classes of Cohomologies of Coverings. Every class of cohomologies α of the symmetric group (i.e., of the space $\mathrm{Sym}_0^n C^\infty$) induces a class of cohomologies $\varphi^*\alpha$ in the base B of any n-sheeted covering: a mapping of the cohomologies φ: H*($\mathrm{Sym}_0^n C^\infty$) → H*(B) is induced by means of the classification mapping φ: B → $\mathrm{Sym}_0^n C^\infty$. We will refer to this class as the class of cohomologies $\alpha[p]$ of the investigated covering p.

Example. Let f be an algebraic n-valued function, assuming n equal values at least for one point (see paragraph A). For such a function we can assign an n-sheeted covering. The classification mapping φ: $G_f \to \mathrm{Sym}_0^n C^\infty$ can be resolved as the composition of a polynomial mapping p: $G_f \to \mathrm{Sym}_0^n C^1$ and an imbedding i: $\mathrm{Sym}_0^n C^1 \to \mathrm{Sym}_0^n C^\infty$. Therefore φ^* can be omitted by means of the cohomologies of the braid group:

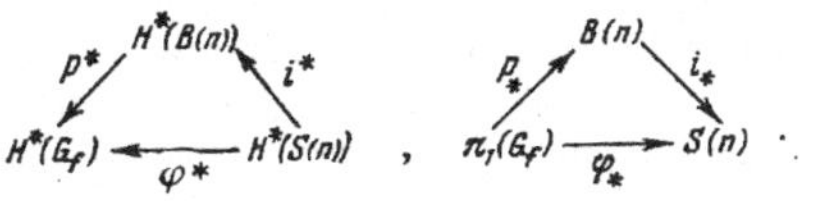

Thus the class of cohomologies of the covering corresponding to the algebraic function is that cohomology class of an algebraic function (see paragraph B), which is induced in the cohomologies of the braid group by means of the cohomologies of the symmetric group for the mapping i*.

In exactly the same way the monodromy group of the algebraic function $\varphi^*\pi_1(G_f)$ is the image of the braid group of the function under the natural homomorphism i_* of the braid group B(n) onto the symmetric group S(n).

E. A Covering and a Vector Stratification. There exists a natural imbedding of the symmetric group into the orthogonal S(n) → O(n) in the form of a permutation of the coordinate axes. This imbedding enables us to construct a vector stratification of any n-sheeted covering. It induces as well a mapping of the classification spaces† j : BS(n) → BO(n) and hence, a mapping of the cohomologies j*: H* (BO(n))→H*(BS(n)). Any class of cohomologies α of the Grassmann manifold BO(n) induces the class $j^*\alpha$ in the cohomologies of the symmetric group.

Let an n-sheeted covering be given by means of a mapping φ of its base into BS(n). Then the class of cohomologies of the covering $\varphi^* j^* \alpha$ is the characteristic class α of vectors of the O(n) stratification corresponding to the given covering.

Example. Let f be an algebraic n-valued function which for at least one point assumes n distinct values. Then among the classes of cohomologies of the complement to its branch manifold we may distinguish the subring of cohomology classes of the function in which the subring of the classes of cohomologies for the corresponding function is a covering and in which the subring of the cohomology classes for the corresponding covering is a stratification:

$$H^*(G_f) \supset p^* H^*(B(n)) \supset p^* i^* H^*(BS(n)) \supset p^* i^* j^* H^*(BO(n)).$$

The following two remarks will not be used by us hereafter.

F. Remark 1. The chain of groups and mappings B(n) → S(n) → O(n) can be extended to the left and to the right:

$$H(n) \to B(n) \to S(n) \to O(n) \to U(n).$$

By means of the reverse mapping from the braid group we obtain the infinite-dimensional group H(n) of homeomorphisms of the plane ‡ leaving n given points fixed. The operation of the entire group of such homeomorphisms of the plane H on the space of collections of n points $\mathrm{Sym}_0^n C$ defines the principal stratifi-

† This mapping $\mathrm{Sym}_0^n C^1 \to$ BO(n) can be realized with the aid of the following "Vandermonde spiral." Let $z \in C$. Let us set $\xi(z) = e_1 + ze_2 + \ldots + z^{n-1}e_n$, where e_k is a basis of the space C^n. For a choice $(z_1, \ldots, z_n) \in \mathrm{Sym}_0^n C$ we compare the plane $\mathbf{R}^n$ with the space $\mathbf{R}^{2n} = C^n$ spanned by $\xi(z_1), \ldots, \xi(z_n)$. We obtain in this way the "Vandermonde mapping" w: $\mathrm{Sym}_0^n C^1 \to G_{n,2n}$.

‡ Stationary in a neighborhood of ∞.

cation H $\to \mathrm{Sym}_0^n\mathbf{C}$ with the structure of the group H(n). It is known that the group H (with respect to a weaker measure) is homotopically trivial. Hence the indicated stratification is a classification under a stratification by means of the structure group H(n), which (in a weaker measure) is homotopically equivalent to the braid group B(n). Thus, the complement to the branch manifold of the universal algebraic function (1) is classified by the space under the group of homeomorphisms of the plane leaving n points fixed, and stationary in a neighborhood of infinity: $\mathrm{Sym}_0^n\mathbf{C} = BH(n)$. In general we can associate with every algebraic function, a stratification by means of the structure group H(n). A polynomial mapping p: $G_f \to \mathrm{Sym}_0^n\mathbf{C}$ (see paragraph A) is simply a classification mapping yielding this stratification.

On the other hand the group O(n) can be imbedded in U(n). However, the image of $H^*(BU(n))$ in $H^*(BB(n))$ is trivial; this stems from the triviality of the tangential stratification to $\mathrm{Sym}_0^n\mathbf{C}$. The latter is seen from the representation of $\mathrm{Sym}_0^n\mathbf{C}$ as the complement to the discriminant of the manifold of the algebraic function (1). See [5].

G. Remark 2. To any two coverings ξ_1, ξ_2 with a general base B we can assign two new coverings by means of the sum $\xi_1 \oplus \xi_2$ and the product $\xi_1 \otimes \xi_2$. These coverings originate from the natural imbeddings of the symmetric groups $S(n_1) \times S(n_2) \to S(n_1 + n_2)$ and $S(n_1) \times S(n_2) \to S(n_1 n_2)$, since the sum of an n_1-sheeted covering has $n_1 + n_2$ sheets, and the product, $n_1 n_2$. We raise the natural problem of how to find the class of cohomologies of a sum and product.

Let us denote by H^{**} the ring of cohomologies with coefficients in the field factored by the ideal spanned by the factor elements (of the form $\alpha = \alpha_1\alpha_2$, $\dim \alpha_i > 0$). Let $\alpha \in H^{**}(S(\infty))$ be the standard class of cohomologies of the symmetric group. Then in $H^{**}(B)$ we have

$$\alpha[\xi_1 \oplus \xi_2] = \alpha[\xi_1] + \alpha[\xi_2], \qquad \alpha[\xi_1 \otimes \xi_2] = n_2\alpha[\xi_1] + n_1\alpha[\xi_2].$$

The composition operation arises for the superposition of coverings; this operation is related to the wreath product of the structure groups.

Let $p_1\colon E_1 \to B_1$, $p_2\colon E_2 \to B_2$ be coverings such that the base of the second coincides with the space of the first. Then we may define a third covering $p_1 \circ p_2 : E_2 \to B_1$. It is of interest to explain the connection between the classes of cohomologies of all three coverings. In the simplest case the result is given by the following formula. Let $\alpha \in H^1(S(\infty), Z_2)$ be a unique nontrivial element of these groups. Then

$$\alpha_1[p_1 \circ p_2] = n_2\alpha_1[p_1] + p_{1!}\,\alpha_1[p_2],$$

where n_2 is the number of sheets of p_2 and the mapping $p_{1!}: H^*(E_1) \to H^*(B_1)$ has a dual preimage p_1^{-1} : $H_*(B_1) \to H_*(E_1)$.

§ 4. PROOF OF THE NON-REPRESENTABILITY

We now present a proof of the theorem of paragraph A, section 2, based on the fact that functions of a small number of variables do not have multivariate cohomologies.

A. LEMMA. Let $f\colon \mathbf{C}^k \underset{n}{\to} \mathbf{C}^1$ be an algebraic function of k variables. Then every class of cohomologies of the function f of dimension greater than k, is equal to 0.

Proof. We will examine the complement G_f to the branch manifold of the function f in $\mathbf{C}^k$. This complement is a k-dimensional Stein manifold; it is given in $\mathbf{C}^k$ by the equation $\Delta \neq 0$, where Δ is the discriminant. Since for any k-dimensional Stein manifold $H^i(G_f) = 0$ for $i > k$, the theorem is proved.

Example. Let $f\colon \mathbf{C}^k \underset{n}{\to} \mathbf{C}^1$ be an n-valued algebraic function. Then every class of cohomologies of the function f whose dimension is greater than or equal to n is null.

In fact, this is so for n-valued functions of $n-1$ variables (3), according to the lemma. In agreement with paragraph B of section 3 the truth of the assertion follows for the universal function of n-values (1) and therefore for any n-valued functions.

B. Invariance of the Classes of Cohomologies of the Covering under a Tschirnhausen Transformation. Let $\varphi\colon \mathbf{C}^l \underset{n}{\to} \mathbf{C}$ and $f\colon \mathbf{C}^k \underset{n}{\to} \mathbf{C}$ be n-valued algebraic functions, connected by relation (5), such that f assumes n distinct values for at least one point. Then the complement G_f to the branch manifold of the function f is nonempty and the mapping P transforms G_f into the complement G_φ to the branch manifold of the function φ.

The function φ assigns an n-sheeted covering of $G\varphi$. The mapping $P : Gf \to G\varphi$ over Gf induces an n-sheeted covering on Gf. Formula (5) proves that this covering is equivalent (as a covering) to that which furnishes the algebraic function f. Thus we have proved the

LEMMA. The cohomologies of the covering of an algebraic function f are induced by the cohomologies of the covering of the function φ under the mapping $P : G_f \to G\varphi$:

$$\begin{array}{ccc} & H^*(BS(n)) & \\ \swarrow & & \searrow \\ H^*(G_f) & \overset{P^*}{\leftarrow} & H^*(G_\varphi). \end{array}$$

Example. Let P be the identity mapping. Then the mapping from φ to f is a Tschirnhausen transformation. Thus under Tschirnhausen transformations the classes of cohomologies of an algebraic function which are cohomology classes of the corresponding covering remain invariant. The invariance under Tschirnhausen mappings was proved in [3] for one- and three-dimensional classes α_1, α_3. The deduction of the general formulation was done by D. B. Fuks using arguments of the work [3].

C. Stiefel-Whitney Classes. In the group of k-dimensional cohomologies the Grassmann manifold $H^k(BO(n), \mathbf{Z}_2)$ is a Stiefel-Whitney class w_k. This class is a major barrier to the construction of $n-k+1$ intersecting $O(n)$-stratifications.

Let ξ be an $O(n)$ stratification, and $S(n)$ a stratification or an n-sheeted algebraic function. We will denote by $w_k[\xi]$ the image of w_k in the group of k-dimensional cohomologies for the base of the covering or stratification (corresponding to the complement to the branch manifold of the algebraic function).

In particular the universal $S(n)$-stratification and universal algebraic function (1) have the Stiefel-Whitney classes

$$w_k \in H^k(S(n), \mathbf{Z}_2); \qquad w_k \in H^k(B(n), \mathbf{Z}_2).$$

Example. For $n = 2^r$ the class $w_{n-1} \in H^{n-1}(B(n), \mathbf{Z}_2)$ is non-null. This follows from the calculations of D. B. Fuks [5].

For $r = 1, 2$ we obtain the classes $w_1 \in H^1(B(2), \mathbf{Z}_2)$ and $w_3 \in H^3(B(4), \mathbf{Z}_2)$, which in [3] are denoted by α_1, α_3.

Every class $w_k[\xi]$ of an algebraic function ξ is invariant under a Tschirnhausen transformation. This follows from the lemma of section B.*

D. Completion of the Proof of the Theorem. Assume that the function of (3) having $n = 2^r$ values and of $k = n-1$ variables is represented in the form (5) with the aid of an algebraic function φ of less variables. By the lemma of section A, the k-dimensional Stiefel-Whitney class $w_k[\varphi] \in H^k(G\varphi, \mathbf{Z}_2)$ of the function φ is empty. By the lemma of part B the k-dimensional Stiefel-Whitney class $w_k[f] \in H^k(Gf, \mathbf{Z}_2)$ of the function f is also empty.

On the other hand, in accordance with section C the k-dimensional Stiefel-Whitney class of the universal function (1) is not 0. By section B of §3, it also differs from 0 for the homotopically equivalent function (1) to the function f given by formula (3). This contradiction completes the proof of the theorem.

E. Remark. The non-triviality of the k-th Stiefel-Whitney class of the function does not exclude the possibility of representing this function in the form of one of the branches of the superposition of algebraic functions of less than k variables.

For example, it is known that for any n the universal algebraic function (3) can be represented in the form of one of the branches of the superposition of algebraic functions of a smaller number of variables (A. Wiman [6]). In particular the equation of fourth order $z^4 + a_1z^2 + a_2z + a_3 = 0$ is solvable by means of radicals.

The proof of the theorem requires only that we obtain the superfluous root.

---

*From the results of D. B. Fuks [5] we also see that every class of cohomologies of an algebraic function with coefficients $\mathbf{Z}_2$ remains invariant under a Tschirnhausen transformation.

We also note that algebraic investigations of the functions are not really essential for our arguments. For example the algebraic functions φ_i in (2) can be replaced by germs of algebraic functions and the polynomials p_i by germs of holomorphic transformations.

LITERATURE CITED

1. V. I. Arnol'd, "Braids of algebraic functions and cohomologies of swallowtails," Usp. Matem. Nauk., 23, No. 4, 247-248 (1968).
2. V. I. Arnol'd, "On some topological invariants of algebraic functions," Trud. Mosk. Matem. Obshch., 21, 28-43 (1970).
3. V. I. Arnol'd, "On classes of cohomologies of an algebraic function, invariant under Tschirnhausen transformations," Funktsional'.Analiz. i Ego Prilozhen., 4, No. 1, 84-85 (1970).
4. E. Fadell and L. Neuwirth, "Configuration spaces," Math. Scand., 10, No. 1, 119-126 (1962).
5. D. B. Fuks, "Cohomologies of a braid group with respect to modulo two," Funktsional' Analiz. i Ego Prilozhen., 4, No. 2, 62-73 (1970).
6. A. Wiman, Über die Anwendung der Tschirnhausen Transformation auf die Reduktion algebraischer Gleichungen," Nova Acta, Uppsala, 10, 3-8 (1927).

ALGEBRAIC UNSOLVABILITY OF THE PROBLEM OF LYAPUNOV STABILITY AND THE PROBLEM OF TOPOLOGICAL CLASSIFICATION OF SINGULAR POINTS OF AN ANALYTIC SYSTEM OF DIFFERENTIAL EQUATIONS

V. I. Arnol'd

Further on, it is proved that the problem of stability, and therefore the problem of topological classification as well, is not a trivial problem in the sense of [1]: i.e., it does not permit an algebraic solution, even if cases of codimensionality infinity (and even codimensionality 103) are neglected.

The solution of both problems for singular points having a general position is well known: the topological type of the singular point is determined by the quantities of eigennumbers of the linear part of the equation in the left and right half-planes. These quantities can be found by means of a finite number of arithmetic operations on the coefficients of the linear part of the equation. Thus, algebraic criteria exist for membership of a singular point having a general position in some topological type. (In particular, an algebraic stability criterion exists.)

In considering singular points which do not have a general position a large number of different cases develops. It is natural to classify them according to the codimensionality of the corresponding submanifold* in the funct... al space of all vector fields (see [1, 2, 3]). For example, the case of one zero eigennumber and the c... ne pair of purely imaginary eigennumbers are cases of codimensionality one.

Numerous investigations have been carried out on singular cases having a small codimensionality.† In these cases, as in the case of a general condition (i.e., the case of codimensionality zero), it is possible to give the algebraic stability criteria (corresponding to the topological classification). In other words, in the cases considered the stability of the singular point (correspondingly, its membership in some topological type) is recognized by a finite number of arithmetic operations on Taylor coefficients of finite order.

The example given in the present paper shows that criteria of this kind cannot exist when singular cases of sufficiently large (but finite) codimensionality are considered.

In order to explain the cause of this phenomenon we shall consider a space of streams (of fixed order) of vector fields at a singular point. Having chosen the coordinates in the neighborhood of the singular point, we can identify the space of streams with the space of segments (of fixed length) of the Taylor series in the right sides of the differential equation. Thus, the space of streams considered is finite-dimensioned.

The singular point of a vector field having a stipulated stream of fixed order can be:

I. stable (in the Lyapunov sense) for any continuation of the Taylor series;

*For the sake of brevity we shall call it the "codimensionality of the case."
†Regrettably, it is difficult to understand the codimensionality up to which exhaustive stability criteria exist at present. The same applies to the problem of topological classification.

Moscow State University. Translated from Funktsional'nyi Analiz i Ego Prilozheniya, Vol. 4, No. 3, pp. 1-9, July-September, 1970. Original article submitted December 18, 1969.

Originally publ. in: Funkts. Anal. Prilozh. 4:3, 1-9, © Russ. Acad. Sciences 1970
English transl.: Funct. Anal. Appl. 4, 173-180, © Consultants Bureau, New York 1971
Reprint by Springer-Verlag Berlin Heidelberg 2014 with kind permission, DOI 10.1007/978-3-642-31031-7_25

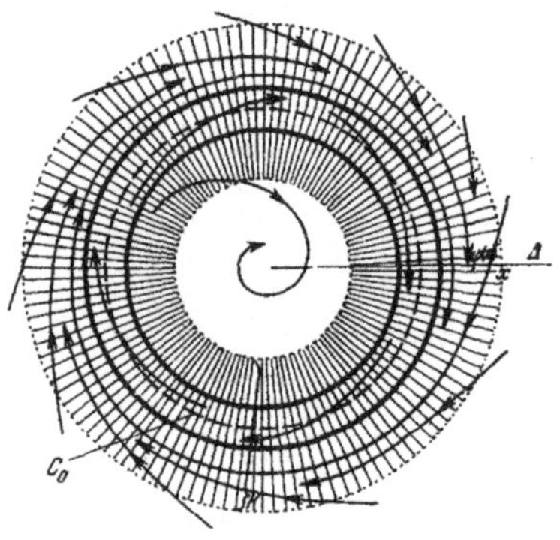

Fig. 1

II. unstable, regardless of the continuation;

III. stable or unstable, depending on the continuation.

In accordance with this we call a stream stable (I), unstable (II), or neutral (III).

Thus, the space of streams of fixed order can be partitioned into three parts: I + II + III. It turns out that the subdivision of the space of streams is not semialgebraic* if the dimensionality of the phase space is larger than two and the order of the stream is larger than four.

From this it is clear that the algebraic criterion of Lyapunov stability (and therefore the criterion for recognizing the topological type also) cannot exist in singular cases having a sufficiently large codimensionality.

The author has not attempted to provide a precise definition of least codimensionality where this phenomenon occurs for the first time. From the example given below it follows that no algebraic stability criterion exists beginning with codimensionality 102.

§1. Stability of Singular Points in Space and Limit Cycles on a Plane

Let us begin by investigating the stability of the singular point zero of a system in the three-dimensional Euclidian space $\mathbf{R}^3$

$$\dot{x}_i = a_i(x_1, x_2, x_3), \qquad i = 1, 2, 3, \tag{1}$$

whose right sides are homogeneous polynomials of degree k.

1.1. Lowering the Order. As is well known, the system (1) allows the order to be lowered. This means the following. A natural mapping p: $\mathbf{R}^3 \setminus 0 \to \mathbf{RP}^2$ of the complement to the point 0 on the projective plane exists: the point in space is associated with a straight line connecting it with 0. As a consequence of the homogeneity of the system (1) it follows that in representing p the partition of $\mathbf{R}^3 \setminus 0$ into trajectories of the system (1) is mapped on a certain (completely defined) partition of the projective plane into curves. We shall call it a factor-partition.

The association between the geometric properties of the original system (1) and the curves of the factor-partition of the projective plane is the following: the singular points on $\mathbf{RP}^2$ correspond to the directions of entry into $\mathbf{R}^3$, the cycles in $\mathbf{RP}^2$ correspond to invariant cones in $\mathbf{R}^3$, etc.

In affine coordinates on the projective plane $(x : y : 1 = x_1 : x_2 : x_3)$ the factor-partition is the partitioning into phase curves of the system of differential equations

$$x' = P(x, y), \qquad y' = Q(x, y), \tag{2}$$

where, as is easily calculated,

$$P = A_1 - xA_3, \qquad Q = A_2 - yA_3. \tag{3}$$

Here $A_i = a_i(x, y, 1)$ are inhomogeneous polynomials of x, y of degree no higher than k.

In order to define radial motion of a point in $\mathbf{R}^3$ we consider the rate of change of the square of the distance from 0. This rate $(r^2)^{\cdot}$ is a homogeneous polynomial of degree k + 1. We shall assume that k is odd (in our example k = 5). Then the character of radial motion (attraction or repulsion) is determined by the sign of the function $(r^2)^{\cdot}/r^{k+1}$, which is defined on the projective plane. In affine coordinates, as is easily calculated,

$$(r^2)^{\cdot}/r^{k+1} = (1+\rho^2)^{-\frac{k+1}{2}} R(x, y), \tag{4}$$

*A set is called semialgebraic if it can be stipulated by a finite number of polynomial equations and inequalities.

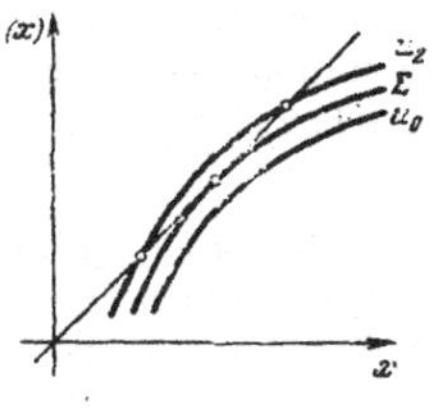

Fig. 2

where

$$R(x, y) = (\rho^2)' + 2A_3(1+\rho^2), \qquad \rho^2 = x^2 + y^2, \tag{5}$$

and where, in accordance with (2), $(\rho^2)' = 2Px + 2Qy$. The function R is an inhomogeneous polynomial of x, y of degree no higher than k + 1. The sign of the radial component (4) in a finite portion of the x, y plane is determined by the sign of the function R, and on an infinitely distant straight line it is determined by the sign of the homogeneous part of R having the degree k + 1.

Equations (2), (3) and (4), (5) reduce the investigation of the system (1) to the problem of differential equations with a polynomial right side on a plane. Note that this reduction can be partially reversed.

Namely, assume P and Q are arbitrary polynomials of the variables x, y having a degree no higher than k, and assume A_3 is any polynomial of degree no higher than k−1. Then Eqs. (3) determine polynomials A_1, A_2 of degree no higher than k. From the polynomials A_1, A_2, and A_3 it is possible to formulate homogeneous polynomials a_i and the system (1). For the formulated system (1) the factor-partition has the form (2), (3), and the radial motion has the form (4), (5) with chosen P, Q, A_3.

Our example of the system (1) will be formulated precisely by this method: we shall choose the polynomials P, Q, A_3 so that the system (1) has the required properties.

1.2. The Change of Stability During the Generation of a Cycle. When the coefficients of the polynomials P and Q change, the topological properties of the factor-partition (2) can change. In particular, the creation of a pair of limit cycles "out of thin air" (i.e., far from the singular points and the separatrixes; Fig. 1) can occur.

More precisely, we can denote a pair of polynomials (P_0, Q_0) and a closed trajectory C_0 which corresponds to the system (2) in such a way that the following proposition is satisfied.

PROPOSITION A. In the space of pairs of polynomials (P, Q) of degree no higher than k there exists a vicinity U of the point (P_0, Q_0) and an analytical hypersurface Σ separating U into two parts $(U = U_2 + \Sigma + U_0)$ which are such that

1) systems (2) belonging to the domain U_2 have two limit cycles in the vicinity of the curve C_0: a stable cycle and an unstable cycle;

2) systems which do not have cycles near C_0 lie in the domain U_0;

3) systems having one twofold (semistable) cycle near C_0 lie on the surface Σ (these include the pair P_0, Q_0 itself).

We shall call the surface Σ the generation surface.

Let us recall that the sequence function is defined as the mapping φ of the segment Δ which is transversal relative to the cycle C_0 on itself (or on its continuation), which is achieved by the trajectories of the system (2). The shape of the graphs of the sequence functions of systems from U_2, Σ, and U_0 is clear from Fig. 2.

Let us consider a ring-shaped vicinity K_0 of the cycle C_0, which is bounded by two smooth curves. This vicinity can be chosen in such a way that the following proposition will be satisfied.

PROPOSITION B. The trajectories of the system (2) corresponding to $(P, Q) \in U$, enter K_0 through one of its boundaries, while they exit from K_0 through its other boundary. Under these conditions if there are no cycles ($(P, Q) \in U_0$), then all the trajectories cross K_0 without being delayed. However, if there are cycles ($(P, Q) \in U_2$), then the entering trajectories are wound on a stable cycle, while the exiting trajectories are unwound from an unstable cycle (see Fig. 1).

Note that if the transversality condition B is satisfied for the polynomials $P = P_0$, $Q = Q_0$ and the ring K_0, then it remains valid if we replace P_0, Q_0 by some sufficiently close polynomials P, Q and the ring K_0 by some sufficiently C^1-close ring K.

Let us now consider the radial component of the motion (1) [more precisely, the function (4)] on the projective plane. The polynomials $P = P_0$, $Q = Q_0$, and $A_3 = A_{30}$ can be chosen in such a way that the following proposition is satisfied.

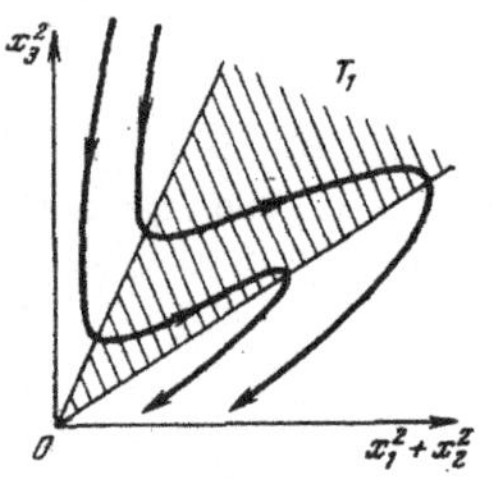

Fig. 3

PROPOSITION C. There exists a ring K_0 such that the function $(r^2)^{\cdot}/r^{k+1}$ is positive inside K_0 and negative throughout the entire remaining part of the projective plane; under these conditions its gradient on the boundary of K_0 does not vanish anywhere.

Note that then the function (4), as calculated for the values of (P, Q) (and the previous $A_3 = A_{30}$) which are sufficiently close to (P_0, Q_0), will have the same properties in a certain ring-shaped region K = K(P, Q) which is close to K_0.

Let us now assume that the vicinity U of the pair of polynomials (P_0, Q_0) is chosen to be so small that all three propositions A, B, and C are satisfied for the pairs of polynomials $(P, Q) \in U$ and domains K = K(P, Q). It is easy to prove the following lemma (see 1.3).

LEMMA 1. The system (1) is stable as long as any trajectory of the factor-system (2) traverses the ring K without being delayed (i.e., for $(P, Q) \in U_0$), and is unstable if the ring K contains two limit cycles (i.e., $P, Q \in U_2$).

From this it is evident that under the conditions indicated a change in stability occurs at precisely the instant at which the cycle is generated. In other words, the boundary of the stability domain in the space of the coefficients of the polynomials a_i defining the system (1) contains a piece whose equation has the form

$$(P, Q) \in \Sigma. \tag{6}$$

On the other hand, in §3 the following lemma is proved.

LEMMA 2. The generation surface Σ in the domain U of the space of the polynomial pairs (P, Q) is not algebraic but transcendental.

From Lemmas 1 and 2 it is evident that the partitioning of the space of the system (1) into stable and unstable domains cannot be semialgebraic if propositions A, B, and C cannot be satisfied. The possibility of satisfying conditions A, B, and C is verified in §2.

1.3. Independence of Stability from Completion of the Process of Writing the Leading Terms. Let us prove Lemma 1 in a somewhat intensified form: For the same conditions the zero equilibrium condition of any system close to (1),

$$\dot{x}_i = a_i + o(r^k), \qquad i = 1, 2, 3. \tag{7}$$

is stable (unstable).

Proof. Assume that there are no cycles. Let us use K_1 to designate the vicinity of the ring K on whose boundary the field (2) is still transverse with respect to the boundary, while the function (4) is already negative.

Let us consider the conical region $p^{-1}K_1$ corresponding to the ring K_1 in $\mathbf{R}^3$ (Fig. 3). Let us designate it by T_1. Any trajectory of the system (1) which enters T_1 through one of the conical surfaces bounding it exits from T_1 through the other surface (since the trajectories of the factor-system (2) pass through K_1). In a sufficiently small vicinity of the origin the same thing also applies to the system (7). Outside of T_1 the trajectories of the system (1) approach 0. Since the function (4) is away from zero on the projective plane outside of K_1, the same thing applies to the system (7) in a sufficiently small vicinity of the origin.

Thus, for $t \to +\infty$ any trajectory of the system (1) and any trajectory of the system (7) having initial conditions sufficiently close to 0 always exit from T_1 and tend to 0. Therefore, the singular point 0 of the system (1) (or, (7)) is asymptotically stable.

Let us assume that cycles exist. Let us use K_2 to designate the vicinity of the stable cycle on whose boundary the field (2) is directed into the interior of K_2, while the function (4) is positive.

Let us consider the conical region $p^{-1}K_2$, corresponding to the ring K_2 in $\mathbf{R}^3$ (Fig. 4). Let us designate it by T_2. Inside T_2 the field (1) is directed away from the center everywhere. Therefore, in a sufficiently small vicinity of 0 the field (7) is similarly directed away from the center.

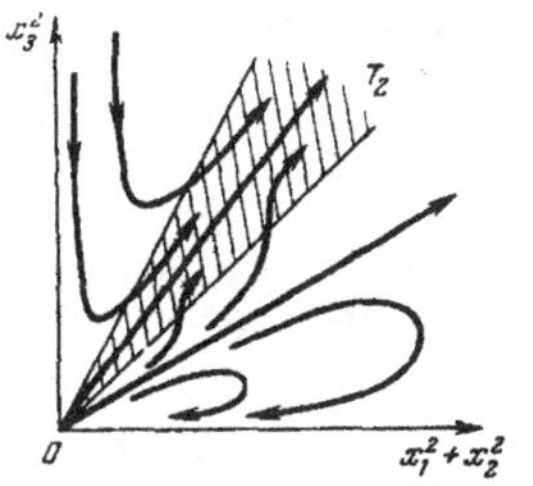

Fig. 4

On the conical surfaces which bound T_2 the field (1) is directed into the interior of T_2. Therefore, in a sufficiently small vicinity of the point 0 the same thing is also valid for the system (7).

Therefore, the singular point 0 of the system (1) (or (7)) is unstable (in the Lyapunov sense).

The lemma has been proved. It remains for us to 1) designate the polynomials P_0, Q_0, A_{30}, the cycle C_0, and the ring K_0 which satisfy propositions A, B, and C of 1.2; 2) prove Lemma 2.

§2. Example

Propositions A, B, and C are in the nature of inequalities: if they are satisfied for a certain system (1), then they are also satisfied for all close systems of the same kind.

Let us indicate the domain in which all three conditions are satisfied in the space of triplets of homogeneous polynomials a_i of degree $k = 5$ of three variables x_i.

2.1. A Degenerate System. Our example is formulated by means of a small perturbation of the following simplest system. Let us give the polynomials P, Q, A_3 in Eqs. (2), (3) of §1 the following values:

$$P_{00} = y, \quad Q_{00} = -x, \quad A_{30} = 1-4(\rho^2-1)^2, \text{ где } \rho^2 = x^2+y^2. \tag{8}$$

The system (2) then has the circles ρ = const as trajectories. The function A_{30} is negative everywhere except in the ring K_{00}, where $0.5 \le \rho^2 \le 1.5$. The radial motion is determined according to (5) by the quantity $R = 2A_{30}(1+\rho^2)$. The homogeneous portion of R (sixth degree) is negative-definite, while the sign of R coincides with the sign of A_3. Thus, condition C (see 1.2) is satisfied for the ring K_{00} and the values of P, Q, A_3 indicated in (8).

Therefore, condition C remains satisfied if P_{00}, Q_{00} are replaced by any sufficiently close polynomials of no higher than the fifth degree, and the ring K_{00} is replaced by an appropriate close ring.

2.2. Dissipative Perturbations. Let us replace P_{00}, Q_{00} by the close polynomials

$$P_0 = y + \varepsilon x f, \qquad Q_0 = -x + \varepsilon y f, \tag{9}$$

where $f = \alpha + \beta\rho^2 + \gamma\rho^4$. By choosing α, β, γ, we can achieve satisfaction of conditions A and B of 1.2. Actually, according to (2) and (9) the function $u = \rho_2$ satisfies the equation

$$u' = 2\varepsilon u f(u), \tag{10}$$

from which it is easy to find (or to investigate) the sequence function. Let us assume, for example, that $f(u) = -(u-1)^2$. Then the circle $C_0 = \{x, y : u = 1\}$ is a semistable cycle of the system (2). It is easy to verify the fact that conditions A and B are satisfied. Moreover, it is possible to choose the ring K_0 from 2.1 as the ring vicinity of the cycle C_0 which is incorporated in these conditions.

Therefore, for a sufficiently small ε all three conditions A, B, and C are satisfied for (A_{30}, P_0, Q_0) in a certain ring K_0 which is close to K_{00}.

Let us specify such a small ε. Then the conditions A, B, and C remain satisfied if P_0, Q_0 are replaced by any polynomials of degree no higher than the fifth which are sufficiently close to the chosen ones, and K_0 is replaced by the appropriate close ring.

2.3. A Hamilton Perturbation. In order to prove the transcendental nature of the generation surface it is sufficient to verify the fact that its intersection with some algebraic submanifold of the space of polynomial pairs (P, Q) is transcendental. It is convenient to take a four-dimensional plane obtained from a three-dimensional plane of systems of the form (9) by means of a one-parameter "Hamiltonian perturbation" (11) as such a submanifold. The intersection of the generation surface with its four-dimensional plane $\mathbf{R}^4$ will be called a generation surface as before, and will be designated by Σ. It is easy to check the fact that in the vicinity of the point (9), where $f = -(u-1)^2$, the surface Σ is an analytical manifold (both in $\mathbf{R}^4$ and in the entire space of polynomial pairs).

Let us assume that

$$P = H_x + \varepsilon x f, \qquad Q = -H_y + \varepsilon y f, \tag{11}$$

where the "Hamilton function" H is a polynomial of degree no higher than the sixth which differs only slightly from $(x^2 + y^2)/2$ for small values of the parameter μ:

$$H = \frac{1}{2}(x^2 + y^2) + \mu H_1(x, y).$$

The choice of H_1 does not have any special significance; it is convenient to take $H_1 = x^4$. For sufficiently small $|\mu|$ the conditions A, B, and C will be satisfied for the system (2) having the right side (11) for the εf chosen in 2.2.

2.4. Explicit Equations for the Four-Parameter Family. In accordance with Eqs. (11), (9), (8), and (3), let us consider the family of systems (2) which depend on four parameters a, b, c, μ:

$$x' = y + x(a + b\rho^2 + c\rho^4), \quad y' = -x - 4\mu x^3 + y(a + b\rho^2 + c\rho^4), \tag{12}$$

and the corresponding four-parameter family of the system (1):

$$\begin{cases} \dot{x}_1 = x_2 x_3^4 + x_1(F + G), \\ \dot{x}_2 = -x_1 x_3^4 + x_2(F + G) - 4\mu x_1^3 x_3^2, \\ \dot{x}_3 = x_3 G, \end{cases} \tag{13}$$

where $F = ax_3^4 + b(x_1^2 + x_2^2)x_3^2 + c(x_1^2 + x_2^2)^2$, $G = x_3^4 - 4(x_1^2 + x_2^2 - x_3^2)^2$.

In accordance with 2.2 and 2.3, all conditions A, B, and C of Lemma 1 (and therefore its conclusion) are satisfied for a sufficiently small ε in the vicinity of the point M: $a = -\varepsilon$, $b = 2\varepsilon$, $c = -\varepsilon$, $\mu = 0$ in the four-dimensional parameter space. On the other hand, § 3 proves the following lemma.

LEMMA 3. The analytical branch of the generation surface Σ passing through the point M of the parameter space is transcendental.

Applying Lemmas 1 and 3 in the vicinity of the point M, we arrive at the following conclusions.

2.5. Results. THEOREM 1. The boundaries separating Lyapunov-stable and Lyapunov-unstable systems (13) in the four-dimensional parameter space (a, b, c, μ) are not a semialgebraic set.

Further, let us consider the space of r-streams of vector fields at the point 0 of three-dimensional space $\mathbf{R}^3$. Assume $r \geq 5$. Then in the space of r-streams there is a submanifold N formed by streams for which the Taylor series begin with terms of degree no higher than the fifth. This submanifold N has codimensionality 102 in the space of all r-streams.

THEOREM 2. The boundary separating Lyapunov-stable and Lyapunov-unstable streams from N is not a semialgebraic subset.

Theorem 2 derives from Theorem 1, since the boundary indicated is an analytical hypersurface in the space N in the vicinity of one of its points, while the four-dimensional plane of the systems (13) intersects this hypersurface transversally.

§3. Proof of Nonalgebraicity

Here Lemma 3, which was formulated in 2.4, is proved.

3.1. The Reduction Plan. The generation surface in four-dimensional parameter space $\mathbf{R}^4$ has an analytical branch Σ passing through the point M (the point M corresponds to the system (9) with $f = -(u-1)^2$ and with small ε).

In order to prove the transcendental nature of Σ, it is sufficient to prove the transcendental nature of Σ_1 which is the analytical continuation of this branch (the continuation Σ_1 may also not belong to the generation surface). In order to establish the transcendental nature of the continuation Σ_1 it is sufficient to prove that its intersection Σ_2 with some algebraic manifold (for example, with the plane $\mathbf{R}^3$) is transcendental.

In order to prove the transcendental nature of the intersection Σ_2 it is sufficient to consider the cone Σ_3 which is tangent to Σ_2 at some point of $\mathbf{R}^3$ and to establish the fact that it is transcendental in $\mathbf{R}^3$.

The latter check reduces to checking whether the curve Σ_4 corresponding to the cone Σ_3 in the projective plane $\mathbf{RP}^2$ is transcendental.

It turns out that this entire construction can be carried out in such a way that explicit equations (17) are written out in terms of Abelian integrals (16) for the curve Σ_4. From these equations it is evident that the mapping of Σ_4 on an affine map is closer to the graph of a logarithm than an algebraic curve can allow. It is this which proves the transcendental nature of the generation surface Σ.

3.2. Reduction to Abelian Integrals. The R^3 plane is stipulated by the equation $\mu = -1/4$; i.e., we choose the Hamiltonian $H = (2x^2 + 2y^2 - x^4)/4$. The point $a = b = c = 0$ of this plane R^3 corresponds to the Hamilton equation (12). We shall study the cone Σ_3 which is tangent to the intersection Σ_2 of the analytical continuation Σ_1 of the generation surface with the R^3 plane at the point $a = b = c = 0$.

The Hamilton equation has the first integral H. Therefore, the sequence function φ of the system (2) having the right side (11) is an identical transformation for $\varepsilon = 0$. Let us consider the behavior of the sequence function φ for small ε. We take Δ as the coordinate on the transversal segment $h = H(x, y)$. We use ψ to designate the increment of the sequence function in such a way that $\varphi(h) = h + \psi(h)$. Easy calculation shows that if

$$x' = H_y + \varepsilon A, \quad y' = -H_x + \varepsilon B, \tag{14}$$

then for $\varepsilon \to 0$ we have

$$\psi(h) = \varepsilon \int_{H=h} -A\,dy + B\,dx + O(\varepsilon^2). \tag{15}$$

We use I to designate the integral in the right side of (15). I is a linear function of the integrand form.

In particular, we shall move along the straight line $a = \varepsilon\alpha$, $b = \varepsilon\beta$, $c = \varepsilon\gamma$ in our three-dimensional plane; substituting the values of A and B from (12) into (15), we find

$$I = \alpha I_0(h) + \beta I_1(h) + \gamma I_2(h),$$

where

$$I_m = \int_{H=h} (x^2 + y^2)^m (y\,dx - x\,dy). \tag{16}$$

We recall now that the cycles correspond to the zeros of the function ψ. The line $H = h$ will be the limit of the cycles of Eqs. (12) (where $a = \varepsilon\alpha, \ldots$) for $\varepsilon \to 0$, provided that $\alpha I_0(h) + \beta I_1(h) + \gamma I_2(h) = 0$, and in the limit of twofold cycles if, in addition, the derivative with respect to h is equal to zero:

$$\alpha J_0(h) + \beta J_1(h) + \gamma J_2(h) = 0, \text{ where } J_m(h) = \frac{d}{dh} I_m(h).$$

From this it is evident that the vector product of the vectors I and J (with the components I_m, J_m) for each h is a direction vector **a** (with the components α, β, γ) tangent to the generation surface:

$$\mathbf{a}(h) = [\mathbf{I}, \mathbf{J}]. \tag{17}$$

It is this which is the parametric equation of the direction cone Σ_3 or the homogeneous equation of the curve Σ_4. It is easy to check the fact that the vector a(h) is tangent precisely to that component Σ_1 of the generation surface which is obtained for the analytical continuation of the portion Σ of the generation surface of interest to us which lies near the point M, provided only that $0 < h < 1/4$.

3.3. Calculations. It is necessary to show that the curve on the projective plane, which is parametrically stipulated by Eqs. (17) by means of the elliptical integrals (16) over the cycle $2x^2 + 2u^2 - x^4 = 4h$, is transcendental. In our calculations we can replace the free integrals I_m by their more convenient linear combinations. Moreover, the multiplication of the vectors I and J by numbers does not affect the curve Σ_4. As the linear combinations replacing I_m it is convenient to take

$$U_0 = \frac{1}{4\sqrt{2}} I_0, \quad U_1 = \frac{1}{8\sqrt{2}} I_1, \; U_2 = \frac{1}{12\sqrt{2}} I_2.$$

Then we introduce the substitutions

$$4h = 1 - u^2, \; \Lambda = \ln \frac{4}{\sqrt{2u}}, \quad M = \Lambda + \frac{1}{4}.$$

179

It is not difficult to check the fact that the asymptotic relationships (for $u \to +0$) for the integrals written above and their derivatives with respect to the parameter 2h are given by the following equations:

$$U = x - \frac{1}{2}(Me - y)u^2 + [u^3], \quad U' = \Lambda e - y + [u],$$

where the vectors **x**, **e**, **y** are linearly independent

$$x = \left(\frac{2}{3}, \frac{22}{105}, \frac{962}{10395}\right), \quad e = (1, 1, 1), \; y = \left(0, \frac{2}{3}, \frac{38}{35}\right).$$

Here $|u^{\nu}|$ designates any small quantity of the form $O(u^{\nu}\log^{\varkappa} u)$ for some $\varkappa > 0$.

From this we find the asymptotic relationships for the components of the vector product $[U, U'] = X[x, e] + Y[y, x] + Z[e, y]$:

$$X = \Lambda + [u], \quad Y = 1 + [u], \; Z = \frac{u^2}{8} + [u^3].$$

Therefore, in the affine coordinates $\xi = 8\,ZY^{-1}$, $\eta = XY^{-1}$, the parametric equation of the curve Σ_4 has the form: $\xi = u^2 + [u^3]$, $\eta = \Lambda + [u]$.

For $\xi \to 0$ we find that $\eta = -1/4 \ln \xi + O(1)$. This means the curve Σ_4 is not algebraic, and this is what it was required to prove.

The author thanks V. M. Alekseev who indicated the errors in calculating the asymptotic relationships for the integrals I_m, and Yu. S. Il'yashenko who showed that these calculations can be replaced by reference to the Riemann-Rojas theorem.

LITERATURE CITED

1. V. I. Arnol'd, "On local problems in analysis," Vestnik Mos. Gos. Univ. (1970).
2. V. I. Arnol'd, "Comments on the singularities of finite codimensionality in complex dynamic systems," Funktsional'.Analiz i Ego Prilozhen., 3, No. 1, 1-6 (1969).
3. V. I. Arnol'd, "Singularities of smooth mappings," Uspekhi Matem. Nauk, 23, No. 1, 3-44 (1968).
4. V. I. Arnol'd, "Algebraic unsolvability of the problem of Lyapunov stability and the problem of topological classification of singular points of an analytic system of differential equations," Uspekhi Matem. Nauk, 25, No. 2, 265-266 (1970).

On the Arrangement of Ovals of Real Plane Algebraic Curves, Involutions of Four-Dimensional Smooth Manifolds, and the Arithmetic of Integral Quadratic Forms

V.I. Arnol'd

There is an interesting connection between the arrangement of branches of a real algebraic curve on the projective plane, on the one hand, and the topology of certain complex algebraic surfaces, on the other. In the present paper this connection is used for extracting, from simple considerations of four-dimensional topology and the arithmetic of integral quadratic forms, information about the arrangement of ovals of a real plane algebraic curve.

§1. Formulation of the Result

Consider a nonsingular real algebraic curve of degree n on the real projective plane $\mathbb{RP}^2$. Such a curve is a compact smooth one-dimensional manifold without boundary. Its connected components are diffeomorphic to the circle. They are called *ovals*.

A. Harnack [1] proved that the number of ovals does not exceed $g+1$, where $g=(n-1)(n-2)/2$ is the genus of the curve. Curves with $g+1$ ovals do exist, and are called *M-curves*.

The question of possible arrangements of ovals was considered by many authors (in particular, D. Hilbert [2], K. Rohn [3], I.G. Petrovsky [4], and D.A. Gudkov [5]), but has been answered only for curves of degree 6 and less (see a survey in [5]).

For formulation of the result, we need the division of ovals into positive and negative introduced by I.G. Petrovsky. Assume that the curve has even degree: $n = 2k$ (we retain this notation throughout the paper). Then the ovals lie two-sidedly in $\mathbb{RP}^2$, and each of them bounds an interior domain (diffeomorphic to a disk) and an exterior (diffeomorphic to a Möbius strip). We shall call an oval *positive* (or *even*) if it lies inside (in the interior domain) of an even number of other ovals, and *negative* (or *odd)* if it lies inside of an odd number of other ovals. For example, the usual circle $x^2+y^2=1$ is an even oval.

Moscow State University. Translated from Funktsional'nyi Analiz i Ego Prilozheniya, Vol. 5, No. 3, pp. 1–9, July–September, 1971. Original article submitted March 22, 1971.

Originally publ. in: Funkts. Anal. Prilozh. 5:3, 1-9, © Russ. Acad. Sciences 1971
Original English transl.: Funct. Anal. Appl. 5, 169-176, © Consultants Bureau, New York 1971
Translated by O.Viro. Springer-Verlag Berlin Heidelberg 2014 with kind permission, DOI 10.1007/978-3-642-31031-7_26

Theorem 1. *Let p be the number of positive and m the number of negative ovals of an M-curve of degree $2k$. Then, the following congruence holds:*

$$p - m \equiv k^2 \bmod 4 . \tag{1}$$

Note that congruence (1) does not exhaust all the restrictions on the arrangement of the ovals. For example, I.G. Petrovsky [4] proved the inequality

$$|2(p-m) - 1| \leq 3k^2 - 3k + 1 \tag{2}$$

for any curve of degree $2k$ (with not necessarily maximal number of ovals), and D.A. Gudkov [5] proved, for M-curves of degree 6, a congruence of the form of (1) but modulo 8. For other restrictions, see §9, Remarks 4, 5, and 6.

The proof of congruence (1) given below is based on a consideration of the action of involutions of a certain four-dimensional manifold Y on the two-dimensional homology of it.

§2. Construction of Four-Dimensional Manifold Y

The manifold Y is a two-fold covering of the complex projective plane, ramified along the set of complex points of the curve under consideration.

Let $f(x,y) = 0$ be the affine equation of the curve. Here x and y are real variables and f is a non-homogeneous polynomial of degree $2k$ with real coefficients. The corresponding homogeneous equation has the form $F(x_0,x_1,x_2) = 0$, where $f(x,y) = F(1,x,y)$.

Here F is a homogeneous polynomial of degree $2k$ with real coefficients. Since the degree of F is even, the sign of F is the same at all the points (x_0,x_1,x_2) corresponding to the same point $(x_0 : x_1 : x_2)$ of the projective plane $\mathbb{RP}^2$ which does not lie on the curve. We can choose F so that, at the points corresponding to the *non-orientable* component of the complement to the curve in $\mathbb{RP}^2$, the values of F will be *negative*. Fix such a polynomial F.

Now consider the equation

$$z^2 = F(x_0,x_1,x_2) , \tag{3}$$

where x_0, x_1 and x_2 are complex variables that do not vanish all simultaneously. This equation defines a compact complex algebraic surface Y embedded in the three-dimensional complex space E of a one-dimensional vector bundle over the complex projective plane

$$P' : E \to \left(\mathbb{CP}^2 = \{(x_0 : x_1 : x_2)\}\right)$$

whose sections are homogeneous functions of degree k of the variables x_0, x_1, and x_2. From the real point of view, Y is a four-dimensional compact smooth orientable connected manifold without boundary.

In the affine coordinate system in E corresponding to the affine coordinate system (x,y) in $\mathbb{CP}^2$, the surface Y is defined by the equation $z^2 = f(x,y)$.

The restriction P of the mapping P' to Y gives a two-fold ramified covering of the complex projective plane. The ramification locus is the intersection of Y with the zero section of P'. Denote this ramification locus by A. It is clear from Eq. (3) that PA is precisely the set of complex points of our algebraic curve $F = 0$. Thus, A is a smooth orientable connected compact submanifold without boundary of the four-dimensional manifold Y. The real dimension of A (and its codimension in Y) equals 2.

§3. Involution τ and Form Φ_τ

Multiplication of z by -1 determines a smooth involution τ of Y. The set of fixed points of τ is just our complex curve A.

On all our complex manifolds we choose the natural orientations (defined by the form $\operatorname{Re} z_1 \wedge \operatorname{Im} z_1 \ldots \operatorname{Re} z_S \wedge \operatorname{Im} z_S$), so that the intersection numbers of any complex manifolds are non-negative. Observe that the involution τ preserves the orientation of Y.

Denote by $H_2(Y) = H_2(Y,\mathbb{Z}) \bmod \mathrm{Tors}$ the two-dimensional integer homology group of Y factorized by the torsion. The intersection number $(\,,\,)$ of two-dimensional cycles defines on $H_2(Y)$ a bilinear integral nonsingular form (the Poincaré duality). The involution $\tau : Y \to Y$ induces an isomorphism $\tau_* : H_2(Y) \to H_2(Y)$. Define on $H_2(Y)$ the bilinear form Φ_τ by the formula

$$\Phi_\tau(a,b) = (\tau_* a, b), \quad a,b \in H_2(Y)\,. \tag{4}$$

Lemma 1. *Form Φ_τ is symmetric and nonsingular* ($\det \Phi_\tau = \pm 1$).

Proof. Since involution τ preserves the orientation of Y, we have: $(\tau_* a,b) = \tau_*(\tau_* a,b) = (a,\tau_* b) = (\tau_* b,a)$. Nonsingularity of Φ_τ follows from the Poincaré duality.

§4. Arithmetic Lemma

Let $\Phi : \mathbb{Z}^r \times \mathbb{Z}^r \to \mathbb{Z}$ be an integral symmetric nonsingular ($\det \Phi = \pm 1$) bilinear form. The following lemma is well known (see [6, 7, 8]).

Lemma 2. *There exists an element $w \in \mathbb{Z}^r$ such that, for all $x \in \mathbb{Z}^r$*

$$\Phi(x,x) \equiv \Phi(w,x) \bmod 2\,. \tag{5}$$

The element w is defined uniquely modulo $2\mathbb{Z}^r$. *The congruence class* $\Phi(w,w) \bmod 8$ *does not depend on the choice of w, and coincides, modulo 8, with the signature of the form.*

In what follows we use only that Eq. (5) implies

$$\Phi(w',w') \equiv \Phi(w,w) \bmod 8 \quad \text{for} \quad w' = w + 2x\,. \tag{6}$$

Proof of (6). We have

$$\Phi(w',w') = \Phi(w,w) + 4\Phi(w,x) + 4\Phi(x,x)\,,$$

where, according to Eq. (5), the numbers $\Phi(w,x)$ and $\Phi(x,x)$ have the same parity.

Element w (defined modulo 2) will be called the *fundamental class of form* Φ.

§5. Calculation of the Fundamental Class of Φ_τ

Let τ: $Y^{4s} \to Y^{4s}$ be an orientation-preserving involution of an oriented compact smooth $4s$-dimensional manifold without boundary with the $2s$-dimensional oriented manifold of fixed points A^{2s}. Define a form Φ_τ on $H_{2s}(Y^{4s})$ by formula Eq. (4).

Lemma 3. *The homology class* $[A^{2s}]$ *represented by the cycle* A^{2s} *is a fundamental class of* Φ_τ.

Proof. To each intersection point Q of the cycles a and τa, which does not lie on A^{2s}, there corresponds another intersection point τQ. At an intersection point of a and A^{2s}, one can achieve pairwise transversality of A^{2s}, a, and τa by a small move of a. Therefore, the intersection numbers of a with τa and with A^{2s} coincide modulo 2.

Now, let $A \subset Y$ be the manifolds from §3.

Lemma 4. *The self-intersection number of A in Y equals the half of square of the degree of algebraic curve PA:*

$$\Phi_\tau([A],[A]) = (A,A) = 2k^2\,.$$

Proof. Consider the projection PA of A in $\mathbb{CP}^2$. Let A' and A'' be curves, homologous to PA, in $\mathbb{CP}^2$ intersecting each other transversally, and not on PA. Then, the intersection number of A' and A'' in $\mathbb{CP}^2$ equals the square of the degree of curve PA, i.e., $4k^2$. Consequently, the intersection number of the cycles $P^{-1}A'$ and $P^{-1}A''$ in Y equals $8k^2$. But in Y we have $P^{-1}A' \sim 2A \sim P^{-1}A''$, therefore $(A,A) = 2k^2$.

§6. The Real Part Π of Y

Now let us study the set of real points of Y, i.e., points at which the coordinates z and $x_0 : x_1 : x_2$ take real values.

Consider, on the real projective plane $\mathbb{RP}^2$, the set C of points at which $F \geq 0$. Set C is a smooth two-dimensional submanifold in $\mathbb{RP}^2$ whose boundary is formed by the ovals we study. Generally speaking, manifold C is not connected; due to the assumption about the sign of F made in §2, it is orientable. Observe that the Euler characteristic of C equals the difference $p - m$ between the numbers of positive and of negative ovals.

Lemma 5. *The set Π of real points of Y is a smooth compact two-dimensional orientable manifold without boundary, smoothly embedded in Y. Manifold Π is diffeomorphic to the sum of two copies of C attached to each other by the identity map of the boundary.*

The proof follows from Eq. (3) and triviality of the fibration P over C; the latter is guaranteed by the choice of sign of F.

Lemma 6. *The self-intersection number of a real orientable compact smooth analytic manifold M of dimension $2s$ in its complexification equals $(-1)^s\chi$, where χ is the Euler characteristic of M.*

Proof. Let us construct on M a tangent vector field with simple singular points, and multiply this vector field by i. The sign $(-1)^s$ corresponds to the difference between the two orientations of $\mathbb{C}^{2s}$: one can take all real base vectors and then all the corresponding imaginary vectors, or one can choose the orientation as prescribed in §3.

Lemma 7. *The Euler characteristic of Π is expressed in terms of the numbers of positive and negative ovals by the formula*

$$\chi = 2(p-m)\,.$$

For the proof, notice that $p - m$ is the Euler characteristic of C, and then use Lemma 5.

By combining Lemmas 6 and 7, we obtain the following result.

Lemma 8. *The self-intersection number of Π in Y equals $(\Pi,\Pi) = 2(m-p)$.*

§7. A Homology between the Cycles A and Π

Lemma 9. *The homology classes represented by surfaces A and Π in Y coincide modulo 2; more precisely,*

$$[A] = [\Pi] \in H_2(Y)\otimes\mathbb{Z}_2\,.$$

Proof. Denote by ∞ the complex line "at infinity" defined by the equation $x_0 = 0$ in $\mathbb{CP}^2$ with coordinates $x_0 : x_1 : x_2$ from formula Eq. (3). We shall assume that this line intersects the curve PA transversally in $n = 2k$ points. We can consider the line ∞ as a cycle (with coefficients in $\mathbb{Z}$ or in $\mathbb{Z}_2$).

Denote by $[\infty]$ the homology class of cycle ∞ in $H_2(\mathbb{CP}^2)$. Each cycle c in $\mathrm{H}_2(\mathbb{CP}^2)$ is homologous to $(c,\infty)[\infty]$. For example, $PA \sim 2k[\infty]$.

Further, denote by ∞_Y the cycle $P^{-1}\infty$ in Y, and by $[\infty_Y]$ its homology class in $H_2(Y)$. Lemma 9 follows from the following two relations.

$$[A] = k[\infty_Y] \in H_2(Y)\,, \tag{7}$$

$$[\Pi] = k[\infty_Y] \in H_2(Y,\mathbb{Z}_2)\,. \tag{8}$$

To prove the relation (8), notice that the integer cycle $PA - 2k\infty$ is the boundary of some integer three-dimensional chain K_3 in $\mathbb{CP}^2$. Set $K_3' = P^{-1}K_3$. Then, $\partial K_3' = 2A - 2k\infty_Y$, i.e., $[A] - k[\infty_Y]$ is an element of order 2 in $H_2(Y,\mathbb{Z})$, which proves Eq. (8) (recall that $H_2(Y)$ is the homology group *factorized by torsion*).

To prove the relation (8), note that the ovals divide M-curve PA into two parts, the complex conjugation taking one part into the other (this is established in the proof of Harnack's Theorem). Let B be one of these parts. B is a connected compact oriented two-dimensional manifold with boundary. The boundary of B consists of the ovals of the curve $F = 0$ on $\mathbb{RP}^2$.

Consider the surface $B + C$, where C is the submanifold in $\mathbb{RP}^2$ bounded by the ovals of the curve $F = 0$ (manifold C was defined in §6). The nonsmooth surface $B + C$ is a combinatorial, not necessarily orientable, connected compact two-dimensional submanifold without boundary in $\mathbb{CP}^2$. Therefore, $B + C$ is a cycle in $\mathbb{CP}^2$ with coefficients in $\mathbb{Z}_2$.

Let us show that the *intersection number of cycles $B + C$ and ∞ in $\mathbb{CP}^2$ equals k modulo 2*.

Indeed, if the ovals do not intersect the line at infinity, then exactly half of the $2k$ intersection points of PA and ∞ belong to B (the complex conjugation interchanges the interiors of B and $PA - B$). If the ovals do intersect the line at infinity, then, by putting this line in general position with respect to $B + C$, one can make it intersect only with B, transversally, and at exactly k points.

For the proof, let us choose the affine coordinates (x, y) so that the line ∞ would have the equation $x = 0$, and its point at infinity would lie outside C. The line $x = i\varepsilon$, where $\varepsilon > 0$ is sufficiently small, is homologous to the line $x = 0$, does not intersect C, and transversally intersects B in precisely k points. Indeed, orient by a tangent vector field ξ each oval. Then, upon movement in the (x, y) plane in the direction of ξ along an oval, we cross the line $x = 0$ from left to right as many times as from right to left. Intersections of one type correspond to points of intersection of PA with line $x = i\varepsilon$ lying in B and, the other, in $PA - B$ (this follows from the fact that $i\xi$ is a normal vector field to the oval in B). The numbers of nonreal intersection points of PA with $(x = i\varepsilon)$ that lie in B and $PA - B$ are equal if ε is sufficiently small (because this is the case when $\varepsilon = 0$).

Thus, our assertion italicized above is proven. It implies that $[B+C] = k[\infty] \in H_2(\mathbb{CP}^2, \mathbb{Z}_2)$. Therefore, the cycle $B+C-k\infty$ is homologous to 0 as a $\mathbb{Z}_2$ chain: there exists a three-dimensional $\mathbb{Z}_2$ chain L_3 in $\mathbb{CP}^2$ such that

$$\partial L_3 = B + C - k\infty\,.$$

Setting $L_3' = P^{-1}L_3$ we find

$$\partial L_3' = 2\left(P^{-1}B\right) + \Pi - k\infty_Y, \quad \text{i.e.,} \quad [\Pi] = k[\infty_Y] \in H_2(Y, \mathbb{Z}_2)\,.$$

Lemma 9 is proven.

§8. Proof of Theorem 1

Apply Lemma 2 to the form Φ_τ and homology classes $w = [A] \in H_2(Y)$, $w' = [\Pi] \in H_2(Y)$.

According to Lemma 1, the form is symmetric and nonsingular; according to Lemma 3, the class w is fundamental. By Lemma 9, the class w' differs from w by an even element. Thus, Lemma 2 is applicable, and we find from Eq. (6) that

$$\Phi_\tau([A],[A]) \equiv \Phi_\tau([\Pi,\Pi]) \bmod 8\,.$$

Notice that $\tau_* w = w$, $\tau_* w' = -w'$ (change of the sign of z changes the orientation of Π). Thus, $\Phi_\tau([A],[A]) = (A,A)$, $\Phi_\tau([\Pi,\Pi]) = -(\Pi,\Pi)$. Therefore, $(A,A) + (\Pi,\Pi) \equiv 0 \bmod 8$. Substituting the values of the self-intersection indices $(A,A) = 2k^2$ from Lemma 4 and $(\Pi,\Pi) = 2(m-p)$ from Lemma 8, we find

$$2k^2 + 2(m-p) = 0 \bmod 8\,,$$

q.e.d.

§9. Remarks

1. Assertion (1) (modulo 8 rather than 4) was formulated by D.A. Gudkov as a conjecture supported by a large number of examples. Although the proof of congruence (1) does not use the results of D.A. Gudkov, the present paper could not have been produced if D.A. Gudkov had not communicated his conjecture to the author.

2. Manifold Y was studied by V.A. Rokhlin in a recent work [9] using of the Hirzebruch–Atiyah–Singer signature formulas ([10], §6). By combining Rokhlin's computations with ours and with the Lefschetz–Dold–Atiyah–Bott formulas given by Hirzebruch in [11], one can obtain an additional information on manifold Y and its involutions.

Notice that the real differential type of manifold Y with involution τ depends only on the degree of curve A, for this the curve A need not to be real. However, the complex conjugation involution σ: $Y \to Y$ depends on the arrangement of the real ovals of A.

Involutions σ and τ commute, so that the group $H = \mathbb{Z}_2 + \mathbb{Z}_2$ of the four elements $1, \sigma, \tau, \sigma\tau$ acts on Y. Denote by Φ_h (where $h \in H$) the form $\Phi_h(x,y) = (h_* x, y)$ on $H_2(Y)$. Denote by Π' the submanifold consisting of those points of Y for which the point $x_0 : x_1 : x_2$ is real, while z is purely imaginary. Then, for any real curve $F = 0$ (not necessarily with maximal number of ovals), the relations shown in Table 1 hold:

Table 1

h	1	σ	τ	$\sigma\tau$
Fixed point set of h;	Y	Π	A	Π'
Fundamental class of Φ_h	$(k+1)[\infty_Y]$	$k[\infty_Y]$	$k[\infty_Y]$	$(k+1)[\infty_Y]$
Trace of h_*.	$2+2g$	$2(p-m)-2$	$-2g$	$2(m-p)$
Signature of Φ_h	$2-2k^2$	$2(m-p)$	$2k^2$	$2(p-m)-2$

None of these results was used in the proof of congruence (1) presented above, but without the computations of V.A. Rokhlin this proof would hardly have been found. To be more specific, the formulation of Lemma 9 has been guessed from a comparison of Table 1 with the D.A. Gudkov conjecture.

3. The connection between the arrangement of ovals and the involutions of four-dimensional manifold Y that we have discovered can also be used in the opposite direction, for obtaining information on the involutions of manifold Y on the basis of information about the real curve. For example, from relationships (1) and (2) we find that

$$2 + \operatorname{Tr}\sigma_* \equiv 2k^2 \bmod 8 \quad (\text{for } M \text{ curves}); \qquad |1 + \operatorname{Tr}\sigma_*| \le 3k^2 - 3k + 1 .$$

From Table 1 it follows also that *the forms Φ_σ and Φ_τ have the same parity as the number k, while the forms Φ_1 and $\Phi_{\sigma\tau}$ have the parity of $k+1$.* Indeed, for even k, the signature of form Φ_1, and for odd k the signature of Φ_τ, is not divisible by 8. Consequently, for all k there exists a cycle whose intersection number with ∞_Y is odd. By combining this information with the form of the fundamental classes, we obtain our assertion.

From Table 1, congruence (1), and inequality (2), it is clear that forms Φ_1 and $\Phi_{\sigma\tau}$ cannot be definite, while forms Φ_σ and Φ_τ are definite only for $k = 1$ and $k = 2$. Theorems on the structure of a quadratic form ([8], Chapter 5), therefore permit a complete reconstruction of the canonical forms for Φ_1, Φ_σ, Φ_τ, $\Phi_{\sigma\tau}$ for all k. For example, when $k = 2$ (i.e., for curves of degree 4), in the notation of [8] the form $-\Phi_\sigma = \Phi_\tau$ is Γ_8 corresponding to the Lie algebra E_8.

4. It is interesting that consideration of the involution σ gives a simple proof of the I.G. Petrovsky inequality (2). To this end, consider the linear space $E = H_2(Y, \mathbb{R})$ with inner product given by the intersection pairing. Involution σ_* acts in E, preserving the inner product, so that E splits into a direct sum of two orthogonal eigenspaces E_1 and E_{-1}, corresponding to eigenvalues 1 and -1 of the operator σ_*.

Observe that the inner product is non-degenerate. Therefore, each of the spaces E_1 and E_{-1} can be presented in the form of the direct sum of orthogonal subspaces on which the inner square is positive (negative) definite:

$$E_1 = E_1^+ + E_1^-, \quad E_{-1} = E_1^+ + E_{-1}^- .$$

Let us introduce now the following notation: $\dim E_1^+ = a$, $\dim E_1^- = b$, $\dim E_{-1}^+ = c$, $\dim E_{-1}^- = d$. In this notation,

$$\operatorname{Tr} \sigma_* = a + b - c - d, \qquad \operatorname{Tr} 1_* = a + b + c + d,$$
$$\operatorname{Sgn} \Phi_\sigma = a - b - c + d, \qquad \operatorname{Sgn} \Phi_1 = a - b + c - d.$$

Substitute into the left hand sides of these equations the values of traces and signatures from Table 1. Adding and subtracting the equations thus obtained, we find

$$a = (k-1)(k-2)/2, \; c = a + 1, \; b + d = 3k^2 - 3k + 1, \; b - d = 2(p - m) - 1 .$$

Since b and d are non-negative, $|b - d| \leq b + d$, that proves the inequality (2).

5. Our constructions give rise also to new restrictions on the arrangement of ovals. In order to formulate these restrictions, we partition all ovals into three classes according to the sign of the Euler characteristic of the component of the curve's complement for which the oval is the exterior boundary. Denote the numbers of positive ovals bounding from outside domains with positive, zero, and negative Euler characteristic by p_+, p_0, and p_- respectively, and the numbers of negative ovals by m_+, m_0, and m_-, so that

$$p = p_+ + p_0 + p_-, \quad m = m_+ + m_0 + m_- .$$

For example, p_+ is the number of positive ovals containing no other ovals inside themselves.

Theorem 2. *For any curve of degree* $2k$,

$$p_- \leq \frac{(k-1)(k-2)}{2}, \quad m_- \leq \frac{(k-1)(k-2)}{2}, \quad p_+ \leq b, \quad m_+ \leq d,$$

where the numbers b and d are defined in Remark 4.

Proof. Consider the connected component Π_i of surface Π, projected to the region of $\mathbb{RP}^2$ bounded externally by the given oval γ_i. Out of p surfaces Π_i, p_- have negative Euler characteristics.

The homology classes $[\Pi_i] \in H_2(Y)$ represented by these p_- surfaces (of whatever orientation) are pairwise Φ_1-orthogonal, and the quadratic form Φ_1 on the p_- classes $[\Pi_i]$ takes positive values. Besides, $\sigma_*[\Pi_i] = [\Pi_i]$.

It follows from this that the p_- classes $[\Pi_i]$ are linearly independent, and that on the subspace L spanned by them, the form Φ_1 is positive definite, while $\sigma_* = 1$. Therefore,

$$p_- = \dim L \leq \dim E_1^+ = \frac{(k-1)(k-2)}{2},$$

which proves our first inequality.

The other three inequalities are proven similarly (in considering m_- it is necessary to take into account the nonorientable component of the complement).

6. Linear independence of *all* Π_i would imply the following stronger inequalities

$$p_- + p_0 \leq \frac{(k-1)(k-2)}{2}, \quad m_- + m_0 \leq \frac{(k-1)(k-2)}{2},$$
$$p_+ + p_0 \leq b, \quad m_+ + m_0 \leq d,$$

and, for M-curves of degree $2k$, the following lower bound on the number of empty ovals: $p_+ + m_+ \geq k^2$.

According to D.A. Gudkov, in all the known examples of M-curves, $p_+ + m_+ \geq k^2 + (k-1)(k-2)/2$. Furthermore, for any odd k there exists an M-curve of degree $2k$ with $p_+ = k^2$, $p_0 = m_+ = (k-1)(k-2)/2$.

7. Note also that the quotient space $X = Y/\tau\sigma$ from the naturally arising commutative diagram of the two-fold ramified coverings

$$\begin{array}{ccc} Y & \rightarrow & X \\ \downarrow & & \downarrow \\ \mathbb{CP}^2 & \rightarrow & S^4 \end{array}$$

is a closed connected simply-connected oriented four-dimensional smooth manifold for which

$$\operatorname{Sgn} X = \frac{1}{2}(\operatorname{Sgn} Y + \operatorname{Sgn} \Phi_{\tau\sigma}) = p - m - k^2 = a - d,$$

and for all known M-curves

$$a \leq d, \quad c \leq b, \quad a - d \equiv 0 \pmod 8.$$

From validity of these relations for all M-curves it would follow both the validity of the Gudkov conjecture modulo 8 and the inequality

$$|p - m - 1| \leq k^2 - 1,$$

which is a strengthening of the Petrovsky inequality for M-curves.

8. For the first proof of Lemma 9 the author is indebted to A.N. Varchenko, to whom the author communicated this lemma in the form of a conjecture. In spite of some flaws in A.N. Varchenko's proof, it convinced the author in the validity of the lemma, and without this the present work could not be done. The proof of Lemma 9 presented above uses some ideas from A.N. Varchenko's arguments. Thus, the proof of congruence (1) is the result of joint efforts of the author and A.N. Varchenko. Unfortunately, A.N. Varchenko did not agree to consider himself a co-author of this work.

The author is grateful to A.N. Varchenko, D.A. Gudkov, V.A. Rokhlin, and D.B. Fuchs for numerous discussions and support.

References

[1] A. Harnack, „Über Vieltheiligkeit der ebenen algebraischen Curven", Math. Ann., **10**, 189–199 (1876).

[2] D. Hilbert, „Über die reellen Züge algebraischer Kurven", Math. Ann., **38**, 115–138 (1891).

[3] K. Rohn, „Die Maximalzahl und Anordnung der Ovale bei der ebenen Kurve 6. Ordnung und bei der Fläche 4. Ordnung", Math. Ann., **73**, 177–229 (1913).

[4] I. Petrovsky, "On the topology of real plane algebraic curves", Ann. Math., **39**, No. 1, 187–209 (1938).

[5] D.A. Gudkov, "Topology of sixth-order curves and fourth-order surfaces", Uchenye Zapiski Gor'k. Un-ta, **87**, 3–153 (1969) (Russian). English translation: Nine Papers on Hilbert's 16th Problem D. A. Gudkov and G. A. Utkin, American Math. Soc. Translations– Series 2, Volume 112 (1978); 172 pp.

[6] F. Hirzebruch and K.H. Mayer, „O-Mannigfaltigkeiten, exotische Sphären und Singularitäten", Lecture Notes Math., **57**, 89–91, J. Springer (1968).

[7] J. Milnor, "On simply-connected 4-manifold", in: Symposium Internacional de Topología Algebraica, 122–128, Mexico (1958).

[8] J.-P. Serre, "Cours d'arithmétique", Presses Universitaires de France, Paris (1970).

[9] V.A. Rokhlin, "Two-dimensional submanifolds of four-dimensional manifolds", Funktsional. Analy. i Ego Prilozhen., **5**, No. 1, 48–60 (1971) (Russian). English translation: Funct. Analysis and Its Appl., 1971, 5:1, 39-48.

[10] M.F. Atiyah and I.M. Singer, "Indices of elliptic operators. III", Usp. Matern. Nauk., **24**, No. 1, 127–128 (1969) (Russian). English original: The Annals of Mathematics, Second Series, Vol. 87, Issue 3 (1968) 546-604.

[11] F. Hirzebruch, "Elliptic differential operators on manifolds", Usp. Matem. Nauk, **23**, No. 1, 191–209 (1968) (Russian). German original: Elliptische Differentialoperatoren auf Mannigfaltigkeiten, Veröff. Arbritsgemein. Forsch. Land. Nordrhein – Westfalen. Natur-und Ing. Giesellschaftswiss., 157 (1965), 33–60.

Topology of real algebraic curves (works of I.G. Petrovsky and their development)*

V.I. Arnol'd
Translated by Oleg Viro

1. Classical results (Apollonius, Descartes, Newton, Harnack). The sixteenth Hilbert problem.

2. Theorems by I.G. Petrovsky on upper bounds for the Euler characteristic of a sublevel set of a polynomial with a fixed degree and for the Euler characteristic of a level surface of a polynomial. The Euler-Jacobi-Kronecker formula (Lefschetz-Atiyah-Bott)

$$\sum \frac{p(x_k)}{\det|\partial^2 f/\partial x^2|_{x_k}} = 0, \quad x \in \mathbb{C}^m, \ \deg f = n, \ \deg p < m(n-2)$$

(the summation runs over all $(n-1)^2$ critical points of the polynomial f).

Applications by I.G. Petrovsky of this formula to proving of lower bound for the sum of the number of critical points of even indices with positive critical values and the number of critical points of odd indices with negative critical values.

3. Estimates for the Betti numbers of real algebraic varieties: the works by I.G. Petrovsky, O.A. Oleinik, J. Milnor, R. Thom.

4. Theorem by D.A. Gudkov on arrangements of ovals of sextic curves. D.A. Gudkov's conjecture about congruence modulo 8, its proof (V.I. Arnold, A.N. Varchenko, V.A. Rokhlin) and generalizations (V.A. Rokhlin, V.M. Kharlamov, D.A. Gudkov, A.D. Krakhnov).

5. The recent achievements in the topology of real algebraic surfaces (G.A. Utkin, V.M. Kharlamov). New proofs of I.G.Petorvsky's inequalities for curves and their generalizations for varieties. (V.I. Arnold, V.M. Khalamov).

6. The conjecture about algorithmic insolvability of the topological classification problem for real algebraic varieties that are non-singular complete intersections of hypersurfaces of fixed degrees defined by equations with rational coefficients.

* A talk at the meeting of Moscow Mathematical Society on March 7, 1973.

Originally published in: Usp. Mat. Nauk 28:5, 260-262 (Russian), © Russ. Acad. of Sciences 1973
Translated by O.Viro. Springer-Verlag Berlin Heidelberg 2014 with kind permission, DOI 10.1007/978-3-642-31031-7_27

General A.A. Markov's results and the approximation J. Nash's theorem seem to imply only impossibility of such an algorithm depending algorithmically on the dgrees of the equations.

7. Estimations for the number of connected components of an algebraic hypersurface and the vibration theory.

The nodal curve of the mth Eigenfunction divides a membrane into at most m parts (Courant – Hilbert, vol. I, Ch. 6, §6). Courant claimed that the zero curve of any linear combination of the first m Eigenfunctions (of the Laplace operator on a compact Riemann manifold) has the same property, i.e., divides the manifold into at most m parts.

Eigen oscllations of the sphere with the standard metric are described by spherical functions, i.e., polynomials. Therefore the Courant statement cited above implies the following estimate

$$\dim_{\mathbb{R}} H_0(\mathbb{R}P^N - V_n, \mathbb{R}) \leq C^N_{N+n-2} + 1 \qquad (1)$$

for the number of connected components of the complement to an algebraic hypersurface of degree n in the N-dimensional projective space.

For planar curves ($N = 2$), the estimation (1) is exact (it turns into equality on a configuration of n lines in general position) and can be proven independently of the Courant statement. For smooth surfaces of degree 4 in $\mathbb{R}P^3$ the estimation is also exact and proved (by V.M. Kharlamov).

In general case, the Courant statement is false (a counter-example can be constructed by a small perturbation of the standard metric on the sphere). Nonetheless the estimation (1) seems to be plausable: for proving it one has to verify the Courant statement only for oscilations of the sphere (or the projective space) with the standard metric.[1]

References

[1] I. Petrovsky, "On the topology of real plane algebraic curves", Ann. of Math. **39**, No. 1, 187–209 (1938).

[2] I.G. Petrovsky, O.A. Oleinik, "On topology of real algebraic surfaces", Izvestia Acad Sci. **13**, 389–402 (1949).

[3] D.A. Gudkov, "Topology of sixth-order curves and fourth-order surfaces", Uchenye Zapiski Gor'k. Un-ta, **87**, 3–153 (1969) (Russian). English translation: Nine Papers on Hilbert's 16th Problem D. A. Gudkov and G. A. Utkin, American Mathematical Society Translations – Series 2, Volume 112 (1978); 172 pp.

[1] Translator's remark: the inequality (1) does not hold true for smooth surfaces of any even degree ≥ 6 in $\mathbb{R}P^3$. Counter-examples to (1) were constructed in the paper of O. Viro, "Construction of multicomponent real algebraic surfaces", Soviet Math. Dokl. **20**, No. 5, 991-995 (1979).

[4] V.I. Arnold, "On the arrangement of ovals of real plane algebraic curves, involutions of four-dimensional smooth manifolds, and the arithmetic of integral quadratic forms", Funktsional. Analiz i Ego Prilozhen., **5**, No. 3, 1–9, (1971), see also this volume, pp 239–249.
[5] V.A. Rokhlin, "Proof of the Gudkov conjecture", Funktsional Analiz i Ego Prilozhen., **6**, No. 2, 62–64 (1972).
[6] V.A. Rokhlin, "Congruences modulo 16 in Hilbert's sixteenth problem", Funktsional. Analiz i Ego Prilozhen. **7**, No. 2, 58–64 (1973) (Russian).
[7] V.M. Kharlamov, "The maximal number of components of a 4th degree surface in $\mathbb{R}P^3$," Funktsional. Analiz i Ego Prilozhen. **6**, No. 4, 101 (1972).
[8] V.M. Kharlamov, "New congruences for the Euler characteristic of real algebraic varieties", Funktsional. Analiz. i Ego Prilozhen. **7** (1973), 74–78
[9] D.A. Gudkov and A.D. Krakhnov, "On the periodicity of the Euler characteristic of real algebraic M-1-manifolds", Funktsional Analiz i ego Prilozhen. **7**, 15–19 (1973).
[10] O.A. Oleinik, "Bounds for Betti numbers of real algebraic hypersurfaces", Mat. Sb., **28**, (1951), 635–640.
[11] R. Thom, "Sur l'homologie des varietes algebriques reeles", Diff. and comb. top., A symposium in honor of M. Morse, Prinston Univ. Press, 1965, pp. 255–265
[12] J. Milnor, "On the Betti numbers of real varieties", Proc. Amer. Math. Soc. **15**, No. 2, 275–280 (1964).

To Ivan Georgievich Petrovskii
on his 70th birthday.

ON MATRICES DEPENDING ON PARAMETERS

V. I. Arnold

Given a family of matrices smoothly depending on parameters of endomorphisms of a complex linear space, it is shown that there is a normal form to which the family can be reduced by the choice of a base smoothly depending on the parameters. The formulae obtained are applied to the investigation of bifurcation diagrams of families of matrices.

Contents

Introduction

The reduction of a matrix to its Jordan normal form is an unstable operation. For both the normal form itself and the reducing mapping depend discontinuously on the elements of the original matrix. Therefore, if the elements of a matrix are known only approximately, then it is unwise to reduce it to its Jordan form.

Furthermore, when investigating a family of matrices smoothly depending on parameters, then although each individual matrix can be reduced to a Jordan normal form, it is unwise to do so, since in such an operation the smoothness (and also the continuity) relative to the parameters is lost.

Thus, *the problem arises of finding the simplest possible normal form* to which not only one specific matrix, but an arbitrary family of matrices close to it can be reduced by means of a mapping smoothly depending on the elements of the matrices.

In the present note this problem is solved for a family of matrices depending holomorphically on parameters. In other words, we investigate

Originally publ. in: Usp. Mat. Nauk 26:2, 101-114, © Russ. Acad. Sciences 1971
English transl.: Russ. Math. Surv. 22, 29-43, © London Mathematical Society, 1971
Reprint by Springer-Verlag Berlin Heidelberg 2014 with kind permission, DOI 10.1007/978-3-642-31031-7_28

the problem of a normal form of matrices over the ring of germs of holomorphic functions of several complex variables.

Obviously, it is enough to consider a deformation of a Jordan matrix, that is, a family containing a matrix that reduces to the Jordan normal form when its parameters have the value zero. Theorem 4.4 gives a form to which all close matrices holomorphically depending on parameters can be reduced by a change of base.

This normal form differs from the original Jordan form in that some entries instead of being zero are holomorphic functions of parameters that vanish when the value of the parameters is zero. This is the simplest form in the sense that the number of such entries is as small as possible.

The reduction to this normal form is stable: if the matrix is known only approximately, then both the elements of the normal form and the reducing mapping can be found approximately.

The construction of the normal form is based on a well known theorem on commuting matrices and on standard arguments of the theory of singularities of differentiable mappings [2]. The result obtained has obvious applications in the theory of the bifurcations of non-linear differential and integral equations (see [3]–[7]).

The author is grateful to N. N. Brushlinskaya, D. A. Kazhdan and S. G. Krein for useful discussions.

§ 1. Versal and universal deformations

1.1. Deformations. We consider square complex matrices of order n. Let A_0 be such a matrix. A *deformation* of A_0 is a matrix $A(\lambda)$ of the same order as A_0 with entries that are power series of an arbitrary number k of complex variables λ_i, convergent in a neighbourhood of $\lambda = 0$, with $A(0) = A_0$. A deformation is also called a family, the variables λ_i *parameters*, and the parameter space $\Lambda = \{\lambda\}$ a *base* of the family.

Two deformations $A(\lambda)$ and $B(\lambda)$ of the matrix A_0 are called *equivalent* if there exists a deformation $C(\lambda)$ of the identity matrix with the same base such that $A(\lambda) = C(\lambda)B(\lambda)C^{-1}(\lambda)$.

Let φ be a germ of a mapping $\mathbf{C}^l \to \mathbf{C}^k$ holomorphic at 0, that is, a set of k power series in l complex variables convergent in a neighbourhood of 0. We assume that $\varphi(0) = 0$. The mapping φ of the parameter space $\{\mu\}$ into the base of the deformation $A(\lambda)$ defines a new deformation of A_0 according to the formula

$$(\varphi^*A)(\mu) = A(\varphi(\mu)), \qquad \mu \in \mathbf{C}^l.$$

The deformation φ^*A is said to be *induced* by A under the mapping φ.

1.2. Versal deformations. A deformation $A(\lambda)$ of A_0 is called *versal*[1] if

[1] The term "versal" is obtained from the word universal by discarding the prefix "uni" indicating uniqueness.

every other deformation $B(\mu)$ of A_0 is equivalent to a deformation induced by $A(\lambda)$ under a suitable change of parameters:

$$B(\mu)=C(\mu)A(\varphi(\mu))C^{-1}(\mu), \qquad C(0)=E, \qquad \varphi(0)=0.$$

A versal deformation $A(\lambda)$ is called *universal* if the change of parameters of φ is determined uniquely by $B(\mu)$.

EXAMPLE. Of the three deformations

$$\begin{pmatrix}\lambda_1 & \lambda_2\\ \lambda_3 & \lambda_4\end{pmatrix}, \quad \begin{pmatrix}\lambda_1 & 1+\lambda_3\\ 0 & \lambda_2\end{pmatrix}, \quad \begin{pmatrix}1+\lambda_1 & 0\\ 0 & \lambda_2\end{pmatrix}$$

the first is versal but not universal, the second is not versal, and the third is not only versal but also universal.

§2. Versality and transversality

2.1. **Transversality.** Let $N \subset M$ be a smooth submanifold of a manifold M. We consider a smooth mapping $A\colon \Lambda \to M$ of another manifold Λ into M, and let λ be a point in Λ such that $A(\lambda) \in N$.

The mapping A is called *transversal to N at λ* if the tangent space to M at $A(\lambda)$ is the sum

$$TM_{A(\lambda)}=A_*T\Lambda_\lambda+TN_{A(\lambda)}.$$

2.2. **Orbits.** We consider the set $M = \mathbf{C}^{n^2}$ of all $n \times n$ matrices and the Lie group $G = GL(n, \mathbf{C})$ of all non-singular $n \times n$ matrices. The group G acts on M according to the formula

(1) $$\mathrm{Ad}_g m = gmg^{-1} \quad (m \in M,\ g \in G).$$

Consider the orbit of an arbitrary fixed matrix $A_0 \in M$ under the action of G. This smooth submanifold of M is denoted by N. Thus, the orbit N of A_0 consists of all the matrices similar to A_0.

2.3. **The versality condition.** A deformation $A(\lambda)$ of a matrix A_0 may be regarded as a mapping $A\colon \Lambda \to M$ of the base of the family into the space M of matrices. The following lemma is obvious and well known.

LEMMA. *A deformation $A(\lambda)$ is versal if and only if the mapping A is transversal to the orbit of A_0 at $\lambda = 0$.*

We shall prove that the versal deformation A is transversal. Let $B(\mu)$ be any deformation of A_0. Then by the versality of A

$$B(\mu)=C(\mu)A(\varphi(\mu))C^{-1}(\mu),$$

hence

$$B_{*0}=A_*\varphi_*+[C_*, A_0].$$

Consequently, for every tangent to the base of B at the origin of a vector ξ we have $B_*\xi = A_*\varphi_*\xi + [C_*\xi, A_0]$. Hence, any vector in the space TM_{A_0}

is the sum of a vector in the image of A_* and a vector tangent to the orbit. Thus, A is transversal to the orbit.

We precede the proof of the versality of a transversal deformation by the following construction.

2.4. Orbit and centralizer. We consider the mapping $\alpha: G \to M$, given by (1) for a fixed $m = A_0$. The mapping α carries the group G into the orbit of A_0. The stationary subgroup H of A_0 is mapped to A_0; it consists of all the elements of G that commute with A_0.

The mapping α induces a mapping of the tangent space at the unit element of the group into the tangent plane to the orbit

$$\alpha_*: TG_e \to TM_{A_0}, \quad \alpha_* C = [C, A_0].$$

The kernel Ker α_* of α_* is the tangent plane to the stationary subgroup H. The subalgebra Ker α_* of the Lie algebra TG_e consists of all the matrices that commute with A_0 and is called the *centralizer* of A_0.

Note that the dimension of the matrix group G is the same as that of the matrix manifold M (namely n^2). Consequently, *the codimension of the orbit of A_0 is equal to the dimension of its centralizer.*

2.5. Decomposition into a direct product. By means of the unit element of G we introduce a submanifold V, transversal to the centralizer and of complementary dimension (equal to the dimension of the orbit). Let $A: \Lambda \to M$, $A(0) = A_0$, be an arbitrary deformation of A_0 transversal to the orbit of A_0, where the dimension of the base Λ is equal to the codimension of the orbit (that is, has the minimum value for transversal deformations).

Let $\beta: V \times \Lambda \to M$ be the mapping given by the formula $\beta(v, \lambda) = \mathrm{Ad}_v A(\lambda)$. Note that the dimension of the direct product $V \times \Lambda$ is the same as that of M (namely n^2). From the transversality of A to the orbit and the transversality of V to the centralizer it follows that the derivation $\beta_*|_{(e, 0)}$ is non-singular. By the inverse functions theorem the mapping β determines a holomorphic diffeomorphism between a sufficiently small neighbourhood of $(e, 0)$ in $V \times \Lambda$ and a sufficiently small neighbourhood of A_0 in M. Replacing V and Λ by smaller manifolds, we may assume that a neighbourhood of A_0 in M splits into the direct product $V \times \Lambda$.

2.6. Transversality $\Rightarrow$ versality. Let A be a transversal deformation. If the dimension of the base is greater than the minimum (that is, the codimension of the orbit), we replace the base by a submanifold whose dimension is equal to the codimension of the orbit and such that the restriction of A to this subspace is still transversal to the orbit. Let us prove that the resulting family is versal. If we obtain more, so much the better.

Let $B(\mu)$ be an arbitrary deformation of A_0. For sufficiently small μ the matrix $B(\mu)$ has a unique representation in the form $B(\mu) = \beta(v, \lambda)$, $v \in V$, $\lambda \in \Lambda$ (see 2.5). Hence, B can be written in the form

$$B(\mu) = C(\mu) A(\varphi(\mu)) C^{-1}(\mu),$$

where $\varphi(\mu) = \pi_2 \beta^{-1} B(\mu)$, $C(\mu) = \pi_1 \beta^{-1} B(\mu)$, and where π_1 and π_2 are the projections of $V \times \Lambda$ onto V and onto Λ. Thus, A is versal, and the lemma of 2.3 is proved.

Note that in spite of the uniqueness of this decomposition, the deformation A need not be universal. The fact is that uniqueness holds only for a fixed submanifold V and the latter can be chosen in different ways.

§ 3. An example

3.1. A Sylvester family. Consider an n-parameter family of $n \times n$ matrices

$$A(\alpha) = \begin{pmatrix} 0 & 1 & & & & \\ & 0 & 1 & & & \\ \cdot & \cdot & \cdot & \cdot & \cdot & \\ & & & 0 & 1 & \\ \alpha_1 & \alpha_2 & & \dots & \alpha_n \end{pmatrix}.$$

We call this a *Sylvester* family.[1]

3.2. Versality of a Sylvester family. By a direct computation of the commutators it is easy to verify that a Sylvester family is transversal to the orbit of each of its matrices. Lemma 2.3 therefore has the following consequence.

COROLLARY. *A Sylvester family determines an n-parameter versal deformation of each of its matrices.*

If a matrix A_0 can be reduced to a Sylvester normal form, then also matrices close to A_0 can be reduced to a Sylvester form without losing their smooth dependence on the parameters.

3.3. Universality of Sylvester deformations. Note that the elements α_i of a Sylvester matrix are (to within the sign) the coefficients of its characteristic polynomial. Hence, the elements of the Sylvester normal form of a given matrix are uniquely determined.

COROLLARY. *A Sylvester family defines a universal deformation of each of its matrices.*

Thus, the problem of a normal form of a deformation of a matrix A_0 is solved for the case when A_0 can be reduced to Sylvester form, that is, when only one Jordan block corresponds to each eigenvalue.

3.4. Universal deformation of a Jordan block. In particular, we consider the case when A_0 is a Jordan block with the eigenvalue λ. From Lemma 2.3 and Corollary 3.3 we obtain:

COROLLARY. *A versal (and even a universal) deformation of a Jordan block is an n-parameter deformation*

$$D(\alpha_1, \dots, \alpha_n) = \lambda E + A(\alpha),$$

where $A(\alpha)$ is a Sylvester matrix.

[1] This can also be regarded as the family of all ordinary linear differential equations of order n with constant coefficients.

If a matrix A_0 is a Jordan block, then matrices close to A_0 can be reduced to the form just given, and the reducing mapping and the parameters α_i of the Sylvester matrix depend holomorphically on the elements of the original matrix.

3.5. Codimensions of orbits. From Corollary 3.3 we also derive:

COROLLARY. *For any Sylvester matrix, irrespective of the multiplicity of the eigenvalues, the codimension of the orbit (which is equal to the dimension of the centralizer and also to the number of parameters of a universal deformation) is one and the same, namely the order n of the matrix.*

§4. The construction of universal deformations

4.1. The orthogonal complement. We introduce in the space of matrices $M = \mathbf{C}^{n^2}$ a Hermitian scalar production $(A, B) = \mathrm{Tr}\,(AB^*)$, where B^* is the adjoint of B. The corresponding scalar square is simply the sum of the squares of the moduli of the matrix elements.

LEMMA. *The vector $B \in TM_{A_0}$ is perpendicular to the orbit of the matrix A_0 if and only if $[B^*, A_0] = 0$.*

PROOF. The tangent vectors to the orbit of A_0 are the matrices that can be represented in the form $[C, A_0]$. Since B is orthogonal to the orbit, for any matrix C we have $([C, A_0], B) = 0$. In other words, for any C we have $0 = \mathrm{Tr}([C, A_0]B^*) = \mathrm{Tr}(CA_0B^* - A_0CB^*) = \mathrm{Tr}([A_0, B^*]C) = ([A_0, B^*], C^*)$. Since C is arbitrary, this condition is equivalent to $[B^*, A_0] = 0$.

Thus, the lemma is proved: *the orthogonal complement to the orbit of A_0 is the adjoint of its centralizer.*

4.2. Centralizers of Jordan matrices. It is not difficult to compute the centralizers of matrices reduced to Jordan normal forms. We assume at first that the matrix A_0 has only one eigenvalue λ and a sequence of upper Jordan blocks of dimensions $n_1 \geqslant n_2 \geqslant \ldots$

LEMMA.[1] *The matrices that commute with A_0 are precisely those described in Figure 1 below.*

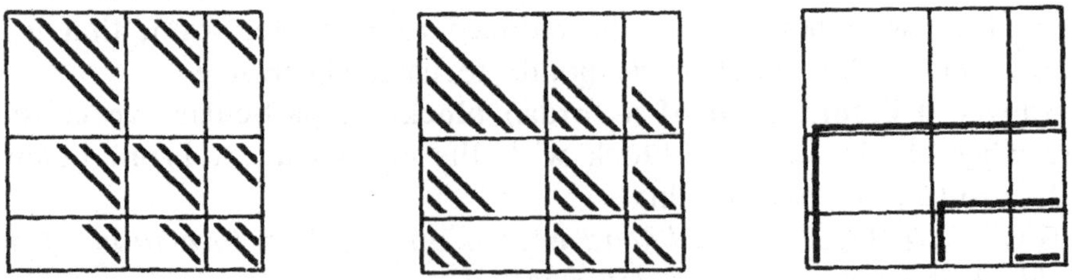

a) A matrix in the centralizer

b) A vector orthogonal to the orbit

c) A normal form

Fig. 1.

[1] The proof is in [8], 199–207.

In Figure 1a each oblique segment denotes a sequence of identical numbers, and the blank entries denote zero. Thus, the number of oblique segments is equal to the dimension of the centralizer.

COROLLARY. *The dimension of the centralizer of the matrix* A_0 (*which is equal to the codimension of the orbit and to the minimum dimension of a versal deformation*) *is given by the formula* $d = n_1 + 3n_2 + 5n_3 + \ldots$

If a Jordan matrix A_0 has several eigenvalues, we divide it into blocks corresponding to these eigenvalues. Then matrices commuting with A_0 are block diagonal, where a block corresponding to an eigenvalue λ has the form described in Figure 1a.

The formula for the dimension of the centralizer (the codimension of the orbit, the dimension of a minimal versal family) is obtained from the above by summation over all the distinct eigenvalues.

4.3. Selection of the normal form. For a transversal deformation of a matrix A_0 we may choose the family of the form $A_0 + B$, where B belongs to the orthogonal complement of the orbit of A_0 described above. So we get a versal deformation of A_0 with the minimum number of parameters (equal to the codimension of the orbit).

If A_0 has only one eigenvalue, B has the form described in Figure 1b. Here on each oblique segment is a sequence of identical numbers, which are the parameters of the deformation. The number of parameters is equal to the number of segments and is given by the formula of Corollary 4.2.

The matrix B has many non-zero entries. We can suggest another form of the versal family $A_0 + B$ in which the number of non-zero entries in B is minimal (and equal to the number of parameters).

To this end we choose a base of the centralizer in the following natural way: with each oblique segment in Figure 1a we associate a matrix with 0's and 1's, the 1's being on the given oblique segment.

By writing down the condition of orthogonality in Lemma 4.1, we can verify that the system of independent equations of the tangent space to the orbit consists of the following equations: for each oblique segment of Figure 1b the sum of the corresponding elements of the matrix is equal to zero.

Thus, to obtain a family $A_0 + B$ transversal to the orbit it is sufficient to take for the family of matrices B matrices for which on each oblique segment in Figure 1b one entry is an independent parameter, and all the other entries are zero.

The non-zero element can be chosen on each oblique segment at an arbitrary place. For example, the matrix described in Figure 1c is a suitable choice. So we come to the following result.

4.4. THEOREM. *Every matrix* A_0 *has a versal deformation, the number of its parameters is equal to the codimension of the orbit and to the dimension of the centralizer of* A_0. *This number is equal to*

$d = \sum_\lambda (n_1 + 3n_2 + 5n_3 + \ldots)$, *where* $n_1 \geqslant n_2 \geqslant \ldots$ *are the orders of Jordan blocks corresponding to the eigenvalue* λ *and the summation is over all the distinct eigenvalues. A versal deformation with fewer parameters does not exist.*

If A_0 *is in the Jordan normal form, then for a versal deformation with the minimum number of parameters we may take a d-parameter "normal form"* $A_0 + B$, *where the blocks of the block-diagonal matrix B have the form described in Figure* 1c. *The order of a block is equal to the multiplicity of the corresponding eigenvalue. The entries indicated in Figure* 1c *are independent parameters and the other entries are equal to zero.*

In other words *any complex matrix close to the given matrix can be reduced to the above d-parameter normal form* $A_0 + B$ (*where* A_0 *is the Jordan normal form of the given matrix*), *so that the reducing mapping and the parameters of the normal form depend holomorphically on the elements of the original matrix.*

Aliter: *each matrix over the ring of germs of functions holomorphic at zero can be reduced to the above d-parameter normal form* $A_0 + B$, *where* A_0 *is the Jordan form of the value of the germ at zero and B is a matrix whose only non-zero entries are those indicated in Figure* 1c.

4.5. Examples. If all the eigenvalues of A_0 are distinct, then $d = n$ and our normal form is diagonal.

If A_0 is the null matrix, then $d = n^2$ and our versal deformation is simply the family of all matrices.

If A_0 consist of one Jordan block, then $d = n$ and our normal form is the Sylvester form.

We denote a matrix in Jordan normal form by the product of the determinants of its blocks. For example, $\alpha^3\alpha^2$ denotes a matrix having two Jordan blocks (of order 3 and 2) with the same eigenvalue α, and the matrix $\alpha^2\alpha\beta^2$ has blocks of order 2 and 1 with the eigenvalue α and a block of order 2 with the eigenvalue β.

By Theorem 4.4 *minimal versal deformations of the matrices* $\alpha^3\alpha^2$ *and* $\alpha^3\alpha\beta^2$ *can be chosen in the following form*

$$\left|\begin{matrix} \alpha & 1 & 0 & 0 & 0 \\ 0 & \alpha & 1 & 0 & 0 \\ \lambda_1 & \lambda_2 & \alpha+\lambda_3 & \lambda_4 & \lambda_5 \\ \lambda_6 & 0 & 0 & \alpha & 1 \\ \lambda_7 & 0 & 0 & \lambda_8 & \alpha+\lambda_9 \end{matrix}\right|, \quad \left|\begin{matrix} \alpha & 1 & 0 & 0 & 0 \\ \lambda_1 & \alpha+\lambda_2 & \lambda_3 & 0 & 0 \\ \lambda_4 & 0 & \alpha+\lambda_5 & 0 & 0 \\ 0 & 0 & 0 & \beta & 1 \\ 0 & 0 & 0 & \lambda_6 & \beta+\lambda_7 \end{matrix}\right|.$$

The dimensions of the families are, respectively, equal to

$$d = n_1 + 3n_2 = 3 + 3\cdot 2 = 9, \quad d = n_1(\alpha) + 3n_2(\alpha) + n_1(\beta) = 2 + 3\cdot 1 + 2 = 7.$$

Thus, *any family of matrices close to a matrix of order 5 with a unique eigenvalue* α *and Jordan blocks of orders* 3 *and* 2 *can be reduced, without loss of smoothness, to the above nine-parameter normal form. If the matrices of the family depend holomorphically on the parameters then the coefficients* $\lambda_1, \lambda_2, \ldots, \lambda_9$ *of the normal form also depend holomorphically on the parameters.*

Similarly, any family of matrices close to $\alpha^2\alpha\beta^2$ can be reduced smoothly to the above seven-parameter normal form.

§ 5. Remarks

5.1. Generalizations. 1. Similar propositions can also be proved for the real case and for differentiable dependence on parameters, but the formulae are tedious.

2. In all the definitions instead of families of matrices we could have talked of fibre sections of fibre endomorphisms of the given vector fibering. This would have made the statements longer, without changing the proofs.

3. Similar (but simpler) results can be obtained for families of rectangular matrices (mappings from one space into another). Here a minimal versal family has the form

$$\begin{pmatrix} E_r & \\ & \lambda_{11} \ldots \lambda_{1l} \\ \ldots & \ldots \\ & \lambda_{k1} \ldots \lambda_{kl} \end{pmatrix},$$

where r is the rank of A_0, E_r is the identity matrix of order r, and λ_{ij} are independent parameters.

4. In Lemma 2.2 neither the fact that the manifold M consists of matrices, nor that G is a group of matrices are essential: important only is the fact that the Lie group G acts on M.

5.2. Applications. By combining the formulae we have obtained with the transversality theorem [2], we can derive many results on families of matrices "of general form".

For example, from the versality of the family in 5.1.3 we obtain the following corollary.

COROLLARY. *In the space of families of linear mappings* $A : \mathbf{C}^m \to \mathbf{C}^n$ *an open everywhere dense set forms a family of the following kind. The rank of A is everywhere equal to* min (m, n), *except on the "bifurcation diagram", which is formed by a finite number of smooth manifolds* Σ^i *in the parameter space. On* Σ^i *the rank is reduced to i units; the codimension of* Σ^i *is equal to* $(m - r)(n - r)$. *In the neighbourhood of each point the stratification* $\{\Sigma^i\}$ *of the parameter space is diffeomorphic to the direct product of a domain in* $\mathbf{C}^k$ *and the stratification of a neighbourhood of zero in one of the verbal families of* 5.1.3.

EXAMPLE. Let $m = n$ and let the rank of A_0 be $n - 1$. Then the bifurcational diagram of a family of general form in the neighbourhood of

A_0 is a smooth hypersurface Σ^1. If the rank of A_0 is $n - 2$, then the bifurcation diagram consists of a smooth submanifold Σ^2 of codimension 4 and a hypersurface Σ^1 which tends to Σ^2 so that in every plane normal to Σ^2, Σ^1 has close to Σ^2 the form of a quadric cone.

The normal forms of § 3 and § 4 lead to similar results, but to state them we have first to get rid of the continuous parameters – the eigenvalues.

5.3. Orbits of a bundle. We call a *bundle* the set of all matrices whose Jordan normal forms differ only by their eigenvalues, but for which the sets of distinct eigenvalues and the orders of the Jordan blocks are the same. For example, all the diagonal matrices with simple eigenvalues define one bundle.

Each bundle is a semi-algebraic smooth submanifold of the space of matrices, it is a fibre space whose fibres are the orbits. The base of this fibering is the configuration space consisting of sets of coloured eigenvalues (two eigenvalues have the same colour if the corresponding Jordan blocks have the same order).

The splitting into bundles is a finite semi-algebraic stratification of the space $\mathbf{C}^{n^2}$, and we can make use of the transversality theorem.

COROLLARY. *In the space of families of $n \times n$ matrices an open everywhere dense set forms a family transversal to all the bundles of the orbits.*

5.4. Families in general position. Families of matrices that are transversal to all the bundles are said to be in *general position.* Corresponding to the decomposition of the space of matrices into bundles, the parameter space of the family decomposes into submanifolds. In a family in general position almost all the matrices have simple eigenvalues. The exceptional parameter values to which there correspond matrices with multiple eigenvalues define a subset of the parameter space. We call this a *bifurcation diagram.*

The bifurcation diagram of a family in general position is a finite union of smooth manifolds; to each bundle of orbits there corresponds a manifold in the parameter space – a manifold of matrices with fixed orders of the Jordan blocks. The codimension of such a manifold in the parameter space of a family in general position is equal to the codimension of the corresponding bundle in the space of all matrices.

The singularities of the bifurcation diagram of a family in general position are the same as those in bifurcation diagrams of versal deformations.

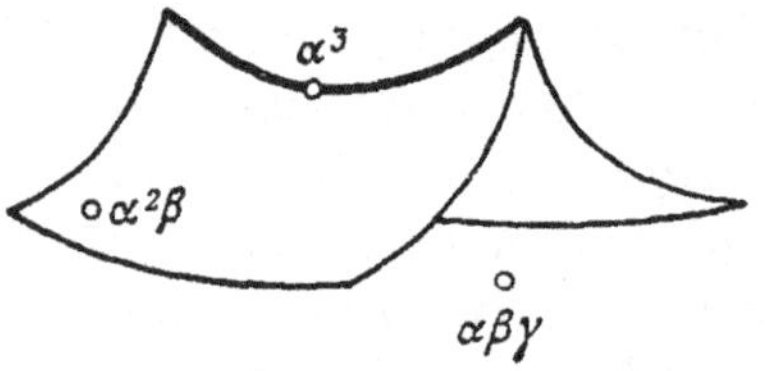

Fig. 2. Bifurcation diagram of a family in general position containing a Jordan block of order 3.

EXAMPLE. We assume that a family in general position contains a Jordan block of order n. Then the bifurcation diagram (in some neighbourhood of the corresponding value of the parameter) has the form of a direct product of a swallow tail[1] of the required dimension and a Euclidean space. For example, Figure 2 represents the bifurcation diagram of a family in general position containing a Jordan block of order 3.

5.5. Codimensions of bundles. It is easy to prove:

LEMMA. *The codimension of a bundle can be expressed by the dimensions of Jordan blocks* $n_1(\lambda) \geqslant n_2(\lambda) \geqslant \ldots$ *according to the formula*

$$c = 2\sum_{\lambda}(n_2 + 2n_3 + 3n_4 + \ldots) + \sum_{\lambda}(n_1 + n_2 + \ldots - 1).$$

For $c = d - \nu$, where d is the codimension of the orbit, and ν is the number of distinct eigenvalues λ. By Theorem 4.4

$$d = \sum_{\lambda}(n_1 + n_2 + \ldots) + 2\sum_{\lambda}(n_2 + 2n_2 + 3n_4 + \ldots), \quad \nu = \sum_{\lambda} 1,$$

from which the formula follows.

Note that in the formula for the codimension of a bundle simple eigenvalues make no contribution. In addition, the codimension of a bundle does not depend on the order n of the matrix, but depends only on the sizes of the Jordan blocks corresponding to multiple eigenvalues. Therefore Lemma 5.5 permits us to enumerate quickly singularities of small codimensions in bifurcation diagrams of families in general position irrespective of the order of the matrices.

5.6. Enumeration of singularities of small codimension. The only bundle of codimension 1 in the space of all matrices is that of matrices having a single eigenvalue of multiplicity two. We denote this bundle by α^2, without indicating the simple eigenvalues, which can be arbitrary. For the sake of brevity we say "the codimension of a case" (instead of "the codimension of a bundle in the space of all matrices"). In the case α^2 the bifurcation diagram is a smooth hypersurface.

There are two cases of codimension 2: that of a pair of Jordan blocks each of order 2 with distinct eigenvalues ($\alpha^2\beta^2$) and that of one Jordan block of order 3 (α^3). The corresponding bifurcation diagrams in the plane are a pair of intersecting straight lines and a semi-cubical parabola.

By Corollary 5.3 *a two-parameter family in general position has a bifurcation diagram with singularities only of these two forms.*

Thus, the versal families of §4 allow us to investigate the singularities of bifurcation diagrams of families in general position with a large number of

---

[1] A swallow tail of dimension $k-1$ is a hypersurface in $\mathbf{C}^k$ given by an equation $\Delta(a) = 0$, where Δ is the discriminant of the polynomial $z^{k+1} + a_1 z^{k-1} + \ldots + a_k = 0$. A one-dimensional swallow tail is a semi-cubical parabola; a two-dimensional one is shown on the left of Figure 3.

parameters.[1] With increasing codimension c the number of distinct bundles N grows in the following form:

c	1	2	3	4	5	6	7	8	9
N	1	2	4	7	11	19	30	51	73

All cases of codimension $c < 5$ are enumerated in the following list:

c	1	2	3	4
det	α^2	α^3, $\alpha^2\beta^2$	α^4, $\alpha\alpha$, $\alpha^3\beta^2$, $\alpha^2\beta^2\gamma^2$	α^5, $\alpha^2\alpha$, $\alpha^4\beta^2$, $\alpha\alpha\beta^2$, $\alpha^3\beta^2\gamma^2$, $\alpha^3\beta^3$, $\alpha^2\beta^2\gamma^2\delta^2$.

For example, the column $c = 3$ indicates that *the bifurcation diagrams of three-parameter families in general position have no singularities except those in Fig. 3.*

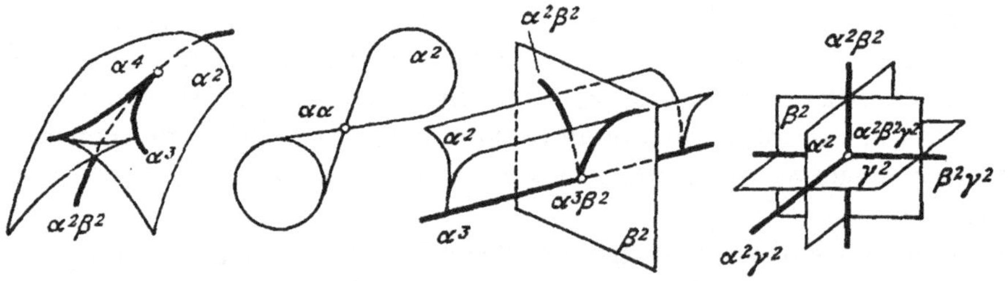

Fig. 3. Singularities of bifurcation diagrams of three-parameter families in general position.

Note that the bifurcation diagram of a family in general position with several distinct multiple eigenvalues is locally diffeomorphic to a direct product of diagrams of families corresponding to cases with one multiple eigenvalue.

We call bundles whose matrices have only one multiple eigenvalue *elementary*. All the other bundles are essentially transversal intersections of elementary ones. In particular, the codimension of a complex singularity is equal to the sum of the codimensions of the elementary components.

[1] Note that a versal d-parameter deformation of the matrix A_0 is transversal to the orbit and consequently to the bundle containing A_0. The codimension c of this bundle is less than d. Therefore, to construct bifurcation diagrams we can consider deformations with fewer parameters transversal to the bundle but not to the orbit.

Such a deformation is, for example, the c-parameter subfamily of the versal family of Theorem 4.4, which is obtained if one of the diagonal parameters of B is equated to zero (Fig. 1b) for each eigenvalue λ. Another c-parameter deformation, transversal to the bundle, is obtained if in the normal form of Theorem 4.4 we restrict ourselves to matrices B in which all the blocks corresponding to distinct eigenvalues have trace zero.

To prove this consider vectors tangent to the bundle but orthogonal to the orbit. Our assertions follow from the fact that if the matrix A_0 has only one eigenvalue, then such vectors and only they are scalar matrices.

The orders of the Jordan blocks of all the elementary bundles of codimension $c < 16$ are given in the following table:

c	1	2	3	...	7	8	...	11	12	...	15
$\{n_i\}$	2	3	4	...	8	9	...	12	13	...	16
			1, 1	...	5, 1	6, 1	...	9, 1	10, 1	...	13, 1
					2, 2	3, 2	...	6, 2	7, 2	...	10, 2
						1, 1, 1	...	4, 1, 1	5, 1, 1	...	8, 1, 1
								3, 3	4, 3	...	7, 3
									2, 2, 1	...	5, 2, 1
											4, 4
											1, 1, 1, 1

The number a_c of elementary bundles of codimension c can be computed by Euler's method [9] with the help of the generating function

$$\sum_{c=-1}^{\infty} a_c x^{c+1} = \prod_{k=1}^{\infty} (1 - x^{k^2})^{-1}.$$

Here we must take $a_{-1} = a_0 = 1$. The first few coefficients are

c	1	2	3	4	5	6	7	8	9	10	11	12	13	14	15
a_c	1	1	2	2	2	2	3	4	4	4	5	6	6	6	8

ADDED IN PROOF. With each matrix we can associate an interesting family of algebraic manifolds. Namely, we assign to each matrix the set of coefficients of its characteristic polynomial. The space of all matrices then becomes the complete space of families of algebraic manifolds (the fibre of a family is the set of all matrices with a fixed characteristic polynomial).

In exactly the same way the base of a deformation of an arbitrary matrix decomposes into subsets corresponding to matrices with common characteristic polynomials. Our versal deformations are algebraic (linear), therefore their bases are complete spaces of algebraic families of algebraic manifolds.

The explicit formulae of versal deformations make it possible to indicate explicitly the equations defining these families. Computations show that even for the simplest matrices we get highly unusual families of algebraic manifolds. Let us show, for example, how to obtain in this way one of Briescorn's results [12] on rational double points.

The dimension of the manifold of a family constructed for a given matrix is expressed in terms of the orders of the Jordan blocks according to the formula

$$D = d - n = \sum_{\lambda} 2n_2(\lambda) + 4n_3(\lambda) + 6n_4(\lambda) + \dots$$

(in the notation of 4.2).

Hence, *the family so constructed is that of algebraic surfaces* ($D = 2$) *if and only if to each eigenvalue there corresponds one Jordan block except for a single eigenvalue to which there correspond two Jordan blocks one of which is of order* 1.

By computing the characteristic polynomial of matrices corresponding to a versal family we can write down explicitly the equations of the resulting family of surfaces. After insignificant transformations these can be put into the form

$$xy + z^{k+1} + a_1 z^{k-1} + \dots + a_k = 0,$$

where $a_1, \dots, a_k$ are the parameters of the family. But this is the standard versal deformation of a surface with a rational double point of type A_k, which was obtained in detail by M. Artin, Briescorn and Tyurina [13]. Thus, *families of algebraic surfaces arising from versal deformations of matrices, and only these, are versal deformations of surfaces with rational double points of type* A_k.

References

[1] G. Frobenius, Ueber die vertauschbaren Matrizen, S.–B. Preuss. Akad. Wiss. Phys. Math. Kl. (1896), 7–16.

[2] Singularities of differentiable mappings. A collection of translations, Izdat. "Mir", Moscow 1968.

[3] V. I. Arnold, Remarks on singularities of finite codimension in complex dynamical systems, Funktsional Anal. i Prilozhen. 3 (1969), 1–6.
= Functional Anal. Appl. 3 (1969), 1–5.

[4] R. J. Sacker, On invariant surfaces and bifurcation of periodic solutions of ordinary differential equations, New York University, 1964.

[5] N. N. Brushlinskaya, A versal family of vector fields of a Poincaré domain, Funktsional Anal. i Prilozhen. 4 (1970), 6–13.
= Functional Anal. Appl. 4 (1970).

[6] S. G. Krein and V. P. Trofimov, On holomorphic operator functions of several complex variables, Funktsional Anal. i Prilozhen **3** (1969), 85–86.
= Functional. Anal. Appl. **3** (1969), 330–331.

[7] F. V. Atkinson, Multiparameter spectral theory, Bull. Amer. Math. Soc. **74** (1968), 1–27.

[8] F. R. Gantmakher, Teoriya matrits, second ed. Izdat. Nauka, Moscow 1967.
Translation: The theory of matrices, Chelsea, New York 1969.

[9] L. Euler, De partitione numerorum, Novi Comment. Acad. Sc. Petrop. **3** (1950/1), 125–135.

[10][1] W. Wasow, On holomorphically similar matrices, J. Math. Anal. Appl. **4** (1962), 202–206.

[11][1] S. Khabbaz and G. Stengle, An application of *K*-theory to global analysis of matrix-valued function, Math. Ann. **179** (1969), 115–122.

[12][1] E. Briescorn, Nice Mathematical Congress, 1970.

[13][1] G. N. Tyurina, The resolution of singularities of plane deformations of rational double points, Funktsional. Anal. i Prilozhen. **4** (1970), 77–83.
= Functional Anal. Appl. **4** (1970).

Translated by H. Freedman.

[1] **Added in proof.**

LECTURES ON BIFURCATIONS IN VERSAL FAMILIES

V. I. Arnol'd

In these lectures we consider the ways in which the disposition of the phase curves of a vector field can alter in a neighbourhood of a singularity as the parameters on which the vector field depends vary. A technical convenience in the study of such changes are certain deformations having a special universality property – the so-called versal families. Our results are presented mainly in the form of explicit formulae for versal families and an analysis of the corresponding bifurcation diagrams. As an application of the general theory we give a classification of the singularities of the decrement of general two-parameter families of linear autonomous systems and a classification of the singularities of the neutral surface (stability boundary) of general three-parameter families of linear systems; we also treat the topologically versal deformations of singular points of non-linear systems of ordinary differential equations for all cases of degeneracy of codimension 1 and for some of codimension 2; we indicate applications to the theory of hydrodynamical stability.

Contents

Introduction

The word bifurcation, meaning some sort of branching process, is widely used to describe any situation in which the qualitative, topological picture of the object we are studying alters with a change of the parameters on which the object depends. The objects in question can be extremely diverse: for example, real or complex curves or surfaces, functions or maps, manifolds or fibrations, vector fields, differential or integral equations. If our object depends on parameters, then we say that we have a family. If we are interested in a family locally, as the parameters vary only slightly in a

Originally publ. in: Usp. Mat. Nauk 27:5, 119-184, © Russ. Acad. Sciences 1972
English transl.: Russ. Math. Surv. 27, 54-123, © London Mathematical Society 1972
Reprint by Springer-Verlag Berlin Heidelberg 2014 with kind permission, DOI 10.1007/978-3-642-31031-7_29

neighbourhood of fixed values, then we speak of a deformation of the object corresponding to these values of the parameters.

It turns out in many cases that the study of all possible deformations leads to that of a single one, from which all others can be obtained. Such a deformation, in some sense the largest one, should give all the essentially distinct bifurcations of the given object; it is called a versal deformation. The study of bifurcations and versal families of various objects of analysis is one of the applications of the theory of singularities of smooth maps. In the present lectures we consider mainly the bifurcation of phase portraits[1] of dynamical systems in a neighbourhood of a position of equilibrium.

This article is based on lectures given in the autumn of 1970 to the Leningrad Mathematical Society, in the winter of 1971 to the Voronezh mathematical winter school, and in more detail in June 1971 to the Katsiveli summer school. The author is grateful to V. M. Alekseev, who took notes of the lectures at Katsiveli, and also to R. I. Bogdanov, N. N. Brushlinskaya, A. D. Bryuno, A. N. Varchenko, S. M. Vishik, N. K. Gavrilov, D. M. Galin, V. M. Zakalyukin, Yu. S. Il'yashenko, D. A. Kazhdan, S. G. Krein, A. G. Kushnirenko, V. P. Palamodov, A. S. Pyartly, A. N. Shoshitaishvili and V. I. Yudovich for useful discussions. I take the opportunity to mention that the study of versal families of vector fields first arose under the influence of the late Galina Tyurina's work on versal deformations of the singularities of algebraic varieties.

§ 1. Local analytic problems

In many problems of analysis it becomes necessary to consider various "singular cases". The study of singularities in diverse problems has a number of common features. One of these is that as the theory becomes deeper, it breaks up into separate branches, devoted to a more and more detailed study of more and more particular cases.

The investigation of the particular cases recalls the computation of the logarithms of individual numbers to more and more significant figures: the general, principal features of the object in question can remain hidden in such a study.

The theory of the singularities of smooth maps (see the articles of Whitney [1], Thom and Levine in the collection [2], or the survey article [3]) gives one general method of setting out these general properties. In the present lectures we show to what results this approach can lead in the study of specific analytical problems. The majority of the problems considered here are connected with the theory of ordinary

[1] The term "phase portrait" is used to describe the disposition of the phase curves of a dynamical system or differential equation. (Transl.)

differential equations, but it is easy to find many problems in other branches of analysis that are amenable to the same study. We begin with the simplest examples.

1.1. A function near a critical point. Consider a smooth (infinitely differentiable) function $y = f(x)$, $x \in \mathbf{R}^n$. A point x_0 is said to be *non-critical* for f if the derivative of f at x_0 is non-zero. The behaviour of a function near a non-critical point is determined by the implicit function theorem: it follows from this theorem that we can choose a coordinate system $x' = (x'_1, \ldots, x'_n)$ in a neighbourhood of a non-critical point such that $f = x'_1$.

How does a function behave near a critical point? Suppose that $df|_{x_0} = 0$, that is, that x_0 is a critical point. If the Hessian $\det \left|\frac{\partial^2 f}{\partial x_i \partial x_j}\right|$ of f does not vanish at x_0, then x_0 is said to be a *non-degenerate* critical point (Fig. 1.1). Morse's Lemma [4] asserts that near a non-degenerate critical point there is a coordinate system in which

$$f = C + x_1^2 + \ldots + x_k^2 - x_{k+1}^2 - \ldots - x_n^2.$$

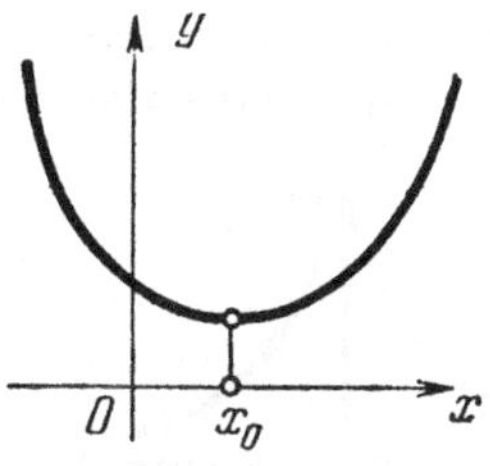

Fig. 1.1.

This normal form permits a detailed analysis of the behaviour of the function in a neighbourhood of a non-degenerate critical point.

If we wish to study more complicated degenerate cases, then we fairly soon get bogged down in a vast classification of particular cases (a fortunate exception is the case of a function of one variable, when the complete classification is easy; see also [90]).

The principal questions arising here are:

1) Is the (differential) topological type of a function in the neighbourhood of a critical point determined by a finite truncation of its Taylor series?

If 1) can be answered positively, then the next question is:

2) Is there a finite algorithm to determine the topological type in terms of a truncation of the Taylor series?

It has been shown that the analytical type of an analytic function in the neighbourhood of a complex-isolated critical point is determined by a finite truncation of its Taylor series; in the smooth (and even in the finitely differentiable) cases, analogous results have been obtained under the assumption that the critical point has finite multiplicity. (A critical point of a function is said to have finite multiplicity if it is isolated in some complex neighbourhood for some finite truncation of the Taylor series.) For critical points of finite multiplicity, the answer to the second question is also positive. For critical points of infinite multiplicity (in the analytic case these are just non-isolated critical points) even the first question has a negative answer. For more details, see the articles of Tougeron [5], Arnol'd [3], Mather [6],

Artin [8], Samoilenko [7], Palomodov [9], Kuiper [10], Kuo [11], [12], Boshnak [13], Takens [14], Varchenko [81], [89].

1.2. Singularities of vector fields. Consider a smooth vector field given by the differential equation

$$\dot{x} = v(x), \quad x \in \mathbf{R}^n,$$

in a neighbourhood of a point $x_0 \in \mathbf{R}^n$.

The point x_0 is said to be *non-singular* if the vector $v(x_0)$ is non-zero. The structure of a vector field near a non-singular point is completely determined by the linearization theorem of the theory of ordinary differential equations [15].

It follows from this theorem that in a neighbourhood of a non-singular point there is a coordinate system $y = (y_1, \ldots, y_n)$ such that the field vectors are all equal and parallel: $\dot{y} = e_1$ (where e_1 is the field with the components $(1, 0, \ldots, 0)$).

What does a vector field look like near a singular point? For simplicity we may translate the singularity to the origin, so that $v(0) = 0$. Let us expand v as a Taylor series in a neighbourhood of the origin: $v = Ax + \ldots$, where $A: \mathbf{R}^n \to \mathbf{R}^n$ is a linear operator. The singular point $x = 0$ is said to be *simple* if the following *non-degeneracy condition* holds: $\operatorname{Re} \lambda_i \neq 0$ $(i = 1, \ldots, n)$ for all eigenvalues λ_i of the operator A. Thus, for a simple singular point a certain number k of eigenvalues of A lie in the right half-plane, and the remaining $n - k$ in the left half-plane of the complex variable λ.

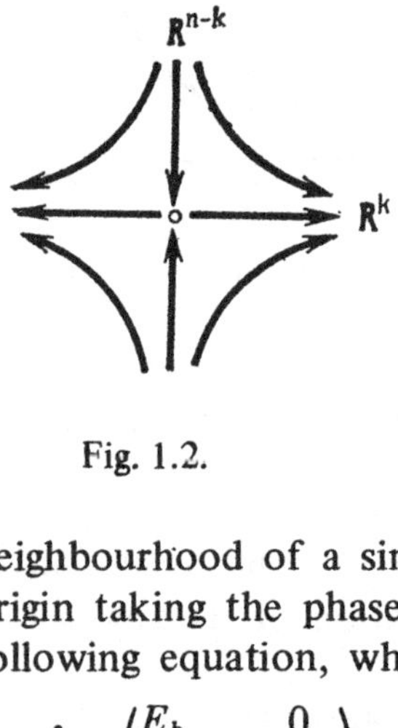

Fig. 1.2.

The Hartman–Grobman theorem ([16]–[19]) asserts that there exists a homeomorphism of a neighbourhood of a simple singular point onto a neighbourhood of the origin taking the phase curves of the given equation into those of the following equation, which is especially easy to study:

$$\dot{y} = \begin{pmatrix} E_k & 0 \\ 0 & -E_{n-k} \end{pmatrix} y \quad (y \in \mathbf{R}^n,\ E_l \text{ is the identity map } \mathbf{R}^l \to \mathbf{R}^l).$$

This normal form (the "multi-dimensional saddle") permits a detailed study of the behaviour of the phase curves in a neighbourhood of a singular point (Fig. 1.2).

If we wish to study more complicated singularities, then we run into a mass of cases which it is difficult to survey. A substantial part of the following lectures is devoted to various approaches to the classification of these cases, to the principal questions arising out of this classification, and to sketches of the few general results that are known in the subject.

1.3. The problem with an oblique derivative. Let D be a domain with a smooth boundary Γ (Fig. 1.3). In the "problem with an oblique derivative" a vector field v is given on the boundary, tangent to D; it is required to determine a function harmonic in D from its directional derivative with respect to v.

In this problem the singularities turn out to be those points of the boundary for which the field is tangent to Γ. In studying this problem one is tempted to look at more and more particular cases. For instance, the set of singular points itself may have complicated singularities, the contact may be of a high order and so on. The principal problem arising in this situation is that of knowing what singularities are to be regarded as being in "general position".

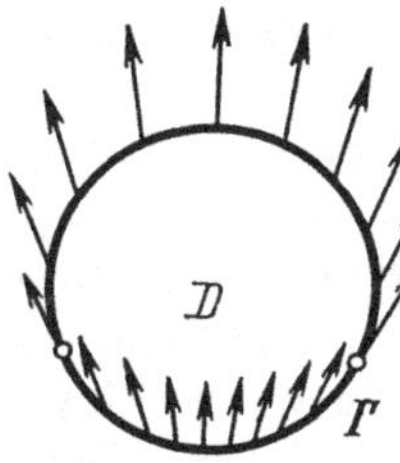

Fig. 1.3.

It can be shown ([20], [21]) that those vector fields which can be written in the form

$$x_2\frac{\partial}{\partial x_1}+x_3\frac{\partial}{\partial x_2}+\ldots+x_k\frac{\partial}{\partial x_{k-1}}+\frac{\partial}{\partial x_k}$$

in a suitable coordinate neighbourhood of any point of the boundary (where Γ has the equation $x_1 = 0$, and k is some integer with $1 \leqslant k \leqslant n = \dim D$) are everywhere dense in the space of all vector fields on Γ tangent to D.

Despite the abundance of investigations of complicated singularities in the problem with an oblique derivative (see the articles of Malyutov [22], Egorov, and Kondratev [23], and Maz'ya [24]), the case of general position has so far only been analyzed in the two-dimensional problem.

1.4. Normal forms of matrices. Consider a linear operator

$$A\colon \mathbf{R}^m \to \mathbf{R}^m.$$

Which operators are "in general position", and which are exceptional?

Here we run into a problem belonging to algebra rather than to analysis: instead of the function spaces of the previous examples we now have the finite-dimensional space of all linear operators from $\mathbf{R}^m$ to $\mathbf{R}^n$, which is, of course, isomorphic to $\mathbf{R}^{mn}$.

The linear operators $\mathbf{R}^m \to \mathbf{R}^n$ of greatest possible rank $r = \min(m, n)$ form an everywhere dense subset of $\mathbf{R}^{mn}$. Such operators will be said to be non-singular; by the theorem on the rank of a matrix, a non-singular matrix is given in a suitable basis by the rxr identity matrix, completed by a rectangle of zeros.

All singular operators form a proper algebraic subvariety of $\mathbf{R}^{mn}$. The subsequent study of special cases includes an investigation of the singularities of this variety.

The group $GL(\mathbf{R}^m) \times GL(\mathbf{R}^n)$ acts on $\mathbf{R}^{mn}$ by interchanging the variable in image and inverse image: $(g, h)A = hAg^{-1}$ (where $g \in GL(\mathbf{R}^m)$ and $h \in GL(\mathbf{R}^n)$). The whole space $\mathbf{R}^{mn}$

is broken up into the orbits of this action: two operators A and B belong to the same orbit if $B = hAg^{-1}$ for some h and g. Thus, each orbit consists of all operators of fixed rank ρ. The operators of rank ρ form a smooth subvariety of $\mathbf{R}^{mn}$, whose codimension is easy to work out: it is just the product of the coranks $(m-\rho)(n-\rho)$.

The decomposition of $\mathbf{R}^{mn}$ into orbits consisting of all operators of fixed rank ρ is a *stratification* – a partition into finitely many subvarieties given by algebraic equations and inequalities. Another example of a stratification is the partition of $\mathbf{R}^3$ into the six pieces shown in Fig. 1.4: the three pieces of the complement of the cone, the two halves of the cone, and the vertex. This is a stratification of the space of quadratic forms on the plane corresponding to the classification of quadratic forms by their inertia indices. Every algebraic variety (and every difference of algebraic varieties) can be stratified – see the articles of Whitney [25], Thom [26], Mather [27] and Lojasievich [28].

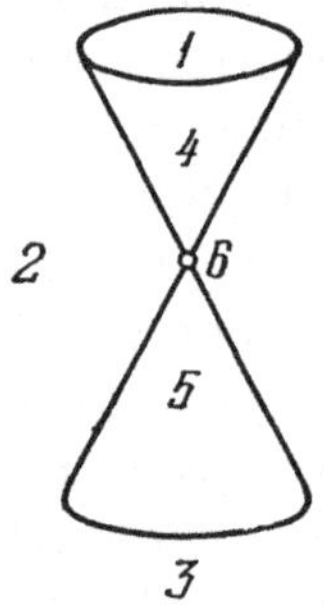

Fig. 1.4.

An analogous situation arises in the study of Jordan forms of matrices of endomorphisms (maps of a linear space into itself). The classification of matrices to within similarity leads to a partition of our space into manifolds; however, even here the number of parts is not finite, but continuous: the eigenvalues are similarity invariants.

Thus, already in this algebraic problem we encounter a difficulty that causes trouble in analytic problems; to deal with it we have to introduce a number of general concepts, to which we now proceed.

§2. Transversality theorems

In formalizing the concept of "general position" it is convenient to use the terminology of spaces of jets; reference to certain standard "transversality theorems" can then often be used to prove the possibility of moving into general position. The present section contains the formulation of these theorems; proofs are in [2], [29].

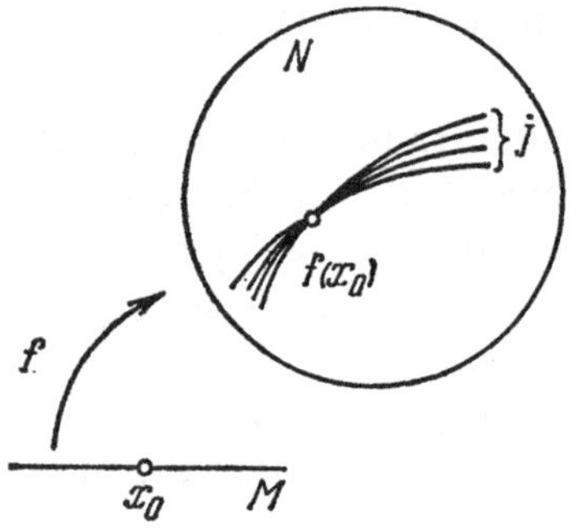

Fig. 2.1.

2.1. Jets and germs. Consider (Fig. 2.1) a smooth map of smooth manifolds

$$f\colon M^m \to N^n$$

(for example, M and N could be domains in Euclidean space). Two such maps are said to be *k-equivalent* at a point $x_0 \in M$ if there is contact of order k:

$$\rho_N(f_1(x), f_2(x)) = o(\rho_M(x, x_0)^k).$$

Here ρ denotes some Riemannian metric (for instance, Euclidean distance); it is easy to see that the k-equivalence of two maps does not depend on the metrics ρ_N and ρ_M.

Two maps are 0-equivalent if and only if they take the same value at x_0. It is easy to check that contact of order k is an equivalence relation (that is, $f_1 \sim f_2 \Rightarrow f_2 \sim f_1$, $f_1 \sim f_2 \sim f_3 \Rightarrow f_1 \sim f_3$ and $f_1 \sim f_1$).

DEFINITION (Ehresman). The *k-jet* of a smooth map f at a point x_0 is the k-equivalence class of maps from M to N to which f belongs. It is denoted by

$$j^k_{x_0}(f) = \{ f_1 : f_1 \text{ is } k\text{-equivalent to } f \text{ at } x_0 \}.$$

Note that if we choose coordinates on M and N in a neighbourhood of x_0 and $f(x_0)$, respectively, then we can identify a k-jet with the set of derivatives of f at x_0 of order up to and including k; under this identification of jets

$$j^0_{x_0}(f) = f(x_0), \quad j^1_{x_0}(f) = \left(f(x_0), \left.\frac{\partial f}{\partial x}\right|_{x_0} \right), \quad j^2_{x_0}(f) = \left(f(x_0), \left.\frac{\partial f}{\partial x}\right|_{x_0}, \left.\frac{\partial^2 f}{\partial x^2}\right|_{x_0} \right), \text{ etc.}$$

Fixing such an identification we can speak of the k-jet of a map at x_0 being "the same thing as" the k-th truncation of the Taylor series of the map.

Apart from k-equivalence of maps there is another important notion of equivalence which leads to the concept of *germs* rather than jets. Two maps f_1 and f_2 given in neighbourhoods U_1 and U_2 of x_0 are said to be equivalent if there is a third neighbourhood U_3 of x_0 on which two maps agree: $f_1|_{U_3} = f_2|_{U_3}$.

The equivalence class of f under this equivalence relation is called the *germ* of f at x_0.

It is clear that we can define the 0-jet, 1-jet, and so on, at x_0 for a germ just as for a map.

2.2. Spaces of jets. The set of all k-jets of germs of smooth maps from M to N at all points of M is called the *space* of *k-jets* and is denoted by

$$J^k(M, N) = \text{the space of } k\text{-jets from } M \text{ to } N.$$

This set has a natural structure of a smooth manifold. For, a point of the set is a k-jet j of a germ of a mapping f at some point $x_0 \in M$. We choose coordinate charts in a neighbourhood of x_0 and $f(x_0)$ in M and N, respectively; then the coefficients of the Taylor polynomial of degree k of f at x_0 form coordinates for j in J^k. Note that, conversely, we can easily construct a germ of a map with any assigned choice of Taylor coefficients: it suffices to take the Taylor polynomial itself.

It is easy to calculate the dimension of $J^k(M, N)$. For example,

$$J^0(M, N) = M \times N, \ \dim J^0(M, N) = \dim M + \dim N,$$

$$\dim J^1(M, N) = \dim M + \dim N + \dim M \dim N.$$

There is a natural map $J^{k+1}(M, N) \to J^k(M, N)$ (a $(k+1)$-jet determines a k-jet, since contact of some order implies contact of a lower order). This smooth map is a fibration, and we obtain a chain of fibrations

$$\ldots \to J^k \to J^{k-1} \to \ldots \to J^1 \to J^0 = M \times N.$$

Each fibre of the fibration $J^k \to J^{k-1}$ is diffeomorphic to a linear space, but does not have a natural linear structure.

2.3. The group of jets of local diffeomorphisms. Consider the space of jets $J^k(M, M)$. Inside this space there lies the submanifold of k-jets of diffeomorphisms. This submanifold is not a group, because two jets can only be composed if the image of the first is the inverse image of the second.

We fix some point $x_0 \in M$ and consider all germs of diffeomorphisms leaving x_0 in place. Their k-jets form a group, which we call the *group of k-jets of local diffeomorphisms* of M at x_0 and denote $J^k_{x_0}(M)$. In particular, the group of 1-jets of local diffeomorphisms is isomorphic to the linear group

$$J^1_0(\mathbf{R}^n) = \mathrm{GL}(\mathbf{R}^n).$$

For $k \geqslant 1$ we obtain a more complicated Lie group. Since a k-jet determines a $(k-1)$-jet, we have a chain of maps

$$\ldots \to J^k_{x_0}(M) \to J^{k-1}_{x_0}(M) \to \ldots \to J^1_{x_0}(M) = \mathrm{GL}(\mathbf{R}^n).$$

It is easy to see that these maps are homomorphisms, and that the kernels of $J^k_{x_0}(M) \to J^{k-1}_{x_0}(M)$ are commutative groups.

For if $f(x) = x + ax^k (\mathrm{mod}\, x^{k+1})$ and $g(x) = x + bx^k (\mathrm{mod}\, x^{k+1})$, then $(f \circ g)(x) = x + ax^k + bx^k\ (\mathrm{mod}\, x^{k+1})$.

2.4. Jets of vector fields. A vector field on a manifold M is a section of the tangent bundle $p\colon TM \to M$, that is, a map $v\colon M \to TM$ such that the diagram

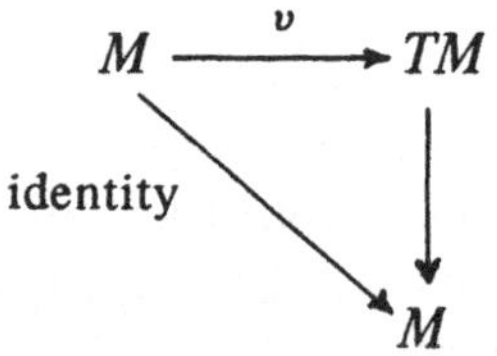

is commutative.

The definitions of germs, jets, and jet spaces of vector fields are just repetitions of the preceding definitions.

The group of diffeomorphisms of M acts on the set of all vector fields on M.

The group of k-jets of local diffeomorphisms of M at x_0 acts on the space of $(k-1)$-jets of vector fields on M at x_0; this action is linear.

For example, let $y = a_1 x + a_2 x^2 + \ldots$ be a 2-jet of a local diffeomorphism. The image of the 1-jet of the vector field $v(x) = v_0 + v_1 + \ldots$ is given by the formula

$$w(x) = w_0 + w_1 x + \ldots, \text{ where } w_0 = a_1 v_0, w_1 = a_1 v_1 a_1^{-1} + 2a_2 a_1^{-1} v_0.$$

This formula is obtained on writing the equation $\dot{x} = v(x)$ in terms of the coordinates y.

2.5. Transversality. DEFINITION. Two linear subspaces L_1 and L_2 of a linear space L are said to be *transversal* if

$$L = L_1 + L_2.$$

For example, two planes of $\mathbf{R}^3$ meeting at a non-zero angle are transversal (Fig. 2.2).

Now suppose that A, B, C are three smooth manifolds, and that two smooth maps are given (Fig. 2.3).

$$B \xrightarrow{f} A \xleftarrow{g} C.$$

The maps f: $B \to A$ and g: $C \to A$ are said to be *transversal at* $(b, c) \in B \times C$ if either $f(b) \neq g(c)$ or if $f(b) = g(c) = a \in A$ and the images of the tangent planes to B and C at b and c under the action of f

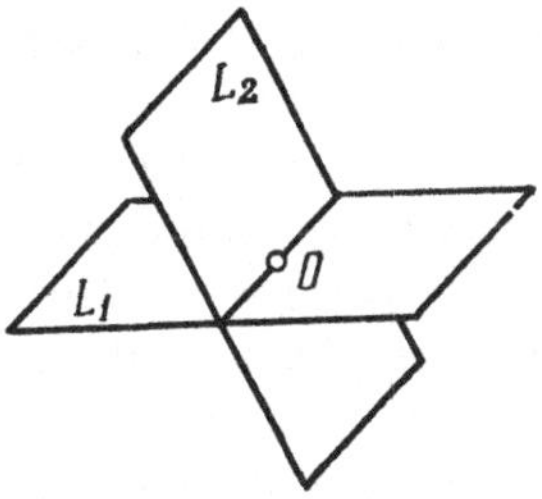

Fig. 2.2.

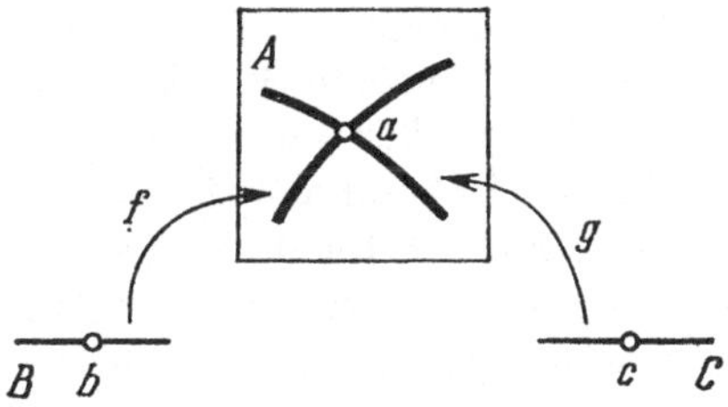

Fig. 2.3.

and g respectively, are transversal in the tangent space to A at a:

$$f_* T_b B + g_* T_c C = T_a A.$$

The two maps f: $B \to A$ and g: $C \to A$ are said to be transversal if they are transversal at any point $(b, c) \in B \times C$.

For example, two lines in $\mathbf{R}^3$ are transversal only if they do not intersect.

Suppose that we are given a smooth map f: $B \to A$ and a smooth submanifold $C \subset A$. Then f is said to be *transversal to the submanifold* C if it is transversal to the embedding i: $C \to A$.

Note that a map of a line into a plane can fail to be transversal to a given line in the plane even when the image is normal to the given line (the image of the tangent space and the tangent space to the image are not one and the same thing).

If $C \subset A$ is a stratified submanifold of A, being given as a finite union of smooth disjoint submanifolds (the *strata*), then a map $f: B \to A$ is said to be transversal to the stratification of C if it is transversal to each stratum.

2.6. The weak transversality theorem. This theorem asserts that a map which is not transversal to a fixed submanifold can be made transversal by an arbitrarily small shift (Fig. 2.4).

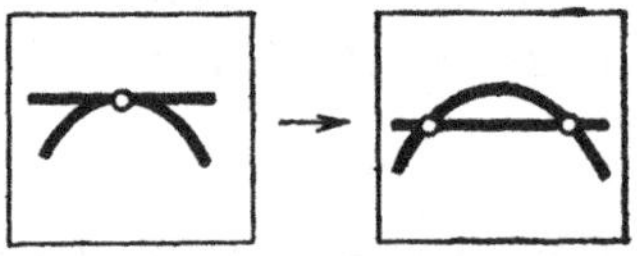

Fig. 2.4.

THEOREM. *Let $f: B \to A$ be a smooth map of a compact manifold B to a manifold A containing a compact submanifold $C \subset A$. Then the maps f that are transversal to C form an open everywhere dense subset of the function space of all maps $f: B \to A$ (with the C^r-topology, $s \leqslant r \leqslant \infty$, where $s >$ max (dim B − dim A + dim C, 0)).*

For a proof see, for example, Thom and Levine [2].

REMARK. An analogous theorem holds when B and C are not compact. If C is non-compact, then the words "open subset" must be replaced by "intersection of a countable family of open sets" (an example: A is a torus and C an everywhere dense helix on the torus). For the maps transversal to C to be everywhere dense it suffices that C is a regular submanifold (that is, each point of C should have a neighbourhood U in A such that the pair $(U, U \cap C)$ is diffeomorphic to $(\mathbf{R}^n, \mathbf{R}^k)$).

If B is non-compact, then it is convenient to give the space of maps the "fine topology". In this topology a neighbourhood of a map $f: B \to A$ is defined as follows: we fix an open set G in the jet space $J^k(B, A)$ for any k. The set of C^∞-maps $f: B \to A$ whose k-jets at any point belong to G is open in the fine topology; these open sets form a basis for the open sets of the fine topology in the space of infinitely differentiable maps $f: B \to A$.

Thus, to say that two maps are close together in the fine topology means that these two maps (and any number of their derivatives) approach each other arbitrarily quickly "at infinity"; in particular, the graph of a map sufficiently close to f lies in a neighbourhood of the graph of f that narrows down arbitrarily quickly "at infinity".

The transversality theorem has an obvious extension to the case of a stratified submanifold C whose strata are all regular submanifolds.

EXAMPLE. Let A be the space of linear operators $M: \mathbf{R}^m \to \mathbf{R}^n$, and let C be the stratified variety of operators of less than maximal rank (see § 1.4). A map $f: B \to A$ is a family of operators depending smoothly on the parameters (that is, on a point of B). The transversality theorem has the following consequence.

COROLLARY. *In the space of smooth families $f: B \to A$ of $m \times n$ matrices, those families that are transversal to the stratified variety C form an everywhere dense set. In particular, the values of the parameters corresponding to matrices of rank r form in general (that is, for families belonging to an everywhere dense intersection of open sets in the space of all families) a smooth submanifold of the parameter space B of codimension $(m - r)(n - r)$.*

For example, in a 5-dimensional family of 2×3 matrices in general position, the rank is 1 on a 3-dimensional smooth submanifold of the parameter space and never vanishes; if for a given family this is not so, then it can be moved to general position by an arbitrarily small deformation.

2.7. Thom's transversality theorem. Thom's transversality theorem is a generalization of the weak transversality theorem in which the part of the submanifold C is played by a submanifold of a jet space.

With every smooth map $f: M \to N$ there is associated its "k-jet extension"

$$\hat{f}: M \to J^k(M, N), \hat{f}(x) = j_x^k(f)$$

(sending a point $x \in M$ to the k-jet of f at x).

THEOREM. *Let C be any regular submanifold of the jet space $J^k(M, N)$. Then the set of maps $f: M \to N$ whose k-jet extensions are transversal to C is an everywhere dense intersection of open sets in the space of smooth maps.*

For a proof see, for example, Thom and Levine [2] or Vishik [29].

We remark that the weak transversality theorem is obtained by putting $k = 0$ in the statement of the theorem. The strong theorem cannot be deduced directly from the weak one for the following reason.

We could apply the weak theorem to the map $\hat{f}: M \to J^k$ and obtain a transversal map near to $\hat{f}$. However, the deformed map need not be the k-jet extension of any map from M to N.

Thom's transversality theorem asserts that one can choose the transversalized deformation of a k-jet extension from a narrower class of deformations: it is sufficient to restrict ourselves to deformations of our map in the space of jets that are induced by genuine deformations of the map $M \to N$. Thus, the integrability conditions (which distinguish deformations of $\hat{f}$ induced by deformations of f) are not an obstruction to obtaining transversality.

The proof of Thom's theorem reduces to a construction of the required deformations by means of explicit polynomial formulae with sufficiently many parameters. The possibility of choosing parameter values so as to achieve transversality is guaranteed by Sard's lemma (the measure of the set of critical values of a smooth map is zero).

Thom's transversality theorem extends in an obvious way to the case of a stratified variety C, whose strata are all regular submanifolds of a jet space. Generalizations to the case of symplectic, or contact, or volume preserving maps can be found in [29].

EXAMPLE. Consider the space of smooth functions $f: M \to \mathbf{R}$ on a smooth compact manifold M.

Let C be the set of 1-jets with vanishing linear term ($df|_{x_0} = 0$).

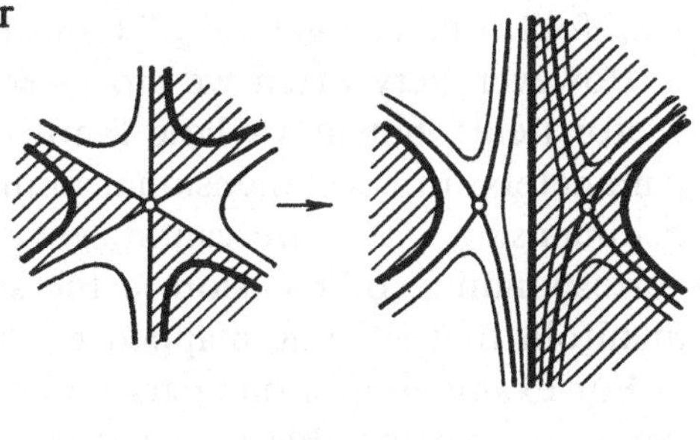

Fig. 2.5.

Then C is a smooth regular submanifold of the jet space $J^1(M, \mathbf{R})$ (whose codimension is equal to the dimension of M).

It is easy to see that transversality of a map $\hat{f}$: $M \to J^1(M, \mathbf{R})$ at a point x_0 of C just means non-degeneracy of the second differential $d^2f|_{x_0}$. Therefore the transversality theorem has the following consequence.

COROLLARY. *In the space of smooth functions* f: $M \to \mathbf{R}$, *those functions whose critical points are all non-degenerate form an everywhere dense intersection of open sets.*

If a function f has a degenerate critical point, then by an arbitrarily small shift of the function it can be ensured that the complicated singularity is dispersed into several non-degenerate ones.

Fig. 2.5. shows how the critical point at the origin of the function $f = x^3 - 3xy^2$ can be broken up into two non-degenerate critical points by the deformation $\tilde{f} = x^3 - 3xy^2 - \varepsilon x$. (The shaded area is the set of points where the function takes negative values.)

§ 3. Versal deformations

From the preceding it is clear that a most important characteristic of "degenerate" or "singular" cases is the codimension of the corresponding submanifold of the space of jets. The transversality theorem allows us to move a given function into general position by an arbitrarily small shift. In particular, if the codimension of the submanifold in question in the space of jets (which for brevity we call the *codimension of the singularity*) is greater than the dimension of the manifold from which we are mapping, then we can arrange, by an arbitrarily small shift, that the singularity does not occur at all: it is split up into a number of simpler ones.

For example, in the study of singularities of a vector field on an n-dimensional manifold the subvariety of the space of jets of vector fields corresponding to degenerate singular points has codimension $n + 1$. This number is greater than the dimension of manifold M that is being mapped. Therefore by a small deformation of any vector field we can achieve that all the singular points become non-degenerate.

From what we have said one might come to the conclusion that one never has to consider singularities whose codimensions exceed the dimension of the manifold: one can get rid of them by an arbitrarily small shift.

However, very often we are interested not in just a single object (say, a vector field), but in a whole family of them, depending on a number of parameters. In a k-dimensional family we can come across singularities of codimension k that we cannot get rid of by a small shift (Fig. 3.1) (the corresponding codimension in the space of jet exceeds the dimension of the manifold that is being mapped by k).

For example, in a one-parameter family of vector fields on M^n we can meet (and not be able to avoid by any small shift of the family) the simplest degenerate singularities (belonging to a subvariety of the space of jet of codimension $n + 1$).

Thus, the discussion of all possible k-parameter families make it necessary for us to consider singularities of all finite codimensions k in the corresponding function space. Clearly, the simpler and more important cases of degeneracy correspond here to small codimensions k.

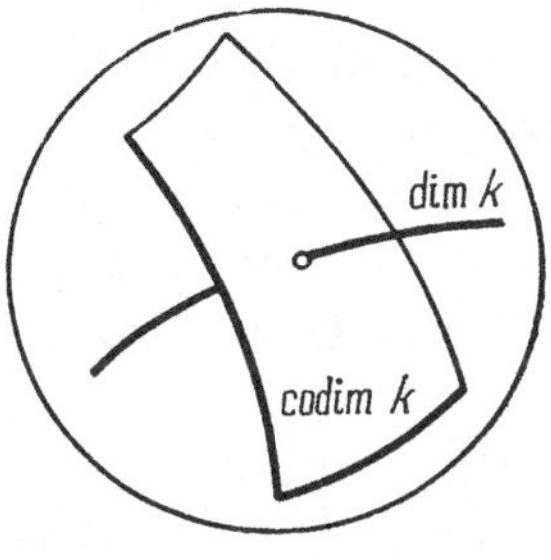

Fig. 3.1.

For example, the sets only of codimension 1 can divide up the function space. Therefore an investigation of the singularities of codimension 1 allows us to go continuously from any "general" point of the function space to any other, and to describe the "bifurcations" that occur as we cross the singularities of codimension 1.

The greater the codimension of a singularity, the harder it is to study it, and (as a rule), the less use there is in such a study.

The study of singularities of large codimension k is reasonable only if we are interested in k-parameter families. But then the natural object of study is not the individual singularity itself (say a vector field with a complicated singular point), but a sufficiently large family so that the singularity of the type in question does not vanish away under a small deformation of the family.

Such a "sufficiently large" family is called a *versal family*. The present section is devoted to the definition of this notion and its application to a problem on the canonical form of matrices.

Sometimes we also have to study singularities of infinite codimension: for instance, Hamiltonian dynamical systems or systems with one particular group of symmetries or another have infinite codimension in the space of all dynamical systems. In such cases one can often reduce the function space such that the case under discussion has only finite codimension.

However, it is not always easy to achieve this. Thus, boundary-value problems for partial differential equations can be regarded from the point of view of transversality of the intersection of the manifold of solutions with the manifold of functions satisfying the boundary conditions. The analysis of this situation requires the ability to distinguish between various infinite codimensions: the condition for a function of one variable constructed from a given object to vanish identically defines in the function space a submanifold of smaller (infinite) codimension than the vanishing of a function of two variables.

The problem with a directional derivative (see § 1.3) seems to be one of the simplest models where such a computation of infinite codimensions is needed, corresponding to kernels and cokernels consisting of functions on manifolds of different dimensions.

The local structure of these manifolds (on which boundary conditions have to be given or, conversely, one kind of discontinuity or another in the solutions is to be allowed) can be understood, for fields in general position, from the local models indicated in § 1.3.

3.1. Versal families. The reader familiar with final objects in categories will recognize in the definition of (uni)versal deformations the final object of the category of deformations. Unfortunately the formal definition is a long one. We begin with an informal description (according to taste, the reader can skip this and proceed to § 3.2, where the formal definition is given; this may also be omitted if the reader wishes to reach the final result of this section immediately: the normal form theorem of § 3.3).

Let $A(\lambda)$ be a family of any kind of objects, and λ the parameter of the family. Let $B(\mu)$ be another such family. We assume that some class of such families has been fixed (in what follows A and B will be matrices, vector fields, or smooth maps).

We can often speak of maps of one family into another, $B(\mu) \to A(\lambda)$. By this we mean that to each value of the parameter μ of the family $B(\mu)$ there corresponds a definite value $\lambda = \varphi(\mu)$ of the parameter of the family $A(\lambda)$ and that some class of such maps has been fixed.

A *versal family* is one into which one can map every other family of the given class (by means of a map of the given class).

A *universal family* is a versal family $A(\lambda)$ such that in mapping a family $B(\mu)$ into it, the "change of parameters" $\lambda = \varphi(\mu)$ is uniquely determined by the family $B(\mu)$.

The term "versal" is formed from the word "universal" by dropping the prefix "uni", which denotes the uniqueness of the map φ. It is also useful to remark that "versal" is the intersection of the two words "universal" and "transversal". For more on the topic of versal families of algebraic and analytic varieties with singularities see the articles of Douadi [30], Grauert and Kerner [31], Tyurina [32], Grauert [33], and Pham [34].

The formal definition of a versal family is given below in a simple special case: the discussion will be on families of matrices of maps of a linear space into itself.

3.2. Example: deformations of matrices of endomorphisms. We denote by

$$\mathrm{End}(\mathbf{C}^n) = \mathbf{C}^{n^2}$$

the set of $n \times n$ complex matrices.

The relation of similarity partitions $\mathbf{C}^{n^2}$ into semi-algebraic varieties: two matrices lie in one subvariety if they have the same eigenvalues and the same size Jordan blocks. On account of the eigenvalues this is a partition into a continuum of parts. As a rough model one can imagine the partition of 3-space into the strata of the varieties $x^2 = y^2 - z^2 + c$ (Fig. 3.2).

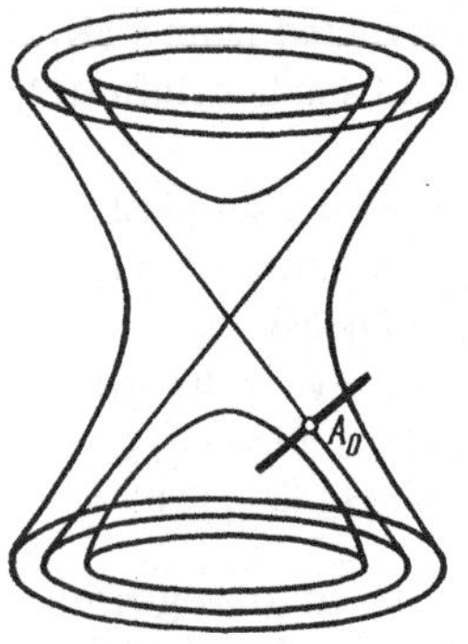

Fig. 3.2.

Every individual matrix $A_0 \in \mathbf{C}^{n^2}$ can be put into the Jordan normal form (by a similarity transformation). However, this process is unstable: under a small alteration of the elements of the matrix both the Jordan form and the similarity transformation leading to it change discontinuously, in general. The versal deformation of a matrix A_0 is a "normal form" into which not only A_0, but also any family of nearby matrices can be put.

Of course, this new normal form must itself contain parameters; its advantage is that both the elements of the new normal form and the similarity transformation leading to it can be chosen depending smoothly on the elements of the original matrix (varying in a neighbourhood of A_0). In Fig. 3.2 a versal deformation of the matrix A_0 is illustrated in the form of a line segment transversal to the cone at A_0.

We define a *family* of matrices to be a holomorphic map $A: \Lambda \to \mathbf{C}^{n^2}$ where Λ is a neighbourhood of 0, the origin of coordinates in some parameter space $\mathbf{C}^l$. The germ of the family A at 0 is called a *deformation* of the matrix $A(0) = A_0$.

A deformation A' of A_0 is said to be *equivalent* to A if there is a deformation C of the identity such that

$$A'(\lambda) = C(\lambda)A(\lambda)(C(\lambda))^{-1}.$$

Let $\varphi: M \to \Lambda$ be a holomorphic map ($M \subset \mathbf{C}^m$, $\Lambda \subset \mathbf{C}^l$). The *family induced from A by φ* is

$$(\varphi^*A)(\mu) = A(\varphi(\mu)),\ \mu \in M.$$

The *induced deformation* φ^*A of the matrix A_0 is defined in the same way with reference to a deformation A of A_0 and a germ of a map $(\mathbf{C}^m, 0) \to (\mathbf{C}^l, 0)$.

A deformation A of a matrix A_0 is said to be *versal* if any deformation A' of A_0 is equivalent to a deformation induced from A; a deformation A is said to be *universal* if the map φ inducing A' is uniquely determined by A'.

A trivial example of a versal deformation is the family $\mathbf{C}^{n^2}$ of all matrices, which defines a n^2-parameter versal deformation of any of its matrices. We would like to have a versal deformation with fewer parameters. For example, if all the eigenvalues $\alpha_1, \ldots, \alpha_n$ of A_0 are distinct, then there is an n-parameter versal deformation of A_0. This is defined by the family of operators having diagonal form with elements $\alpha_i + \lambda_i$ in the same basis as for A_0; it is also universal.

3.3. Explicit form of versal deformations of matrices. Denote by α_i the eigenvalues of A_0 and let $n_1(\alpha_i) \geqslant n_2(\alpha_i) \geqslant \ldots$ be the size of the Jordan blocks belonging to the α_i, in decreasing order.

THEOREM. *The smallest number of parameters of a versal deformation of a matrix A_0 is* $\sum_i [n_1(\alpha_i) + 3n_2(\alpha_i) + 5n_3(\alpha_i) + \ldots]$.

For a proof see [35]. There still remains some element of arbitrariness in the choice of one versal family or another.

In particular, the three normal forms described in the following theorem are versal deformations of a matrix reduced to upper triangular form.

THEOREM. *Let A be a family of linear operators from $\mathbf{C}^n$ into itself depending holomorphically on parameters $\lambda \in \mathbf{C}^l$, and suppose that for some value λ_0 of the parameters the operator $A(\lambda_0)$ has eigenvalues α_i and*

Jordan blocks of order $n_1(\alpha_i) \geqslant n_2(\alpha_i) \geqslant \dots$

Then there exists a basis of $\mathbf{C}^n$ *depending holomorphically on the parameters* λ *as they vary in some neighbourhood of* λ_0, *and such that in this basis the matrix of the operator* $A(\lambda)$ *has the diagonal block form*

$$A_0 + B(\lambda),$$

where A_0 *is the Jordan upper triangular matrix of* $A(\lambda_0)$ *and* $B(\lambda)$ *is a diagonal block matrix whose blocks correspond to the eigenvalues of* A_0. *The block*

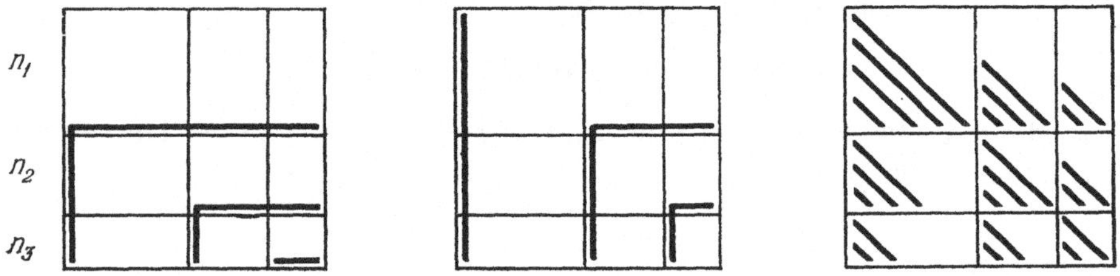

Fig. 3.3.

B_i *corresponding to the eigenvalue* α_i *has only zero entries apart from the places indicated in Fig.* 3.3, *where the entries are holomorphic functions of the parameters* λ.

Three normal forms are illustrated in Fig. 3.3. In the first two the number of non-zero entries in B_i is $n_1(\alpha_i) + 3n_1(\alpha_i) + \dots$; in the third all the elements on any one of the indicated diagonals are equal. A versal deformation of A_0 is obtained if the indicated elements of the B_i are regarded as independent parameters; in any of the three cases there are $\sum_i [n_1(\alpha_i) + 3n_2(\alpha_i) + \dots]$ of them. The first two normal forms have the advantage that the matrix has the fewest possible non-zero entries. The advantage of the third form is that the versal deformation is orthogonal to the corresponding orbit (in the sense of the element-wise scalar product of matrices). Thus, the "regularization" of the unstable problem of reducing matrices to the Jordan normal form leads to normal forms of the third type. A proof of this theorem is in [35].

3.4. Examples. We denote an upper triangular Jordan matrix by the product of the determinants of its blocks. For instance, α^2 denotes a 2×2 Jordan block, and $\alpha\alpha$ a 2×2 block that is a scalar multiple of the unit matrix.

The first normal form of the theorem of § 3.3 leads to the following versal deformations with the minimal number of parameters.

A) A versal (and universal) 2-parameter deformation of the 2×2 Jordan block α^2:

(1) $$\begin{pmatrix} \alpha & 1 \\ 0 & \alpha \end{pmatrix} + \begin{pmatrix} 0 & 0 \\ \lambda_1 & \lambda_2 \end{pmatrix}.$$

B) A versal (but not universal) 4-parameter of the scalar 2×2 matrix $\alpha\alpha$:

$$\begin{pmatrix} \alpha & 0 \\ 0 & \alpha \end{pmatrix} + \begin{pmatrix} \lambda_1 & \lambda_2 \\ \lambda_3 & \lambda_4 \end{pmatrix}.$$

C) A versal and universal 3-parameter deformation of the 3×3 Jordan block α^3:

$$\begin{pmatrix} \alpha & 1 & 0 \\ 0 & \alpha & 1 \\ 0 & 0 & \alpha \end{pmatrix} + \begin{pmatrix} 0 & 0 & 0 \\ 0 & 0 & 0 \\ \lambda_1 & \lambda_2 & \lambda_3 \end{pmatrix}.$$

D) A versal 5-parameter deformation of the matrix $\alpha^2\alpha$:

$$\begin{pmatrix} \alpha & 1 & 0 \\ 0 & \alpha & 0 \\ 0 & 0 & \alpha \end{pmatrix} + \begin{pmatrix} 0 & 0 & 0 \\ \lambda_1 & \lambda_2 & \lambda_3 \\ \lambda_4 & 0 & \lambda_5 \end{pmatrix}.$$

For example, every holomorphic family of matrices having α^2 as a Jordan block when the parameters are zero can be put into the normal form (1) for small values of the parameters, where λ_1 and λ_2 are holomorphic in the parameters.

In the investigation of many problems relating to the behaviour of operators depending on parameters, our construction of normal forms allows us to restrict our attention to special families: versal deformations with the minimal number of parameters. One such problem is the question of the structure of bifurcation diagrams.

3.5. Bifurcation diagrams. A *bifurcation diagram* of a family of matrices is a partition of the parameter space Λ according to the Jordan types of the matrices. The family is a map $A: \Lambda \to \mathbf{C}^{n^2}$ of the parameter space into the space of matrices, so that in studying the bifurcation diagram we have to investigate the partition of the space of all matrices into matrices with Jordan forms of various types. In this partition we will put together all the matrices with the same size Jordan blocks that differ only in their eigenvalues. The partition so obtained is a finite stratification of the space $\mathbf{C}^{n^2}$ of all matrices.

Each stratum of this stratification is determined by a list of all the sizes $n_1(\alpha_i) \geqslant n_2(\alpha_i) \geqslant \ldots$ of the Jordan blocks corresponding to the ν distinct eigenvalues ($1 \leqslant i \leqslant \nu$). The codimension c of such a stratum in $\mathbf{C}^{n^2}$ is less than the codimension d of the corresponding orbit by the number of distinct eigenvalues, that is, by ν:

$$c = d - \nu = \sum_{i=i}^{\nu} [n_1(i) + 3n_2(i) + \ldots - 1].$$

We remark that simple eigenvalues contribute nothing to this sum. Applying the weak transversality theorem we arrive at the following result:

THEOREM. *In the space of all families of $n \times n$ matrices the families transversal to the stratification according to the Jordan type form an everywhere dense set.*

This theorem together with the formulae for versal families of § 3.3 allows us to describe the bifurcation diagrams of families in general position. In particular, for families with only a few parameters we arrive at the following results.

A. ONE-PARAMETER FAMILIES. From $c = 1$ we deduce that the matrix has only one two-fold eigenvalue corresponding to a 2×2 Jordan block. This stratum will be denoted by α^2.

COROLLARY. *In a one-parameter family of matrices of general form the only matrices that can turn up are those with only simple eigenvalues, and for particular isolated values of the parameters, matrices of type* α^2 *(with one* 2×2 *Jordan block). If a family has matrices of a more complicated Jordan structure, then this can be avoided by an arbitrarily small shift of the family.*

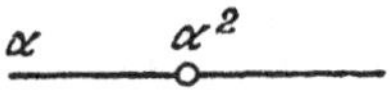

Fig. 3.4.

Thus, the bifurcation diagram of a general one-parameter family has the form indicated in Fig. 3.4.

B. TWO-PARAMETER FAMILIES. There are precisely two different Jordan types with $c = 2$: α^3 (a single 3×3 Jordan block) and $\alpha^2\beta^2$ (two 2×2 blocks with distinct eigenvalues).

COROLLARY. *The bifurcation diagram of a general two-parameter family of matrices has the form of a plane curve whose only singularities are cusps and nodes* (Fig. 3.5). *The cusps correspond to matrices of type* α^3 *with a* 3×3 *Jordan block, and the nodes to matrices of type* $\alpha^2\beta^2$ *with two* 2×2 *Jordan blocks with distinct eigenvalues. Other points on the curve correspond to matrices having a single* 2×2 *Jordan blocks, and points off the curve to matrices with distinct eigenvalues.*

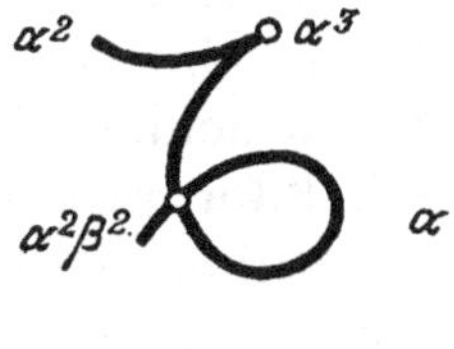

Fig. 3.5.

If the family has matrices of a more complicated type, or if the bifurcation diagram has worse singularities, then they can be avoided by an arbitrarily small shift in the family.

C. THREE-PARAMETER FAMILIES. There are four strata with $c = 3$: $\alpha^2\beta^2\gamma^2$ (three 2×2 blocks), $\alpha\alpha$ (two 1×1 blocks with the same eigenvalue), $\alpha^2\beta^3$ (a 2×2 and a 3×3 block), and α^4 (a 4×4 block).

Consequently, the bifurcation diagrams of general three-parameter families have at worst singularities of the forms shown in Fig. 3.6.

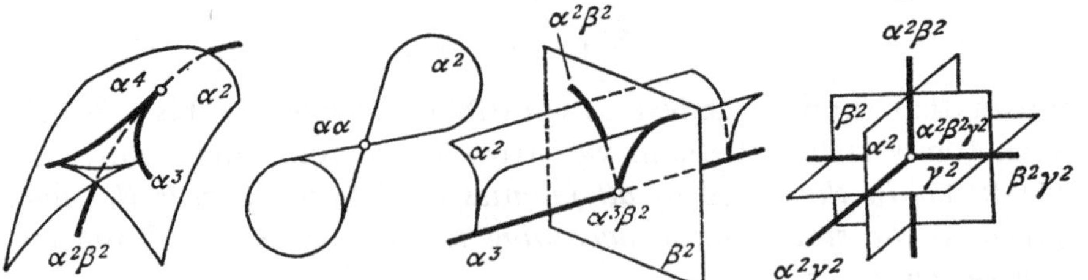

Fig. 3.6.

The singularity corresponding to α^4 is called the swallow's tail: this surface is given by the equation $\Delta(a, b, c) = 0$, where Δ is the discriminant of the polynomial $z^4 + az^2 + bz + c$. Strictly speaking, all that has been said above refers to the complex case, so that the surfaces illustrated in Fig. 3.6 should be regarded as complex surfaces.

Versal deformations of real matrices are constructed by Galin in [36]. The construction goes as follows. Suppose first that the real operator on $\mathbf{R}^{2n}$ whose versal deformation we look for has just one pair of complex conjugate eigenvalues $x \pm iy$ ($y \neq 0$), with Jordan blocks of order $n_1 \geqslant n_2 \geqslant \ldots$, so that $n_1 + n_2 + \ldots = n$. Then for some real basis of $\mathbf{R}^{2n}$ the matrix of the operator has the same form as the matrix of the realification of the complex Jordan operator $\hat{A}_0: \mathbf{C}^n \to \mathbf{C}^n$ with a unique eigenvalue $x + iy$ and Jordan blocks of order $n_1 \geqslant n_2 \geqslant \ldots$, that is,

$$A_0 = \begin{pmatrix} X & -yE \\ yE & X \end{pmatrix}, \qquad (2)$$

where X is an upper triangular real Jordan matrix with the eigenvalue x and blocks of order $n_1 \geqslant n_2 \geqslant \ldots$, and E is the $n \times n$ unit matrix.

It turns out that *one can take the realification of a minimal complex versal deformation of the complex matrix $\hat{A}_0$ as a minimal versal deformation of a real matrix A_0*.

For example, a real 4×4 matrix with two 2×2 Jordan blocks corresponding to eigenvalues $x \pm iy$ has a 4-parameter versal deformation obtained by realifying the complex versal deformation

$$\begin{pmatrix} z & 1 \\ 0 & z \end{pmatrix} + \begin{pmatrix} 0 & 0 \\ \lambda_1 & \lambda_2 \end{pmatrix},$$

that is, the following deformation with parameters $\rho_1, \rho_2, \tau_1, \tau_2$:

$$\begin{pmatrix} x & 1 & -y & 0 \\ 0 & x & 0 & -y \\ y & 0 & x & 1 \\ 0 & y & 0 & x \end{pmatrix} + \begin{pmatrix} 0 & 0 & 0 & 0 \\ \rho_1 & \rho_2 & -\tau_1 & -\tau_2 \\ 0 & 0 & 0 & 0 \\ \tau_1 & \tau_2 & \rho_1 & \rho_2 \end{pmatrix}, \qquad \begin{aligned} z &= x + iy, \\ \lambda_k &= \rho_k + i\tau_k. \end{aligned}$$

Every real matrix is similar over the real field to a matrix of block diagonal form in which every real eigenvalue has a corresponding real Jordan matrix, and every complex conjugate pair has a corresponding block of the form (2).

A versal real deformation (with the minimal number of parameters) of such a matrix is obtained on replacing each block by its minimal versal deformation. The minimal number of parameters of a real versal deformation is thus given by the formula

$$d = \sum_{\lambda} [n_1(\lambda) + 3n_2(\lambda) + 5n_3(\lambda) + \ldots],$$

where the summation extends over all ν eigenvalues, real and complex.

Explicit formulae for versal deformations and tables of bifurcation diagrams are given in [36] for $d - \nu \leqslant 3$.

For applications to mechanics one should also produce tables of versal deformations of symplectic and Hamiltonian (infinitesimally symplectic) matrices (deformations are to preserve the symplecticity).

One use of the bifurcation diagrams we have obtained is the following. Suppose that in the study of some phenomenon or other a bifurcation diagram of a different type from those listed here turns up. Then it is likely that one of the following two is happening: either in our idealization of the phenomenon we have omitted something essential, which can qualitatively alter the structure of the diagram, or else there is some special reason for additional multiplicity of the spectrum or for non-transversality to the Jordan stratification (for example, symmetry, or Hamiltonian character of the problem).

§ 4. Application: the classification of singularities of decrement diagrams

As one of the applications of versal deformations of matrices we consider here the solution of the following problem. Suppose that we are given a family of linear homogeneous autonomous ordinary differential equations. As is well known, the asymptotic behaviour of solutions is determined by the eigenvalue with the greatest real part of the operator specifying the equation. How does this real part depend on the parameters?

In engineering the real part (with a minus sign) is known as the decrement. Thus, our problem consists in the investigation of the behaviour of the decrement as the parameters vary.

A convenient way of describing the behaviour of the decrement as the parameters vary is to indicate in the parameter space (plane, 3-space, . . .) the level curves (surfaces, . . .) of the decrement. The family of curves of level decrement, drawn in the parameter plane, is called the *decrement diagram*.

The appearance of the decrement diagram varies a lot from family to family; in some cases it can have very complicated singularities. It turns out, however, that for families in general position the decrement diagram can only have certain very simple singularities: all the more complicated singularities break up under a small shift of the family.

In the present section we describe all the singularities of the decrement diagrams of 2-parameter families in general position.

The classification of singularities of decrement diagrams in general position serves much the same purpose in the study of the dependence of the system on parameters as the classification of general position singularities in the study of phase portraits.[1]

The appearance on a decrement diagram of singularities not in general position should cause concern: it can be explained by special symmetry properties of the family, or they can show up the inadequacy of the idealization ("uncorrectness"), in which the neglect of small effects in the equations (for instance "parasite connections" in electronics) can change the picture qualitatively.

The classification of the singularities of 2-parameter decrement diagrams in general position includes, in particular, a study of the singularities of the stability boundary of 3-parameter families of linear equations in general position (null decrement surfaces).

The results we obtain can also be applied to non-linear systems having stationary points that depend smoothly on the parameters: the decrement of the linearization of the non-linear system at such a point has, as a function of the parameters, only the simplest singularities (in the case of a family in general position).

---

[1] See transl.'s footnote on p. 55.

However, in applying the results to non-linear systems we have to exclude the part of the stability boundary corresponding to the eigenvalue zero, since the smooth dependence of the stationary point on the parameters is lost here. Thus, a description of the singularities of the stability boundary for non-linear systems in general position (and of the decrement diagram in the neighbourhood of points of this boundary) demands a further study. We return to this problem in §§ 5 and 6.

In the study of iterations of maps, and also of equations with periodic coefficients or motions in a neighbourhood of a periodic trajectory, the greatest modulus of the eigenvalues plays the role of the decrement. If this modulus is non-zero, then its singularities (as function of the parameters in a family in general position) are just the same as those of the decrement in a family in general position. For this reason henceforth we only consider the decrement.

In the study of the moduli of eigenvalues in non-linear problems of the types just referred to, the results of the present section are applicable outside the stability boundary and at those points of the stability boundary for which 1 is not an eigenvalue.

4.1. Decrement diagrams. Consider a family A of linear differential operators in a Euclidean space $\mathbf{R}^n$, depending smoothly on a point λ of the parameter space Λ,

$$A(\lambda)\colon \mathbf{R}^n \to \mathbf{R}^n.$$

DEFINITION. The *increment*[1] of the family A is defined as being that function f of the parameter whose value at λ is the greatest real part of eigenvalues of the operator $A(\lambda)$:

$$f(\lambda) = \lim_{t\to+\infty} \frac{1}{t} \ln \| e^{A(\lambda)t} \|.$$

The function f is continuous, but not necessarily differentiable. Our task is to study the singularities of f for two-parameter families in general position. Thus, we may assume that the parameter space Λ is either $\mathbf{R}^2$ or some plane domain.

The family of level curves of f on the plane Λ is called the *decrement diagram.* A bar perpendicular to the level lines ("Bergstrich" in topography) indicates the direction of the slope, that is, the direction in which f decreases. In other words, the direction bar is on the side of increasing stability.

EXAMPLE. Consider the differential equation

$$\ddot{z} = xz + y\dot{z},$$

depending on the parameters x and y. The matrix of the corresponding system has the form

$$A(x,y) = \begin{pmatrix} 0 & 1 \\ x & y \end{pmatrix}.$$

---

[1] Engineers refer to the absolute value $|f|$ as the decrement if $f < 0$, and as the increment if $f > 0$.

The decrement diagram is sketched in Fig. 4.1. The parabola $4x + y^2 = 0$ divides the (x, y)-plane into two parts in each of which the increment is a continuous function. To the left of the parabola the eigenvalues are complex and $f = y/2$. To the right the eigenvalues are real, and

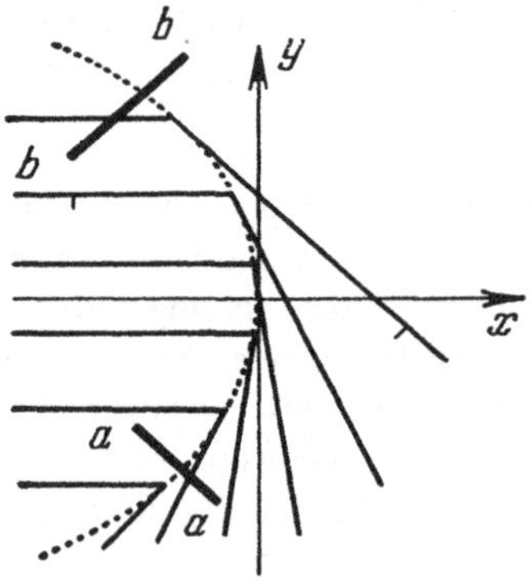

Fig. 4.1.

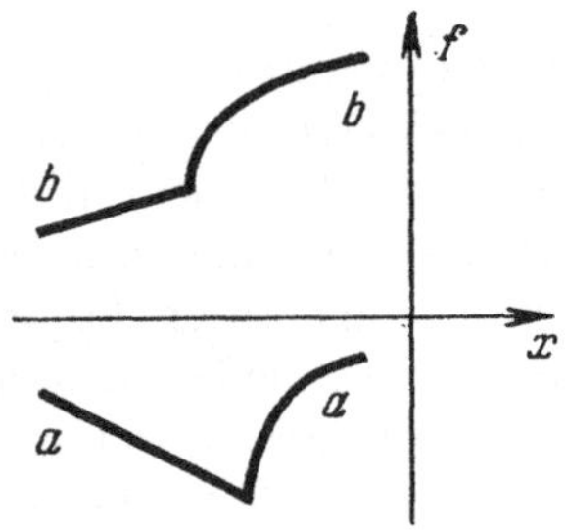

Fig. 4.2.

$f = [y + \sqrt{(4x + y^2)}]/2$. The level lines of the increment are rays tangent to the parabola.

All the points on the parabola are singularities of the decrement diagram. They correspond to matrices A having a 2×2 Jordan block. As we cross the parabola from left to right, the increment changes from linear to square root behaviour (Fig. 4.2).

It is clear that the singularity indicated here cannot be avoided by a small shift of the family. There are other unavoidable singularities, and our aim is to list them completely.

4.2. Strata of codimension one and two in the space of matrices. If the maximal real part of an eigenvalue of $A(\lambda_0)$ belongs to a single real eigenvalue, or to a single complex conjugate pair,[1] then the increment is a smooth function in a neighbourhood of the parameter value λ_0 in question.

Smoothness can only be lost when an eigenvalue with the maximal real part is not unique. Matrices for which the maximal real part belongs to several eigenvalues at once form a closed semi-algebraic subvariety F in the space $\mathbf{R}^{n^2}$ of all $n \times n$ matrices. F has codimension 1, and its complement consists of two open components.

***D_1*. The stratum (α).** The maximal real part belongs to precisely one real eigenvalue.

***D_2*. The stratum ($\alpha \pm i\omega$).** The maximal real part belongs to precisely one complex conjugate pair.

It is easy to stratify F. The strata of maximal dimension (codimension 1) can be listed as follows:

***F_1*. The stratum (α^2).** The maximal real part belongs to precisely two coincident eigenvalues; these are real, and correspond to a 2×2 Jordan block.

***F_2*. The stratum (α, $\alpha \pm i\omega$).** The maximal real part belongs to precisely three eigenvalues, one real, and one complex conjugate pair.

[1] Here and in what follows the numbers of a complex conjugate pair are assumed to be non-real.

F_3. **The stratum** $(\alpha \pm i\omega_2)$. The maximal real part belongs to two distinct complex conjugate pairs.

It is clear that the strata F_1, F_2 and F_3 are smooth regular non-closed non-intersecting subvarieties of codimension 1 in the space $\mathbf{R}^{n^2}$ of all matrices. The remainder $F \setminus (F_1 \cup F_2 \cup F_3)$ of F (the variety of matrices with non-unique eigenvalues of maximal real part) is a closed semi-algebraic sub-variety of codimension 2 in the space $\mathbf{R}^{n^2}$ of all matrices. The strata of maximal dimension in $F \setminus (F_1 \cup F_2 \cup F_3)$ have codimension 2 in $\mathbf{R}^{n^2}$, and it is easy to list them:

G_1. **The stratum** (α^3). The maximal real part belongs to precisely three coincident real eigenvalues corresponding to a 3×3 Jordan block.

G_2. **The stratum** $((\alpha \pm i\omega)^2)$. The maximal real part belongs to precisely two coincident complex conjugate pairs of eigenvalues corresponding to a 2×2 Jordan block.

G_3. **The stratum** $(\alpha^2, \alpha \pm i\omega)$. The maximal real part belongs to precisely four eigenvalues, two of which are real and coincident and correspond to a 2×2 Jordan block; the other two are complex conjugate.

G_4. **The stratum** $(\alpha, \alpha \pm i\omega_1, \alpha \pm i\omega_2)$. The maximal real part belongs to precisely five eigenvalues: one real, and two distinct complex conjugate pairs.

G_5. **The stratum** $(\alpha \pm i\omega_1, \alpha \pm i\omega_2, \alpha \pm i\omega_3)$. The maximal real part belongs to precisely three distinct complex conjugate pairs.

The strata $G_1 - G_5$ are regular non-closed, non-intersecting submanifolds[1] of codimension 2 in the space $\mathbf{R}^{n^2}$ of all matrices. The remainder $F \setminus \cup F_i \setminus \cup G_i$ is a closed semi-algebraic subvariety of $\mathbf{R}^{n^2}$ of codimension 3.

The following is a corollary of the weak transversality theorem of § 2.6:

COROLLARY. *In 2-parameter families of matrices in general position there are no matrices having sets of eigenvalues with maximal real part other than the types* (D_i, F_i, G_i) *listed above; these collections occur only transversally.*

Thus, in a family in general position collections of eigenvalues of codimension 1 (F_i) occur in the parameter plane along smooth curves having singularities only at those points of the parameter plane where collections of codimension 2 (G_i) occur. The latter can only happen at isolated points of the parameter plane.

The segments F_1 and F_2, together with the singular points corresponding to the G_i, form curves that divide the parameter plane into parts of the two types D_1 and D_2. It is not hard to see that the segments F_3 all lie inside the part D_2.

---

[1] All the manifolds D_i, F_i, G_i are connected for sufficiently large n. Exceptions are: D_2 and F_1 for $n = 2$, F_3, G_2 and G_3 for $n = 4$, and G_5 for $n = 6$ all have two components.

Furthermore,

the points $G_1(\alpha^3)$ lie at the join of $F_1(\alpha^2)$ and $F_2(\alpha, \alpha \pm i\omega)$,
$G_2((\alpha \pm i\omega)^2)$ adjoin $F_3(\alpha \pm i\omega_1, \alpha \pm i\omega_2)$,
$G_3(\alpha^2, \alpha \pm i\omega)$ lie at the join of $F_1(\alpha^2)$, $F_2(\alpha, \alpha \pm i\omega)$ and $F_3(\alpha \pm i\omega_{1,2})$,
$G_4(\alpha, \alpha \pm i\omega_{1,2})$ at the join of $F_2(\alpha, \alpha \pm i\omega)$ and $F_3(\alpha \pm i\omega_{1,2})$,
$G_5(\alpha \pm i\omega_{1,2,3})$ adjoin $F_3(\alpha \pm i\omega_{1,2})$

Fig. 4.3 shows a (hypothetical) example of a configuration which these curves could form on the parameter plane of a family in general position.

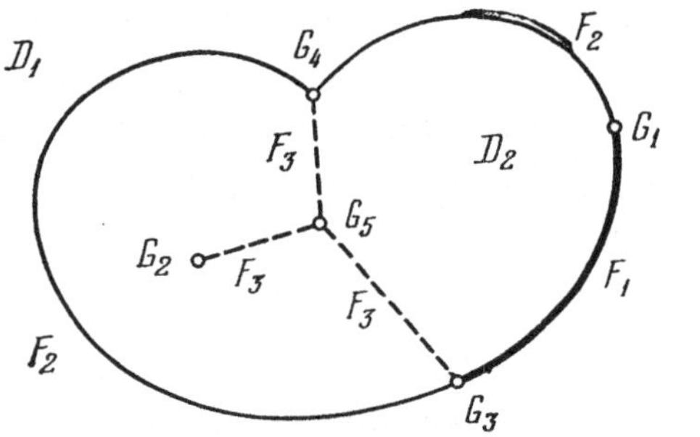

Fig. 4.3.

4.3. The structure of the decrement diagrams near points of a strata of codimension 0 or 1. On the complement to the set F of singularities f is a smooth function of the parameters. However, the decrement diagram can have singularities at some points of this complement: these are the critical points of f.

From Thom's transversality theorem and Morse's lemma we deduce that outside F the increment of a family in general position has only simple critical points, that is, the three types of points shown in Fig. 4.4 (six types if we distinguish between the case of

Fig. 4.4.

real eigenvalues (D_1) and complex eigenvalues (D_2)):

D_i^0. **Minimum.** Near the point under discussion we can choose smooth coordinates x and y in the parameter plane so that the increment takes the form

$$f = \text{const} + x^2 + y^2.$$

D_i^1. **Saddle point.** In suitable coordinates, $f = \text{const} + x^2 - y^2$.

D_i^2. **Maximum.** $f = \text{const} - x^2 - y^2$.

Now let us investigate the behaviour of f near non-singular points of F, that is, near the interior points of the curves F_i in the parameter plane. Here we must distinguish two cases: the point of F_i can be non-critical for the increment, regarded as a smooth function along this curve, but it can also be critical.

From Thom's transversality theorem it follows that in families in general position critical points of the restriction of the increment to the curves F_i can only be non-degenerate maxima or minima.

Combining this information with the explicit formulae of § 3 for versal families of matrices, we easily arrive at the following normal form of the increment near points of the strata of codimension 1.

THEOREM *In a neighbourhood of a non-critical point of the restriction of the increment f of a family in general position to a curve F_i we can choose smooth coordinates x and y in the parameter plane such that f takes one of the following three forms:*

The case F_1^0 (*a Jordan block; Fig.* 4.5):

$$f = \text{const} + y + \begin{cases} \sqrt{x} \ \textit{if} \ x \geqslant 0, \\ 0 \ \textit{if} \ x \leqslant 0. \end{cases}$$

The cases F_2^0 and F_3^0 (*simple fishbone; Fig.* 4.6):

$$f = \text{const} + x + |y|.$$

The curves F_1^0 and F_2^0 divide up the domains of real and of complex eigenvalues D_1 and D_2. The level increment lines approach the F_1 curve tangentially from the side of real roots, and transversally from the side of complex roots. They approach the curves F_2 and F_3

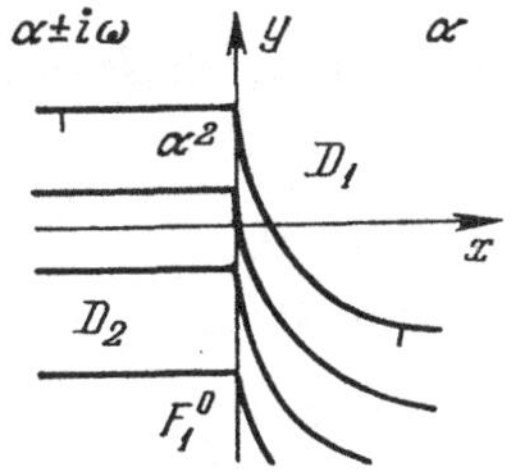

Fig. 4.5.

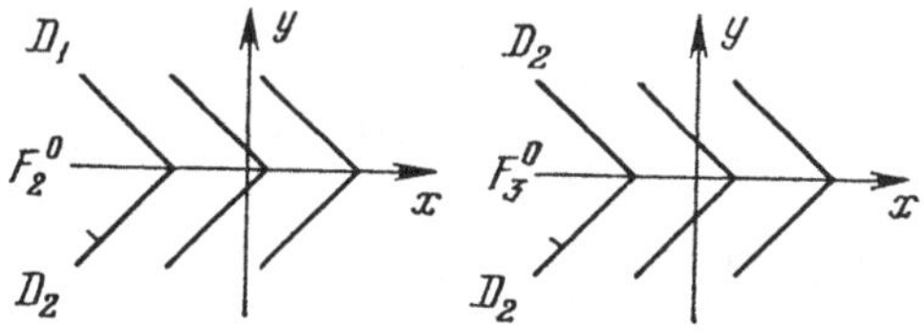

Fig. 4.6.

transversally from either side. In all cases the level curves have an angle of less than 180° at the break line on the side of decreasing f.

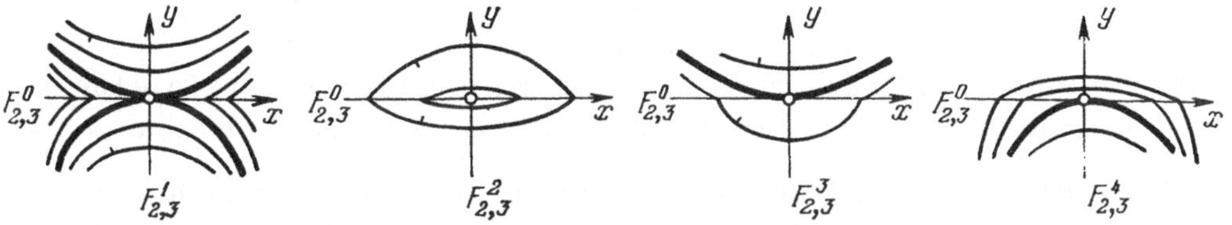

Fig. 4.7.

In a neighbourhood of a critical point of the restriction of the increment in a family in general position, we can choose coordinates x and y such that the increment takes one of the following twelve forms:

The cases F_2^k and F_3^k, $k = 1, \ldots, 4$ (*a conditional extremum along a fish-bone; Fig.* 4.7)

$$f = \text{const} + \varepsilon x^2 + \varphi(y) + |y|, \quad \varepsilon = (-1)^k,$$

where $\varphi(y) = ay + \ldots$ *is a smooth function,* $a > 0$, $a \neq 1$.

The four values of k are obtained as combinations of the two signs of ε and the two possibilities for a:

k	1, 2	3, 4
a	(0, 1)	$(1, +\infty)$

Odd values of k correspond to a conditional maximum and even ones to a minimum. To obtain an explicit idea of the form of the decrement diagram it suffices to consider the case $\varphi(y) = ay$: in this case the level curves of f consist of pieces of two parabolas, shifted along the y-axis.

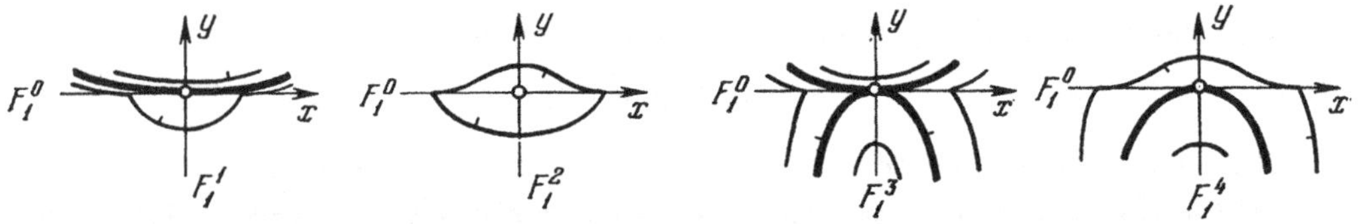

Fig. 4.8.

The cases F_1^k, $k = 1, \ldots, 4$ (*a conditional extremum with Jordan block* α^2; *Fig.* 4.8):

$$f = \text{const} + \varepsilon x^2 + \varphi(y) + \begin{cases} \sqrt{y}, & \textit{if } y \geqslant 0, \\ 0, & \textit{if } y \leqslant 0. \end{cases}$$

Here $\varepsilon = \pm 1$, $\varphi(y) = ay + \ldots$ *is a smooth function,* $a \neq 0$.

The four values of k are obtained as combinations of the signs ε and a:

k	1	2	3	4
sign ε, sign a	$--$	$-+$	$+-$	$++$

Odd k correspond to a conditional maximum, and even ones to a minimum. To obtain an explicit idea of the form of the decrement diagram, it suffices to consider the cases $\varphi(y) = \pm y$.

Our theorem asserts that the increment of a 2-parameter family in general position has no singularities at internal points of the curves F other than the 15 listed types F_i^k ($15 = 12 + 3$): if some particular family has other singularities, then these can be avoided by an arbitrarily small shift in the family. The singularities F_i^k, however, are obviously unavoidable.

4.4. The structure of decrement diagrams near strata of codimension 2. In the study of singularities of the strata of codimension 2 in 2-parameter families in general position we can restrict our attention to the "least degenerate" cases, since any further degeneracy increases the codimension, and the singularity becomes avoidable.

Combining the transversality theorem and the explicit formulae of § 3 for versal deformations of families of matrices we arrive at the following normal forms of the increment near points of strata of codimension 2.

THEOREM. *In a neighbourhood of any point of a stratum of codimension* 2 (G_i in the notation of § 4.2) *we can choose smooth coordinates x and y in the parameter plane of a family in general position such that the increment f takes on one of the following* 18 *forms:*

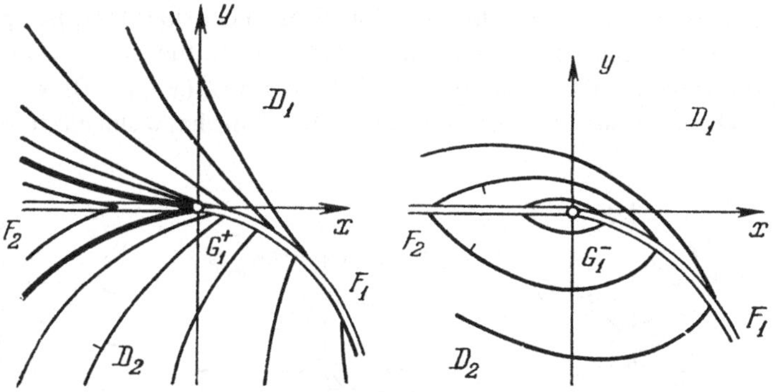

Fig. 4.9.

The cases $G_1^{\pm}$ (*a* 3 × 3 *Jordan block; Fig.* 4.9):

$$f = \varphi(x, y) + \lambda(x, y),$$

where λ is the greatest real part of the roots of the cubic equation $\lambda^3 = \lambda x + y$, and φ is a smooth function for which $(\partial\varphi/\partial x)(0, 0) = a \neq 0$. The form of the decrement diagram is determined by the sign of a.

The signs + and − in $G_1^{\pm}$ correspond to $a > 0$ and $a < 0$, respectively. To obtain an explicit idea of the form of the decrement diagram it suffices to consider the cases $\varphi = \pm x$. Two curves approach tangentially the point $x = y = 0$: the ray $F_2 (y = 0, x < 0)$ and one half of the semi-cubical parabola $F_1 (4x^3 = 27y^2, y < 0)$. These two curves divide the domain of complex conjugate roots D_2 (the convex one) from the domain D_1 of real eigenvalues. On moving along this boundary between D_1 and D_2 the increment f varies monotonically in the case $a > 0$, but has a minimum at the point G_1^- in the case $a < 0$. From the side of D_1 the level increment curves touch the semi-cubical parabola F_1.

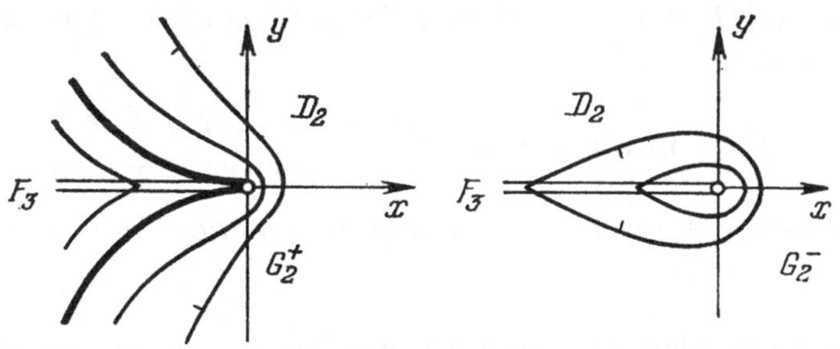

Fig. 4.10.

The cases $G_2^{\pm}$ (*a complex pair of* 2×2 *Jordan blocks; Fig.* 4.10):

$$f = \varphi(x, y) + |\operatorname{Re}\sqrt{(x + iy)}|.$$

Here Re *denotes the real part,* φ *is a smooth function for which* $(\partial\varphi/\partial x)(0, 0) = a \neq 0$. *The form of the decrement diagram is determined by the sign of a.*

The signs + and – in $G_2^{\pm}$ correspond to $a > 0$ and $a < 0$, respectively. To get an explicit idea of the form of the decrement diagram it suffices to consider the cases $\varphi = \pm x$. The ray $F_3(y = 0, x < 0)$ approaches $x = y = 0$ (and stops there). In the case $a < 0$ the function f has a minimum at $G_2^-(x = y = 0)$. In the case $a > 0$ the point $G_2^+(x = y = 0)$ is topologically nonsingular for f. The level line of f passing through this point has a singularity of the semicubical type.

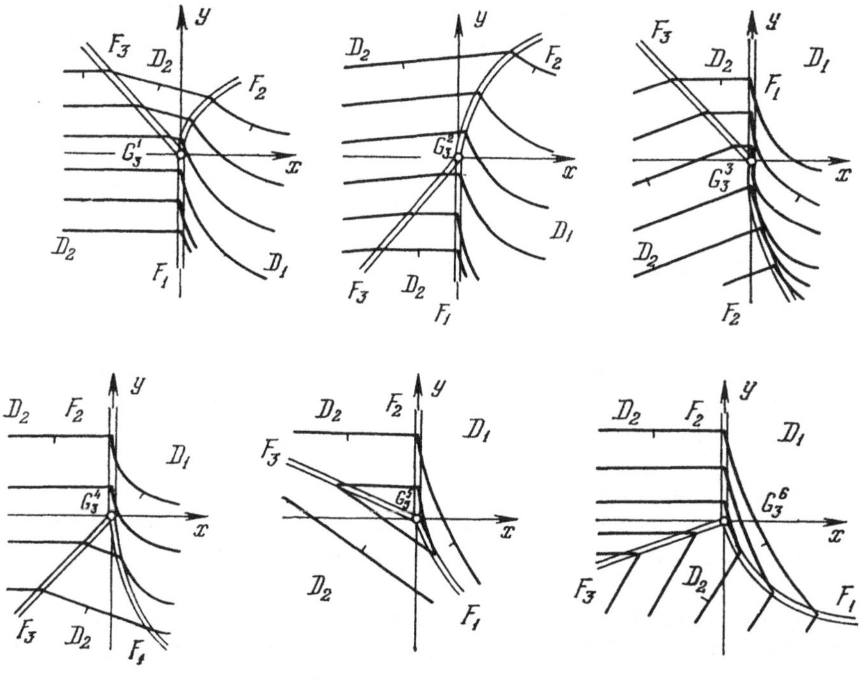

Fig. 4.11.

The cases $G_3^k (k = 1, \ldots, 6)$ (*a coincident complex pair and Jordan block; Fig.* 4.11):

$$f = \text{const} + y + \max \begin{cases} \sqrt{x}, \ \varphi(x, y) \ \text{if } x \geqslant 0, \\ 0, \ \varphi(x, y) \ \text{if } x \leqslant 0. \end{cases}$$

Here $\varphi(x, y) = ax + by + \ldots$ *is a smooth function,* $a \neq 0$, $b \neq 0$, *and* $b \neq -1$.

The six values of k are obtained as combinations of the two possible signs of a and the three intervals in which b can lie:

k	1	2	3	4	5	6
sgn a, interval of b	$+$, $(0, +\infty)$	$-$, $(0, +\infty)$	$-$, $(-1, 0)$	$+$, $(-1, 0)$	$-$, $(-\infty, -1)$	$+$, $(-\infty, -1)$

To obtain an explicit idea of the form of the decrement diagram, it suffices to take φ to be linear. Three smooth rays F_1, F_2, and F_3 approach the point, F_1 and F_2 meet head-on with first order tangency; F_3 approaches transversally from the side of D_2 corresponding to complex roots. In the case G_3^5 (that is, when $a < 0$, $b < -1$) the increment has a minimum at $x = y = 0$; in all other cases $G_3^k (k \neq 5)$ is a topologically non-singular point of f.

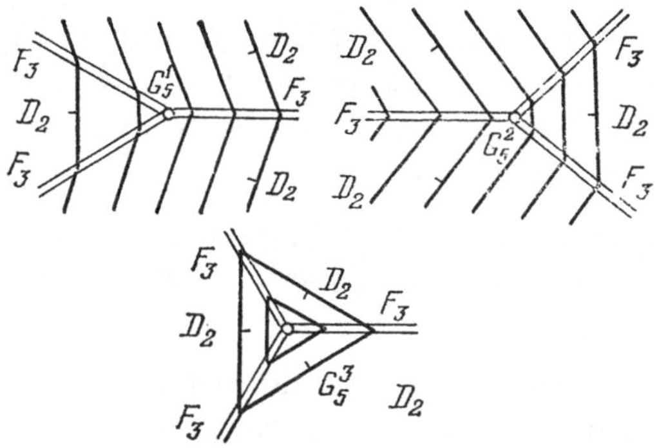

Fig. 4.12.

The cases G_5^k, $k = 1, 2, 3$ (*double fish-bone; Fig.* 4.12)

$$f = \text{const} + x + \max(|y|, \varphi(x, y)),$$

where $\varphi(x, y) = ax + by + \ldots$ *is a smooth function,* $a < 0$, $b > 0$, $a + 1 \neq \pm b$.

The three values of k for G_5^k correspond to the three intervals in which a can lie:

k	1	2	3
condition on a	$b - 1 < a$	$-b - 1 < a < b - 1$	$a < -b - 1$

To obtain an explicit idea of the form of the decrement diagram it suffices to take φ to be linear.

In each of the three cases ($k = 1, 2, 3$) three smooth branches of F_3 approach G_5^k transversally. In the last case the point is a minimum of the increment, and in the other two it is a topologically non-singular point. On approaching G_5^k along three of the curve segments the increment increases, and along the others it decreases.

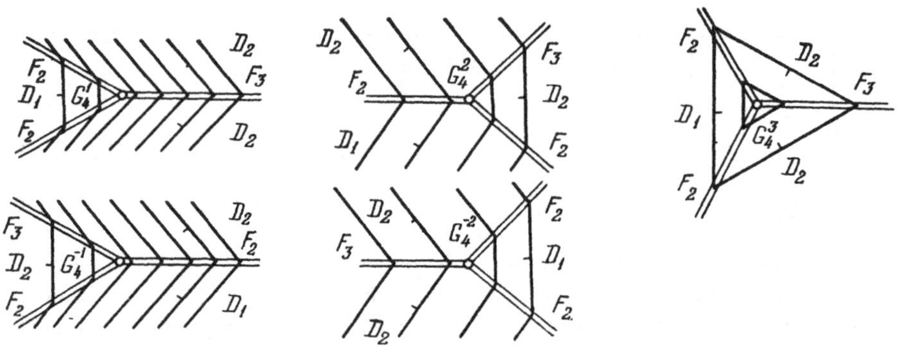

Fig. 4.13.

The cases G_4^k, $k = \pm 1, \pm 2, 3$ (*a double fish-bone together with a real root; Fig.* 4.13). *The increment is given by the same formula as in the cases G_5^k, but we have to distinguish more variants according to which of the sectors correspond to the real root.*

Negative k corresponds to the cases in which the decrement increases on approaching G_4^k along the line F_3 (along which the complex pairs coincide). The other two rays are branches of F_2.

4.5. Discussion. An analysis of the normal forms listed above leads to a number of conclusions of a general character on the structure of the decrement diagram both locally and globally. First of all we note that the theorems of § 4.3 and § 4.4 have the following consequence:

COROLLARY. *The increment f: $\Lambda \to \mathbf{R}$ of a 2-parameter family in general position is topologically equivalent to a smooth function having only simple critical points.*

For the minima are just the points of type D_i^0, F_i^2, $G_{\bar{1},2}$, G_3^5, $G_{4,5}^3$. Points D_i^1 and F_i^3 are topologically equivalent to saddle points. In a neighbourhood of the maxima (D_i^2 points) the increment is a smooth function. Points of all other types are topologically non-singular for the increment.

This corollary obviously implies Morse's inequalities for the number of singular points of the various types. In particular, *if some closed level increment curve bounds a simply-connected domain, then the total number of points of type $D_i^{0,2}$, F_i^2, $G_{\bar{1},2}$, G_3^5 and $G_{4,5}^3$ inside this domain exceeds by* 1 *the number of points D_i^1 and F_i^3*. I do not know whether the assertion of the corollary holds good for l-parameter families with $l > 2$.[1]

From the fact that the segments F_1 and F_2 together form closed curves, and from the above description of the singularities at the ends of the

[1] Note that in the case $l \geqslant 2$ the singularities of the increment are the same as the singularities of the greatest real part of a root of an algebraic equation whose coefficients are functions in general position of l parameters. For $l \geqslant 3$, this is no longer true: the increment can have more complicated singularities.

segments F_3 we deduce the following result.

COROLLARY. *If the parameter space Λ is a closed 2-dimensional manifold, then the number of points of type G_1 and G_3 have the same parity, and the total number of points G_2, G_3, G_4 and G_5 is even.*

If the parameter space Λ is a compact domain with a boundary that intersects the F_i transversally and does not pass through points G_i, then the result has to be altered as follows: the total number of points G_1 and G_3 has the same parity as the total number of intersections of the boundary with curves F_1 and F_2, and the total number of points G_2, G_3, G_4 and G_5 has the same parity as the number of intersections of the boundary with F_3.

The study of the increment we have made allows us, in particular, to investigate the singularities of the stability boundary (that is, the curve of zero increment) in the parameter plane of a two-parameter system in general position. The following is a corollary to the theorems of § 4.3 and § 4.4.

COROLLARY. *The stability boundary of a general two-parameter family of matrices consists of smooth arcs intersecting transversally at their ends.*

Note that the breaks in the stability boundary, according to the classification of § 4.3 and § 4.4, can only be of the types F_1^0 (2×2 Jordan block) or F_2^0 or F_3^0 ("simple fish-bone"). Each arc of the boundary can therefore be extended beyond its end-points without loss of smoothness.

Here the total number of break points of type F_1^0 and F_2^0 is even on each closed component of the stability boundary.

Note also that the preceding analysis of the singularities of the increment of *two-parameter* families suffices for the study of the stability boundary of *three-parameter* families.

For by the transversality theorem the singularities of the strata of codimension 3, as well as the critical points of the restrictions of the increment to the strata of codimension 0, 1 or 2 can be moved away from the stability boundary by small shifts of the family. Thus, the stability boundary of a general family consists of smooth surfaces, and its singularities lie along the curves of intersection with the surfaces of type F_i, and at the points of intersection with the strata G_i (the latter appear as curves in general three-parameter families).

Moving along such a curve G_i we can consider our three-parameter family as a one-parameter family of two-parameter families (the two parameters are the coordinates in a plane transversal to G_i, and the one a coordinate t along G_i). In considering the normal forms of § 4.3 and § 4.4 we must now assume that all the constants and the derivatives of φ depend smoothly on t. Furthermore, in the case of general position we may take for the parameter z these functions $\varphi(x, y, t)$ themselves. So we come to the following conclusions:

COROLLARY. *The stability boundary of a general three-parameter family of matrices has the same singularities as the increment graphs of general two-parameter families.*[1] *These singularities can be listed to within diffeomorphism as follows:*

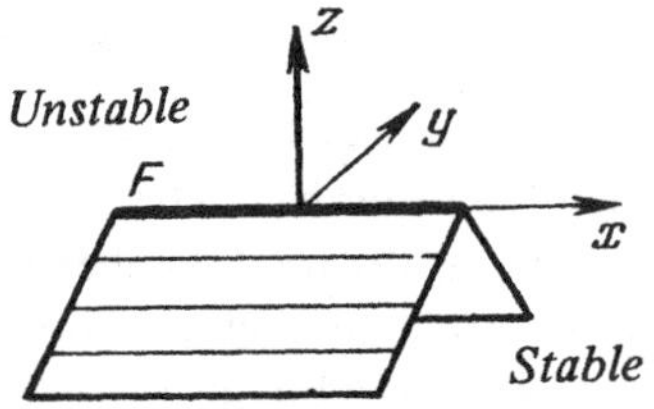

Fig. 4.14.

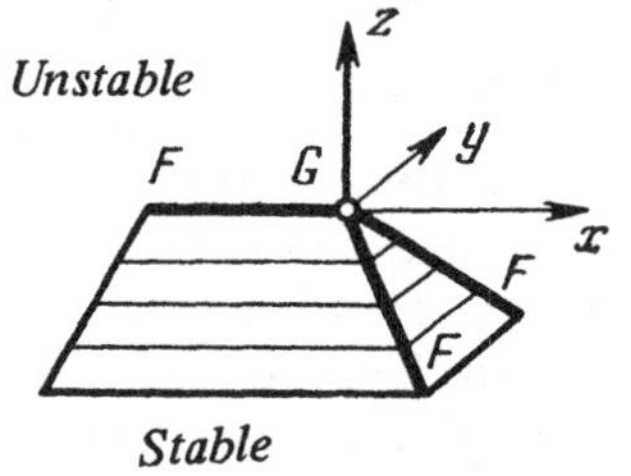

Fig. 4.15.

Two faces meeting along a ridge (F_i; Fig. 4.14): $z + |y| = 0$.

Three faces meeting at a corner ($G_{3,4,5}$; Fig. 4.15): $z + \max(x, |y|) = 0$.

Cuspidal point on a ridge (G_2; Fig. 4.17): $z + |\mathrm{Re}\sqrt{(x + iy)}| = 0$. (*This surface in* $\mathbf{R}^3$ *is diffeomorphic to that given by the equations* $XY^2 = Z^2$, *where* $Y \geqslant 0$.)

Node on a ridge (G_1; Fig. 4.16): $z + \lambda(x, y) = 0$, *where* λ *is the greatest real part of the roots of the equation* $\lambda^3 = x\lambda + y$. (*This surface in* $\mathbf{R}^3$ *is diffeomorphic to that given by* $X^2 Y^2 = Z^2$, $X \geqslant 0$, $Y \geqslant 0$.)

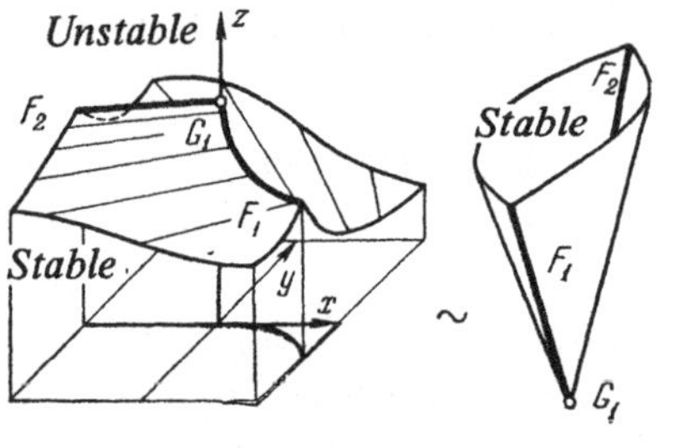

Fig. 4.16.

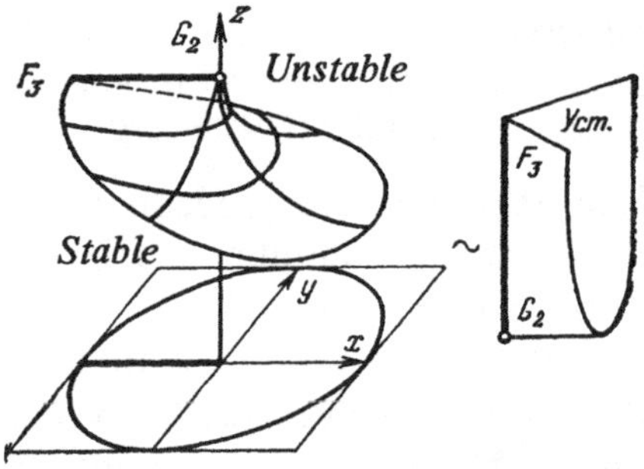

Fig. 4.17.

The acute angles of the stability boundary always point into the domain of instability.

From what we have said we can also deduce some global properties of the stability boundary. For example, if this boundary is closed, then the total number of vertices of all the types (G_i) is even and so is the number of vertices of type G_1 and G_3 together.

4.6. The construction of normal forms. The proofs of the theorems of § 4.3 and § 4.4 demand an exhausting analysis of many special cases; here we only touch on the basic methods on which the study is based.

Let $f_1 : X \to \mathbf{R}$ and $f_2 : X \to \mathbf{R}$ be smooth functions on an n-dimensional manifold X. We construct the new function $f: X \to \mathbf{R}$

[1] In just the same way the stability boundary of an $(n + 1)$-parameter family has the same singularities as the increment graphs of n-parameter families.

$$f(x) = \max_{i=1,2} f_i(x).$$

The function f is not necessarily everywhere differentiable. The question arises: what singularities can it have? The following lemma answers this for a function f in general position.

LEMMA. For almost all[1] (f_1, f_2) we can choose smooth coordinates in a neighbourhood of any point $x_0 \in X$ such that f takes one of the following forms:

a) $f = \text{const} + x_1$ (*non-singular point*),

b) $f = \text{const} + x_1^2 + \ldots + x_k^2 - x_{k+1}^2 - \ldots - x_n^2$ (*a Morse point*),

c) $f = \text{const} + x_1 + |x_2|$ (*simple ridge*),

d) $f = \text{const} + x_1^2 + \ldots + x_k^2 - x_{k+1}^2 - \ldots - x_{n-1}^2 + |x_n| + \varphi(x_n)$, *where* $\varphi(x_n) = ax_n + \ldots$ *is a smooth function,* $a > 0$, $a \neq 1$ (*a conditional extremum*).

PROOF. If $f_1(x_0) \neq f_2(x_0)$, then the maximum in a neighbourhood of x_0 is just one of the functions, and is smooth. According to Morse theory we may assume that f_i has one of the forms a) or b) in a neighbourhood of x_0.

If $f_1(x_0) = f_2(x_0)$, then we may assume by the (weak) transversality theorem that $df_1(x_0) \neq df_2(x_0)$. Hence f_1 and f_2 coincide (in a neighbourhood of x_0) along a smooth hypersurface Y. We must now distinguish two cases:

c) *the differentials of f_1 and f_2 are linearly independent.* This condition means precisely that x_0 is a non-critical point of the restriction of the f_i to the manifold Y, where they coincide. In this case we can set $2x_1 = f_1 + f_2 - 2f(x_0)$, $2x_2 = f_1 - f_2$, and obtain $f = \text{const} + x_1 + |x_2|$.

d) *the differentials of f_1 and f_2 are linearly dependent.* This condition means that x_0 is a critical point of the restriction of the f_i to Y.

We can deduce from the transversality theorem that for functions f_i in general position all the critical points x_0 are non-degenerate. Now consider the one-parameter family of manifolds Y_t given in a neighbourhood of x_0 by the equations $f_1 - f_2 = 2t$.

The function $2g = f_1 + f_2$ restricted to Y_t has a non-degenerate critical point on Y_t depending smoothly on t (as t varies in a small neighbourhood of zero). An easy generalization of Morse's lemma to the case of a one-parameter family of functions allows us to choose a system of coordinates on Y_t in a neighbourhood of the critical point such that g takes the canonical form

$$g_{|Y_t} = c + x_1^2 + \ldots + x_k^2 - x_{k+1}^2 - \ldots - x_{n-1}^2,$$

and that the coordinates vary smoothly with t (for small $|t|$).

The constant c in this equation depends on t; setting $c(t) = c(0) + \varphi(t)$ and choosing $x_n = \pm t$ we obtain a coordinate system on X in which f has the form indicated in the statement of the lemma.

Almost all the assertions of the theorem of §4.3 are obtained by applying the lemma we have just proved.

---

[1] That is, for all except a countable union of nowhere dense closed set in the function space of pairs (f_1, f_2).

The normal form for the case F_1^0 indicated in the list in a neighbourhood of a 2×2 Jordan block is obtained from the versal family

$$A(\lambda_1, \lambda_2) = \begin{pmatrix} \alpha & 1 \\ 0 & \alpha \end{pmatrix} + \begin{pmatrix} 0 & 0 \\ \lambda_1 & \lambda_2 \end{pmatrix}$$

on making the change of variables $x = \lambda_1 + \lambda_2^2/4$, $y = \lambda_2$.

The normal form for the cases F_1^k can be obtained as follows. Since the deformation A is versal, there is a map of the parameter plane onto the x, y-plane, where x and y are the coordinates just indicated. In the case F_1^k (see [3]) this map has a fold that cuts the line of Jordan blocks ($x = 0$) transversally; we may assume that this intersection takes place at the origin. Let $y = \varphi(x)$ be the equation of the fold. Then we can take as coordinates on the parameter plane of the considered family in question the functions

$$Y = x,\ X = \sqrt{[\varepsilon(y - \varphi(x))]}\ \text{(where } \varepsilon = 1 \text{ or } -1).$$

In these coordinates the eigenvalues are given by the formula

$$\alpha + y \pm \sqrt{x} = \alpha + \varepsilon X^2 + \varphi(Y) \pm \sqrt{Y},$$

which leads us to the normal form F_1^k.

The study of points of type G_1, when 3×3 Jordan blocks appear, is based on the consideration of the versal family with parameters φ, x, y:

$$\begin{pmatrix} \varphi & 1 & 0 \\ 0 & \varphi & 1 \\ 0 & 0 & \varphi \end{pmatrix} + \begin{pmatrix} 0 & 0 & 0 \\ 0 & 0 & 0 \\ y & x & 0 \end{pmatrix}.$$

We denote by $\lambda(x, y)$ the greatest real part of the roots of the cubic equation $\lambda^3 = x\lambda + y$. We must study the function $f = \lambda + \varphi$, where φ is a smooth function in general position in a neighbourhood of the origin.

It is easy to check that the level curves of the function λ consist of segments tangent to the semi-cubical parabola $4x^3 = 27y^2$ (Fig. 4.18). It is convenient to carry out the investigation of the function $\lambda + \varphi$ on the plane of roots of the equation.

In the smaller of the two domains bounded by the semi-cubical parabola the roots are real; we denote them by $\alpha > \beta > -\alpha - \beta$. Then $\lambda(x, y) = \alpha$, and x and y vanish at the origin O of the (α, β)-plane faster than to the first order. Hence the level lines of the function $\lambda + \varphi$ on the (α, β)-plane behave like the straight lines $\alpha = \text{const}$ (Fig. 4.19).

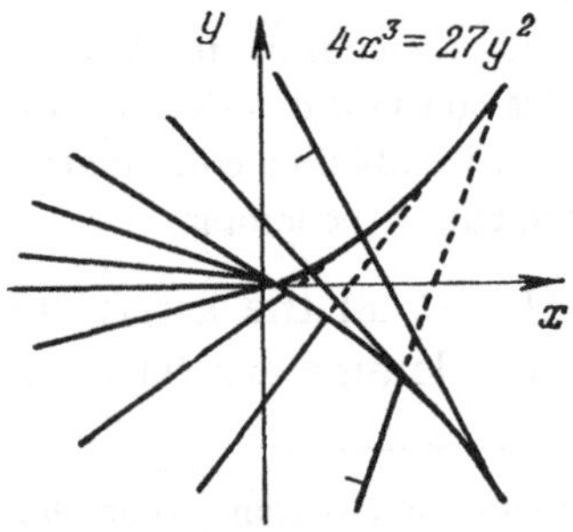

Fig. 4.18.

In the larger region we denote the roots by $2\alpha, -\alpha \pm i\omega\ (\omega > 0)$. When $\alpha > 0$ we find $\lambda = 2\alpha$, and when $\alpha < 0$ we have $\lambda = -\alpha$. The form of the level curves of $\lambda + \varphi$ on the (α, ω)-plane is determined by the behaviour of that part of φ on the line $\alpha = 0$ that is quadratic in ω. By Vietà's formula $x = 3\alpha^2 - \omega^2$, and $y = 2\alpha(\alpha^2 + \omega^2)$. Hence the behaviour of φ on the ray $\alpha = 0$ is determined by the sign of $\partial\varphi/\partial x$ at zero. Depending on this sign we obtain on the (α, ω)-plane one of the two pictures of Fig. 4.20, which gives us the set-up of Fig. 4.9 on the (x, y)-plane.

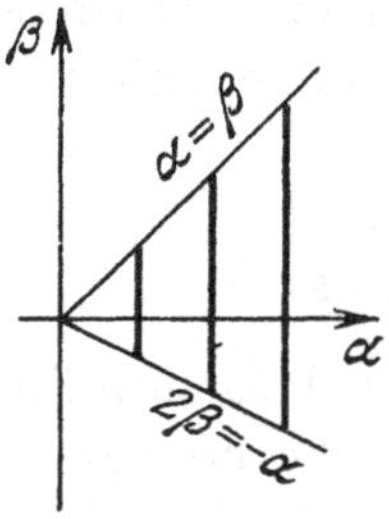

Fig. 4.19.

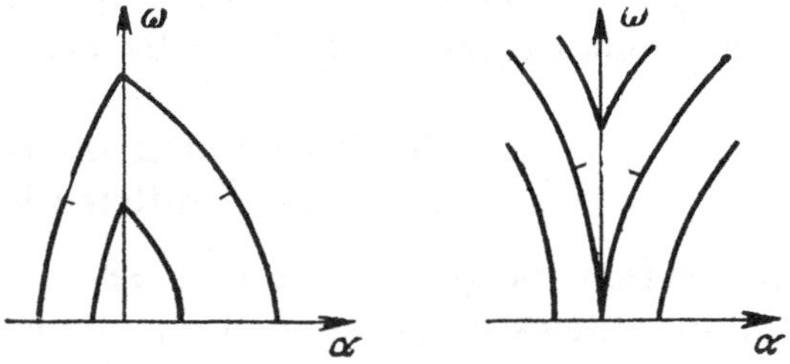

Fig. 4.20.

The study of the points of type G_2 (two complex conjugate 2×2 Jordan blocks) is based on the consideration of the (real four-parameter) versal family with complex parameters p and q

$$\begin{pmatrix} \alpha & 1 \\ 0 & \alpha \end{pmatrix} + \begin{pmatrix} 0 & 0 \\ p - q^2/4 & q \end{pmatrix}.$$

Setting $q = x + iy$, $p = p(x, y)$ we arrive at the problem of studying the real function $f = \varphi(x, y) + |\operatorname{Re}\sqrt{(x + iy)}|$, where $\varphi = \operatorname{Re}(\alpha + p(x, y))$ is a function in general position.

It is again convenient to carry out the investigation on the plane of the roots. Let $x + iy = (u + iv)^2$, $u \geqslant 0$. Then $|\operatorname{Re}\sqrt{(x + iy)}| = u$. The behaviour of f on the half-plane $\{u, v \colon u \geqslant 0\}$ depends on the behaviour of the part of φ that is quadratic in v on the line $u = 0$. This quadratic part in v is defined by the part that is linear in x and y: if $\varphi = \text{const} + ax + by + \ldots$, then $\varphi = \text{const} + a(u^2 - v^2) + 2buv + \ldots$.

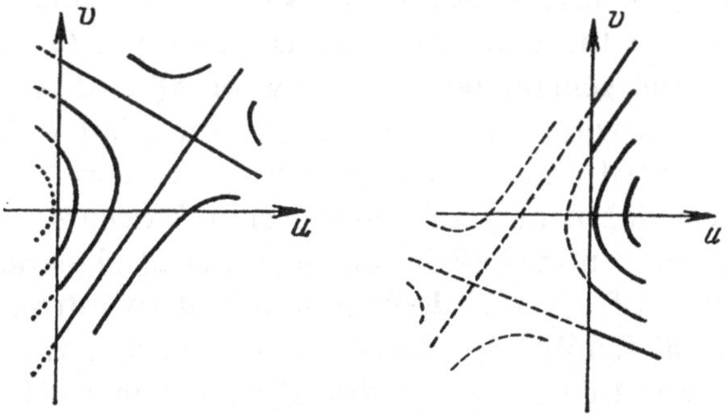

Fig. 4.21.

Depending on the sign of a we obtain one of the two set-ups of Fig. 4.21 on the half-plane $\{ u, v: u \geqslant 0\}$. Returning to the (x, y)-plane, we obtain Fig. 4.10.

The cases $G_3 - G_5$ are simpler and their study demands only patience. The form of the decrement diagram of general two-parameter families becomes perhaps more lucid if we remark that the level decrement lines have the form of sections of the stability boundary of general three-parameter families (Figs. 4.14–4.17) by planes in general position.

§5. The simplest bifurcations of singularities of differential equations

In this section the general ideas of §§1–3 are applied to non-linear autonomous systems of differential equations. We shall be concerned with the neighbourhood of a singularity at which the vector field defining the system vanishes. In local coordinates with the origin at the singular point the system can be written in the form $\dot{x} = Ax + \ldots, x \in \mathbf{R}^n$, where $A: \mathbf{R}^n \to \mathbf{R}^n$ is a linear operator.

In this situation the case of general position is that in which all the eigenvalues of A have non-zero real parts. If we denote by n_- and n_+ the number of eigenvalues in the left and right half-planes, respectively (so that $n_- + n_+ = n$), then the integer n_- is the only topological invariant and it determines the system in a neighbourhood of zero to within a homeomorphism.

Thus, for the topological classification in the case of general position we obtain altogether $n + 1$ classes of singularities in $\mathbf{R}^n$.

The smooth classification is quite another matter. First of all, the eigenvalues $\lambda_k(A)$ are diffeomorphism invariants, so that there is at any rate a continuum of classes. Apart from this, a more detailed analysis shows that so-called "resonances" or effects connected with "small denominators" make the differentiable classification very complicated and cause "pathological" phenomena.

We shall return to the smooth classification in §6, restricting ourselves here to a topological investigation, but we consider apart from the cases of general position also the cases of small codimension (1 or 2). In accordance with the general point of view of §§1–2 we shall study not individual equations with degenerate singularities, but k-dimensional families of differential equations in which for certain "bifurcational" values of the parameters degeneracy of codimension k occurs.

Since we are only interested in the topological classification we have to modify the notion of a versal deformation and to introduce that of a topologically versal family. A differentiable determination of versal families of vector fields was first given by Brushlinskaya in [58].

5.1. Topologically versal deformations. We recall that a *topological* *orbital equivalence* of the systems

$$\dot{x} = v(x),\ x \in X;\ \dot{y} = w(y),\ y \in Y$$

is a homeomorphism h: $X \to Y$ that takes the phase curves (orbits) of the first system into those of the second, with preservation of the direction of time.

Agreement of the speed of motion is not required, so that h does not necessarily take solutions of the first system into solutions of the second.

Two systems are said to be *topologically orbitally equivalent* if there is a topological orbital equivalence between them.

Consider a family of differential equations depending smoothly on parameters ε:

$$(1) \qquad \dot{x} = v(x, \varepsilon).$$

Topological orbital equivalence determines a partition of the parameter space into classes. We call this partition the *bifurcation diagram of the family*.

A family of differential equations depending smoothly on parameters ε

$$(2) \qquad \dot{y} = w(y, \varepsilon),$$

is said to be (topologically) *equivalent* to the family (1) if there exists a homeomorphism $h(\varepsilon)$, depending continuously on ε, that takes the (oriented) phase curves of the system (1_ε) into those of (2_ε).

In other words, a topological equivalence of families is a topologically orbital equivalence of systems, depending continuously on the parameter.

A family of differential equations depending smoothly on parameters μ

$$(3) \qquad \dot{x} = u(x, \mu),$$

is said to be *induced* from (1) if it is obtained from (1) by a continuous change of parameters, that is, if there exists a continuous map $\varepsilon = \varphi(\mu)$ such that

$$u(x, \mu) = v(x, \varphi(\mu)).$$

The germ of the family (1) at $\varepsilon = 0$ is called a *deformation* of the equation $\dot{x} = v(x, 0)$. The deformation (1) is said to be a *topologically versal* deformation of the equation $\dot{x} = v(x, 0)$ if every other deformation of the same equation is equivalent to one induced from (1).

5.2. Topologically versal deformations of germs. We shall be concerned with the local problem, that is, the study of germs of vector fields at a point and the study of the behaviour of the phase curves in a neighbourhood of the point, as the parameter passes through a bifurcational value. For this purpose we have to localize the preceding notions, replacing vector fields by their germs everywhere.

However, in this localization procedure we must take a little care lest we throw out the baby with the bathwater.

In the classical theory of singularities of differential equations for each value of the parameter a neighbourhood of the singular point is constructed in which the phase portrait is studied. As the parameter approaches a bifurcational value, this neighbourhood contracts to 0, and at the bifurcational value goes up with a jump. As a result, we may not be able to sense the change-over (say the approach of a neighbouring singular point) that occurs at the bifurcational value (Fig. 5.1).

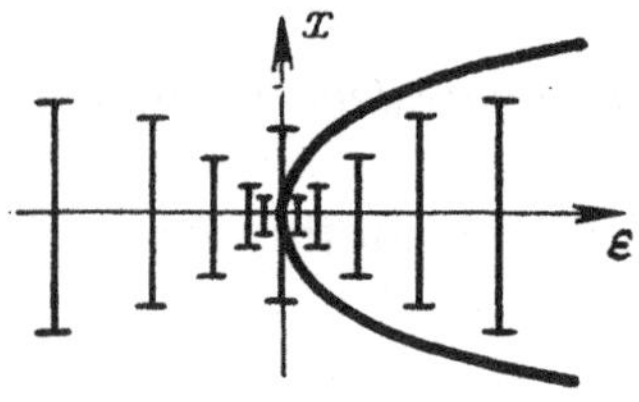

Fig. 5.1.

This reasoning explains why it is impossible in the localization simply to replace the family $v(x, \varepsilon)$ by the germ of a family of germs of vector fields, and why we have to introduce the following definitions.

By a *local family* (v, x_0, ε_0) we mean the germ of the map v (see (1)) at the point (x_0, ε_0) of the direct product of the phase space with the parameter space. Thus, each representative of this germ is given in a whole neighbourhood of (x_0, ε_0) in the product space (and not just in a neighbourhood of x_0 in the phase space).

By an *equivalence of the local families* (1) (v, x_0, ε_0) *and* (2) (w, y_0, ε_0) we mean a germ (at (x_0, ε_0)) of a continuous map $y = h(x, \varepsilon)$ for the representatives of which $h(\cdot, \varepsilon)$ for each ε is a homeomorphism, taking the phase curves of (1) (within the domain of definition of h) into those of (2), and for which $h(x_0, \varepsilon_0) = y_0$.

We note that the germs of the homeomorphisms $h(\cdot, \varepsilon)$ do not necessarily take x_0 into y_0 for $\varepsilon \neq \varepsilon_0$.

A local family (3) (u, x_0, μ_0) is *induced* from (1) by means of the germ at μ_0 of the continuous map $\varepsilon = \varphi(\mu)$, where $\varphi(\mu_0) = \varepsilon_0$.

A local family (v, x_0, ε_0) is said to be a *topologically versal deformation* of the germ of the field $v_0 = v(\cdot, \varepsilon_0)$ at x_0 if every other local family containing this germ is equivalent to a family induced from the given one.

We should emphasize that the existence of a topologically versal deformation of a given family is not in the least obvious; it is easy to give examples of fields that do not have such a deformation with finitely many parameters (for example, the zero field).

5.3. Simplest examples. EXAMPLE 1. Let x_0 be a non-singular point of a field v_0 (that is, $v_0(x_0) \neq 0$). Then we can take the trivial 0-parameter deformation as a versal deformation. In a neighbourhood of a non-singular point all vector fields have the same structure (even in the smooth sense).

EXAMPLE 2. Let x_0 be a non-degenerate singular point of a field v_0 (that is, $v_0(x_0) = 0$, but none of the eigenvalues of the matrix $(\partial v_0/\partial x)(x_0)$ is purely imaginary). Then we can still take the trivial 0-parameter family as a versal deformation (whereas one might require infinitely many parameters to construct "differentially versal" deformations). Thus, for non-singular and non-degenerate singular points the bifurcation diagrams are trivial: the topological type of the field does not change on deformation, and the topologically versal deformation consists just of the one original field.

Things turn out otherwise when x_0 is a degenerate singular point of v_0. In this case the minimal dimension that a topologically versal family can have is the minimal number of parameters of families in which singularities of the given topological type occur so that they cannot be avoided by small shifts.

We then have a bifurcation diagram in the parameter space of the family (Fig. 5.2): to every class of points into which the bifurcation diagram decomposes the parameter space, there corresponds a certain phase portrait. In the large, the collections of phase portraits obtained for various parameter values, form something like a fibration: as long as we stay in one class of points of the bifurcation diagram we have a genuine fibration, but to each contiguity of one class onto another there corresponds a certain degeneration of the phase portrait.

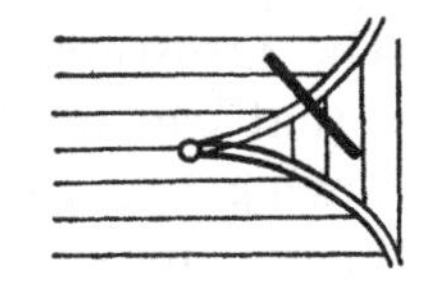

Fig. 5.2.

The construction of versal deformations reduces the study of all possible deformations of a given field to that of a single – versal – one. In particular, to investigate the degeneration of the phase portrait as we approach strata of smaller dimension on the bifurcation diagram, we construct a plane transversal to this part (Fig. 5.2). The family induced on the transversal plane by its inclusion mapping into the parameter space is a versal deformation of its singular point, or at any rate, it can be investigated by means of versal deformations with no more parameters than the dimension of the transversal.

The program sketched above for studying versal deformations has so far only been completely worked out for cases of codimension 1, and in part for cases of codimension 2. Already in certain cases of codimension 3 finite-dimensional/topologically versal deformations apparently do not exist. (this was pointed out to me by N. K. Gavrilov).

5.4. A remark on codimensions. In defining the codimensions of degenerate singularities of vector fields there are two possible ways of starting the count, so that a more detailed discussion is in order.

Let J^k be the manifold of k-jets of vector fields on a manifold M of dimension n. Amongst these jets there are the jets of vector fields with singularities; all such jets form in J^k a submanifold of codimension n.

However, we wish to regard this case to be of codimension zero, because it is the case of general position: the vector field determines a map of M into J^k, and the transversal intersections of M with the submanifold of singular jets cannot be avoided by small shifts.

Consequently we define the "codimension of a case" as the codimension of the corresponding submanifold of J^k minus n. Thus, to say that a singularity of a vector field has codimension c is to say that the k-jet of the vector field at the point in question belongs to some submanifold of codimension $c + n$ in the space of jets J^k.

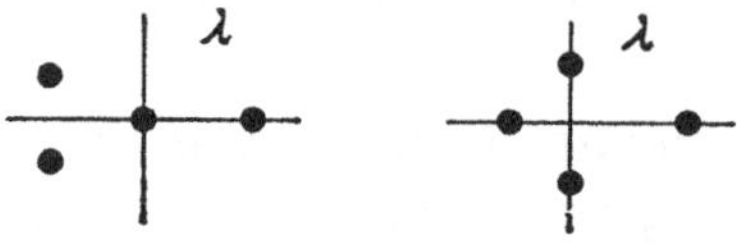

Fig. 5.3.

In what follows, by the codimension of a singularity we always understand the codimension of the submanifold of the space of jets minus n. According to the transversality theorem, singularities of codimension c occur unavoidably in families with not fewer than c parameters. We begin with cases of codimension 1, that is, with the unavoidable singularities of one-parameter families.

The transversality theorem guarantees that the only degenerate singularities that can occur in one-parameter families in general position are of the following two types (Fig. 5.3):

1) the linear part of the field at the singular point has one eigenvalue zero;

2) the linear part of the vector field at the singular point has one pair of purely imaginary eigenvalues.

We present topologically versal deformations for the germs of these two types. After this it turns out that in a general one-parameter family all the singular points are non-degenerate points for all values of the parameter except for a discrete set of bifurcational parameter values in a neighbourhood of which the situation can be studied by means of versal families.

5.5. The case of one eigenvalue zero. To begin with we consider the one-parameter family of vector fields on the line

(4) $\dot{x} = \pm x^2 + \varepsilon, \quad x \in \mathbf{R}, \; \varepsilon \in \mathbf{R}.$

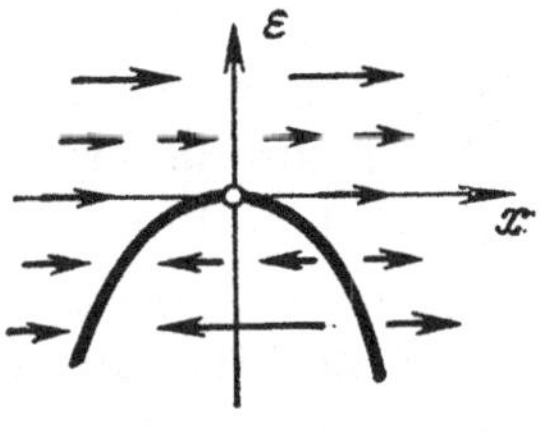

Fig. 5.4.

This family gives the topologically versal deformation of the germ of the vector field v at the simplest degenerate singular point ($v(x) = \pm x^2$), which was undoubtedly well known to Poincaré (and possibly earlier). As the parameter passes through the critical value $\varepsilon = 0$ (Fig. 5.4), what happens depends on the sign: either two singular points, one stable and one unstable, annihilate one another, or a pair of singular points is born and they immediately move apart (asymptotically to $\sqrt{|\varepsilon|}$)..

It turns out [37] that a topologically versal deformation of a general singularity in $\mathbf{R}^n$, in the case of one zero root of the characteristic equation, can be obtained from (4) by a simple building-on process, which we will refer to as "suspension":

$$(5)\qquad \begin{cases} \dot{x} = \pm x^2 + \varepsilon, & x \in \mathbf{R},\ \varepsilon \in \mathbf{R}, \\ \dot{y} = -y, & y \in \mathbf{R}^{n_-}, \\ \dot{z} = z, & z \in \mathbf{R}^{n_+}, \end{cases}$$

n_- and n_+ being the number of roots in the left and right half-planes, so that $n = n_- + n_+ + 1$.

For example, for $n = 2$ the system (5) describes the fusion of a saddle point and a sink (Fig. 5.5). At $\varepsilon = 0$ we obtain a so-called saddle-sink.

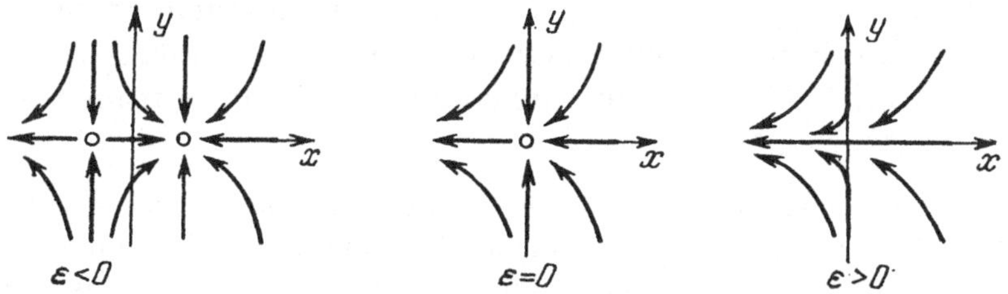

Fig. 5.5.

From the transversality theorem we deduce:

THEOREM. *In the space of one-parameter families of vector fields those families for which the following propositions hold form an open*[1] *everywhere dense set:*

1. *For all parameter values except a discrete set of bifurcational values, all the singular points are non-degenerate.*

2. *If ε_0 is a bifurcational parameter value, then a single degenerate singular point corresponds to it; for this parameter value the linear part of the field at this point has either one eigenvalue zero or a conjugate pair of purely imaginary eigenvalues.*

3. *In the first case the germ of the family at (ε_0, x_0) is topologically equivalent to the germ of the form* (5) *at the origin, and in the second case to the germ* (7) *below.*

We consider the set of all pairs (ε, x) in the product of the parameter space $\{\varepsilon\}$ and the phase space, where x is a singular point of the field corresponding to the parameter value ε.

By the transversality theorem this set, swept out by the singular points of the field as the parameter varies, is a smooth curve for a family in general position. This curve might have the form shown in Fig. 5.6: as the parameter passes through bifurcational values, singular points can swallow each other up or be born in pairs.

---

[1] Under the assumption that the base of the family is compact; in the non-compact case it is the intersection of a sequence of open sets.

The points at which the curve projects badly onto the ε-axis are just the degenerate singularities, and their projections are the bifurcational values of the parameter. For we can determine $x(\varepsilon)$ by means of the implicit function theorem from the equation $v(x, \varepsilon) = 0$, provided that a certain determinant is non-zero, and this determinant is just $|\partial v/\partial x|$, that is, the product of the eigenvalues of the linear part of v at the singular point.

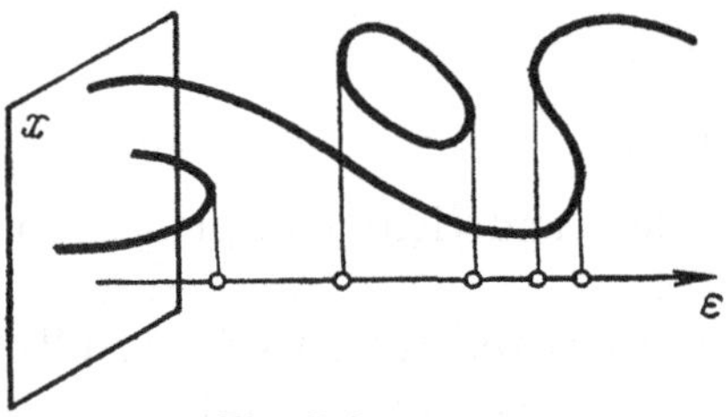

Fig. 5.6.

We note that the curve of singularities in the (ε, x)-space is non-singular even at birfurcational values (this is clear, for example, from (5)): the whole singularity comes from the projection.

From what has been said above it follows that if in some problem a more complicated curve of singularities turns up, for example, the one illustrated in Fig. 5.7 on the left, then this indicates the fact that the family in question is not in general position. This may be connected with some special symmetry of the situation, or it may point to an inadequate

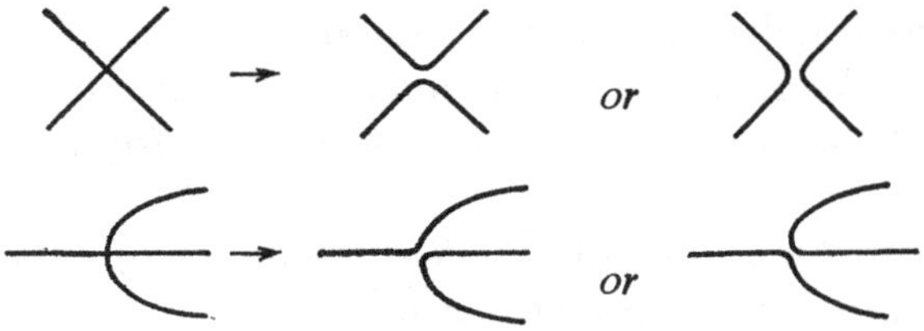

Fig. 5.7.

idealization in which we have neglected some small effect that is nevertheless capable of changing qualitatively the behaviour of the singular points as the parameter varies.

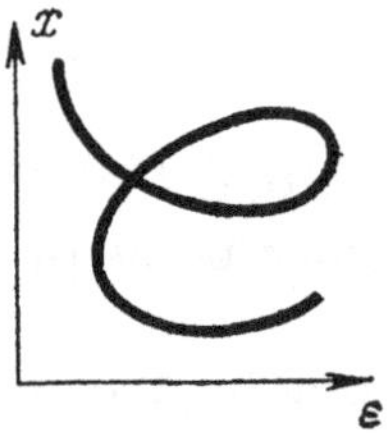

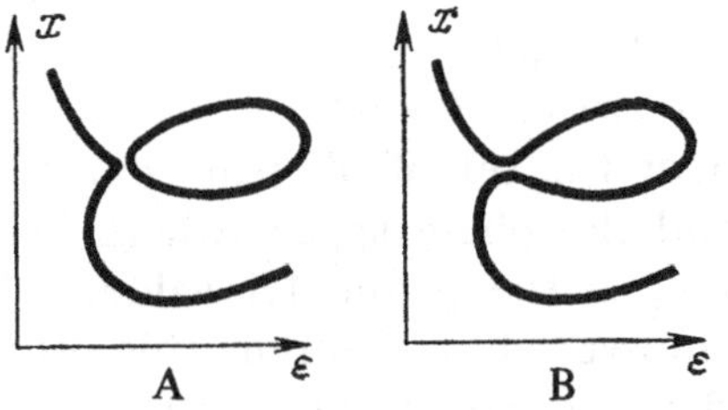

Fig. 5.8. Fig. 5.9.

An example of such a configuration of singular points, which is not possible in a family in general position, is shown in Fig. 5.8, which is taken from the book "The theory of oscillations" by Andronov, Witt and Khaikin.

Under a small shift of the family the singularities of the curve of singular points illustrated in Fig. 5.7 on the left disintegrate, for example, as shown in Fig. 5.7 on the right; Fig. 5.8 must take on the form of Fig. 5.9 A or B.

If we wish to find out which of the cases actually holds in a real system whose idealization leads to a non-general singularity, then we must compute some of the terms discarded in the idealization; formula (5) then suggests which ones.

5.6. The case of a pair of purely imaginary eigenvalues. Let us start with an example due to Poincaré and Andronov of a one-parameter family of vector fields on the plane. We write it in complex form

$$\dot{z} = z(i\omega + \varepsilon + cz\bar{z}), \tag{6}$$

where $z = x + iy$ is a complex coordinate on the plane $\mathbf{R}^2$ regarded as the complex z-plane.

In (6) ω and c are real non-zero constants, which we could take to be ± 1 if so desired; ε is a real parameter.

For all ε the point $z = 0$ is an equilibrium point of the spiral type. This spiral is stable if $\varepsilon < 0$ and unstable for $\varepsilon > 0$. For $\varepsilon = 0$ the linear approximation gives the centre; the character of the singular point for $\varepsilon = 0$ is determined by the sign of c: $c < 0$ corresponds to stability, $c > 0$ to instability.

If we carry out the analysis of the singular points locally in z, we observe that at $\varepsilon = 0$ the singular point loses its stability, but we miss an important phenomenon connected with this loss in stability – the birth of a limit cycle (see Fig. 5.12). To avoid this mistake we have to consider a neighbourhood of zero in the (z, ε)-space, and not just in the z-plane for a fixed ε.

The investigation of a neighbourhood of zero in the (z, ε)-space can conveniently be carried out as follows. Consider the function $\varphi(z) = z\bar{z}$. From (6) we find the equation for φ:

$$\dot{\varphi} = \varphi(2\varepsilon + 2c\varphi), \quad \varphi \geqslant 0.$$

The resulting family of equations on the ray $\varphi \geqslant 0$ is easy to study. Apart from the singularity at $\varphi = 0$, which is present for each ε, we also have (if ε and c have opposite signs) the singular point $\varphi = -\varepsilon/c$. For $c > 0$ the vector field $\dot{\varphi}$ has one

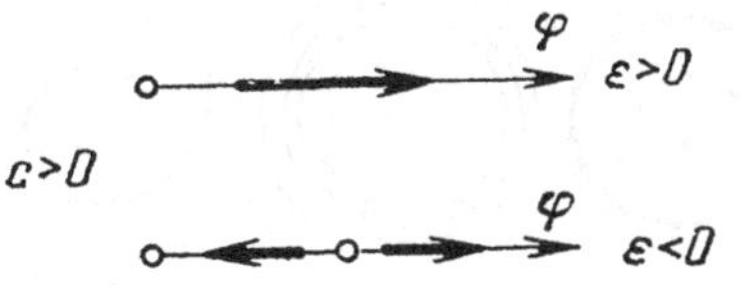

Fig. 5.10.

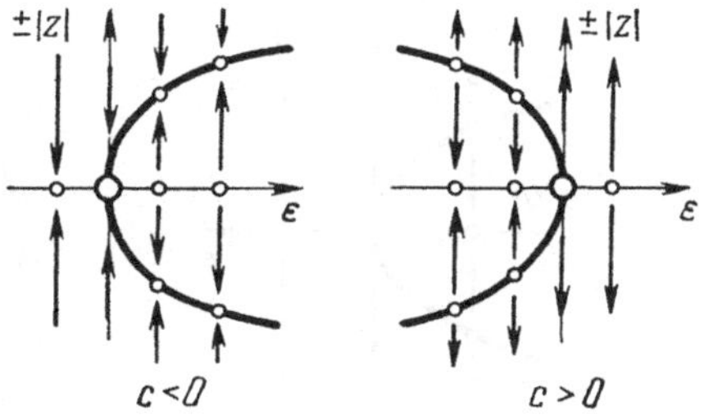

Fig. 5.11.

of the forms indicated in Fig. 5.10, depending on the sign of ε.

To the point $\varphi = 0$ there corresponds the origin of the $z = x + iy$-plane, and to $\varphi = -\varepsilon/c$ a limit cycle (which is real only if ε and c have opposite signs).

To understand the situation better we trace $\pm|z|$ against ε. Then the behaviour of the cycle, as the parameter varies, is depicted by one of the two diagrams of Fig. 5.11, depending on the sign of c. Thus, the radius of the cycle is proportional to $\sqrt{|\varepsilon|}$.

Let us first consider the case $c < 0$. As ε passes through zero, the spiral at the origin loses its stability. When $\varepsilon = 0$, there is still a stable spiral at the origin, but it is only weakly stable: the phase curves no longer approach the origin exponentially.

For $\varepsilon > 0$ the phase curves leave the origin to a distance proportional to $\sqrt{\varepsilon}$ and wind themselves around a stable limit cycle. Thus, the loss in stability as ε passes through zero is accompanied in the case $c < 0$ by the birth of a stable limit cycle whose radius grows like $\sqrt{\varepsilon}$.

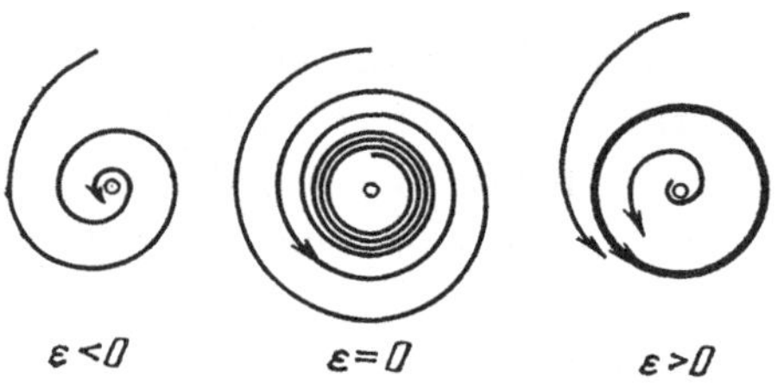

Fig. 5.12.

In other words, the stationary situation becomes unstable and is replaced by a stable periodic regime whose amplitude is proportional to the square root of the distance of the parameter from the critical values. Physicists refer to this situation as *weak perturbation* of auto-oscillations.

In the case $c > 0$ (Fig. 5.13) we have an unstable limit cycle for $\varepsilon < 0$. As ε tends to zero, this cycle settles at the equilibrium position, which for $\varepsilon < 0$ is a stable spiral. At $\varepsilon = 0$ the spiral becomes unstable (weakly rather than exponentially unstable). For positive ε the spiral is unstable even in the linear approximation.

This case is known as *strong perturbation* for the following reason.

Let us imagine that a system is near a position of stable equilibrium and that this equilibrium position becomes unstable as the parameter changes. In the case $c > 0$, as ε tends to zero from the negative side (and even somewhat earlier), the disturbances, which are

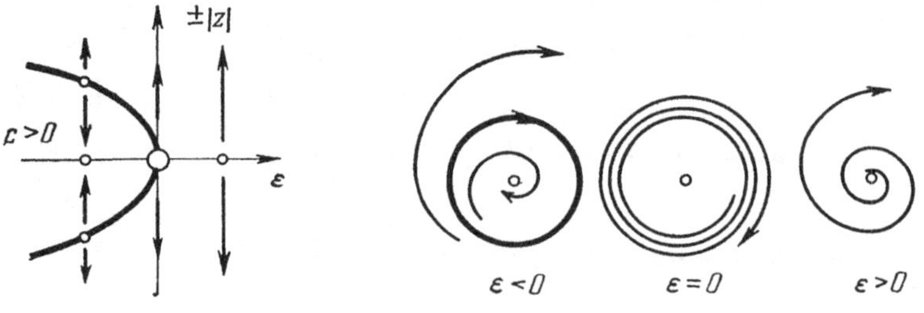

Fig. 5.13.

always present, throw the system out of the neighbourhood of the equilibrium position, and it immediately jumps over to some other regime (for example, to some distant equilibrium position, or limit cycle, or to some more complicated attracting set). Thus, under a continuous change of the parameter the regime of motion changes by a jump, strongly.

In the case $c < 0$ the amplitude of the growing auto-oscillation depends at least continuously (though not smoothly) on the parameter (singularity of the square root type); in this sense the regime of motion changes weakly.

In studying the equation (6) we have made essential use of the "versal" point of view: if instead of a neighbourhood of the origin in the (z, ε)-space we had considered a neighbourhood in the z-space for fixed ε, then we should have missed the limit cycles. This agrees with the principle that we must study degeneracies of codimension k in a k-dimensional family: our case (6) of codimension 1 is contained in a one-parameter family, which is in fact a topologically versal deformation for a general plane field with a pair of purely imaginary eigenvalues.

In a space of an arbitrary number n of dimensions, a topologically versal deformation of a degenerate singular point of general type with a pair of purely imaginary eigenvalues is obtained by a simple process of building on to (6):

$$(7)\qquad \begin{cases} \dot{z} = z(i + \varepsilon \pm z\bar{z}), & z \in \mathbf{C}^1, \ \varepsilon \in \mathbf{R}, \\ \dot{u} = -u, & u \in \mathbf{R}^{n_-}, \\ \dot{v} = v, & v \in \mathbf{R}^{n_+}, \ n_- + n_+ + 2 = n. \end{cases}$$

After what has been said above the study of the topologically normal form (7) does not present any difficulties. We note that there are no singularities on the curve of singular points of the degeneracy in question. We note further that the map h that takes a family in general position in a neighbourhood a degeneracy of the type in question into the form (7) is not only a homeomorphism, but also a combinatorial (P.L.) equivalence, and even has many of the attributes of smooth maps. In particular, a large part of the invariant manifolds obtained on multiplying the equilibrium position and the cycle of the system (6) by the saddle point ($u = 0, v = 0$) is smooth not only for the normal form (7), but also for the non-smooth map of the original system that takes it into the form (7).

The degenerate case considered here has been investigated by many authors (in particular, E. Hopf [38], Neimark [39], Brushlinskaya [40]); however, the final theorem on the topological versality of the family (7) (and with it the theorem of § 5.5) has only been proved recently in the article of Shoshitaishvili (see [37]).

5.7. Applications to the theory of hydrodynamical stability. The phenonema analyzed above often occur in diverse concrete situations: mechanical, physical, chemical and biological, economics systems lose their stability at every step. Here we consider as an example one special problem of this kind – the question of the loss of stability of a stationary current of an incompressible viscous fluid.

Let D be a domain filled with a fluid, and v the velocity field of the fluid. The motion is described by the Navier–Stokes equations

$$(8)\qquad \frac{\partial v}{\partial t} + (v\nabla, v) = \nu \triangle v - \operatorname{grad} p + f, \ \operatorname{div} v = 0,$$

where the coefficient ν denotes the viscosity, f is a field of non-potential mass forces; the pressure p is determined by the condition of incompressibility.

On the boundary of D we impose, say, the adhesion condition ($v|_{\partial D} = 0$).

We assume that the initial velocity field determines the whole subsequent motion, so that the equations (8) determine a dynamical system in the infinite-dimensional space of source-free vector fields vanishing on the boundary of D.

In fact, this has only been proved for the two-dimensional case. A wide literature is devoted to the existence, uniqueness, and properties of solutions of the equations (8), however, the basic problems remain open. See the articles of Gyunter [41], Ladyzhenskaya [42], [44], Ebin and Marsden [43].

Consider, for example, the Poiseuille flow (with parabolic velocity profile; Fig. 5.14) in a plane channel or a cylindrical tube. The Poiseuille flow is a stationary point of our dynamical system in the function space for any value of the viscosity ν. This position of equilibrium is stable for sufficiently large viscosity; however, it loses its stability as the viscosity drops. We can investigate what happens in this process using the theorems of § 5.5 and 5.6.

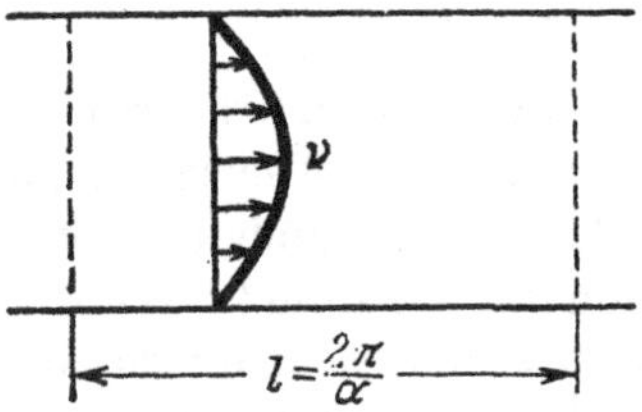

Fig. 5.14.

Of course, we must take special precautions owing to the fact that the problem is infinite-dimensional. This is not very dangerous because viscosity quickly dampens the higher harmonics, so that for any non-zero value ν of the viscosity the system, in fact, reduces to a finite-dimensional one. There is another difficulty in that we cannot be certain that our system is in general position: this has to be checked by computations. It seems natural that the system (8) should be in general position in a domain "of general type" and for general mass forces f, however, the Poiseuille flow is very special, for example, in that it has a large symmetry group.

We restrict ourselves to perturbations whose velocity field repeats itself periodically along the stream, with wave length l. To normalize the velocity of the basic current we change the external forces proportionally to the viscosity so that the outflow Q of the fluid is constant ($f = \text{const} \cdot Q\nu$). In that case we obtain a two-parameter system with the parameters l and ν. One usually takes as parameters the reciprocals $\alpha = 2\pi/l$ (the wave number), and $\mathrm{R} = \text{const} \cdot Q/\nu$ (the Reynolds number). Thus, a decrease in the viscosity, which produces instability, corresponds to an increase in the Reynolds number.

Numerical computations (which in practice cannot be performed without a computer) show that as the Reynolds number grows past a certain critical value $\mathrm{R}_0 = \mathrm{R}_0(\alpha)$, a pair of complex eigenvalues crosses the imaginary axis from the stable half-plane to the unstable one (see, for example, the books of Lin' [45] and of Betchov and Kriminale [46]). Hence we run here into that situation of loss of stability (§ 5.6) in which a limit cycle is born or dies.

The sign of the coefficient c in (6), which determines whether the perturbation is weak or strong, has been computed by Andreichikov and Yudovich [47]. To describe the result it is convenient to draw the stability boundary in the (α, R)-plane. It turns out that it has the "tongue" shape shown in Fig. 5.15; the extreme left-hand tip of the tongue is especially important: its R-coordinate corresponds to the first loss of stability, and its α-coordinate determines the most dangerous wavelength for instability.

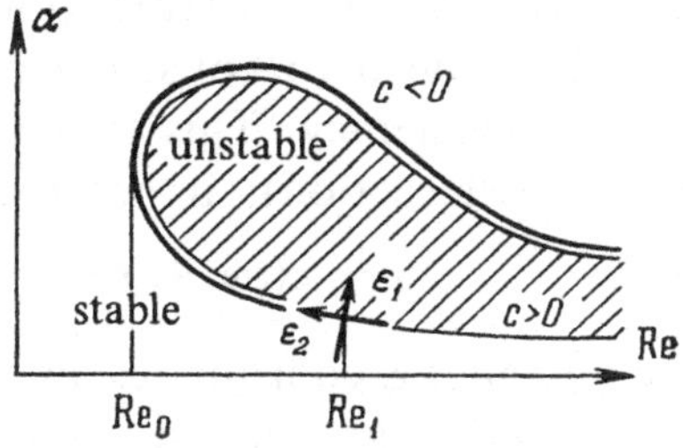

Fig. 5.15.

It turns out (see [47]) that for the whole left-hand and upper part of the tongue-shaped stability boundary the coefficient c is positive, so that *strong* perturbation takes place. Consequently, even before the Reynolds number passes through the critical value R_0, some oscillatory regime[1] arises somewhere in the phase space, away from the stationary point (that is, from the Poiseuille flow), and any small perturbation, as the Reynolds number approaches R_0, throws the system into it. This new regime can be a stable stationary point (in hydrodynamical terms – a stationary current other than the Poiseuille flow), or a limit cycle (a periodic current), but it can also have a more complicated structure; for example, it can turn out to be a conditionally periodic motion on a torus. Furthermore, the regime arising from a strong perturbation can be a Y-system, that is, an attracting set having very irregular, unstable trajectories on it. The spectrum of the corresponding dynamical system can turn out to be discontinuous even despite the fact that the number of degrees of freedom is finite (that is, the attracting set is finite-dimensional). Experimentalists would call such a regime of currents turbulent.

We remark, however, that the determination of the regime into which our concrete system actually goes, as stability of the Poiseuille flow is lost, lies, according to the specialists, on the boundary of what is possible with present-day computers.

In this situation the qualitative predictions that can be made without any calculations, merely on the basis of the above account of the general theory of bifurcations probably must not be neglected.

In the problem considered there are two parameters, α and R. Hence, apart from singularities of codimension 1, those of codimension 2 may also turn up. Let us turn our attention to one of these, namely that connected with the change of sign of c. The computations of Andreichikov and Yudovich [47] show that for a sufficiently large Reynolds number R strong

[1] **Another possibility in systems of arbitrary form is going off to infinity; in our case apparently this does not happen, because at infinity the phase velocity is directed back to the origin in view of the damping action of viscosity.**

perturbation on the lower side of the loss of stability tongue of Fig. 5.15 gives way to weak perturbation. To understand what happens at this instant we need to construct a two-parameter versal family for such a double degeneracy. This family is easy to construct (how, we shall discuss in the next section): it has the form

$$\dot{z} = z(i\omega + \varepsilon_1 + \varepsilon_2 z\bar{z} + c_2 z^2 \bar{z}^2), \quad z = x + iy \in \mathbf{R}^2,$$

(the remaining coordinates in the phase space correspond to stable eigenvalues, and I do not write them down). The meaning of the parameters ε_1 and ε_2 is clear from Fig. 5.15; the character of the change-over at $\varepsilon_1 = \varepsilon_2 = 0$ is determined by the sign of c_2.

If we set $\varphi = z\bar{z}$ as above, we obtain for φ the equation

$$\dot{\varphi} = 2\varphi(\varepsilon_1 + \varepsilon_2\varphi + c_2\varphi^2), \quad \varphi \geqslant 0.$$

Depending on the signs of ε and c we have the following possibilities:

1. Let $c_2 < 0$, $\varepsilon_2 < 0$. Then as ε_1 passes from negative to positive values, the system is weakly perturbed into a stable periodic regime of oscillations (Fig. 5.16).

2. Let $c_2 < 0$, $\varepsilon_2 > 0$. Then as ε_1 passes from negative to positive values (Fig. 5.17), the system strongly goes into a stable periodic regime of auto-oscillations, which comes into being even before the loss of stability by the position of equilibrium (together with an unstable oscillatory regime that settles at the position of equilibrium at the moment of loss of stability).

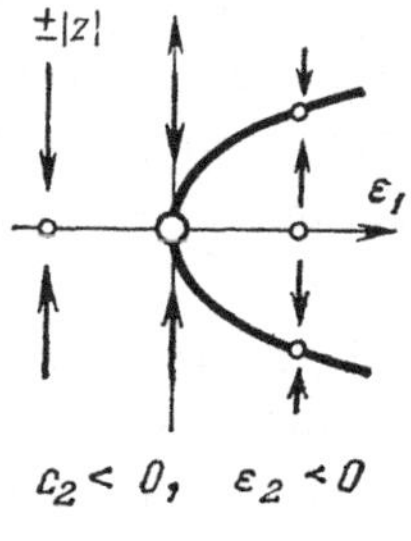

Fig. 5.16.

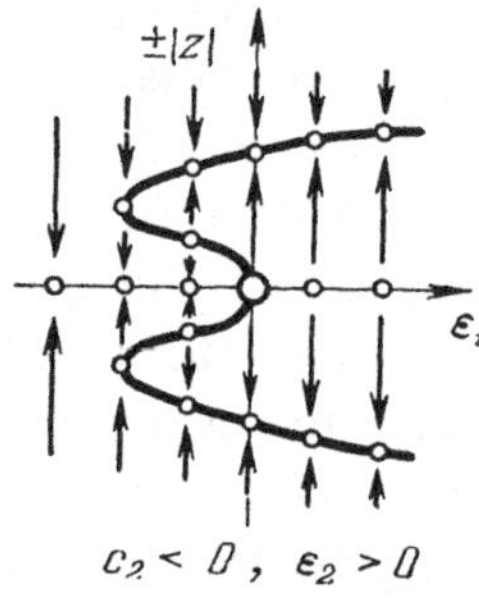

Fig. 5.17.

We were able to study the indicated stable limit cycle near the point at which strong perturbation gives way to weak since here it is near to the equilibrium position. However, an analytic continuation of this cycle can exist (far away from the equilibrium position) also for other values of the parameters (α, R); as we can see, we can look for it as an analytic continuation of the unstable cycle which falls into the equilibrium position when strong loss of stability occurs. This stable cycle is one of the candidates for the role of the regime established as stability is lost.

3. $c_2 > 0$, $\varepsilon_2 < 0$ (Fig. 5.18). There is weak loss of stability, but the limit cycle that appears soon dies off, fusing with an unstable limit cycle which arrives from afar; after this the system is strongly perturbed into a

new regime.

4. $c_2 > 0$, $\varepsilon_2 > 0$ (Fig. 5.19). Ordinary strong perturbation.

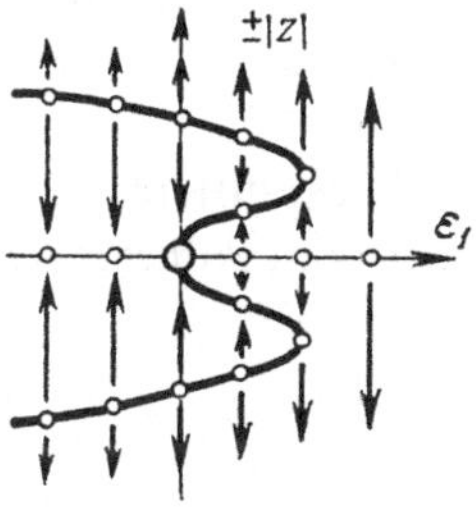

Fig. 5.18.

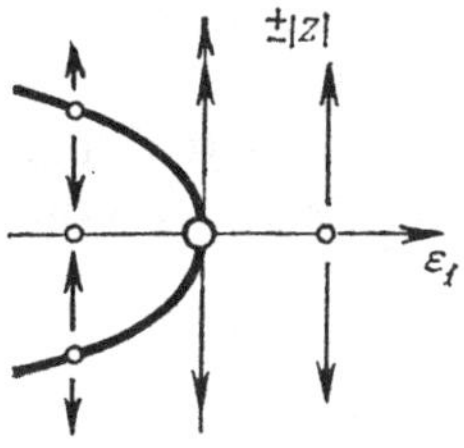

Fig. 5.19.

Thus, whatever the sign of c_2, for a suitable sign of ε_2 our analysis allows us to establish a phenomenon that is qualitatively new by comparison with those of §5.6: for $c_2 < 0$ we find a clearly established regime after strong perturbation; and for $c_2 > 0$ we find a short-lived weak perturbation of the regime. To find out which of the two cases ($c_2 < 0$ or $c_2 > 0$) actually occurs, we have to carry out vast calculations.

We remark in conclusion that in the theory of hydrodynamical stability various singularities of stability boundaries and decrement diagrams occur (see, for example, [46]), so that here we can find applications of the results of §4. To apply the general theory of bifurcations to the theory of hydrodynamical stability it would be important to study the case of general position in problems with various symmetry groups, because in many hydrodynamical problems the domain D of flow has one group of symmetries or another (for example, the group of translations of the channel in the Poiseuille flow problem; representations of this group take part in investigations in the form of the parameter α).

The behaviour of a fluid after the loss of stability of a stationary flow is discussed in many articles (see, for example, the textbook of Landau and Lifshits [48] and the recent articles of Ruelle and Takens [49], [50]). Here one usually assumes a weak regime of auto-oscillations, and the loss of stability is studied by means of a limit cycle. Landau proposed that conditionally periodic regimes with greater and greater frequencies would then arise; this can undoubtedly be explained by the fact that other dynamical systems were not known to him. Recently Ruelle and Takens [49] have constructed examples of the loss of stability of a cycle in which a more complicated regime arises; however, their example has an exotic character, because it corresponds to a metrically very thin (though open) part of the parameter space.

It should be mentioned that the results of the papers quoted can only be applied if the loss in stability originates in a weak regime, whereas the regime of stability loss of the Poiseuille flow turns out to be strong, at least in the plane case.

§6. The construction of versal deformations

Here we describe the arguments by means of which the versal families investigated in §5 were found.

There are two steps in the construction of versal deformations. Firstly, the algebraic investigation of normal forms of individual objects (for example, the reduction to the Jordan normal form in the problem of deformations of matrices) and secondly, the execution of a (trans)versal deformation, the choice of which is dictated largely by aesthetic considerations.

As applied to ordinary differential equations, the first part – the construction of normal forms of vector fields – leads essentially to Poincaré's dissertation [51], although Poincaré used a different terminology. Poincaré's theory was developed by Dulac [52] and Birkhoff [53], and was perfected by many authors.

It is interesting to note that specialists in singularities of linear differential equations with variable coefficients do not seem to be familiar with the theory of Poincaré's normal forms. The equation

$$x^n \frac{dy}{dx} = A(x)\, y$$

can be written in the form of a non-linear autonomous system with a singular point at zero:

$$\dot{y} = A(x)\, y, \quad \dot{x} = x^n.$$

The application of Poincaré's method to this system immediately gives a considerable part of the formal (and sometimes also the analytical) results on singular points; see, for example, the book of Wasow [54], who apparently did not know of the works of Poincaré, Dulac, and Birkhoff.

Poincaré's theory as expounded below has a number of variants: instead of equilibrium positions of autonomous systems one can consider fixed points of maps, or also equilibrium positions and periodic solutions of equations with constant or with periodic coefficients. There are also analogous theories for the Hamiltonian case, but we do not concern ourselves with all the variants here. For more details see the books of Birkhoff [53] and Siegel [55].

6.1. Formal normal forms. All possible vector fields (on a fixed manifold) form the infinite-dimensional Lie algebra $\mathfrak{A}$ of the infinite-dimensional group of diffeomorphisms A. The question of normal forms is that of the orbits of the adjoint action of the group on the algebra:

$$(\mathrm{Ad}_g v)(x) = g_* v(g^{-1}x), \quad v \in \mathfrak{A}, \quad g \in A.$$

In the study of the orbits of this action it is helpful to go over to the infinitesimal case. For convenience we suppose in the subsequent account that the field v has a singular point at 0, so that in a fixed coordinate system near the origin

$$v(x) = \Lambda x + \dots$$

As for the diffeomorphism g, we assume that it leaves the point $x = 0$ fixed and has at this point the identity E as its linear part, so that the Taylor expansion of g in the coordinate system in question has the form

$$g = E + h_2 + h_3 + \dots = E + h.$$

We restrict ourselves for the time being to the formal point of view and do not yet worry about the convergence of series: in actual fact we are only interested in the algebraic problem connected with the behaviour of jets of finite order.

A short calculation leads to the usual formula

$$\text{(1)} \qquad \mathrm{Ad}_{E+h}\, v = v + \mathrm{ad}_h\, v + O(h^2),$$

where $\mathrm{ad}_h v = [h, v]$ is the commutator in the Lie algebra of vector fields, that is, the Poisson bracket.

In coordinates the components of the Poisson bracket of two vector fields a and b is expressed in terms of the components of the fields by the formula

$$\text{(2)} \qquad [a, b]_i = \left(\frac{\partial a}{\partial x} b - \frac{\partial b}{\partial x} a\right)_i = \sum_{j=1}^{n} \frac{\partial a_i}{\partial x_j} b_j - \frac{\partial b_i}{\partial x_j} a_j.$$

Note that if a and b are homogeneous vector polynomials of degree k and l, then $[a, b]$ is a homogeneous vector polynomial of degree $k + l - 1$.

Hence if v is a linear vector field and h is homogeneous, then $[h, v]$ is a homogeneous vector polynomial of the same degree as h.

Now let us study the orbit of the linear field $v(x) = \Lambda x$, that is, let us see how we can write it in other coordinate systems. We must therefore investigate the conditions for the following equation in h to be soluble:

$$\mathrm{Ad}_{E+h}\, v = v + w,$$

where v is a linear field, and w is a non-linear "perturbation". By the implicit function theorem we must linearize this equation with respect to h; according to (1) we obtain the equation

$$\mathrm{ad}_h\, v = w,$$

which we can also write in the form

$$L_v h = w,$$

where L_v is the operator of commutation with the linear field.

The investigation of the linear operator L_v acting from the space of vector fields with zero linear part into itself is made easier by the fact indicated above: it leaves invariant the space of homogeneous vector polynomials of any degree. Suppose for simplicity that the operator Λ giving our linear vector field $v(x) = \Lambda x$ is diagonal. Let λ_i $(i = 1, \ldots, n)$ be its eigenvalues, and $\mathbf{e}_i$ its eigenvectors.

Computing commutators by (2) we obtain the following conclusion, again discovered by Poincaré:

LEMMA. *If the operator Λ of the field $v(x) = \Lambda x$ is diagonal, then the operator L_v of commutation with v is also diagonal. The eigenvectors of L_v are the vector monomials $x^l \mathbf{e}_i$ (where $x^l = x_1^{l_1} \ldots x_n^{l_n}$, x_i being the coordinates in a basis of eigenvectors of Λ). The eigenvalues of L_v are linear functions in the eigenvalues of Λ, namely*

$$-L_v x^l \mathbf{e}_i = [\lambda_i - (l, \lambda)]\, x^l \mathbf{e}_i, \quad |l| = l_1 + \ldots + l_n \geqslant 2.$$

If Λ is not diagonal, then L_v also has Jordan blocks, but the eigenvalues are all given by exactly the same formula.

It is clear from the lemma that if all the numbers $\alpha = [\lambda_i - (l, \lambda)]$ are non-zero (that is, if none of the eigenvalues λ_i is such a linear combination of the eigenvalues including itself with non-negative integral coefficients whose sum exceeds 1), then L_v is invertible.

Suppose that the condition "all the α are non-zero" is fulfilled. Then we say that there is no resonance between the λ_i, or that the field v is resonance-free. The formal result of Poincaré consists in the following theorem.

THEOREM. *The orbit of a resonance-free linear field $v(x) = \Lambda x$ in each space of jets under the action of the group of diffeomorphisms with the identity as the linear part at zero consists of all jets of fields with the linear part Λx at zero.*

In other words, by a polynomial change of variables any equation whose linear part at zero is resonance-free can be linearized to within terms of arbitrarily high degree. From the proof it is also clear that one can construct a "formal change of variables" which completely linearizes the formal series $v(x) = \Lambda x + \ldots$ at zero.

A proof of Poincaré's theorem is obtained from the lemma by induction on the degree of the jet. To linearize the 2-jet $v_1 + v_2$, where $v_1(x) = \Lambda x$, we set $h_2 = L_{v_1}^{-1} v_2$. To linearize the $(k + 1)$-jet $v_1 + v_2 + \ldots + v_{k+1}$ we linearize first the k-jet, that is, we choose a system of coordinates in which the $(k + 1)$-jet has the form $v_1 + w_{k+1}$. After this we set $h_{k+1} = L_{v_1}^{-1} w_{k+1}$; then the diffeomorphism $(E + h_{k+1})$ takes v_1 into $v_1 + w_{k+1}$, and the inverse diffeomorphism linearizes $v_1 + w_{k+1}$.

In the resonance case when some of the eigenvalues of L_v are zero, an analogous procedure can be applied. This does not, however, allow us to get rid of all the non-linear terms of the expansion of v in a Taylor series, but only the non-resonant ones, that is, those for which the eigenvalues α are non-zero. As a result, instead of a linear formal normal form we can use a formal mapping to bring our system of equations to the form

$$\dot{x}_i = x_i \sum_k c_k^i x^k \quad (i = 1, \ldots, n), \tag{3}$$

where k ranges over the integral vectors satisfying the relation $(\lambda, k) = 0$ (resonance), and such that if we insert x_i into the sum, then we obtain the series

$$\lambda_i x_i + \sum b_l^i x^l \quad (l \geqslant 0, \ \sum l_i \geqslant 2).$$

Thus, the coefficients c_k^i in (3) can only be non-zero if $k_i \geqslant -1$, $k_j \geqslant 0$ (for $i \neq j$), $\Sigma k_s \geqslant 1$.

The proof that we can bring our equation to the form (3) is just like that of the formal linearization in the resonance-free case. The only difference is that the operator L_v is not diagonal in this case, but upper triangular. However, the diagonal elements have the same form $\alpha = \lambda_i - (l, \lambda)$ as in the resonance-free case. Hence the application of L_v^{-1} allows us to annihilate all the non-resonant terms of degree 2, 3, and so on.

6.2. Poincaré and Siegel domains. The step from the formal consideration on the level of jets to the study of the actual reduction of our equation to normal form by means of a smooth change of the variables is not a harmless operation, as Poincaré observed (and before him Euler). Even in the analytical case the resulting series can diverge (for example, because in inverting L_v we have to divide by "small denominators" α). To formulate convergency results we need some auxiliary constructions. We consider the space of matrices, for simplicity even diagonal matrices, with diagonal entries λ_i. This is a complex n-dimensional space $\mathbf{C}^n = \{\lambda = (\lambda_1, \ldots, \lambda_n)\}$. The set of all possible vectors λ_i satisfying a fixed resonance relation $(\lambda, k) = 0$ forms a hyperplane in $\mathbf{C}^n$; we denote it by Γ_k and call it a *resonance hyperplane.*

Changing the vector k with integer components we obtain a countable set of resonance hyperplanes corresponding to the countable number of possible values of k in (3). Let us consider now how the whole set of resonance hyperplanes is situated in the space $\mathbf{C}^n$ of eigenvalues.

LEMMA. *If $n > 2$, then the space $\mathbf{C}^n = \{\lambda\}$ is divided by a many-sided cone into two open parts: in one of these the resonance hyperplanes are everywhere dense (this part is called the Siegel domain), and in the other it is discrete so that a sufficiently small neighbourhood of any point meets only finitely many resonance hyperplanes (this part is called the Poincaré domain).*

The point λ belongs to the Poincaré or the Siegel domain according as to whether the convex hull of the points $\lambda_1, \ldots, \lambda_n$ on the complex plane contains the point 0.

In the case $n = 2$ the closure of the resonance hyperplanes has real dimension 3, that is, codimension 1 in $\mathbf{C}^2$. In this case the only singular points that do not fall into the Poincaré domain are those for which the ratio of the eigenvalues is a real negative number. Saddle points and centres belong to this type.

POINCARÉ'S THEOREM. *If the vector λ of eigenvalues of the linear part of an analytic system of differential equations $\dot{x} = \Lambda x + \ldots$ is resonance-free at zero and belongs to the Poincaré domain, then the system is analytically equivalent to its linear part.*

SIEGEL'S THEOREM. *For almost all (in the sense of Lebesgue measure) choices λ of eigenvalues the system is equivalent to its linear part. There exists an everywhere dense subset of the Siegel domain consisting of resonance-free choices of λ for which there is a system with eigenvalues λ that is analytically not equivalent to its linear part.*

Poincaré's theorem can be generalized to the case of resonance; if λ belongs to the Poincaré domain, then there can only be finitely many resonances in the system. A proof of the convergence of the series reducing the system to normal form was given in this case by Dulac [52]. Bryuno [56] has recently concerned himself with sharpening Siegel's theorem.

Proofs of Poincaré's and Siegel's theorems can be found, for example, in the books of Nemytskii and Stepanov [57] or Siegel [55]).

In accordance with the general statements of §§ 1 and 2, for the study of resonances, as for that of every kind of degeneracy, a reasonable formulation of the problem must contain at least as many parameters as the codimension of the degeneracy.

Thus, suppose that the vector fields defining our differential equations depend on parameters. If we try to apply the above construction at once to the whole family of equations, then we have to divide by the quantities $\alpha = \lambda_i - (\lambda, l)$. Each of the $\alpha = \alpha_{i,l}$ is a function of the parameters of the family; on dividing by the α poles appear. In the Siegel domain the zeros of the various $\alpha_{i,l}$ form an everywhere dense set, so that there is no hope that a finite-dimensional (smooth or analytic) versal family exists here.

The situation with topologically versal families turns out to be more pleasant: the point is that the first few terms of the formal series give the necessary normal form, and the remaining terms do not affect the topology of the phase curves of the equations of the family. The topological normal forms introduced in § 5 were obtained in precisely this way; we now explain in more detail how they were found.

6.3. Example: a pair of roots crossing the imaginary axis. When the pair $\lambda_{1,2} = \pm i\omega$ (Fig. 6.1) crosses the imaginary axis, resonances always appear: $\lambda_j = \lambda_j + m\lambda_1 + m\lambda_2$ for all j, m. If there are no other resonances (which we will suppose), then it is natural to look for a one-dimensional (trans)versal family, since crossing the imaginary axis is a phenomenon of codimension 1. Thus, we are concerned with the reduction of the family

$$\dot{x} = v(x, \varepsilon), \quad x \in \mathbf{R}^n, \ \varepsilon \in \mathbf{R}^1,$$

to the topologically normal form, and we may assume that the equilibrium position is fixed as ε varies (this can always be arranged by a translation of the origin depending smoothly on ε). Suppose then that the linear part of v in the equilibrium position does not have multiple eigenvalues (otherwise the system would be even more degenerate). Then our differential equations take the form

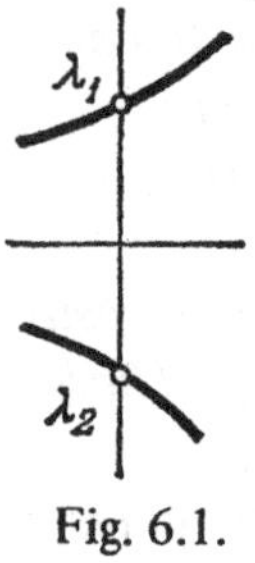

Fig. 6.1.

$$\dot{x}_j = x_j\lambda_j(\varepsilon) + \ldots,$$

where the $\lambda_j(\varepsilon)$ are smooth functions. For $\varepsilon = 0$ the resonance relation $(\lambda(0), k) = 0$ holds. To construct a versal family it is natural to assume transversality to the resonance, that is, to suppose that

$$(\lambda(0), k) = 0, \quad \frac{d}{d\varepsilon}(\lambda(\varepsilon), k)\Big|_{\varepsilon=0} \neq 0.$$

The resonance vector k in our case has the form $k = (1, 1, 0, \ldots, 0)$, since $\lambda_1(0) + \lambda_2(0) = i\omega - i\omega = 0$, and there are no other relations. To annihilate the terms of degree 2 we have to invert L_v on polynomials of degree 2, that is, to divide by $\alpha = \lambda_j - (l, \lambda)$, $|l| = 2$. For our resonance k, with $|l| = 2$, we obtain $\alpha \neq 0$, so that all the terms of degree 2 can be annihilated by a smooth (for example, polynomial) change of variables, depending smoothly on ε .

We can continue in the same way until we run into a resonance term, which we do not annihilate not only for $\varepsilon = 0$ (which is impossible), but also for $\varepsilon \neq 0$ (when it can be annihilated at the cost of a pole in the change of variables, that is, in the final analysis by a restriction of the neighbourhood of the singular point in the x-space).

Thus, a polynomial change of variables depending smoothly (or analytically if we are in the analytical case) on ε allows us to reduce the whole family at once to the form

$$\dot{x}_j = x_j[\lambda_j(\varepsilon) + c_j(\varepsilon)x_1x_2 + d_j(\varepsilon)x_1^2x_2^2 + \ldots, \quad j = 1, \ldots, n, \tag{4}$$

where the dots denote terms of higher degree.

In this way we can always bring the family to its normal form (in which the terms that remain are resonant at $\varepsilon = 0$), to within terms of arbitrarily high degree.

If we start from a point of the Poincaré domain, then this procedure gives an analytic method of reducing the whole family to the polynomial normal form. Brushlinskaya [58] has proved the following result:

THEOREM. *A germ of an analytic (smooth) vector field of Poincaré type has a finite-dimensional analytic (smooth) versal family consisting of polynomial vector fields. A local family of analytic (smooth) vector fields with a singular point O of Poincaré type is analytically (smoothly) equivalent in a neighbourhood of O to a family formed from sufficiently long segments of the Taylor series at O.*

In the Siegel domain, in the presence of resonances, it may not be possible to reduce a family, or even an individual equation to the resonance normal form by an analytic map (for examples, see Euler [59], Siegel [60], Bryuno [56]).

The divergence of the series reducing our system to a normal form does not, however, contradict the *topological* equivalence of a family with its normal form. Thus, for example, we can check that the family (4) obtained above,

$$\dot{x}_j = x_j[\lambda_j(\varepsilon) + c_j(\varepsilon)x_1x_2] + O(|x|^5),$$

after returning to the real space where the family $v(x; \varepsilon)$ was given, is topologically equivalent to the family

$$\dot{x}_j = x_j[\lambda_j(\varepsilon) + c_j(\varepsilon)x_1x_2], \tag{5}$$

(provided that $\frac{d}{d\varepsilon}(\lambda_1+\lambda_2)\Big|_{\varepsilon=0}\neq 0$, $c_1(0)+c_2(0)\neq 0$). The latter family is obviously topologically equivalent to the versal deformation considered in § 5.6 (6); the derivation of this normal form is now complete.

6.4. A discussion of the divergence of the transformations to normal forms. We have seen above that in the Siegel domain difficulties connected with small denominators arise in the reduction to normal form. However, the topological picture can be quite simple, for example, an ordinary saddle has the same topological structure whether the ratio of the eigenvalues is rational or irrational. The same phenomenon also occurs in the Poincaré domain: resonances need not have any effect on the phase portrait.

A natural question is why resonances, which do not appear topologically, can hinder us in the analytic (or even finitely smooth) reduction to normal form. Here it is useful to bear in mind the behaviour of resonances in the theory of perturbations of conditionally periodic motions, for example, in problems of mechanics (see [61]). Consider the differential equation on an n-dimensional torus T^n

$$(6)\qquad \dot{\theta}=\omega+\varepsilon\ldots,\quad (\text{mod } 2\pi)\in T^n \quad \omega\in\mathbf{R}^n,\ \varepsilon\ll 1.$$

The resonance $(\omega, k)=0$ corresponds (at least in the absence of perturbations, that is, when $\varepsilon=0$) to a change in the topological properties of the system: the phase curves are everywhere dense not in the n-dimensional torus, as in the resonance-free case, but in an $(n-1)$-dimensional subtorus. For example, for $n=2$ resonances usually give rise to coarse periodic regimes (stable and unstable limit cycles on the torus). Clearly the existence of such cycles prevents us from reducing the equations to the normal form

$$(7)\qquad \dot{\theta}=\omega,$$

which is the usual one for the resonance-free case.

These arguments lie at the base of Poincaré's proof of the non-existence of first integrals of the three-body problem (see [62], [63]).

One assumes that the influence of resonances on the local problem, which we have discussed above, is of a similar nature, but connected with the change of the topology of the fibration generated by the phase curves in a complex rather than a real domain (see [64]). Such a change, even if it makes absolutely no difference to the real part of phase space necessarily prevents the analytic reduction and can hinder the C^r-smooth reduction.

We note that the system (3) can be brought to the form (6) by the substitution $x=e^{i\theta}$ (real ω correspond to purely imaginary λ). The usual methods of searching for limit cycles for the system (6) lead to the consideration of a first integral $\varphi=e^{i(\theta,k)}$ of the non-perturbed system (7); in the notation of the system (3) this is nothing other than the quantity $\varphi=z^k$, which was used in § 5. The equation of the first approximation for the invariant manifold corresponding to resonance is obtained from relations analogous to (4):

$$\dot{\varphi} = \varphi[(k, \lambda) + (k, c)\varphi + \ldots].$$

We find

$$\varphi \approx -\frac{(k, \lambda(\varepsilon))}{(k, c(\varepsilon))}. \tag{8}$$

For $n = 2$ these arguments can be reinforced by a rigorous proof of the existence of a complex limit cycle which near a resonance has the asymptotic form (8) (such a proof was given by Pyartli; see [65]).

At the instant of resonance, when $(k, \lambda) = 0$, a cycle (a complex not simply-connected phase curve) approaches the complex separatrix of the singular point. A non-contractible path lying on this cycle vanishes under resonance, by fusing with the equilibrium position. A special case is the birth (or death) of a cycle from an equilibrium position under loss of stability (see § 5). In this case $k = (1, 1, 0, \ldots, 0)$, and all the phenomena can be observed in the real domain (see Fig. 5.12). In other cases (even with the same resonance, for example, in the case of a saddle) the topology of the real phase curve need not alter under resonance.

The differences in the topology between the complex phase curves of an equation (or a family) and those of its normal form is one obstruction to the analytic reduction to normal form. Furthermore, if this difference is determined (as it usually is) by a jet of finite order, then it is an obstruction not only to the analytic reduction to normal form but also to the finitely-smooth reduction. For example, the divergence of the reducing series in the case when the ratio of the eigenvalues can be well approximated by rational numbers can be explained by the existence in any neighbourhood of a stationary point of complex limit cycles coming from nearby resonances of high orders: the system in normal form does not have such cycles, so that the transformation to normal form must necessarily diverge. See Pyartli's article [65].

The investigation of the questions touched upon here is far from complete. The existing divergency proofs (Siegel [60] and Bryuno [56]) are based on an estimate of the growth of coefficients and do not explain the essence of the matter (in the same sense that such an estimate of the coefficients of the series for arctan z does not explain the divergence of this series for $|z| > 1$, although it proves the divergence). We should also mention that the completeness of the study of divergence in [56] is illusory, because it is not at all proved that for the normal forms in question divergence is the rule and convergence the exception. In fact, in [56] only very special examples are given of divergence for chosen normal forms. The divergence of the reduction to these normal forms for a system "in general position" (in a class of systems with a fixed normal form) is plausible, but is not proved.

6.5. The reduction theorem. Apart from the construction of the formal theory of Poincaré's normal forms another general theorem was used in the derivation of the versal families of § 5, which allows us not to worry about the eigenvalues with non-zero real part. This theorem consists in the following.

Suppose that we are given a family of equations

(9) $$\dot{x} = v(x, \varepsilon), \quad x \in \mathbf{R}^n, \ \varepsilon \in \mathbf{R}^l.$$

We assume that for $\varepsilon = 0$ the system has an equilibrium position at the origin and that the operator $A = \frac{\partial v}{\partial x}(0, 0)$ of the linear part of the field has k eigenvalues on the imaginary axis (zero and conjugate pairs being counted). Let n_- and n_+ be the number of eigenvalues with negative or positive real part, so that $k + n_- + n_+ = n$.

Shoshitaishvili has proved the following reduction theorem [37]:

THEOREM. *The local family (9) is topologically equivalent to a suspension of some smooth local family with a k-dimensional phase space.*

In other words, the local family (9) is topologically equivalent to a family of the form

(10) $$\begin{cases} \dot{p} = w(p, \varepsilon), & p \in \mathbf{R}^k, \ \varepsilon \in \mathbf{R}^l, \\ \dot{q} = -q, & q \in \mathbf{R}^{n-}, \\ \dot{r} = r, & r \in \mathbf{R}^{n+}. \end{cases}$$

The proof is based on the construction of a k-dimensional invariant submanifold in each system (9) such that the situation in any $(n_- + n_+)$-dimensional transversal direction is "hyperbolic". We then construct, as usual in hyperbolic situations, invariant fibrations under which the whole phase space projects onto the k-dimensional submanifold. There are five such fibrations in the present case, rather than two as in the ordinary hyperbolic theory (see the book of Anosov [66] or Smale's survey [67]): stretching, contraction, neutral, neutral-stretching or neutral-contracting. These fibrations were also constructed, independently of Shoshitaishvili, by Tikhonova [68], and by Hirsch, Pugh and Shub [69].

The reduction theorem gives useful results not only in the theory of bifurcations for which it was devised, but also in the study of individual equations. Indeed, it follows from this theorem that the topological type of a singular point in $\mathbf{R}^n$ is determined by the integers n_-, n_+ and by the behaviour of the phase curves on some invariant submanifold whose dimension k is equal to the number of eigenvalues on the imaginary axis. Thus, the study of the topological type of a singular point in $\mathbf{R}^n$ reduces to the analogous problem in a space of, as a rule, much smaller dimension k. For a special case of the problem of topological classification, the stability problem (either in the sense of Lyapunov or asymptotic), this reduction was introduced earlier by Pliss [70].

We also note that actually Shoshitaishvili has proved a little more than what was stated above. In particular, it follows from his results that if the system (10) is in "general position" in some sense, then the reduced system

(11) $$\dot{p} = w(p, \varepsilon), \qquad p \in \mathbf{R}^k$$

is in general position. In particular, in many cases one can deduce from the topological versality of the family (11) that of the suspended system (10). This is just how the versal deformations (5), (7) in §5 were obtained.

6.6. Cases of codimension 2. Degenerate singular points of codimension 1 were analysed in § 5: they correspond to a single eigenvalue zero or to a pair of purely imaginary ones, and we have given and studied topologically versal deformations for them.

The cases of codimension 2 have so far not been fully analysed. There are five such cases:

1. A single zero root with additional degeneracy. An example is the following:

$$\dot{x} = \pm x^3 + \varepsilon_1 x + \varepsilon_2, \quad x \in \mathbf{R}$$

(Fig. 6.2). It is easy to check that the above deformation is (topologically) versal; in the higher-dimensional case a versal deformation can be obtained by suspension of a saddle (see [37]).

The bifurcation diagram (for the case $+x^3$) is illustrated on the left of Fig. 6.2. The semi-cubical parabola divides the $(\varepsilon_1, \varepsilon_2)$-plane into two parts. In the smaller part the system has three equilibrium positions near $x = 0$, and in the greater part one only. The change-over in the phase portrait as the parameters move around the point $\varepsilon_1 = \varepsilon_2 = 0$ in the small circle of

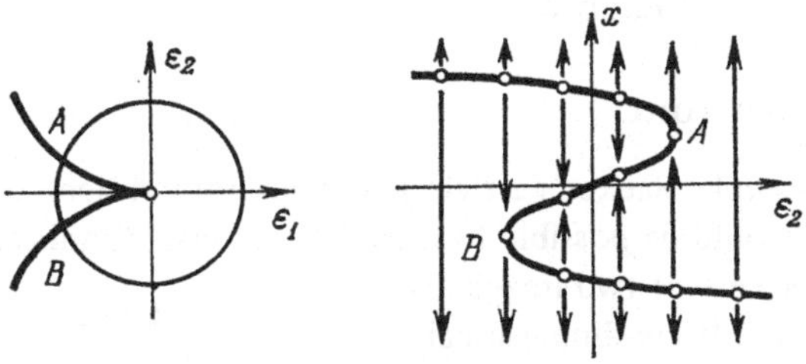

Fig. 6.2.

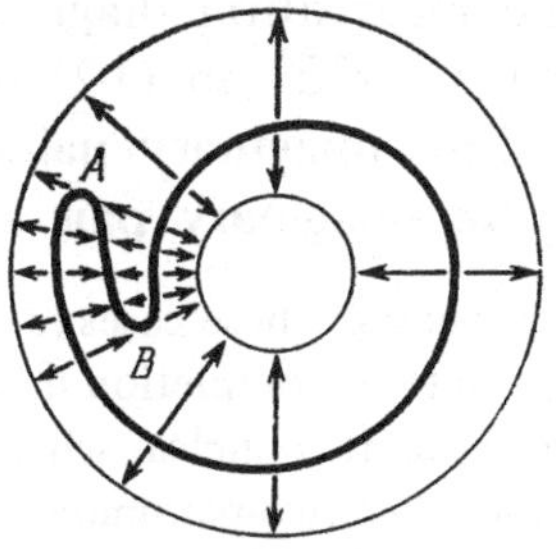

Fig. 6.3.

Fig. 6.2 is shown in Fig. 6.3. The product of this circle with the (one-dimensional) phase space is an annulus on which the equilibrium positions form a smooth curve; the behaviour of the vector field is clear from Fig. 6.3.

2. An imaginary pair with additional degeneracy. An example is the following:

$$\dot{z} = z[i\omega + \varepsilon_1 + \varepsilon_2 z\bar{z} \pm z^2\bar{z}^2], \; z = x + iy \in \mathbf{R}^2. \tag{12}$$

We have already met this deformation in § 5.7. The bifurcation diagram consists of the line $\varepsilon_1 = 0$ and half of a parabola touching it at zero; it is shown in Fig. 6.4 for the case when we have $+z^2\bar{z}^2$ in (12).

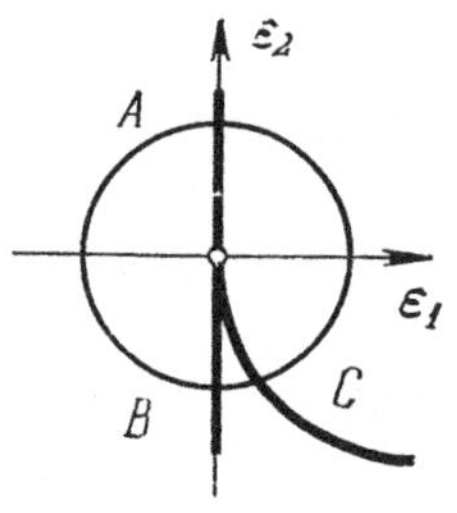

Fig. 6.4.

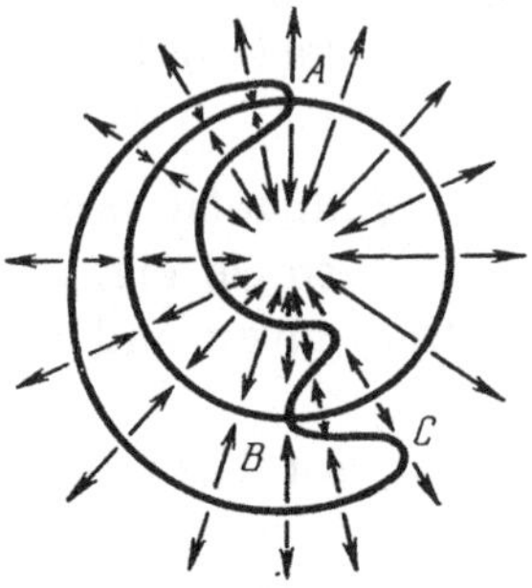

Fig. 6.5.

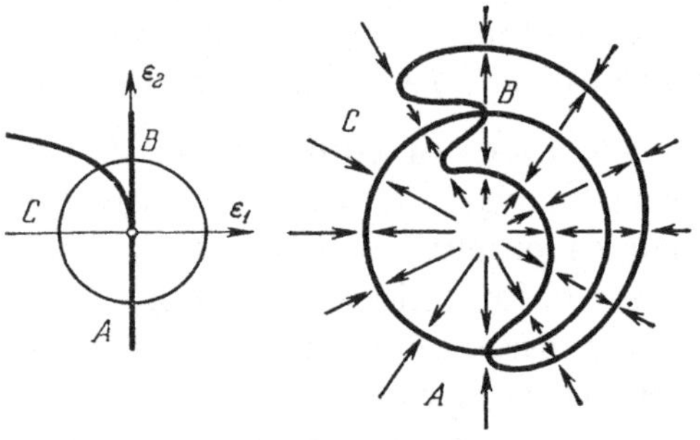

Fig. 6.6.

The change-over in the phase diagram as we move in a small circle around 0, is shown in Fig. 6.5. The annulus in the figure is the product of the circle on the parameter plane with the line on which $\pm |z|$ is plotted. The circle in this figure corresponds to the equilibrium position

$$z = 0,$$

and each limit cycle is represented by the two points of intersection of the radius with the curve

$$\varepsilon_1 + \varepsilon_2 |z|^2 + |z|^4 = 0.$$

The bifurcation diagram and the family over the circle for the case when we have $-z^2\bar{z}^2$ in (12) is shown in Fig. 6.6.

3. Two imaginary pairs.

4. An imaginary pair and a zero root.

The study of these cases (see, for example, Sacker's article [71]) has not yet been carried to that state of completion in which it would be possible to describe the versal families; furthermore, it is unclear whether in the case of two imaginary pairs there is a two-parameter (or at least a finite-dimensional) topologically versal family (even under the assumption of normal incommensurability of the ratios of the frequencies as they cross simultaneously from one half-plane to the other).

Finally, there remains one last case of co-dimension 2:

5. Two zero roots. An example is the family of equations on the plane

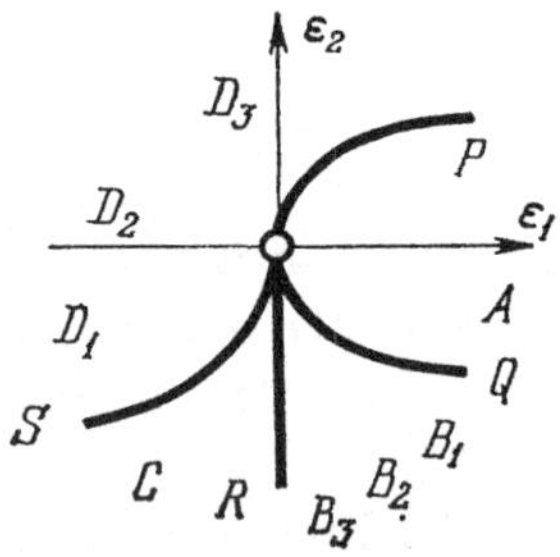

Fig. 6.7.

$$(13) \qquad \begin{cases} \dot{x}_1 = x_2, \\ \dot{x}_2 = \varepsilon_1 + \varepsilon_2 x_1 + x_1^2 \pm x_1 x_2 \end{cases}$$

with parameters ε_1 and ε_2. The bifurcation diagram divides the $(\varepsilon_1, \varepsilon_2)$-plane into the four parts denoted A, B, C, and D in Fig. 6.7, which corresponds to the choice $+x_1x_2$ in (13) (the case with $-x_1x_2$ is just the same if we reverse the direction of time, and we do not consider it).

The phase portraits corresponding to each of the four parts of the $(\varepsilon_1, \varepsilon_2)$-plane are shown in Fig. 6.8. The branches P, Q, R, and S of the

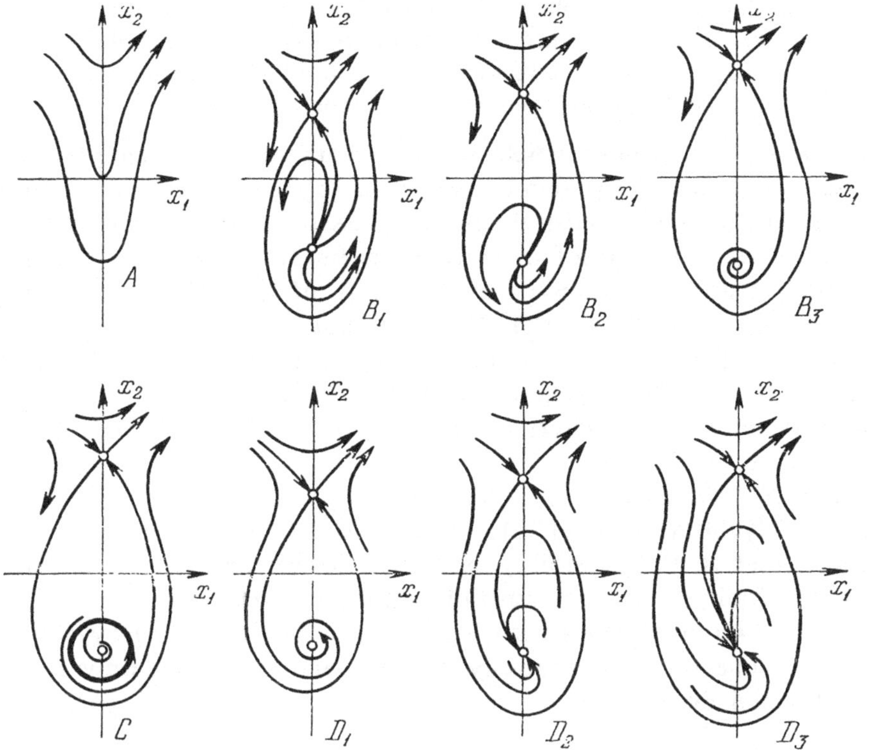

Fig. 6.8.

bifurcation diagram correspond to the systems of codimension 1 illustrated in Fig. 6.9.

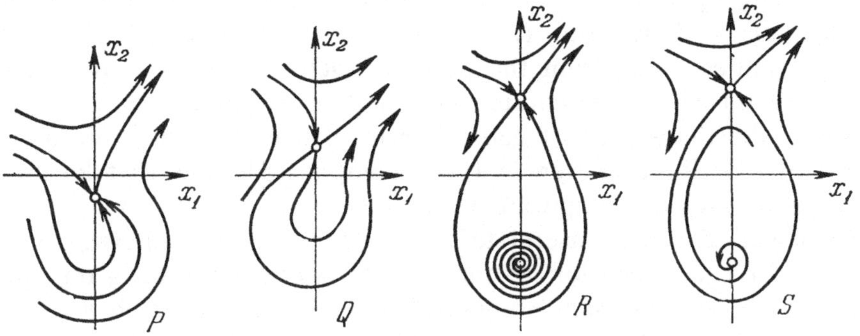

Fig. 6.9.

Finally the singular point $\varepsilon_1 = \varepsilon_2 = 0$ itself corresponds to the phase portrait shown in Fig. 6.10.

We note that the bifurcation on the S branch – the birth of a cycle from a node of a separatrix – does not occur in our classification of singularities of codimension 1 since it is a global rather than a local phenomenon (near the singular point). So we see that as the number of parameters of a family grows, in the local study of the bifurcations of singular points global bifurcations of smaller codimensions begin to play a role. It then follows that with a sufficient number of parameters we run into a local problem with the same difficulties that arise in the global problem on vector fields on a manifold from the fact (discovered by Smale [72]) that the structurally stable systems are non-dense.

Bogdanov, from whose article we have borrowed the example we have analysed, constructed the family (13) as a topologically versal deformation for a system in general position in the class of plane systems with two eigenvalues zero. The proof of versality is far from simple; the principal difficulty consists in the study of the limit cycles: it has to be shown that there are not more than one both for the system (13) and for any member of a two-parameter family in general position in which for some value of the parameters a singular point with two eigenvalues zero appears.

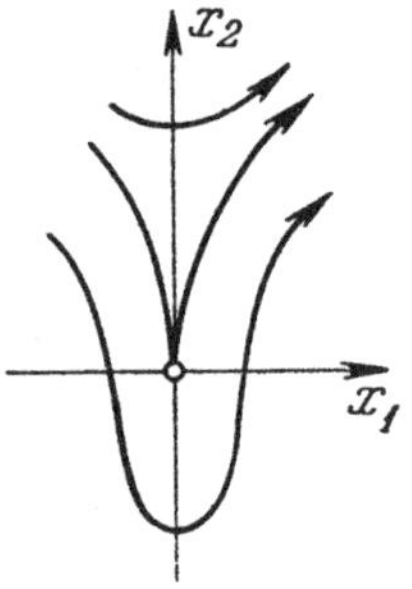

Fig. 6.10.

The system (13) has been studied by many authors, especially Rychkov ([73]–[75]), but a proof of its versality (that is, the possibility of putting every two-parameter family in general position with a singular point with two eigenvalues zero locally into the form (13) by a homeomorphism depending continuously on the parameters) has only recently been obtained by Bogdanov.

§ 7. The classification of singular points

In this section we abandon our "universal" point of view and consider not families, but individual systems of differential equations in a neighbourhood of a singular point of the vector field, and we allow degeneracies of arbitrarily large codimension. The study of such complicated singularities has a very limited value from the point of view of §§ 1–6, since complicated degeneracies have large codimension and occur rarely.

However, the knowledge of general, principal features of arbitrary singularities is of interest even in those complicated cases that lie outside the bounds of present computing techniques.

In particular, it can be useful to know what sort of pathologies can occur in cases of large codimension if only not to waste our efforts on non-essential things. Among such non-essential objects upon which much effort has been wasted are, for example, algebraic criteria for stability (in the sense of Lyapunov or asymptotic), and also algebraic criteria in the problem of the spiral centre (when there are eigenvalues zero).

To show what our principal objects of discussion are we first consider a very simple problem that can be easily studied in complete detail.

7.1. Singular points of functions on the line. Let $f(x)$, $x \in \mathbf{R}$, be a smooth real function in a neighbourhood of $x = 0$. If $x = 0$ is not a critical point, then the function is smoothly equivalent in this neighbourhood to a linear function ($y = x + c$). What happens at a critical point is also well known: if $f'(0) = 0$, then the behaviour of f is determined by the sign of $f''(0)$, and so on.

Consider now for definiteness the problem of finding conditions for f to have a minimum at 0. The answer can be presented in the following way: the space J^k of k-jets of functions at 0 splits up into three parts,

$$J^k = \mathrm{I} \cup \mathrm{II} \cup \mathrm{III};$$

where I consists of the jets that guarantee a minimum,
II consists of the jets guaranteeing the absence of a minimum, and
III consists of the jets from which we cannot decide whether there is a minimum or not.

Jets of types I and II are said to be *sufficient*, and those of type III to be *indeterminate.*

The sets I, II, and III in our problem satisfy the following two conditions.

1) SEMI-ALGEBRAICITY. Each of the sets I, II, and III is a semi-algebraic submanifold of the space of jets J^k.

A *semi-algebraic set* of $\mathbf{R}^N$ is defined as a finite union of subsets each of which is given by a finite system of polynomial equations and inequalities ($P_i = 0, Q_j > 0, R_k \geqslant 0$). If the inequalities are not needed, then the set is said to be algebraic. The following theorem is a useful property of semi-algebraic sets (for a proof, see [76], [77]).

THE TARSKII–SEIDENBERG PRINCIPLE. *The image of a semi-algebraic set under a polynomial map is semi-algebraic.*

A formally weaker, but equivalent statement is the following proposition:

The projection of a semi-algebraic set onto a subspace is semi-algebraic.

We remark that even the projection of an algebraic set need not be algebraic (for example, the projection of a sphere onto a plane).

2. ALMOST FINITE DETERMINACY. As $k \to \infty$ the codimension of the set $\mathrm{III} \subset J^k$ of indeterminate jets tends to infinity.

In other words, the indeterminate jets of J^k are determined by a number of conditions that increases with k. As a result it turns out that the set of functions for which the question whether 0 is a local minimum is not decidable by any number of terms of the Taylor series is very slim: it has infinite codimension in the function space.

7.2. Other examples. The analogous problem for functions of several variables does not have such a simple algorithm: if the second differential is degenerate, we have to appeal to the following ones, and we arrive at a problem of the same type as that of classifying algebraic curves, surfaces, and so on. All the same, the decomposition $J^k = \mathrm{I} \cup \mathrm{II} \cup \mathrm{III}$ of the space of k-jets of functions on $\mathbf{R}^n$ has the properties of semi-algebraicity and almost finite determinacy, although there is no hope of writing out explicitly the equations and inequalities on the Taylor coefficients when n and k are large. The existence of these equations and inequalities can be deduced from the Tarskii–Seidenberg theorem, whose proof also contains an algorithm to compute these equations and inequalities (generalizing the theory of Sturm).

The next example is the problem of the topological classification of germs of smooth mappings. In 1964 Thom announced a theorem on semi-algebraicity and almost finite determinacy in this problem, but he did not give a complete proof (see [79]). The recent articles of Mather ([80], [27]) and Varchenko ([81], [82]) allow us to hope that Thom's theorem will soon at last be proved. Varchenko has proved an analytic variant of Thom's theorem by the same method by which he previously proved Zariski's conjecture on the topological equivalence of algebraic varieties (see [88], [89]).

Following [20], we call problems in which the two properties of semi-algebraicity and almost finite determinacy are satisfied trivial. Another example of a trivial problem is that of computing the index of a singular point of a vector field. The triviality of this problem was proved by Zakalyukin in [83], where an algorithm is indicated which either computes the index by a finite number of arithmetical and logical operations, or goes on working indefinitely. The latter only happens in cases of infinite codimension (for example, in the analytic situation only if the singularity is non-isolated).

7.3. Singular points of vector fields. We return to the problem of the topological classification of singular points of vector fields. At first sight the situation appears to be as simple as that of a function. Non-degenerate singular points are classified by the number of eigenvalues in the left half-plane. The space of 1-jets splits into finitely many parts corresponding to the number of roots in the left half-plane. Each of these parts is a semi-algebraic set in the space of jets; the polynomial inequalities defining it can even be written out explicitly (the Routh–Hurwitz conditions).

The indeterminate 1-jets form a semi-algebraic submanifold of codimension 1 dividing up the domains corresponding to different numbers of eigenvalues in the left half-plane. In the preceding sections we have discussed a number of examples of investigations into what happens in these degenerate cases as we go over to 2-jets, and so on. Thus, the impression is formed that here too we can go as far as we like, and that only the complexity of the computations and the number of variants prevents us from giving an algebraic classification of cases of arbitrarily large codimension. It turns out, however, that things just don't happen this way, and that we are, in fact, up against a non-trivial problem (see [84]). Furthermore, the condition of semi-algebraicity of the decomposition of the jet space ceases to hold even in such a simple problem as that of distinguishing a centre and a spiral (true, only for eigenvalues zero: the triviality of this problem for non-zero eigenvalues was already established by Poincaré).

The conjecture as to the transcendency of the conditions determining the character of a degenerate spiral was put forward by Bryuno and proved by Il'yashenko in [85]. In particular, it follows from this result that no algebraic algorithm can determine the problem of stability of an equilibrium position (in the sense of Lyapunov or asymptotic), and hence also that of the topological classification of singular points (of sufficiently large codimension).

There still remains some hope for the existence of a non-algebraic algorithm, that is, one for which the condition of almost finite determinacy still holds: the set of germs whose topological type (or whose stability) is not determined by any finite segment of the Taylor series may possibly have infinite codimension. The question whether this is so is very difficult; it must be made more precise by indicating the exact meaning of the word codimension: the sets in the space of k-jets whose codimensions have to be determined are non-algebraic, and set-theoretical difficulties are possible. Thom has conjectured in [87] that the answer to this question is in the negative.

7.4. Structure of the sets of indeterminacy. The question of the almost finite determinacy refers to the behaviour of the sets of indeterminate jets in the space J^k of k-jets as k tends to infinity. Questions on the structure of the sets of indeterminacy for a fixed k lend themselves better to investigation. Let us fix an indeterminate $(k - 1)$-jet of a vector field at 0 and consider the set J of all possible k-jets with the given $(k - 1)$-jet. For the sake of definiteness we consider the problem of asymptotic stability. Then the space J splits up into two (possibly empty) parts: I (the stable k-jets) and II (the unstable k-jets), and the remainder III consists of the indeterminate jets (in the problem of the topological classification there are more parts). A reasonable form of the problem of a stability criterion consists in asking what properties the parts I, II, and the boundary between them have. For example, the transcendence of the boundary points to the non-existence of an algebraic stability criterion. One could ask how complicated the structure of this boundary can be. For example, can it (or either of the open parts I or II) have infinitely many connected components? And could points of part I mix in with points of part II like rational and irrational numbers?

Examples of this kind are not known, but one might fear that this is just the sort of thing that happens in singular cases of sufficiently large codimension in a many-dimensional space.

There is a close connection between the local problem on the behaviour of phase curves near a singular point in $\mathbf{R}^n$ and the global problem on differential equations given by a polynomial system in the projective space $\mathbf{RP}^{n-1}$ of dimension one less. In [84] this connection is used to deduce the transcendence of the stability boundary in the space of jets for a local problem in $\mathbf{R}^3$ from the transcendence of the surface of births of limit cycles in the space of coefficients of polynomial systems on the projective plane. But in the many-dimensional global situation much more complicated phenomena are possible than limit cycles, for example, systems on the torus with intermingled commensurable and incommensurable winding numbers, or the open domains in a function space discovered by Smale [72], which are free from structurally stable systems. All these phenomena can be realized in polynomial systems in a projective space, and each of them can make its contribution to the entanglement of the stability boundary in J.

References

[1] H. Whitney, On singularities of mappings of Euclidean spaces, I, Mappings of the plane into the plane, Ann. of Math. (2) **62** (1955), 374–410.

[2] R. Thom and H. Levine, Singularities of differentiable mappings, Bonn. Math. Schr. **6**, Bonn 1959.
= Singularities of differentiable maps, Mir, Moscow 1968, 164–182.

[3] V. I. Arnol'd, Singularities of smooth mappings, Uspekhi Mat. Nauk **23**:1 (1968), 3–44.
= Russian Math. Surveys **23**:1 (1968), 1–43.

[4] J. Milnor, Morse theory, Ann. of Math. Studies **51**, Univ. Press, Princeton, N.J., 1963. Translation: *Teoriya Morsa*, Mir, Moscow 1965.

[5] J. C. Tougeron, Idéaux de fonctions différentiables, Thèse, Université de Rennes 1967.

[6] J. N. Mather, Stability of C^∞-mappings, III, Finitely determined map-germs, Inst. Hautes Études Sci. Publ. Math. no. 35 (1968), 279–308.
= Matematika **14** (1970), 145–175.

[7] A. M. Samoilenko, On the equivalence of a smooth function to its Taylor polynomial in a neighbourhood of a critical point of finite type, Functsional Anal. i Prilozhen. **2**:4 (1968), 63–69.
= Functional Anal. Appl. **2** (1968), 318–323.

[8] M. Artin, On the solutions of analytic equations, Inventiones Math. **5** (1968), 277–291.

[9] V. P. Palomodov, Remarks on differentiable maps of finite multiplicity, Functsional Anal. i Prilozhen. **6**:2 (1972), 52–61.

[10] N. H. Kuiper, C^1-equivalence of functions near isolated critical points, Sympos. infinite-dimensional topology, Baton Rouge 1967, University Press, Princeton, N.J., 1968.

[11] T. C. Kuo, A complete determination of C^0-sufficiency in $J^r(2, 1)$, Invent. Math. **8** (1969), 226–235.

[12] T. C. Kuo, On C^0-sufficiency of jets of potential functions, Topology **8** (1969), 167–171.

[13] J. Bochnak, Jets suffisants et germes de determination finie, Preprint, Inst. Hautes Études Sci.

[14] F. Takens, A note on sufficiency of jets, Invent. Math. **13** (1971), 225–231.

[15] V. I. Arnol'd, Obyknovennye differentsial'nye uravneniya (Ordinary differential equations), Nauka, Moscow 1971.

[16] D. M. Grobman, Homeomorphism of systems of differential equations, Dokl. Akad. Nauk SSSR **128** (1959), 880–881.

[17] D. M. Grobman, Topological classification of a neighbourhood of a singular point in n-dimensional space, Mat. Sb. **56** (1962), 77–94.

[18] Ph. Hartman, A lemma in the theory of structural stability of differential equations, Proc. Amer. Math. Soc. **11** (1960), 610–620.

[19] Ph. Hartman, On the local linearization of differential equations, Proc. Amer. Math. Soc. **14** (1963), 568–573.

[20] V. I. Arnol'd, On local problems of analysis, Vestnik Moscov Gos. Univ. Ser. Mat. Mekh. 1970, no. 2, 52–55.

[21] S. M. Vishik, On the oblique derivative problem, Vestnik Moscov. Gos. Univ. Ser. Mat. Mekh. 1972, no. 1, 21–28.

[22] Yu. V. Egorov and V. A. Kondrat'ev, On the oblique derivative problem, Mat. Sb. **78** (1969), 148–176.
= Math. USSR Sb. 7 : 1 (1969), 139–169.

[23] M. B. Malyutov, On the problem with oblique derivative in three-dimensional space, Dokl. Akad. Nauk SSSR **172** (1962), 283–286.
= Soviet Math. Dokl. **8** (1967), 87–90.

[24] V. G. Maz'ya, A degenerating problem with an oblique derivative, Uspekhi Mat. Nauk **25** : 2 (1970), 275–276.

[25] H. Whitney, Tangents to analytical varieties, Ann. of Math. (2) **81** (1964), 496–549.

[26] R. Thom, Ensembles et morphismes stratifiés, Bull. Amer. Math. Soc. **75** (1969), 240–284.

[27] J. Mather, Stratifications and mappings, preprint, Harvard, 1971, 1–68.
= Uspekhi Mat. Nauk **27** : 5 (1972), 85–118.

[28] S. Lojasiewicz, Triangulation of semi-analytic sets, Ann. Scuola Norm. Sup. Pisa (3) **18** (1964), 449–474.

[29] S. M. Vishik, Some transversality theorems, Mat. Sbornik **85** (1971), 140–156.

[30] A. Douady, Variétés et espaces mixtes, Familles d'espaces complexes et fondement de la géométrie analytique, Séminaire H. Cartan 1960/1961.

[31] H. Grauert and H. Kerner, Deformationen von Singularitäten komplexer Räume, Math. Ann. **153** (1964), 236–260.

[32] G. N. Tyurina, Local semi-universal plane deformations of isolated singularities of complex spaces, Izv. Akad. Nauk SSSR Ser. Mat. **33** (1969), 1026–1058.
= Math. USSR Izv. 3 : 5 (1969), 967–999.

[33] H. Grauert, Über die Deformationen isolierter Singularitäten analytischer Mengen, Invent. Math. **15** (1972), 171–198.

[34] F. Pham, Remarque sur l'équisingularité universelle, Preprint 1970.

[35] V. I. Arnol'd, On matrices depending on parameters, Uspekhi Mat. Nauk **26** : 2 (1971), 101–114.
= Russian Math. Surveys **26** : 2 (1971), 29–43.

[36] D. M. Galin, On real matrices depending on parameters, Uspekhi Mat. Nauk **27** : 1 (1972), 241–242.

[37] A. N. Shoshitaishvili, On the bifurcation of topological type of a vector field near a singular point, Funktsional. Anal. i Prilozhen. **6** : 2 (1972), 97–98.
= Functional Anal. Appl.

[38] E. Hopf, Abzweigung einer periodischer Lösung von einer stationären Lösung, Sächs. Akad. Wiss. Leipzig Sber. Math. Phys. Kl. **94** (1942) no: 19, 15–25.

[39] Yu. I. Neimark, On some cases of dependence of periodic motions on parameters, Dokl. Akad. Nauk SSSR **129** (1959), 736–739.

[40] N. N. Brushlinskaya, Qualitative integration of one system of differential equations, Dokl. Akad. Nauk SSSR **139** (1961), 9–12.
= Soviet Math. Dokl. **2** (1961),
[41] N. M. Gyunter, On a basic problem of hydrodynamics, Izv. Fiz.-Mat. Inst. Steklov **2** (1927), 1–168.
[42] O. A. Ladyzhenskaya, *Matematicheskie voprocy dinamiki vyazkikh nezhimaemykh zhidkostei* (Mathematical problems of the dynamics of viscous incompressible fluids), Nauka, Moscow 1970.
[43] D. G. Ebin and J. E. Marsden, Groups of diffeomorphisms and the motion of an incompressible fluid, Ann. of Math. (2) **92** (1970), 102–163.
[44] O. A. Ladyzhenskaya, On the solubility in the small of non-stationary problems for incompressible ideal and viscous fluids and for vanishing viscosity, Zap. Nauchn. Sem. LOMI **21** (1971), 65–78.
[45] Ts. Ts. Lin', *Teoriya gydrodynamicheskoi ustoichivosti* (The theory of hydrodynamical stability), Izdat. Inost. Lit., Moscow 1971.
[46] R. Betchov and V. Kriminale, *Voprosi gydrodynamicheskoi ustoichivosti* (Questions of hydrodynamical stability), Mir, Moscow 1971.
[47] I. P. Andreichikov and V. I. Yudovich, On auto-oscillatory regimes branching out from the Poiseuille flow in a plane channel, Dokl. Akad. Nauk SSSR **202** (1972), 791–794.
= Soviet Math. Dokl. **13** (1972),
[48] L. D. Landau and E. M. Lifshits, *Mekhanika sploshnoi sredy,* second ed. Gostekhizdat. Moscow 1953 (§27, The onset of turbulence).
Translation: Mechanics of a continuous medium, Pergamon Press, Oxford 1958.
[49] D. Ruelle and F. Takens, On the nature of turbulence, Comm. Math. Phys. **20** (1971), 167–192.
[50] D. Ruelle and F. Takens, Note concerning our paper "On the nature of turbulence", Comm. Math. Phys. **21** (1972).
[51] H. Poincaré, Sur les propriétés des fonctions définies par les équations aux differéntielles partielles, Thèse, 1879, Oeuvres complètes, vol. I. Gauthier–Villars, Paris 1928.
[52] H. Dulac, Solutions d'un système d'équations différentielles dans le voisinage des valeurs singulières, Bull. Soc. Math. France **40** (1912), 324–383.
[53] G. D. Birkhoff, Dynamical systems, second ed. Amer. Math. Soc. New York 1948.
Translation: Dinamicheskie sistemy, Gostekhizdat, Moscow–Leningrad 1941.
[54] W. R. Wasow, Asymptotic expansions for ordinary differential equations, John Wiley & Sons, Inc. New York, 1966.
Translation: *Asimptoticheskie razlozheniya reshenii obiknovennykh differentsial'nykh uravnenii,* Mir, Moscow, 1968.
[55] C. L. Siegel, Vorlesungen über Himmelsmechanik, Springer–Verlag, Berlin–Heidelberg–Göttingen 1956.
Translation: *Lektsii no nebesnoi mekhanike,* Izdat. Inost. Lit. Moscow 1959.
[56] A. D. Bryuno, The analytic form of differential equations, Trudy Moskov. Mat. Obshch. **25** (1971), 119–262.
= Trans. Moscow Math. Soc.
[57] V. V. Nemytskii and V. V. Stepanov, *Kachestvennaya teoriya obiknovennykh differentsial'nykh uravnenii* (The qualitative theory of ordinary differential equations), Gostekhizdat. Moscow–Leningrad 1947.

[58] N. N. Brushlinskaya, A finiteness theorem for families of vector fields in a neighbourhood of a singular point of Poincaré type, Funktsional. Anal. i Prilozhen. **5**:3 (1971). 10–15.
[59] L. Euler, De seriebus divergentibus, Opera omnia, Ser. 1 **14**:247, 585–617 (see p. 601).
[60] C. L. Siegel, On the integrals of canonical systems, Ann. of Math. (2) **42** (1941), 806–822.
= Matematika **5** (1961), 103–117.
[61] V. I. Arnol'd and A. Avez, Problèmes ergodiques de la mécanique classique, Gauthier–Villars, Paris, 1967.
[62] H. Poincaré, Les méthodes nouvelles de la méchanique céleste, Ch. V., Oeuvres complètes, vol. I., Gauthier–Villars, Paris 1928.
[63] C. L. Siegel, Über die existenz einer Normalform analytischer Hamiltonischer Differentialgleichungen in der Nähe einer Gleichgewichtslösung, Math. Ann. **128** (1954), 144–170.
= Matematika **5** (1961), 129–155.
[64] V. I. Arnol'd, Remarks concerning singularities of finite codimension in complex dynamical systems, Funktsional. Anal. i Prilozhen. **3**:1 (1969), 1–6.
= Functional. Anal. Appl. **3** (1969), 1–3.
[65] A. S. Pyartli, The birth of complex invariant manifolds near a singular point of a vector field which depends on parameters, Funktsional. Anal. i Prilozhen. **6**:4 (1969), 1–6.
= Functional Anal. Appl.
[66] D. V. Anosov, Geodesic flows on closed Riemannian manifolds of negative curvature, Trudy Mat. Inst. Steklov **90** (1967).
= Proc. Math. Inst. Steklov. **90** (1967).
[67] S. Smale, Differentiable dynamical systems, Bull. Amer. Mat. Soc. **73** (1967), 747–817.
= Uspekhi Mat. Nauk **25**:1 (1970), 113–185.
[68] E. A. Tikhonova, Analogy and homeomorphism of a perturbed and an unperturbed system with a block triangular matrix, Differentsial'nye Uravneniya **6** (1970), 1221–1229.
[69] M. W. Hirsch, C. C. Pugh, and M. Shub, Invariant manifolds, Bull. Amer. Math. Soc. **76** (1970), 1015–1019.
[70] V. A. Pliss, A reduction principle in the theory of stability of motion, Izv. Akad. Nauk SSSR Ser. Mat. **28** (1964), 1297–1324.
[71] R. J. Sacker, On invariant surfaces and bifurcation of periodic solutions of ordinary differential equations, New York University, New York 1964.
[72] S. Smale, Structurally stable systems are not dense, Amer. J. Math. **88** (1966), 491–496.
= Matematika **11** (1967), 107–112.
[73] N. N. Bautin, *Povedenie dinamicheskikh sistem vblizi granits oblasti ustoichivosti* (The behaviour of dynamical systems near the boundary of the domain of stability), Gostekhizdat, Moscow 1949.
[74] A. Andronov, E. Leontovich, I. Gordon, and A. Maier, *Teoriya bifurcatsii dinamicheskikh sistem na ploskosti* (The theory of bifurcation of plane dynamical systems), Nauka, Moscow 1967.

[75] G. S. Rychkov, A complete investigation of the number of limit cycles of the equation $(b_{10}x + y)\,dy = \sum_{i+j\leqslant 2} a_{ij}x^i y^j\,dx$. Differentsial'nye Uravneniya **6** (1970), 2193–2199.

[76] A. Seidenberg, A new decision method for elementary algebra, Ann. of Math. (2) **60** (1954), 356–374.

[77] E. A. Gorin, On asymptotic properties of polynomials and algebraic functions of several variables, Uspekhi Mat. Nauk **16**: 1 (1961), 91–118.
= Russian Maths. Surveys **16**: 1 (1961), 93–119.

[78] R. Thom, Local properties of differentiable mappings, Differential Analysis, Bombay 1964, 191–202.
= "Singularities of differentiable maps", Mir, Moscow 1968, 164–182.

[79] R. Thom, The bifurcation subset of a space of maps, Manifolds, Amsterdam 1970, Springer Lecture Notes **197** (1971), 176–193.
= Uspekhi Mat Nauk **27**: 5 (1972), 51–57.

[80] J. Mather, Notes on topological stability, Preprint, Harvard 1970.

[81] A. N. Varchenko, Theorems on the topological equisingularity of families of algebraic varieties and families of polynomial mappings, Izv. Akad. Nauk SSSR Ser. Mat. **36** (1972).

[82] A. N. Varchenko, A theorem on the equisingularity of families of algebraic varieties, Uspekhi Mat. Nauk **26**: 1 (1971), 217–218.

[83] V. M. Zakalyukin, The algebraic computability of the index of a singular point of a vector field, Funktsional. Anal. i Prilozhen. **6**: 1 (1972), 77–78.

[84] V. I. Arnol'd, The algebraic insolubility of the problem of Lyapunov stability and the problem of the topological classification of singular points of an analytic system of differential equations, Funktsional Anal. i Prilozhen **4**: 3 (1970), 1–9.
= Functional Anal. Appl.

[85] Yu. S. Il'yashenko, The algebraic insolubility and the almost algebraic solubility of the spiral centre problem, Funktsional. Anal. i Prilozhen. **6**: 3 (1972), 30–37.

[86] N. N. Brushlinskaya, A versal family of vector fields in the Poincaré domain, Funktsional. Anal. i Prilozhen. **4**: 1 (1970), 6–13.

[87] R. Thom, Problem 5 concerning manifolds, Manifolds, Amsterdam 1970, Springer Lecture Notes **197** (1971), 229.

[88] O. Zariski, Some open questions in the theory of singularities, Bull. Amer. Math. Soc. **77** (1971), 481–491.
= Uspekhi Mat. Nauk **27**: 5 (1972), 59–68.

[89] A. N. Varchenko, On germs of analytic maps whose topological type is determined by a finite jet, Funktsional. Anal. i Prilozhen. **6**: 3 (1972), 63–64.

[90] V. I. Arnol'd, Some forms of functions near degenerate critical points, the Weyl groups A_k, D_k and E_k, and Lagrangian singularities, Funktsional. Anal. i Prilozhen. **6**: 4 (1972).

Translated by M. Reid.

ВЕРСАЛЬНЫЕ СЕМЕЙСТВА И БИФУРКАЦИИ ДИФФЕРЕНЦИАЛЬНЫХ УРАВНЕНИЙ

В.И.АРНОЛЬД

Речь идет о приложениях идей и методов теории особенностей гладких отображений к анализу бифуркаций в локальных задачах теории обыкновенных дифференциальных уравнений. Преимущество общей точки зрения теории особенностей состоит, главным образом, в том, что она указывает правильную постановку вопросов; к их решению сделаны лишь первые шаги, о которых и рассказывалось в лекциях.*/

С простейшими понятиями теории особенностей (особенности общего положения и особенности конечной коразмерности, версальные деформации и бифуркационные диаграммы) можно познакомиться при рассмотрении совсем простых примеров.

Рассмотрим в качестве такого примера функциональное пространство гладких функций на окружности. Будем считать две функции эквивалентными, если они устроены одинаково, то есть переходят одна в другую при подходящих гладко обратимых заменах координат в прообразе и образе. Можно сказать то же иными словами: прямое произведение групп диффеоморфизмов окружности и прямой действует на функциональном пространстве функций на окружности, и эквивалентны друг другу те функции, которые принадлежат одной орбите этого действия.

---

*/ Более подробно лекции изложены в журнале "Успехи математических наук" (т.27, вып.5 (1972)).

Originally published in: Izd. Inst. Akad. Nauk Ukrain. SSR, Kiev, 42-49 (Russian), © Naukova Dumka, Kiev 1972
Reprint by Springer-Verlag Berlin Heidelberg 2014 with kind permission, DOI 10.1007/978-3-642-31031-7_30

Как устроены орбиты (т.е. класс эквивалентных функций) и как они подходят друг к другу в функциональном пространстве? Этот вопрос является основным в теории бифуркаций (вместо пространства функций могло бы стоять пространство векторных полей или иных объектов анализа). Ниже вводятся термины, позволяющие точно сформулировать поставленный вопрос, и с помощью которых на него можно ответить.

Прежде всего, выделим орбиты "общего положения". Они состоят из функций, все критические точки которых невырождены (второй дифференциал не нуль), а все критические значения (значения в критических точках) различны. Эти орбиты открыты и образуют все вместе открытое всюду плотное множество в функциональном пространстве – множество функций общего положения. Такие функции обладают важным свойством структурной устойчивости: при малой деформации функция общего положения остается эквивалентной себе.

Остальные орбиты составлены из вырожденных функций и на первый взгляд представляют гораздо меньший интерес. Действительно, при сколь угодно малой деформации вырожденные особенности могут распадаться на простейшие, невырожденные.

Простейшие вырожденные орбиты имеют очевидное значение границ, разделяющих разные орбиты общего положения. Типов таких простейших орбит 2: в них входят функции с двумя равными критическими значениями и функции с одной кубической критической точкой (второй дифференциал нуль, третий не нуль). Орбиты этих двух типов имеют в функциональном пространстве коразмерность один.

Имеет ли смысл рассматривать орбиты большей коразмерности?

Ответ отрицателен, если нас интересует отдельная, случайно выбранная функция, мало вероятно, что такая функция будет иметь сложные вырожденные критические точки. Однако, представим себе, что мы исследуем целое семейство функций, зависящих от некоторого количества параметров. Каждую отдельную функцию семейства можно малым шевелением превратить в функцию общего положения. Однако, сделать функциями общего положения все функции семейства сразу удается не всегда.

Действительно, рассмотрим, например, однопараметрическое семейство функций. Можем представлять его себе как кривую в функциональном пространстве, и эта кривая может трансверсально (под ненулевым углом) пересекать орбиту коразмерности 1, разделяющую орбиты общего положения (это произойдет, например, если при изменении параметра функция деформируется так, что один локальный максимум обгоняет другой, или так, что локальный максимум и локальный минимум умирают слившись).

В таком случае ясно, что всякая близкая кривая в функциональном пространстве также пересекает указанную орбиту коразмерности один , и притом в близкой точке. Следовательно, всякое семейство, близкое к данному, имеет функцию вырожденного типа при некотором значении параметра, близком к тому значению, при котором имело вырожденную точку исходное семейство.

Точно также в k -параметрических семействах могут встречаться при отдельных значениях параметров неустранимые малым шевелением семейства-вырожденные случаи, принадлежащие орбитам коразмерности не выше k .

Итак, вырождения всех конечных коразмерностей имеет смысл рассматривать, но лишь в том случае, когда нас интересует не индивидуальная функция а семейство; при рассмотрении семейств основной объект изучения не особенности отдельных членов семейства (они могут быть очень простыми, например, невырожденными), а бифуркации, то есть перестройки, происходящие при изменении параметра.

Удобным техническим средством для изучения этих перестроек являются версальные семейства, определенные ниже.

Проведем через точку орбиты конечной коразмерности конечномерное многообразие, трансверсальное к этой орбите, минимально возможной размерности (равной коразмерности орбиты). Иными словами, включим данную функцию в конечнопараметрическое семейство функций так, чтобы выполнялось указанное выше условие трансверсальности. (Например, в случае орбиты функций с двумя равными максимумами условие трансверсальности заключается в том, что производные максимумов по параметру различны)

Теперь легко сообразить, что всякое другое семейство, содержащее данную функцию при одном из значений параметров, эквивалентно семейству, полученному из данного заменой параметра. А именно, всякое семейство $G(x,\mu)$ (μ – параметр), эквивалентно семейству, индуцированному из построенного семейства $F(x,\lambda)$, при гладкой замене параметра $\lambda=\lambda(\mu)$, так что

$$G(x,\mu)=Y(F(X(x,\mu),\lambda(\mu)),\mu),$$

где $X(x)$ и $Y(y)$ – гладко зависящие от μ диффеоморфизмы.

Всякое семейство, обладающее указанным выше свойством универсальности, называется версальной деформацией деформируемой функции $F(x,0)=G(x,0)$.

Если версальная деформация построена, то различные вопросы о всевозможных деформациях данной функции достаточно решать для одной единственной деформации – версальной.

Рассмотрим, например, функцию x^4. Легко сосчитать, что двупараметрическая версальная деформация задается локально формулой

$$x^4+ax^2+bx.$$

Рассмотрим плоскость параметров a, b. Разбиение функционального пространства на орбиты высекает на этой плоскости разбиение на 4 области, разделяемые полукубической параболой и идущим внутри ее острия лучом. В частях внутри параболического крючка лежат значения параметров, при которых функция имеет три критические точки, снаружи - одно; на луче совпадают критические значения, на параболе лежат функции с кубическими критическими точками, в вершине (острие) параболы - функция x^4, принадлежащая орбите коразмерности 2.

Указанное разбиение называется бифуркационной диаграммой. Бифуркационные диаграммы можно строить для любых семейств, а не только для версальных. Бифуркационные диаграммы версальных деформаций имеют особое значение, так как они являются локальными моделями бифуркационных диаграмм любых семейств общего положения. Более того, бифуркационная диаграмма любой деформации получается из версальной при гладком отображении в последнюю.

Аналогичному исследованию можно подвергнуть другие аналитические объекты, где встречаются вырождения. Рассмотрим, например, задачу о поведении фазовых кривых системы обыкновенных дифференциальных уравнений в окрестности особой точки. Случаи общего положения здесь разобраны Пуанкаре и Зигелем; дополнительная неприятность (по сравнению с задачей о функциях) состоит здесь в том, что открытых орбит нет, так как собственные числа линейной части поля в особой точке - инварианты диффеоморфизмов.

Если собственные числа лежат в одной полуплоскости (случай Пуанкаре), то конечномерное версальное семейство существует. А именно, как доказано Н.Н.Брушлинской, гладкое векторное поле, зависящее гладко от параметров,с особой точкой типа Пуанкаре, гладкой и гладко зависящей от параметров заменой переменных приводится к полиномиальному виду - эти полиномы с неопределенными коэффициентами и образуют версальную деформацию (хотя и не с минимальным числом параметров).

Вне области Пуанкаре более разумным представляется понятие топологически версального семейства (в определениях диффеоморфизмы заменены гомеоморфизмами). В настоящее время полностью изучены топологически версальные деформации вырождений коразмерности один (их две, во-первых, слияние и смерть или рождение пары неподвижных точек, во-вторых, рождение или смерть предельного цикла в положении равновесия). А.Н.Шошитайшвили, которому принадлежат эти результаты, получает их на основе принципа сведения, согласно которому собственные числа с ненулевой вещественной частью не влияют на бифуркации, так что задача сводится к аналогичной в пространстве такого числа измерений, сколько есть собственных чисел с нулевой вещественной частью.

Случаи коразмерности 2 изучены не полностью. Здесь нужно отметить недавний результат Р.И.Богданова, построившего и исследовавшего топологически версальные деформации особых точек с двумя нулевыми корнями на плоскости. Наиболее трудным кажется случай двух пар чисто мнимых собственных чисел; можно также думать, что в общем случае коразмерности три (а возможно и в некоторых случаях коразмерности два) топологически версального конечномерного семейства вообще не существует.

Другие результаты, обсуждавшиеся в лекциях: нормальные формы матриц, зависящих от параметров (Д.М.Галин [6]) алгебраическая вычислимость индекса особой точки векторного поля (В.М.Закалюкин [14]) и, алгебраическая неразрешимость проблемы устойчивости и проблемы центр-фокус(А.Д.Брюно [11] Ю.С.Ильяшенко [16]), особенности декремента как функции от параметров, жесткое и мягкое возбуждение автоколебаний в жидкости [8,17] и т.д.

Л и т е р а т у р а

1. Особенности дифференцируемых отображений, Сборник переводов, М., Мир, 1968.

2. В.И.Арнольд, Особенности гладких отображений, УМН, 23 № 1, 1968, 3-44.

3. В.И.Арнольд, О локальных задачах анализа, Вестник МГУ, сер. математика, № 2, 1970, 52-55.

4. С.М.Вишик, Векторные поля в окрестности края многообразия, Вестник МГУ, сер.математика, № 1, 1972, 21-28.

5. В.И.Арнольд, О матрицах, зависящих от параметров, УМН, 26, № 2, 1971, 101-114.

6. Д.М.Галин, О вещественных матрицах, зависящих от параметров, УМН, 27, № 1, 1972, 241-242.

7. А.Н.Шошитайшвили, О бифуркациях топологического типа особых точек векторных полей, Функц.анализ и его приложения, 6, № 2, 1972, 97-98.

8. И.П.Андрейчиков, В.И.Юдович, Об автоколебательных режимах, ответвляющихся от течения Пуазейля в плоском канале, ДАН, 202, № 4, 1972, 791-794.

9. Н.Н.Брушлинская, Версальное семейство векторных полей области Пуанкаре, Функц. анализ и его приложения, 4, № 1, 1970, 6-13.

10. Н.Н.Брушлинская, Теорема конечности для семейств векторных полей области Пуанкаре, Функц. анализ и его приложения, 5, № 3, 1971, 10-15.

11. А.Д.Брюно, Аналитическая форма дифференциальных уравнений, Труды ММО, 25, 1971, 119-262.

12. В.И.Арнольд, Замечания об особенностях конечной коразмерности в комплексных динамических системах, Функ.анализ и его приложения, 3, № 1, 1969, 1-6.

13. А.С.Пяртли, Рождение комплексных инвариантных многообразий вблизи особой точки векторного поля, зависящего от параметра, Функц. анализ и его приложения, 6, № 4, 1972, 95-96.

14. В.М.Закалюкин, Алгебраическая вычислимость индекса особой точки векторного поля, Функц.анализ и его приложения, 6, № 1, 1972, 77-78.

15. В.И.Арнольд, Алгебраическая неразрешимость проблемы устойчивости по Ляпунову, Функц.анализ и его приложения, 4, № 3, 1970, 1-9.

16. Ю.С.Ильяшенко, Алгебраическая неразрешимость и почти алгебраическая разрешимость проблемы центр-фокус, Функц.анализ и его приложения, 6, № 3, 1972, 30-37.

17. D.Ruell, F.Takeus, On the na ure of turbulence, Comm.Math.Phys., 20 /1971/,167-192.

Translated by J. J. D.

UDC 532.51

NOTES ON THE THREE-DIMENSIONAL FLOW PATTERN OF A PERFECT FLUID IN THE PRESENCE OF A SMALL PERTURBATION OF THE INITIAL VELOCITY FIELD

PMM Vol. 36, №2, 1972, pp. 255-262
V. I. ARNOL'D
(Moscow)
(Received March 31, 1971)

Euler equations of the three-dimensional motion of a perfect incompressible fluid, linearized for a nearly stationary flow are considered and the class of stationary flows for which these linearized equations admit exact explicit solutions is indicated. The analysis of derived equations shows that in some stationary flows the perturbation buildup considerably differs from that obtaining in cases generally considered in the theory of hydrodynamic stability: there appears an infinitely great number of unstable configurations, the flow pattern is difficult to predict (since an approximate determination of perturbation development with time necessitates a rapidly increasing amount of information about initial conditions, etc). These differences are due to the different geometry of stationary flows. In the recently constructed models of stationary flows the assumption is made that a fluid particle in motion stretches into a filament or ribbon whose length exponentially increases with time, while in the usually considered flows the length is assumed to be a linear function of time. In two-dimensional flows the phenomenon of exponential stretching of particles is impossible. It is shown that this is, also, impossible in three-dimensional flows in which the vectors of velocity and viscosity are not collinear.

1. The linearized Euler equation. The shortened equation. Let us write Euler's equation in the form of a vortex equation

$$\partial \mathbf{r} / \partial t = \{\mathbf{v}, \mathbf{r}\} \qquad (\mathbf{r} = \operatorname{rot} \mathbf{v}) \tag{1.1}$$

where the Poisson's bracket of the two vector fields is defined by the condition

$$D_{\{\mathbf{a},\mathbf{b}\}} = D_{\mathbf{b}} D_{\mathbf{a}} - D_{\mathbf{a}} D_{\mathbf{b}}$$

in which $D_{\mathbf{q}}$ denotes integration in the direction of field $\mathbf{q}$. Let us consider a small perturbation $\mathbf{u}$ of the stationary flow $\mathbf{v}$. Let $\mathbf{s}$ be the vortex perturbation field: $\operatorname{rot}(\mathbf{v} + \mathbf{u}) = \mathbf{r} + \mathbf{s}$. Equation (1.1) linearized in the neighborhood of flow $\mathbf{v}$ is of the form

$$\partial \mathbf{s} / \partial t = \{\mathbf{v}, \mathbf{s}\} + \{\operatorname{rot}^{-1} \mathbf{s}, \mathbf{r}\} \tag{1.2}$$

The operation rot^{-1} is understood as the restitution of a nondivergent vector field over

Originally published in: Prikl. Mat. Mekh. 36:2, 255-262, © Nauka, Moscow 1972
English transl.: J. Appl. Math. Mech. 36:2, 236-242, © Elsevier, Oxford, 1972
Reprint by Springer-Verlag Berlin Heidelberg 2014 with kind permission, DOI 10.1007/978-3-642-31031-7_31

its vortex field. In the multiply-connected case it is necessary to consider instead of the vortex field the totality of velocity field circulations over all possible closed contours (not necessarily homologous to zero), i.e., the vortex field together with velocity field circulations over basic univariate cycles. If the flow region has a rim, the velocity field is to be assumed tangent to it.

Let us examine the bahavior of solutions of this equation which is linear with respect to s . Note that the first term in the right-hand side of (1.2) is a more powerful linear operator over s than the second. Hence the second term may be considered as a perturbation of the first. In this way we obtain the shortened equation

$$\partial \mathrm{s} / \partial t = \{\mathrm{v}, \mathrm{s}\} \tag{1.3}$$

If the stationary flow is potential ($\mathbf{r} = 0$), the second term in Eq. (1.2) vanishes, and in that case the shortened equation (1.3) is the same as the linearized Euler equation (1.2). In accordance with the theory of perturbations [1] it is reasonable to assume that the shortened equation defines the continuous part of the spectrum of Eq. (1.2).

The shortened equation (1.3) implies that vector s is carried by the stationary flow. If the geometry of the stationary flow $\mathbf{v}$ is known, this equation can be solved explicitly. Let $\{g^t\}$ be a one-parameter group of diffeomorphisms induced by the stationary flow, hence $g^t\ (x)$ is the solution of the system of ordinary differential equations

$$\frac{d}{dt} g^t (x) = \mathrm{v}\,(g^t (x)), \qquad g^0 (x) = x \tag{1.4}$$

Solution for s of the shortened equation (1.3) can now be expressed in terms of its initial conditions by formula

$$\mathrm{s}\,(t, x) = g_*{}^t \mathrm{s}\,(0, g^{-t} (x)) \tag{1.5}$$

where $g_*{}^t$ is the derivative of the image of g^t.

2. The action – angle variables. The geometry of stationary flows of perfect fluid was examined in [2]. It is shown there that, when the fields v and r are not identically collinear in any region, the space filled with fluid becomes divided into cells in each of which the stream- and vorticity-lines lie on torus surfaces (*). Curvilinear coordinates, similar to the action – angle variables in conventional mechanics. We denote these coordinates by φ and z. The coordinates $\varphi = (\varphi_1, \varphi_2) \bmod 2\pi$ are angular coordinates along the tori and z is the "action variable" numbering the latter. The coordinates φ_1, φ_2 and z can be chosen so that a volume element is defined by $d\varphi_1 d\varphi_2 dz$ and the fields v and r by

$$\mathrm{v} = v_1 (z) \frac{\partial}{\partial \varphi_1} + v_2 (z) \frac{\partial}{\partial \varphi_2}, \qquad \mathbf{r} = r_1 (z) \frac{\partial}{\partial \varphi_1} + r_2 (z) \frac{\partial}{\partial \varphi_2}$$

Equations (1.4) are integrable in the system of coordinates φ, z . For the components of field s in coordinates φ, z

$$\mathrm{s}\,(t; \varphi, z) = s_1 \frac{\partial}{\partial \varphi_1} + s_2 \frac{\partial}{\partial \varphi_2} + s_3 \frac{\partial}{\partial z}$$

from (1.5) we obtain the expressions

*) Cells of a different kind in which all streamlines are closed are also possible in the case of flow in a manifold with a rim; this case is not considered here.

$$s_{1,2}(t;\varphi,z)=s_{1,2}(0;\varphi_0,z)+tv_{1,2}s_3(0;\varphi_0,z),\ s_3(t;\varphi,z)=s_3(0;\varphi_0,z)$$
$$(\varphi_0=\varphi-\mathbf{v}t) \tag{2.1}$$

where the prime denotes a derivative with respect to z.

Formulas (2.1) imply that solutions of the shortened equation (1.3) (for $\mathbf{v}'\neq 0$) usually increase linearly with time. Hence the conventional (exponential) instability of linearized Euler equation can only be due to the second term in formula (1.2). In accordance with the theory of perturbations it is reasonable to expect the appearance of a finite number of unstable discrete eigenvalues (there is no rigorous proof of this).

An interesting exception is the instability of the Couette flow between two cylinders (this was brought to the author's attention by V. I. Iudovich).

In a Couette flow the velocity component of the basic flow along the cylinder axis is zero, hence invariant with respect to Bernoulli's constant. This results in the degeneration of a whole segment of the continuous spectrum into a single point. The longitudinal velocity component $v_{\mathbf{m}}$ in formula (3.1) is for certain values of the wave vector $\mathbf{m}$ independent of z.

The assumption of finiteness of the number of unstable configurations relates to a nondegenerate continuous spectrum, when the longitudinal velocity component varies with Bernoulli's constant, i.e., $v_{\mathbf{m}}'\neq 0$ in (3.1). For such nondegeneracy to exist it is, for example, sufficient for the curvature of the plane curve $v_1=v_1(z)$, $v_2=v_2(z)$ to be nonzero and for the curve to be regular.

The question of retention of the detected above slow instability when passing from the shortened equation (1.3) to the complete equation (1.2) is discussed in Sect. 4 below. The other possibility of exponential instability is related to the collinearity of $\mathbf{v}$ and $\mathbf{r}$, when the action – angle variables cannot be introduced and the stationary flow geometry differs from the described above (cf. [3]). This form of instability is examined in Sect. 5.

3. Spectrum of the shortened equation. For a more detailed analysis of solutions of Eq. (1.3) we expand $\mathbf{s}$ into a Fourier series in terms of φ, using the following notation. Let $\mathbf{m}$, which we shall call the wave vector, be a pair of integers m_1 and m_2. We denote $m_1\varphi_1+m_2\varphi_2$ by $(\mathbf{m},\varphi)$ the number $\sqrt{m_1^2+m_2^2}$ by m and the pair $n_1=-m_2$ and $n_2=m_1$ by $\mathbf{n}$.

For each wave vector we determine the "longitudinal", "transverse", and "normal" vector fields

$$\mathbf{e}_{\mathbf{m}}=\frac{m_1}{m}\frac{\partial}{\partial\varphi_1}+\frac{m_2}{m}\frac{\partial}{\partial\varphi_2},\qquad \mathbf{e}_{\mathbf{n}}=-\frac{m_2}{m}\frac{\partial}{\partial\varphi_1}+\frac{m_1}{m}\frac{\partial}{\partial\varphi_2},\qquad \mathbf{e}_z=\frac{\partial}{\partial z}$$

(For $m=0$ we assume, e.g., $\mathbf{e}_{\mathbf{m}}=\partial/\partial\varphi_1$ and $\mathbf{e}_{\mathbf{n}}=\partial/\partial\varphi_2$).

The Fourier expansion of field $\mathbf{s}$ can now be written as

$$\mathbf{s}=\sum_{\mathbf{m}}(A_{\mathbf{m}}\mathbf{e}_{\mathbf{m}}+B_{\mathbf{m}}\mathbf{e}_{\mathbf{n}}+C_{\mathbf{m}}\mathbf{e}_z)\,e^{i(\mathbf{m},\varphi)}$$

where $A_{\mathbf{m}}$, $B_{\mathbf{m}}$ and $C_{\mathbf{m}}$ are functions of z.

It can be readily verified that the divergence of fields $\mathbf{e}_{\mathbf{m}}$, $\mathbf{e}_{\mathbf{n}}$ and $\mathbf{e}_z$ is zero (this is the result of the form $d\varphi_1\,d\varphi_2 dz$) of the volume element). Hence

$$\operatorname{div}\mathbf{s}=\sum_{\mathbf{m}}(imA_{\mathbf{m}}+DC_{\mathbf{m}})\,e^{i(\mathbf{m},\varphi)}\qquad \left(D=\frac{d}{dz}\right)$$

Consequently the nondivergent fields are determined by the condition " $imA_{\mathbf{m}}+$

$+ DC_{\mathbf{m}} = 0$ for all $\mathbf{m}$".

In accordance with this condition the set of functions $B_{\mathbf{m}}$ and $C_{\mathbf{m}}$ (for $\mathbf{m} = 0$ we have $C_0 =$ const but A_0 is to be added) can be taken as the "coordinates" in the space of convergent fields. In such system of coordinates Eq. (1.3) is decomposed into a series of triangular systems

$$\begin{cases} \dot{B}_{\mathbf{m}} = -imv_{\mathbf{m}}B_{\mathbf{m}} + v_{\mathbf{n}}'C_{\mathbf{m}} \\ \dot{C}_{\mathbf{m}} = -imv_{\mathbf{m}}C_{\mathbf{m}} \end{cases} \tag{3.1}$$

where $\mathbf{v} = v_{\mathbf{m}}\mathbf{e}_{\mathbf{m}} + v_{\mathbf{n}}\mathbf{e}_{\mathbf{n}}$ is the velocity field of the stationary flow (for $\mathbf{m} = 0$ we add the equation $\dot{A}_0 = v_0' C_0$); the prime and the dot denote differentiation with respect to z and t. respectively.

Formula (3.1) again implies the nonexponential instability of Eqs. (1.3). Furthermore, it contains the definition of the spectrum of Eq. (1.3): to each wave vector $\mathbf{m}$ corresponds a segment of the continuous spectrum along the imaginary axis. The related "frequencies" $mv_{\mathbf{m}}$ are equal to all kinds of frequencies $(\mathbf{m}, \mathbf{v})$ of the stationary flow at various tori corresponding to various values of the z-coordinate. The multiplicity of each segment is not less than two (the B- and C-components have the same frequencies).

4. The theorem of Squire for shear flows. The coordinates introduced above are suitable for analyzing the shortened equation (1.3), however, since in curvilinear coordinates the operator rot^{-1} is of a complicated form, analysis of the complete equation (1.2) is generally difficult. A particular case in which the analysis can be reduced to a one-dimensional problem is that of flow with straight streamlines. All plane rectilinear flows, as well as the more general ones in which the fluid flows in parallel planes at constant velocity which varies in magnitude and direction when passing from one plane to another, belong to this class. Study of the latter may be considered as an approximate analysis of general flow in the torus geometry, in which the torus curvature is neglected, while shear (variation of the direction of streamlines from one torus to another) is taken into consideration.

Let φ_1, φ_2 and z be Cartesian coordinates and $dl^2 = d\varphi_1^2 + d\varphi_2^2 + dz^2$. In this case it is expedient to consider periodic flows of not necessarily 2π periodicity (e.g. we can assume the periods of φ_1 and φ_2 to be $2\pi X_1$ and $2\pi X_2$, respectively). The only alteration to be introduced in formulas in Sect. 3 is that now the wave vector $\mathbf{m}$ does not run through a grid of whole points but through grid $\{(m_1 / X_1, m_2 / X_2)\}$.

On these assumptions the expansion of the vortex field $\mathbf{r}$ in terms of unit vectors $\mathbf{e}_{\mathbf{m}}$, $\mathbf{e}_{\mathbf{n}}$ and $\mathbf{e}_z$ is of the form $\mathbf{r} = -v_{\mathbf{n}}'\mathbf{e}_{\mathbf{m}} + v_{\mathbf{m}}\mathbf{e}_{\mathbf{n}}$. The matrices of operator rot in coordinates $B_{\mathbf{m}}$, $C_{\mathbf{m}}$ and the operator of Poisson's bracket containing $\mathbf{r}$ are, respectively,

$$im\begin{pmatrix} 0 & -E + m^{-2}D^2 \\ E & 0 \end{pmatrix}, \quad -\begin{pmatrix} imv_{\mathbf{n}}' & v_{\mathbf{m}}'' \\ 0 & imv_{\mathbf{n}}' \end{pmatrix}$$

where E is an identical transformation. Hence in our coordinates the linearized Euler's equation (1.2) is decomposed into a set of systems of equations corresponding to various $\mathbf{m}$. After calculation we obtain for $\mathbf{m} \neq 0$ the triangular system

$$\begin{aligned} \dot{B}_{\mathbf{m}} &= \left[imv_{\mathbf{m}} + \frac{v_{\mathbf{m}}''}{im} (E - m^{-2}D^2)^{-1} \right] B_{\mathbf{m}} \\ \dot{C}_{\mathbf{m}} &= imv_{\mathbf{m}}C_{\mathbf{m}} + v_{\mathbf{n}}' (E - m^{-2}D^2)^{-1} B_{\mathbf{m}} \end{aligned} \tag{4.1}$$

and for $\mathbf{m} = 0$ we have the system $\dot{A}_0 = \dot{B}_0 = \dot{C}_0 = 0$. The first equation becomes separated and, if the B-component does not have exponential instability, the latter is also absent in the C-component (this is implied by the nonhomogeneous linear equation obtaining for C_m).

Note that the equation for B_m contains only the longitudinal velocity component v_m. Hence this equation is the same as that derived in the analysis of two-dimensional flow of a perfect fluid, whose velocity profile is the component $v_m(z)$ of the velocity vector of a three-dimensional flow in the direction of the wave vector **m**.

Thus a rectilinear three-dimensional flow is exponentially unstable when, and only when, at least one of the two-dimensional flows of a perfect fluid, obtained by the substitution for the velocity vector v of its longitudinal component v_m, is exponentially unstable. The problem of exponential instability of the considered class of three-dimensional flows of perfect fluid is thus reduced to a similar problem for a series of two-dimensional flows corresponding to various values of the wave vector.

In the particular case of flow free from shear (constant direction of v) all velocity profiles are proportional to each other and the obtained result conforms to the theorem of Squire for a perfect fluid [4].

The Jordan form of system (4.1) tends to indicate that in three-dimensional flows, unlike in two-dimensional ones, the linear increase of vortex perturbations with time is the rule, even in the absence of exponential instability.

5. Stationary flows with exponential stretching of particles. The flow region is a three-dimensional compact manifold M constructed in the following manner. (*)

Let us, first, consider a conventional three-dimensional space in coordinates x, y, z and determine the following three diffeomorphisms of that space:

$$T_1(x, y, z) = (x+1, y, z), \quad T_2(x, y, z) = (x, y+1, z)$$

$$T_3(x, y, z) = (2x+y, x+y, z+1)$$

Each of these transforms into itself the lattice of points with complete coordinates x, y, z. Let us identify all points of the xyz-space which can be obtained from each other by successive application of T_i and T_i^{-1} (in any order). As the result a compact analytic manifold M is created which may be visualized as the product of multiplication of a two-dimensional torus $\{(x, y) \bmod 1\}$ by segment $0 \leqslant z \leqslant 1$ whose end tori are identified by formula $(x, y, 0) \equiv (2x+y, x+y, 1)$. We introduce on the drived manifold a Riemann metric. For this we construct in the xyz-space a Riemann metric invariant with respect to all T_i.

Let us examine the linear transformation of the xy-plane

$$A\begin{pmatrix}x\\y\end{pmatrix} = \begin{pmatrix}2x+y\\x+y\end{pmatrix}, \qquad (A) = \begin{pmatrix}2 & 1\\1 & 1\end{pmatrix}$$

Transformation A has eigenvalues $\lambda_{1,2} = (3 \pm \sqrt{5})/2$. Note that $\lambda_1 > 1 > \lambda_2 > 0$, $\lambda_1\lambda_2 = 1$, and the eigendirections are orthogonal to each other. Let (p, q)

*) This kind of manifold became important in the contemporary qualitative theory of ordinary differential equations after the work of Smile, whose attention was drawn to this example by Thom.

be a Cartesian system of coordinates in the $x\ y$ -plane with the axes p and q directed along eigenvectors with eigenvalues $\lambda_1 > 1$ and $\lambda_2 < 1$, respectively.

Let us set

$$ds^2 = e^{-2\mu z}dp^2 + e^{2\mu z}dq^2 + dz^2 \quad (\mu = \ln \lambda_1) \tag{5.1}$$

Metric ds^2 is invariant with respect to transformations T_i, hence it defines on the three-dimensional compact manifold M an analytic Riemann structure.

Let us now consider the vector field $\partial / \partial z$ in the xyz-space. Since it is invariant with respect to transformations T_i, it defines the vector field v on manifold M. Field v on the Riemann manifold M is harmonic: div v $= 0$, rot v $= 0$. Hence v may be taken as the velocity field of a stationary potential flow of a perfect fluid. Every particle of fluid moving in that field exponentially stretches in the q-direction and contracts in the p direction, as implied by formula (5.1).

6. Analysis of the linearized Euler equation. Since the considered flow is potential, the linearized Euler's equation (1.2) is equivalent to the shortened equation (1.3). Owing to the simple geometry of flow the latter equation is solved by formula (1.5). The solution is conveniently expressed with the use of the following notation. Let us consider in the pqz-space the vector fields

$$\mathbf{e}_p = e^{\mu z}\frac{\partial}{\partial p}, \qquad \mathbf{e}_q = e^{-\mu z}\frac{\partial}{\partial q}, \qquad \mathbf{e}_z = \frac{\partial}{\partial z}$$

These fields are invariant with respect to all transformations T_i, consequently they can be considered as vector fields on the manifold M. The directions of fields $\mathbf{e}_p$, $\mathbf{e}_q$ and $\mathbf{e}_z$ are invariant with respect to the phase stream g^t of field $\mathbf{e}_z$ (in the coordinate form $g^t(p, q, z) = (p, q, z + t)$). Under the action of the stream the fields themselves are transformed by formulas

$$g_*{}^t\mathbf{e}_p = e^{-\mu t}\mathbf{e}_p, \qquad g_*{}^t\mathbf{e}_q = e^{\mu t}\mathbf{e}_q, \qquad g_*{}^t\mathbf{e}_z = \mathbf{e}_z$$

In accordance with this, the direction of field $\mathbf{e}_q$ is called the stretching direction, that of $\mathbf{e}_p$ the compression direction, and that of $\mathbf{e}_z$ the neutral direction. Any vector field w on M can be decomposed in these directions

$$\mathbf{w} = w_p\mathbf{e}_p + w_q\mathbf{e}_q + w_z\mathbf{e}_z$$

where w_p, w_q and w_z are functions on the manifold M.

Formula (1.5) applied to the stationary flow $\mathbf{v} = \mathbf{e}_z$ has with the introduced notation the form

$$s_p(t) = e^{-\mu t}U^t s_p(0), \qquad s_q(t) = e^{\mu t}U^t s_q(0), \qquad s_z(t) = U^t s_z(0) \tag{6.1}$$

where U^t is a linear operator acting on functions on the manifold M by formula "$(U^t f)(\xi) = f(g^{-t}\xi)$ for any point ξ from M". Note that the stream g^t maintains its volume, hence operator U^t is unitary.

Formula (6.1) provides fairly complete answers to all kinds of questions on the growth of perturbations of a stationary flow v. First, it shows that the q-component of vortex perturbation exponentially increases with time, while the p component is exponentially attenuated.

Next, the spectrum of operator U^t can be easily analyzed by a Fourier series expansion in terms of (x, y) with fixed z, and for functions independent of x and y by such expansion in terms of z. This spectrum has a countably-multiple continuous (Lebesgue) component along a unit circle and, also, a discrete set of eigenvalues corresponding to eigenfunctions $\varphi_m = e^{2\pi imz}$ (m are integers). This implies that Euler's equation (1.2) linearized for a close to stationary flow $v = e_z$ has a countable set of unstable eigenvalues $\mu - 2\pi im$ related to the countable set of increasing perturbations of vortex $s = \varphi_m(z)\, e_q$ ($m = \pm 1, \pm 2, \ldots$).

The difficulty of predicting solutions of the linearized Euler equation (1.2) for flows with exponential stretching of particles is also indicated by formulas (6.1): to find an approximate solution in terms of t it is necessary to know with considerable exactitude a number of high order harmonics in the initial perturbation $s(0)$ which rapidly increase with t. Comparison of formulas (6.1) and (2.1) shows that the exponential increase of particle stretching considerably increases the difficulty of predicting the growth of perturbations, as compared with conventional flows with linear stretching of particles considered in Sect. 2 – 4.

Phenomena similar to those brought to light in this example are to be expected also in other flows with exponentially stretched particles, and such flows are possible in regions of conventional three-dimensional space. Experimental confirmation of this can be found in [5, 6]. Computer calculations cited in [5] tend to show that the stationary flow of a perfect fluid specified by formulas [2]

$$v_x = A\sin z + C\cos y, \qquad v_y = B\sin x + A\cos z, \qquad v_z = C\sin y + B\cos z$$

has the property of exponentially stretched particles.

The author thanks L.D. Faddeev and V.I. Iudovich for valuable discussions.

BIBLIOGRAPHY

1. Faddeev L.D., On the theory of stability of plane-parallel stationary flows of perfect fluid. Boundary Value Problems of Mathematical Physics, Vol. 5, Zap. Nauchn. Seminarov LOMI, Vol. 21, pp. 164-172, "Nauka", Moscow, 1971.

2. Arnold V., Sur la géométrie differentielle des groupes de Lie de dimension infinie et ses applications à l'hydrodynamique des fluides parfaits. Annales de l'Institut Fourier, Vol. 16, №1, p. 347, 1966.

3. Grad H., Mathematical problems arising in plasma physics. Actes du congrès international des mathématiciens. Vol. 3, pp. 105-113, Gautier-Villars, Paris, 1971.

4. Squire H.B., On the stability of the three-dimensional disturbances of viscous flow between parallel walls. Proc. Roy. Soc., A, Vol. 142, 1933.

5. Henon M., Sur la topologie des lignes de courant dans un cas particulier. Comptes Rendues de l'Acad. des Sc., Vol. 262, Paris, 1966.

6. Froeschle C., A numerical study of the stochasticity of dynamical systems with two degrees of freedom. Astronom. and Astrophys., Vol. 9, 1970.

Translated by J.J.D.

SEL MATH SOV
VOL. 5, NO. 4, 1986

The Asymptotic Hopf Invariant and Its Applications*

V. I. Arnold

The classical Hopf invariant distinguishes among the homotopy classes of continuous mappings from the three-sphere to the two-sphere and is equal to the linking number of the two curves that are the preimages of any two regular points of the two-sphere.

The asymptotic Hopf invariant is an invariant of a divergence-free vector field on a three-dimensional manifold with given volume element. It is invariant under the group of volume-preserving diffeomorphisms, and describes the "helicity" of the field, i.e., the mean asymptotic rotation of the phase curves around each other. The asymptotic Hopf invariant coincides with the classical Hopf invariant for the unitary vector field that is tangent to the Hopf bundle. In the general case the asymptotic Hopf invariant can have any real value (whereas the classical Hopf invariant is always an integer).

The asymptotic Hopf invariant can also be considered as a quadratic form on the Lie algebra of the volume-preserving diffeomorphisms of the three-dimensional manifold that is invariant under the adjoint action of the group on the algebra.

In this paper we present the definition and simplest properties of the asymptotic Hopf invariant, as well as some of its applications to an unusual variational problem that arises in magnetohydrodynamics which was called to the author's attention by Ya. B. Zel'dovich. In connection with this problem there arise a whole series of unsolved mathematical problems, some of which appear to be difficult. The main object of this paper is to discuss these unsolved problems; all the theorems in the paper are obvious.

Attention was first called to the problems considered here by Voltjer [7] in connection with magnetohydrodynamics. Applications to ordinary hydrodynamics were given by Moffatt [4], [5] and Kraichnam [3].

* Originally published in Materaly Vsesoyuznoi Shkoly po Differentsialnym Uravneniyam s Beskonechnym Chislom Nezavisimyh i po Dinamicheskim Sistemam c Beskonechnym Chislom Stepenei Svobodny, Dilizhane, May-June 1973, Akademiya ArmSSR, Erevan, 1974, pp. 229-256. Translated by R. P. Boas.

Originally publ. in: Proc. All-Union School in Diff. Eq. with Infinite Number of Variables and in Dyn. Syst. with Infinite Number of Degrees of Freedom, Dilhan 1973, 229-256, © Acad. Sc. Arm. SSR 1974
English transl.: Selecta Math. Sov. 5:4, 327-345, © Birkhäuser Boston, Inc. 1986
Reprint by Springer-Verlag Berlin Heidelberg 2014 with kind permission, DOI 10.1007/978-3-642-31031-7_32

1. The problem of the minimum magnetic energy of a frozen-in field

Let M be a three-dimensional closed Riemannian manifold, and ξ a divergence-free vector field on M. The *energy* of the field is the integral

$$E = \tfrac{1}{2}\langle \xi, \xi \rangle = \tfrac{1}{2} \int_M (\xi, \xi)\, dv.$$

We are to find the minimum energy for fields obtained from a given field under the action of volume-preserving diffeomorphisms of the manifold M.

Here the action of a volume-preserving diffeomorphism $g\colon M \to M$ associates with a divergence-free field ξ on M another divergence-free field $g * \xi$ such that the flux of the field ξ across any surface σ is equal to the flux of $g * \xi$ across $g\sigma$. In other words, the field is frozen into a covering of M by an incompressible fluid: the vector field can be thought of as drawn on the elements of the fluid and expanding as these elements expand.

The two-dimensional analog can be formulated as follows:

To find a function that minimizes the Dirichlet integral

$$E = \tfrac{1}{2}\langle \nabla u, \nabla u \rangle = \tfrac{1}{2} \int_M (\nabla u, \nabla u)\, dv$$

among all functions u on the closed two-dimensional Riemannian manifold M obtained from a given function u_0 by the action of an area-preserving diffeomorphism on itself:

$$u(x) = u_0(g^{-1}x).$$

It is clear that similar problems arise for manifolds with boundary, for example, for functions $u(x, y)$ in an ordinary Euclidean disk. The mathematical aspects of investigations of these problems have been highly unsatisfactory.

1.1. The Euler equation

Theorem. *The extremals of the problem stated above are divergence-free fields that commute with their curl.*

Proof. Let η be any divergence-free field. The variation of a field ξ under the infinitesimal diffeomorphism defined by η is given by the Poisson bracket $\delta\xi = \{\eta, \xi\}$ (in terms of the coordinates, $(\eta, \xi) = (\xi\nabla)\eta - (\eta\nabla)\xi$).

Consequently $\delta E = \langle \xi, \delta\xi \rangle = \langle \xi, \{\eta, \xi\} \rangle$. But, by a formula from vector analysis, $\operatorname{curl}[\eta, \xi] = \{\eta, \xi\} - \eta \operatorname{div} \xi - \xi \operatorname{div} \eta$ on any three-dimensional Riemannian manifold. Since our fields ξ and η are divergence-free, $0 = \delta E = \langle \xi, \operatorname{curl}[\eta, \xi] \rangle = \langle \operatorname{curl} \xi, [\eta, \xi] \rangle = \langle \eta, [\xi, \operatorname{curl} \xi] \rangle$. Since η is divergence-free, the vector product $[\xi, \operatorname{curl} \xi]$ is orthogonal to all divergence-free fields. Consequently it is a gradient:

$$[\xi, \operatorname{curl} \xi] = \operatorname{grad} \alpha,$$

whence, taking the curl of both sides, we obtain

$$\{\xi, \operatorname{curl} \xi\} \equiv 0,$$

as was to be proved.

Remark 1. In the two-dimensional case we obtain the equation

$$[\nabla u, \nabla \Delta u] \equiv 0,$$

which says that the gradient of the extremal function is collinear with that of its Laplacian.

Remark 2. A similar calculation leads to the following expression for the second variation:

$$\delta^2 E = \langle \{\eta, \xi\}, \{\eta, \xi\} \rangle + \langle \{\eta, \xi\}, [\operatorname{curl} \xi, \eta] \rangle$$

(where ξ is an extremal whose first and second variations are given by the formula

$$\xi(\epsilon) = \xi + \epsilon\{\eta, \xi\} + \frac{\epsilon^2}{2}\{\eta, \{\eta, \xi\}\} + \cdots, \qquad \varepsilon \to 0,$$

in terms of a divergence-free vector field η).

1.2. *Study of the extremal fields*

Let ξ be a divergence-free field on a three-dimensional closed orientable Riemannian manifold M for which $[\xi, \operatorname{curl} \xi] = \operatorname{grad} \alpha$. All such fields are extremals for our problem. It turns out that the field lines of ξ have a very special topology.

Theorem. *Every noncritical level set of the function α is diffeomorphic to a torus (or union of tori). In a neighborhood of such a torus we can define coordinates $\varphi_1, \varphi_2 \pmod{2\pi}$, z, such that φ is the angular coordinate on the torus, z indexes the torus, and the field ξ (as well as the field* curl ξ*) has components*

$$\xi = \omega_1(z)\frac{\partial}{\partial\varphi_1} + \omega_2(z)\frac{\partial}{\partial\varphi_2}; \qquad \operatorname{curl} \xi = \omega_1'(z)\frac{\partial}{\partial\varphi_1} + \omega_2'(z)\frac{\partial}{\partial\varphi_2}.$$

Here the coordinate z can be chosen so that the volume element has the form $d\varphi_1 \wedge d\varphi_2 \wedge dz$.

Remark. The coordinates φ_1, φ_2, z are analogs of the action-angle variables of classical mechanics. The theorem means, in particular, that both the field lines of ξ and of curl ξ lie on the tori $\alpha = \text{const}$. These lines are either closed (if the relative frequency ω is rational) or dense on the torus.

For the proof see [**1**]. It follows from the theorem that, for example, in the analytic case, when $\alpha \not\equiv$ const the manifold M is divided by the critical level sets of α into a finite number of cells, diffeomorphic to the product of the torus by intervals in each of which the fields ξ and curl ξ are tangent to the torus and generate periodic or conditionally periodic windings of the torus. Consequently, we obtain an explicit description of the topology of the field ξ (or curl ξ).

It remains to consider the case when grad $\alpha \equiv 0$. In this case $[\xi, \text{curl}\, \xi] = 0$, i.e., the fields ξ and curl ξ are collinear at each point. Such fields are called force-free fields in magnetohydrodynamics.

If a force-free field ξ is never zero, then curl $\xi = c\xi$, where $c\colon M \to R$ is a smooth function. But div curl $\xi \equiv 0$; consequently, (grad $c, \xi) \equiv 0$, i.e., the function c is a first integral of the field ξ (and also of curl ξ). Hence it follows that the connected components of the nonsingular level surfaces of c are tori, and the field lines of ξ are windings on these tori (in the corresponding coordinates φ_1, φ_2, z, the constants along the field lines of ξ will be the frequency ratios, $\dot\varphi_1/\dot\varphi_2 = \kappa(z)$). Therefore even in the case of a force-free field the field lines lie on two-dimensional tori, provided that the field does not have zeros and c is not constant.

A force-free field with curl $\xi = \lambda\xi$, where λ is a constant, can have a much more complicated topology. An example of such a field on the three-dimensional torus $\{x, y, z, \text{mod } 2\pi\}$ is given by the components

$$\xi_x = A \sin z + C \cos y, \qquad \xi_y = B \sin x + A \cos z, \qquad \xi_z = C \sin y + B \cos z.$$

The topology of these field lines was investigated experimentally by Henon [**2**], using the computer at the Astrophysical Institute at Paris. As a result he discovered a set of tori filled out by field lines ("magnetic surfaces") together with whole domains of three-dimensional space whose field lines, as far as one can tell from the experimental data, are ergodic, or everywhere dense.

1.3. Discussion

Returning to our extremal problem, we see that a field of minimum energy in a given class of frozen-in fields must either have a very special topology (the field lines fill out tori), or be force-free fields of a special kind. But the topological properties of the field lines are invariant under diffeomorphisms, and therefore if the original field is a general one, then every field obtained from it by a diffeomorphism has the same property. Consequently, a field of minimum energy either does not exist (in the class of smooth fields to which the preceding analysis applies) or is a force-free field of special type.

But force-free fields with curl $\xi = \lambda\xi$ are scarce; they are eigenvectors of the field of the operator curl on the space of divergence-free fields. Hence we must assume that our variational problem apparently does not always admit a smooth solution.

In this connection we consider the following example. Let M be a sphere in three-dimensional Euclidean space, and let the field lines of ξ be horizontal circles with centers on the vertical axis. According to Zel'dovich, the energy of such a field can be made arbitrarily small by means of a suitable diffeomorphism which preserves volumes and is fixed in a neighborhood of the boundary. In fact, let us divide the whole sphere into a number of slender solid tori (doughnuts) formed from the circles of the field, and a remainder of small volume. Then let us deform (preserving its volume) each solid torus (violating the axial symmetry of the field) so that it becomes fat and small, with the hole decreasing almost to zero. Then the field energy in the solid tori is decreased (since the field lines are shortened). It can be seen that the whole construction can be carried out in such a way that the field energy in the remaining small volume is not increased by too much, as a result the total energy remains arbitrarily small.

It would be of interest to carry out this construction precisely.

In connection with this example, there arises the *question of whether it is possible to reduce the energy of an arbitrary field to an arbitrarily small value* by an appropriate volume-preserving diffeomorphism. We shall see below that this is not the case. An obstacle to the complete annihilation of the energy can be constructed by considering two linked doughnuts of field lines. In this case the shortening of the field lines in one doughnut, shrinking its hole, induces a lengthening of the field lines in the other, so that there is an obstacle to the decrease of the energy. The asymptotic Hopf invariant, which measures the linking of the field lines (not necessarily closed) lets us give a qualitative expression for this situation in the form of a lower bound for the energy.

1.4. *Magnetohydrodynamic discussion*

In magnetohydrodynamics the role of ξ is played by the magnetic field H, frozen into a fluid of finite viscosity, but of infinite conductivity, which fills M. With an appropriate choice of units, the velocity field v and the magnetic field H satisfy the system of equations

$$\frac{\partial v}{\partial t}+(\nabla v, v)=-\operatorname{grad} p-\nu\Delta v+[jH], \qquad \operatorname{div} v=0,$$

$$\frac{\partial H}{\partial t}=\{v, H\}, \qquad \operatorname{div} H=0, \qquad \operatorname{curl} H=j.$$

The magnetic field H and the velocity field v are prescribed at the initial time. In the course of time, the kinetic energy is dissipated because of the viscosity, and the motion ceases "in the end," since each particle approaches some terminal position. The magnetic field, being frozen in, then attains some terminal value. The energy of this terminal field must be a minimum; otherwise the magnetic energy would have been converted into kinetic energy and, on account of the Lorenz force, the fluid would move until it dissipated the excess of magnetic energy above the minimum.

This sort of description of the behavior of solutions of the system presented above is usually given by physicists. Unfortunately, the preceding analysis of the topology of the extremal fields holds out little hope that this description is correct for any general initial conditions: in fact, the initial magnetic field can be taken without having magnetic surfaces, and then the limiting field, if there is one, must be a force-free field of special type; but such fields are too scarce, and one would hardly find a field with the prescribed lines of force among them.

It appears that for a correct description of the actual process it is necessary to take account of the magnetic viscosity, which violates the assumption that the field is frozen in, and was not taken into account in our system of equations.

The question of the extent to which one can use the extremal field to study the behavior of H over an extended period of time during which the ordinary viscosity succeeds in extinguishing the motion of the fluid, but the magnetic viscosity does not extinguish H, is an interesting unsolved problem.

Zel'dovich proposed the problem of the minimum magnetic field in connection with the question of the evolution of the magnetic field of a star. In this case M is a sphere in three-dimensional Euclidean space, and the field is propagated over the whole space with the boundary conditions

$$\operatorname{curl} H = 0 \text{ outside } M, \qquad \operatorname{div} H = 0 \text{ outside } M,$$

$$(H, n) \text{ is continuous on } \partial M,$$

and the condition of decrease at infinity. Consequently, the volume-preserving diffeomorphism of M acts on the field H throughout the whole space. It is necessary to minimize the total energy of the field H (i.e., the integral over all space). The minimizing field must provide a minimum of the magnetic energy inside M with respect to fields obtained from the given diffeomorphism and stationary near the boundary.

We will not discuss the question of how close this simple model is to reality. In what follows we restrict ourselves to a more simple system, in which M is a manifold without boundary.

2. Definition of the invariant

We begin with a dogmatic presentation: we consider an *ad hoc* definition of the invariant, and establish its simplest properties. The interesting meaning of the invariant (and an explanation of how the invariant was found) will be discussed in the following sections.

Let M be a three-dimensional manifold that is closed (compact, without boundary), oriented, and connected, and let v be the volume element (i.e., a 3-form defining the correct orientation) on M. It will be convenient to assume that the total volume of M is 1. Notice that we are given a volume element on M, but *we are not given any particular Riemannian metric.*

2.1. *Notation*

Every vector field ξ on M generates a differential 2-form ω_ξ according to the formula

$$\omega_\xi(\eta, \zeta) = v(\xi, \eta, \zeta) \qquad \text{for all } \eta, \zeta,$$

and the correspondence $\xi \mapsto \omega_\xi$ is an isomorphism of the linear spaces of fields and 2-forms. The derivative of the form ω_ξ, as for every 3-form, can be written in the form

$$d\omega_\xi = \varphi v,$$

where $\varphi: M \to R$ is a smooth function. The function φ is called the *divergence* of the field ξ:

$$\varphi = \operatorname{div} \xi.$$

The velocity field of a flow that preserves the volume element on M is divergence-free; and, conversely, every flow with divergence 0 on M is the velocity field of an incompressible flow (i.e., of a flow that preserves the volume element v on M).

A divergence-free vector field ξ on M is said to be *homologous to zero* if the 2-form ω_ξ corresponding to it is the total differential of a 1-form α on M:

$$\omega_\xi = d\alpha.$$

The 1-form α will be called a *form-potential.* A field is homologous to zero if and only if its flux across every closed surface is zero.

Remark. If a Riemannian metric is given on M then the 1-form α can be identified with the vector field a for which

$$\alpha(\eta) = (a, \eta) \qquad \text{for every } \eta.$$

In this case $\xi = \operatorname{curl} a$, and the vector field a is called the *vector potential* of ξ. However, it is essential to observe that the forms ω and α (in contrast to the field a) do not depend on the Riemannian metric, but only on the choice of the volume element v.

2.2. *Definition*

The (*mean*)*Hopf invariant* of a field ξ that is homologous to zero on the three-dimensional manifold M with volume element v is the integral of the product of the form ω_ξ and its form-potential, i.e., the number

$$I(\xi)=\int_M d\wedge d\alpha, \qquad \text{where} \quad \omega_\xi = d\alpha.$$

Let us show that this definition is consistent, i.e., that *the value of I does not depend on the particular choice of the form-potential* α, *but only on the field* ξ.

In fact, if $\beta=\alpha+\gamma$ is another form-potential, then $d\gamma=0$, and therefore

$$\int_M \alpha\wedge d\alpha-\beta\wedge d\beta=\int_M \gamma\wedge d\alpha=\int_M d(\gamma\wedge\alpha)$$
$$=\int_{\partial M}\gamma\wedge\alpha=0.$$

Remark. If a Riemannian metric with volume element v is given on M, then

$$I(\xi)=\int_M (\xi, a)\, dv=\langle\xi, \operatorname{curl}^{-1}\xi\rangle,$$

where a is any vector potential of ξ. Therefore *I is the scalar product of the field with its vector potential.* It is essential to observe, however, that the Riemannian metric *does not enter* into the definition of I.

2.3. *Invariance*

Corollary. *Every volume-preserving diffeomorphism* $g\colon M\to M$ *carries every field* ξ *that is homologous to zero into a field with the same Hopf invariant.*

In particular, on a Riemannian manifold the scalar product of a divergence-free field and its vector potential is preserved when the field is acted on by a volume-preserving diffeomorphism.

Consequently the invariance of I under diffeomorphisms that preserve the volume element follows from the fact that *I can be defined by using no structures other than the smooth structure of M and the volume element v.*

Remark. The question of whether I is preserved under homeomorphisms that preserve the volume element (transforming the phase flow of ξ into the phase flow of another field ξ') is an interesting unsolved problem, as is the closely related problem of *whether one can define the invariant I directly for one-parameter groups of homeomorphisms that preserve the volume element.*

Remark. In the case when M is a manifold with boundary, the number I is preserved under all volume-preserving diffeomorphisms that are stationary in a neighborhood of the boundary. If, however, ξ is tangent to the boundary, then I is preserved under all volume-preserving diffeomorphisms provided that M is simply connected. The question of whether one can define an invariant analogous to I for general divergence-free fields on a manifold with boundary (including a surface term in I) is an interesting unsolved problem.

2.4. *Examples*

If we take ξ to be a magnetic field, we arrive at the conclusion that *the Hopf invariant of a magnetic field frozen into an incompressible fluid that fills a closed manifold does not change during any flow of the fluid.*

If we interpret the field ξ as the vorticity field of a perfect fluid, we obtain the result that *in the flow of a perfect fluid on a closed three-dimensional manifold, the scalar product of the velocity field and the vorticity field does not change with time.*

If we consider the field ξ as an element of the Lie algebra of the group S Diff M of volume-preserving diffeomorphisms of the three-dimensional manifold M, we obtain the result that *on the Lie algebra of the group S* diff M *there is a symmetric bilinear form that is invariant with respect to the corresponding action of the group on the algebra.* If we give a Riemannian metric on M then

$$I(\xi, \eta) = \langle \xi, \operatorname{curl}^{-1} \eta \rangle,$$

where $\operatorname{curl}^{-1} \eta$ is the vector potential of the field η. In particular, *for every divergence-free field η we have*

$$\langle \{\xi, \eta\}, \operatorname{curl}^{-1} \xi \rangle = 0,$$

which is, of course, easily verified by direct calculation.

For a two-dimensional manifold M we obtain a skew-symmetric form instead of a symmetric form.

3. Asymptotics of the coefficient of linking with a curve

Let M be a closed connected oriented and simply connected three-dimensional manifold with volume element v, let γ be a smoothly embedded closed orientable curve in M, and let ξ be a divergence-free vector field on M. We define an asymptotic coefficient of the linking of the field lines of ξ that issue from the point x with the curve γ. Let $\{g^t: M \to M\}$ be the phase flow of the field ξ. Select a 2-chain σ (of smooth simplexes) for which $\partial \sigma = \gamma$.

3.1. *Asymptotic linking coefficient*

For every pair of points x, y of M we introduce a "short curve" $\Delta(x, y)$ that joins these points and has the following properties:

(1) If x and y do not belong to γ, then Δ does not intersect γ.

(2) The number of intersections of $\Delta(x, y)$ with γ is bounded by a constant independent of x and y.

It is easy to construct such a system of "short curves;" the dependence of Δ on x and y can be made measurable (and even piecewise smooth).

We fix a system of curves Δ and consider the segment of the orbit $g^t x$ of x corresponding to $0 \leq t \leq T$. We join the last point $g^T x = y$ with the first by $\Delta(y, x)$; then we have a closed curve $\Gamma_T(x)$. We assume that this curve does not intersect γ.

Let $N_T(x)$ denote the linking coefficient of $\Gamma_T(x)$ with γ (i.e., the index of the intersection of $\Gamma_T(x)$ with σ).

Theorem. *For almost all x in M the limit*

$$\lim_{T \to \infty} \frac{1}{T} N_T(x) = \lambda(x)$$

exists (where T runs through the values for which $\Gamma_T(x)$ does not intersect γ). This limit belongs to $L_1(M, v)$ and, as an element of L_1, is independent of the system of curves Δ.

The limit $\lambda(x)$ is called *the asymptotic linking number of the orbit $g^t x$ with the curve γ.*

To prove the theorem it is convenient to give a different definition of the asymptotic linking number, and then prove that it is equivalent to the definition given above.

3.2. Second definition of the asymptotic linking number

On the manifold $M - \gamma$ we can construct a closed differential 1-form α with the following properties:

(1) The linking number with γ of every closed curve δ in $M - \gamma$ is equal to the integral of α over δ.

(2) There is a diffeomorphic embedding $u: S^1 \times D^2 \to M$ of the direct product of a circumference and a disk into M such that the circumference $S^1 \times 0$ maps to γ and the form α induces, on the complement of this circumference, the standard form $u * \alpha = (1/2\pi) \arctan(y/x)$ (where x, y are the coordinates in D^2).

We select a form α with these properties, and consider the limit

$$\hat{\lambda}(x) = \lim_{T \to \infty} \frac{1}{T} \int_0^T \alpha\left(\frac{d}{dt} g^t x\right) dt.$$

Theorem. *The limit exists for almost all x and is independent of the choice of the 1-form α satisfying hypotheses* (1) *and* (2).

Proof. Consider the function

$$f(x)=\alpha(\xi(x)), \qquad \text{where} \quad \xi(x)=\frac{d}{dt}\bigg|_{t=0} g^t x.$$

This function belongs to the space $L_1(M, v)$ by condition (2) on α. By Birkhoff's ergodic theorem, the time average of f exists almost everywhere. This establishes the first part of the theorem, since $\hat{\lambda}(x)$ is the time average of f.

To prove the second part, we observe that α is defined on $M-\gamma$ up to a differential of a single-valued function. If φ is a smooth function on $M-\gamma$, then

$$\int_0^T d\varphi\left(\frac{d}{dt}g^t x\right)dt=\varphi(g^T x)-\varphi(x).$$

Now we observe that if $g^t x$ does not approach γ asymptotically as $t\to\infty$, we can choose a sequence $T_i\to\infty$ such that the distance of the points $g^{T_i}x$ from γ remains bounded below. But φ is bounded above by a constant $C(\epsilon)$ outside an ϵ-neighborhood of γ. Consequently, for all points x that are not asymptotic to γ there is a sequence $T_i\to+\infty$ along which $\varphi(g^{T_i}x)$ is bounded above. Therefore the limit $\hat{\lambda}(x)$ is the same for any two 1-forms α for all points x except those asymptotic to γ (and those points for which one of the limits does not exist). But the points asymptotic to γ form a set of measure 0 (since the field ξ is divergence-free), and we have established that $\hat{\lambda}$ is independent of α for almost all x.

3.3. Equivalence of the definitions

The theorem of Section 3.1 is a consequence of the following theorem.

Theorem. *For almost all x, the limit $\lambda(x)$ exists and is equal to $\hat{\lambda}(x)$.*

Proof. By property (1) of the form α, it is enough to prove that for almost all x

$$\lim_{T\to\infty}\frac{1}{T}\int_{\Delta(g^T x,x)}\alpha(\xi)\,dt=0.$$

But since the number of intersections of Δ and σ is bounded (see property (2) of the curves Δ), it follows that the integrals of α along Δ are uniformly bounded; consequently, the limit (over a sequence of values of T for which Δ does not intersect γ) is zero; this establishes the theorem.

Remark. We have simultaneously proved that $\lambda(x)$ is independent of the family of short curves Δ.

Remark. It is obvious from the theorem that the asymptotic linking number is invariant under volume-preserving diffeomorphisms, in the sense that if a diffeomorphism h carries the system (M, v, γ, ξ, x) to $(M', v', \gamma', \xi', x')$ then

$$\lambda_{\xi,\gamma}(x) = \lambda_{\xi',\gamma'}(x').$$

The question of *whether the asymptotic linking number is invariant under volume-preserving homeomorphisms is an unsolved problem, as is the related question of the possibility of defining an asymptotic linking number with a topological curve γ for a one-parameter group of volume-preserving homeomorphisms.*

3.4. The mean linking number with a curve

Let $\{g^t\}$ be the phase flow of a divergence-free field ξ on a simply connected three-dimensional manifold M with volume element v. Let $\gamma = \partial\sigma$ be an oriented smooth curve in M, and let σ be a piecewise smooth 2-chain. *The mean linking number of* $\{g^t\}$ *with* γ is the average of the asymptotic linking number with respect to M:

$$\lambda = \int_M \lambda(x) v.$$

Theorem. *The mean linking number λ is equal to the flux of the field ξ through the surface σ.*

Proof. The number $\lambda(x)$ is the time average of $f(x) = \alpha(\xi(x))$. Consequently, the space averages of f and λ are the same, i.e.,

$$\lambda = \int_M \alpha(\xi) v = \int_M \alpha \wedge \omega_\xi.$$

Now the theorem follows from the homology of the 2-chain σ and the 1-form α as de Rham flows in $M - \gamma$ (strictly speaking, we should consider not $M - \gamma$, but the complement in M of an ϵ-neighborhood of γ, and then let $\epsilon \to 0$).

Remark. One can obtain similar results for the case when γ is not smoothly embedded, but is a piecewise smooth curve. In addition, one can assume that M is n-dimensional and that the chain γ is $(n-2)$-dimensional.

4. Asymptotic linking number of a pair of trajectories

Let M be a three-dimensional closed simply connected manifold with volume element v, let ξ be a divergence-free field on M, and let $\{g^t\}$ be its phase flow.

4.1. *Definition of the asymptotic linking number of a pair of trajectories*

We consider a pair x_1, x_2 of points of M. We are going to associate with this pair of points a number that characterizes the "asymptotic linking" of the trajectories of $\{g^t\}$ that issue from them. For this purpose we first join any two points of M by a "short path" connecting the points (the conditions imposed on a short path were described above and are satisfied for "almost any" choice of the short path).

We select two large numbers T_1 and T_2, and close the segment $g^{t_k}x_k$ $(0 \leq t_k \leq T_k)$ of the trajectories issuing from x_1 and x_2 by short paths $\Delta(g^{T_k}x_k, x_k)$ $(k=1,2)$ so that we obtain two closed curves $\Gamma_k = \Gamma_{T_k}(x_k)$. We assume that these curves do not intersect (which is true for almost all pairs x_1, x_2 for almost all T_1, T_2). Then the linking number $N_{T_1,T_2}(x_1, x_2)$ of Γ_1 and Γ_2 is defined as follows.

Definition. The asymptotic linking number of the pair of trajectories $g^t_{x_1}, g^t_{x_2}$ is defined as the limit

$$\lambda(x, y) = \lim_{T_1, T_2 \to \infty} \frac{N_{T_1,T_2}(x_1, x_2)}{T_1 T_2}$$

(T_1 and T_2 are to vary so that Γ_1 and Γ_2 do not intersect).

We are going to prove that this limit exists almost everywhere and is independent of the system of short paths (as an element of $L_1(M \times M)$).

4.2. *Digression on Gauss's formula*

It will be useful to have the formula given by Gauss for the linking number of two closed curves in three-dimensional Euclidean space. There is also a similar formula for a simply connected manifold: see de Rham's book *Variétés différentiables.*

In order to state Gauss's formula, we introduce the following notation.

Let x_1: $S^1_1 \to \mathbf{R}^3$ and x_2: $S^1_2 \to \mathbf{R}^3$ be smooth mappings of a circumference in three-dimensional Euclidean space, with disjoint images. Let $t_1 \pmod{T_1}$ and $t_2 \pmod{T_2}$ be coordinates on the first and second circumference; then we denote by $\dot{x}_1 = \dot{x}_1(t_1)$ the velocity vector of the flow on the first, and by $\dot{x}_2 = \dot{x}(t_2)$ that on the second.

We assume that the circumferences are oriented by the choice of the parameters t_1 and t_2, and we fix an orientation for $\mathbf{R}^3$. Then we can define vector products and triple scalar products in $\mathbf{R}^3$.

Gauss's Theorem. *The linking number of the closed curves* $x_1(S^1)$ *and* $x_2(S^2)$

is equal to

$$N_{1,2}=+\frac{1}{4\pi}\int_0^{T_2}\int_0^{T_1}\frac{(\dot{x}_1,\dot{x}_2,x_1-x_2)}{|x_1-x_2|^3}\,dt_1\,dt_2.$$

Proof. Consider the mapping

$$f\colon T_2\to S^2$$

of the torus on the sphere, making a pair of points on our circumferences correspond to the vector of unit length directed from $x_2(t_2)$ to $x_1(t_1)$: $f=F/\|F\|$, where $F(t_1,t_2)=x_1(t_1)-x_2(t_2)$.

We orient the sphere by the inner normal and the torus by the coordinates t_1,t_2. *The degree of the mapping is equal to the linking number* $N_{1,2}$. In fact, this is true for widely separated small circumferences: both the linking number and the degree of the mapping f are 0. Furthermore, it is easy to verify that under a deformation of a curve by any passage of one curve through another both the linking number and the degree of the mapping change by 1, in the same direction. Now the equation $N_{1,2}=\deg f$ is established, in view of the connectedness of the set of smooth mappings $S^1\to\mathbf{R}^3$.

Let us show that *the degree of the mapping f is given by the integral formula of Gauss.* In fact, by the definition of the degree,

$$\deg f=\frac{1}{4\pi}\iint_T f*\omega^2,$$

where ω^2 is the area element on the unit sphere. By the definition of f, the value of the form $f*\omega^2$ on the pair of vectors ξ_1,ξ_2, tangent to the torus, is equal, at t, to its triple scalar product with the vector $-f=-f(t)$ (we oriented the sphere by the *inner* normal),

$$\omega^1(f*\xi_1,f*\xi_2)=(f*\xi_1,f*\xi_2,-f).$$

Differentiating f, we obtain $f*\xi=F*\xi/\|\xi\|+c(\xi)f$, and therefore

$$\omega^2(f*\xi_1,f*\xi_2)=(F*\xi_1,F*\xi_2,-F)/\|F\|^3.$$

Since $F=x_1-x_2$, we obtain, for an element of the spherical image of the torus, the expression

$$f*\omega^2=+(\dot{x}_1,\dot{x}_2,x_1-x_2)\|x_1-x_2\|^{-3}\,dt_1\wedge dt_2,$$

as was to be shown.

4.3. A second definition of the asymptotic linking number

Let $\{g^t\}$ be a phase flow, defined by a divergence-free field ξ in a three-dimensional compact Euclidean domain M. The field is assumed to be tangent to M on the boundary of M. We set

$$\hat{\lambda}(x_1, x_2) = \lim_{\substack{T_1\to\infty \\ T_2\to\infty}} \frac{1}{T_1 T_2} \frac{1}{4\pi} \int_0^{T_2} \int_0^{T_1} \frac{(\dot{g}^{t_1}x_1, \dot{g}^{t_2}x_2, x_1 - x_2)}{\|x_1(t_1) - x_2(t_2)\|^3}\, dt_1\, dt_2.$$

We shall show that:

(1) the limit $\hat{\lambda}(x_1, x_2)$ exists almost everywhere on $M \times M$;
(2) the number $\hat{\lambda}(x_1, x_2)$ coincides, for almost all x_1, x_2, with the number $\lambda(x, y)$ defined above.

To prove the first statement it is enough to verify that $\hat{\lambda}$ is the "time average" of a summable function on the manifold $M \times M$, on which the commutative group $\{g^{t_1}\} \times \{g^{t_2}\}$ acts.

The integrand

$$\Phi(x_1, x_2) = (\xi_1, \xi_2, x_1 - x_2)/\|x_1 - x_2\|^3 \qquad \left(\xi_k = \frac{d}{dt_k}\bigg|_{t_k=0} g^{t_k}x_k\right)$$

has a singularity on the diagonal of $M \times M$ of order no higher than r^{-2} (where r is the distance to the diagonal): since the codimension of the diagonal is 3, the function Φ belongs to the space $L_1(M \times M)$, as was to be proved.

To compare $\hat{\lambda}$ with λ we represent the linking coefficient of the curves $\Gamma_{T_1}x_1$ and $\Gamma_{T_2}x_2$ by Gauss's integral with $0 \le t_1 \le T_1 + 1$, $0 \le t_2 \le T_2 + 1$, and using the value of the parameter t_k from T_k to T_{k+1} for parametrizing the "short path" that joins $g^{T_k}x_k$ to x_k.

Definition. A system of short paths joining the points $x, y \in M$ is a system of paths, depending in a measurable way on x and y, such that the integrals of Gauss type for every pair of nonintersecting paths of the system, and also for any nonintersecting pairs (paths of the system, segments of phase curves $g^t x$, $0 \le t \le \tau < 1$), are bounded independently of the paths by a constant c.

It is easy to verify that systems of short paths exist (it is useful to keep in mind that an integral of Gauss type for a pair of straight-line segments remains bounded when the segments approach each other).

Now the difference

$$\int_0^{T_2 \cdot 1} \int_0^{T_2 \cdot 1} - \int_0^{T_2} \int_0^{T_1}$$

of integrals of Gauss type can be estimated by the sum of at most $[T_1] + [T_2] + 3$ terms, none of which exceeds c. Consequently,

$$\lambda(x, y) - \hat{\lambda}(x, y) = \lim_{T_{1,2}\to\infty} \frac{1}{T_1 T_2} \left(\int_0^{T_2 \cdot 1} \int_0^{T_1 \cdot 1} - \int_0^{T_2} \int_0^{T_1} \right)$$

(where T_1 and T_2 tend to infinity on any sequence for which the curves $\Gamma_{T_1}x_1$ and $\Gamma_{T_2}x_2$ do not intersect).

4.4. *Average linking number*

Theorem. *The mean value of the asymptotic linking number of a pair of trajectories,*

$$\iint\limits_{M\times M} \lambda(x_1, x_2)\, dv_1\, dv_2 \Big/ \iint\limits_{M\times M} dv_1\, dv_2,$$

is equal to the asymptotic Hopf invariant of the phase velocity field,

$$I(\xi) = \langle \operatorname{curl}^{-1} \xi, \xi\rangle.$$

Proof. Consider the Biot–Savart integral

$$\eta(x_2) = -\frac{1}{4\pi}\int_M \frac{[\xi(x_1), x_1 - x_2]}{\|x_1 - x_2\|^3}\, dv(x_2),$$

where [,] denotes the vector product. Then curl $\eta = \xi$ and therefore

$$\langle \eta, \xi\rangle = \langle \operatorname{curl}^{-1} \xi, \xi\rangle = +\frac{1}{4\pi}\iint\limits_{M\times M} \frac{[\xi(x_1), \xi(x_2), (x_1 - x_2)]}{\|x_1 - x_2\|^3}\, dv(x_1)\, dv(x_2),$$

as was to be proved.

Remark. There is a similar result for any compact simply connected three-dimensional Riemannian manifold M, but the Gauss integral has to be replaced by the integral of de Rham's "double form;" this form cannot be written as explicitly, but has similar properties.

Remark. The question of *whether the asymptotic and mean linking numbers are invariant for a pair of trajectories under homeomorphisms that preserve the volume element* remains open, as does the closely related question of *whether one can define asymptotic and mean linking numbers for trajectories of one-parameter groups of volume-preserving homeomorphisms.*

5. Applications to the variational problem

From the existence of the Hopf invariant there follow some lower bounds for the energy of any field obtained from a given field by a volume-preserving diffeomorphism. In particular, on any three-dimensional Riemannian manifold one can find a field that is minimal in its class. In particular, certain special force-free fields have this property.

5.1. *Minimal force-free fields*

Let M be a three-dimensional closed Riemannian manifold. We consider the operator curl on divergence-free fields that are homologous to zero (i.e., have a single-valued divergence-free potential). By Weyl's lemma on orthogonal projctions, we can define a single-valued inverse of the operator curl on our space, so that there is an inverse (integral) operator curl^{-1} from the space of divergence-free fields that are homologous to zero, onto itself. This operator is symmetric, and its spectrum accumulates at zero on both sides.

Theorem. *The eigenfield of* curl^{-1} *corresponding to the eigenvalue* v *of largest modulus has minimum energy in the class of divergence-free fields obtained from the eigenfield under the action of volume-preserving diffeomorphisms.*

Proof. Let v_- and v_+ be the smallest and largest eigenvalues of the operator curl^{-1}. Then for every field ξ that is homologous to zero we have

$$v_-\langle \xi, \xi\rangle \leq \langle \mathrm{curl}^{-1}\,\xi, \xi\rangle \leq v_+\langle \xi, \xi\rangle, \qquad v_- < 0 < v_+.$$

Consequently, we have the following bound for the energy in terms of the Hopf invariant:

$$\langle \xi, \xi\rangle \geq \langle \mathrm{curl}^{-1}\,\xi, \xi\rangle / v,$$

where v denotes the value v_+ or v_- of larger modulus.

The inequality becomes an equality for the eigenfield with the eigenvalue v. The right-hand side of the inequality is invariant under volume-preserving diffeomorphisms (see Section 2). Consequently, under the action of such a diffeomorphism on the eigenfield with eigenvalue v, the field energy can only increase. This completes the proof of the theorem.

5.2. *Examples*

Let us take M to be the three-sphere with the usual Riemannian metric. The eigenfield of the operator curl^{-1} can be calculated explicitly. The eigenfields with largest and smallest eigenvalues are the Hopf field and its symmetric field (corresponding to Hopf invariant -1). The moduli of these eigenvalues are equal.

Corollary. *The Hopf field on the three-sphere has minimum energy among all fields obtained from it by the action of a volume-preserving diffeomorphism.*

(The field lines of the Hopf field are circles, and the linking coefficient of any two of them is 1.)

As another example, we consider the three-dimensional torus with the usual Riemannian metric. The eigenfields of the operator curl^{-1} with largest and smallest eigenvalues can be written explicitly in terms of sines and cosines. We obtain the following corollary:

Corollary. *Each of the fields*

$$\xi_x = A \sin z + C \cos y, \qquad \xi_y = B \sin x + A \cos z, \qquad \xi_z = C \sin y + B \cos z$$

on the three-dimensional torus has minimum energy among all fields obtained from it under volume-preserving diffeomorphisms.

Consequently, a minimal force-free field can have a complicated topology for its field lines, as is the case for generic nonintegrable systems (some field lines cover two-dimensional tori densely, others do not lie on any two-dimensional surfaces: see the experiment of Henon mentioned in Section 1.2).

In conclusion, we remark that we can extract from the asymptotic linking number $\lambda(x, y)$ more invariants than the mean linking number λ; for example, the measure $m(\lambda_0)$ of the set $\{x, y \in M \times M: \lambda(x, y) < \lambda_0\}$, or the value of the Hopf invariant for various regions that are invariant under the flow of a given field ξ. By using such invariants one can sometimes give lower bounds for the energy of a field obtained from a given field by the action of diffeomorphisms, more precisely than those found by using only the Hopf invariant.

Acknowledgment

The author thanks the participants in Zel'dovich and Novikov's seminar for their information about the existence of papers [3]-[5] and [7] and for valuable discussions.

Note added June 6, 1985. A survey of modern generalizations of the asymptotic Hopf invariant is given in [6]. In the simplest generalization one begins with two closed 2-forms a, b on S^4 such that $as^2 = a \wedge b = b^2 = 0$ and considers $I(a, b) = p \int a \wedge d^{-1}a \wedge d^{-1}b + q \int b \wedge d^{-1}a \wedge d^{-1}b$. Such forms define two foliations of S^4 into surfaces intersecting along lines and the functional I probably has an asymptotic ergodic description similar to that given here for the Hopf invariant.

References

[1] V. I. Arnold, *Sur la géometrie différentielle des groupes de Lie de dimension infinie et ses applications à l'hydrodynamique des fluides parfaits*, Ann. Inst. Fourier (Grenoble) **16** (1966), *fasc.* 1, 319-361.

[2] M. Henon, *Sur la topologie des lignes de courant dans un cas particulier*, C. R. Acad. Sci. Paris **262** (1966), 312-314.
[3] H. K. Moffatt, *The degree of knottedness of tangled vortex lines*, J. Fluid Mech. **35** (1969), 117-129.
[4] H. K. Moffatt, *Report on the NATO Advanced Study Institute on magnetohydrodynamic phenomena in rotating fluids*, J. Fluid Mech. **57** (1973), 625-649.
[5] H. K. Moffatt, *Magnetostatic equilibria and analogous Euler flows of arbitrarily complex topology*, J. Fluid Mech. **159** (1985), 359-378.
[6] S. P. Novikov, *Analytical generalized Hopf invariant. Multivalued functionals* (in Russian), Uspehi Mat. Nauk. **39**:5 (1984), 97-106.
[7] L. Voltjer, *A theorem on force-free magnetic fields*, Proc. Nat. Acad. Sci. U.S.A. **44** (1958), 489-491; *On hydromagnetic equilibrium*, ibid., 833-841.

A magnetic field in a moving conducting fluid

V.I. Arnold, Ya.B. Zel'dovich, A.A. Ruzmaikin and D.D. Sokolov
Translated by Gerald Gould

The problem of the evolution of a magnetic field in a moving conducting medium is of current importance in connection with an explanation of the nature of the Earth's magnetic field, the mechanism of cyclic solar activity and the magnetism of other astrophysical objects. In recent years there have been discovered magnetic fields in several planets of the solar system and intensive cycles of activity of a broad class of stars.

In the absence of foreign sources the magnetic field $H(r,t)$ in a medium moving with velocity $v(r,t)$ is subject to the induction equation

$$\frac{\partial H}{\partial t} = \mathrm{rot}[v,h] - R_m^{-1}\mathrm{rot}\,\mathrm{rot}\,H, \quad \mathrm{div}\,H = 0,$$

where the magnetic Reynolds number R_m in cosmic conditions is usually very large. It is important to answer the question whether exponential growth of the field ($H \sim \exp\gamma t$) is possible for a given velocity (kinematic dynamo problem). A definite answer to this problem can be given when the flow is stationary and two-dimensional: the increment of exponential growth $\gamma \to 0$ as $R_m \to \infty$ or is negative. The dynamo is impossible when the fluid moves in stationary plane or spherical surfaces (Zel'dovich, 1956).

A sufficient condition for the streamlines of an Eulerian stationary flow to lie on surfaces $\psi(r) = \mathrm{const}$, that is, to satisfy $(v(r),\nabla)\psi = 0$, is the non-vanishing of the quantity $[v,\mathrm{rot}\,v]$ (the condition of an Euler stationary flow means that $\mathrm{rot}[v,\mathrm{rot}\,v] \equiv 0$).

A stationary flow is not two-dimensional if the individual streamlines everywhere densely fill spatial domains (stochastic flow). It is natural to expect that in such flows the magnetic field will increase with its velocity that is independent of R_m. A most important property of a stochastic flow is the exponential scattering of close particles. An example of such an Eulerian stationary flow on the compact Riemannian

Talk at the Petrovsky seminar on differential equations and mathematical problems of physics, March 25, 1981; Translation of Uspekhi Mat. Nauk **36**, No. 5, (1981), 220–221.

Originally published in: Usp. Mat. Nauk 36:5, 220-221, © Russ. Acad. Sciences 1981
Translated by G. Gould. Springer-Verlag Berlin Heidelberg2014 with kind permission,DOI 10.1007/978-3-642-31031-7_33

manifold $(x,y,z) \sim (x+1,y,z) \sim (x,y+1,z) \sim (2x+y,x+y,z+1)$ was constructed by V.I. Arnol'd in 1972. The velocity field of such a flow is constant and directed along one of the coordinate axes (z). Special magnetic fields that are periodic in z and perpendicular to the z axis increase exponentially and unboundedly in time. Each Fourier harmonic of the deviation from this increasing field starts to increase sharply with the increase rate, which does not depend on R_m during the time $\sim \ln R_m$, and then sharply decays. The authors assume that there is an analogous increase of the field H for $v = (A\sin z + C\cos y, B\sin x + A\cos z, C\sin y + B\cos x)$. In this case $[v, \operatorname{rot} v] \equiv 0$, and computer simulations of E.I. Korkina detect exponential scattering of particles in certain domains (for example, for $A = B = C$).

A magnetic field in a stationary flow with stretching in Riemannian space

V. I. Arnol'd, Ya. B. Zel'dovich, A. A. Ruzmaikin, and D. D. Sokolov

M. V. Keldysh Institute for Applied Mathematics of the USSR Academy of Sciences
(Submitted 13 July 1981)
Zh. Eksp. Teor. Fiz. **81**, 2052–2058 (December 1981)

The problem of the kinematic dynamo at large magnetic Reynolds numbers R_m is considered using as an example artificial flow with exponential particle stretching, simulating the stationary stochastic flow of a conducting fluid. The magnetic fields which have a periodic dependence on only one coordinate grow in time exponentially and without bound. Each Fourier harmonic of the deviation from this growing field increases initially rapidly with a rate independent of R_m during a time interval $t_* \simeq t_0 \ln R_m$, and then decays very rapidly.

PACS numbers: 47.65. + a

1. INTRODUCTION

In spite of substantial progress in the solution of the problem of the behavior of the magnetic field for a prescribed motion of a conducting fluid (the so-called kinematic dynamo, see, e.g., the book of Moffatt[1]), some qualitative aspects of this problem are still unclear. The most difficult and pressing for applications in cosmic physics is the case of large magnetic Reynolds numbers, $R_m = Lv/\nu_m$ (here L and v are characteristic scales of length and velocity and ν_m is the microscopic magnetic diffusion coefficient of the fluid). A decisive role in the generation of the magnetic field is played by the geometric (topological) structure of the velocity field. We restrict our attention to stationary flows of an incompressible fluid ($\operatorname{div} \mathbf{v} = 0$). Then, from the point of view of the geometric structure the flows fall into two classes: 1) flows for which the streamlines are on stationary surfaces; 2) stochastic flows, in which individual streamlines fill a spatial region everywhere densely.

In principle, a solution of the problem of the kinematic dynamo is known for the flows of the first class. At $R_m^{-1}=0$, only a nonexponential growth of the initial (intrinsic) magnetic field is possible. Taking into account a small diffusion $R_m^{-1}=\varepsilon \ll 1$ one is led, in certain circumstances, to an exponential instability, however the rate of growth of the field (the argument of the exponential) turns our to be small, proportional to some function of ε, say $\varepsilon^{1/3}$ (Ref. 2, 3). An exception are the degenerate flows along a system of parallel planes or spherical surfaces, when an exponential instability is impossible.

The answer is less clear for flows of the second class, although a number of papers[1,5,6] have been devoted to the solution of the problem of the dynamo for stochastic flows that simulate turbulence. The difficulty of the problem is that, one the one hand, the exponential separation of neighboring trajectories which is characteristic for a stochastic flow leads to an exponential growth of the magnetic field with a rate independent of R_m, and on the other hand, there occurs a sharp fragmentation of the size scales of the field and the role of diffusion increases, and does not become small even in the limit as $R_m \to \infty$.

In the present paper we consider an aritficial example of a flow with exponential stretching of particles, proposed earlier[4] by one of the authors. In this case the problem of the kinematic dynamo is amenable to a detailed investigation. Although the discussion of the example requires passing to a compact manifold with a Riemannian metric, this flow simulates the main peculiarities of a stochastic flow in Euclidean space.

Magnetic diffusion plays an important role. At $R_m^{-1}=0$ the magnetic fields depending on the three spatial coordinates may grow indefinitely. The introduction of an arbitrarily small diffusion ($\varepsilon \ll 1$) changes the result qualitatively. In this case an infinite exponential growth is possible only for fields which have a periodic dependence on one of the spatial coordinates. The Fourier harmonics of the deviation from this solution at first increase exponentially, with an exponent independent of ε, and then decay sharply. The possibility of such a temporal growth of the field was first pointed out in a paper of one of the authors,[7] and the succeeding rapid damping was indicated by Saffman.[8]

2. FORMULATION OF THE PROBLEM OF THE KINEMATIC DYNAMO

The behavior of the magnetic field for a given stationary flow of an incompressible conducting fluid is described by the induction equation, which in terms of dimensionless variables has the form

$$\frac{\partial \mathbf{H}}{\partial t} + (\mathbf{v}\nabla)\mathbf{H} = (\mathbf{H}\nabla)\mathbf{v} + R_m^{-1}\Delta\mathbf{H}, \tag{1}$$
$$\operatorname{div} \mathbf{v} = 0, \quad \operatorname{div} \mathbf{H} = 0.$$

Here $\mathbf{H}(\mathbf{r}, t)$ is the magnetic-field pseudovector, $\mathbf{v}(\mathbf{r})$ is the velocity of the flow, R_m is the magnetic Reynolds number, which is assumed to be large

$$R_m^{-1} = \varepsilon \ll 1.$$

The usual condition imposed on the solutions of the induction equation is the absence of external sources for the field. In infinite space this means that $Hr^3 \to 0$ as $r \to \infty$. However, if the velocity does not decrease at infinity, and is periodic, for example, it suffices to require that H should not increase at infinity (or should be periodic).

 0038-5646/81/121083-04$02.40

Originally published in: Zh. Eksp. Teor. Fiz. 81:6, 2052-2056, , © Russ. Acad. Sciences 1981
English transl.: Sov. Phys. JETP 54:6, 1083-1086 © American Institute of Physics 1982
Reprint by Springer-Verlag Berlin Heidelberg 2014 with kind permission, DOI 10.1007/978-3-642-31031-7_34

Time does not enter explicitly into Eq. (1). Therefore one can set in (1)

$$H\sim e^{\gamma t},\quad \gamma\mathbf{H}=\hat{P}\mathbf{H},$$

where, in general, γ is complex since the differential operator $\hat{P}$ is not self-adjoint.

3. STATIONARY FLOW WITH EXPONENTIAL PARTICLE STRETCHING

The problem (1) can be solved to the end for the case of an artifical example proposed by one of the authors,[4] which is a model of the fundamental property of a stochastic flow, namely the exponential stretching of the fluid particles. The domain of the flow is a three-dimensional compact manifold, which in Cartesian coordinates can be constructed as the product of the two-dimensional torus $\{(x, y) \bmod 1\}$ with the segment $0\leqslant z\leqslant 1$, for which the end-tori are identified according to the law

$$(x,y,0)\equiv(2x+y,\ x+y,\ 1).$$

On this manifold one can introduce a Riemannian metric as the metric in R^3 which is invariant under the transformations

$$\begin{aligned}(x,y,z)&\to(x+1,y,z),\\ (x,y,z)&\to(x,y+1,z),\\ (x,y,z)&\to(2x+y,\ x+y,\ z+1).\end{aligned}\tag{2}$$

The last transformation is implemented by the area-preserving matrix

$$A=\begin{pmatrix}2&1\\1&1\end{pmatrix},$$

which has the eigenvalues

$$\begin{aligned}&\lambda_{1,2}=(3\pm\sqrt{5})/2,\quad \lambda_1\lambda_2=1,\\ &\lambda_1\approx 2.11>1,\quad \lambda_2\approx 0.34<1.\end{aligned}\tag{3}$$

Changing from the coordinates x, y, z to the coordinates p, q, z, where p has the direction of the eigenvector with $\lambda_2<1$, and q is along the eigenvector with the eigenvalue $\lambda_1>1$, then the metric given by the line element

$$ds^2=e^{-2\mu z}dp^2+e^{2\mu z}dq^2+dz^2,\quad \mu=\ln\lambda_1\approx 0.75\tag{4}$$

is invariant with respect to the transformations (2) and therefore determines an analytic Riemannian structure on the compact three-dimensional manifold.

On this Riemannian manifold we consider a flow with the stationary velocity field

$$\mathbf{v}=(0,0,v),\tag{5}$$

where v=const, so that div $\mathbf{v}=0$ and curl $\mathbf{v}=0$. When moving in this velocity field, each fluid particle is exponentially stretched in the q-direction and exponentially contracted along the p-axis (cf. Ref. 4).

In the space with the metric (4) and coordinates p, q, z the differential operations have the form

$$\nabla=(e^{\mu z}\nabla_p,\ e^{-\mu z}\nabla_q,\ \nabla_z),$$

$$\operatorname{div}\mathbf{H}=e^{\mu z}\frac{\partial H_p}{\partial p}+e^{-\mu z}\frac{\partial H_q}{\partial q}+\frac{\partial H_z}{\partial z},$$

$$\operatorname{rot}_p\mathbf{H}=e^{-\mu z}\left(\frac{\partial H_z}{\partial q}-\frac{\partial e^{\mu z}H_q}{\partial z}\right),$$

$$\operatorname{rot}_q\mathbf{H}=e^{\mu z}\left(\frac{\partial e^{-\mu z}H_p}{\partial z}-\frac{\partial H_z}{\partial p}\right),$$

$$\operatorname{rot}_z\mathbf{H}=e^{\mu z}\frac{\partial H_q}{\partial p}-e^{-\mu z}\frac{\partial H_p}{\partial q},$$

$$\Delta=e^{2\mu z}\frac{\partial^2}{\partial p^2}+e^{-2\mu z}\frac{\partial^2}{\partial q^2}+\frac{\partial^2}{\partial z^2}.$$

The Laplacian $\Delta\mathbf{H}$ is identified with $-$curl curl $\mathbf{H}$.

Projecting Eq. (1) with the velocity field (5) onto the directions $e^{-\mu z}\nabla_p$, $e^{\mu z}\nabla_q$, ∇_z we obtain

$$\begin{aligned}&\frac{\partial H_p}{\partial t}+v\frac{\partial H_p}{\partial z}=-\mu vH_p+\varepsilon\left[(\Delta-\mu^2)H_p-2\mu e^{\mu z}\frac{\partial H_z}{\partial p}\right],\\ &\frac{\partial H_q}{\partial t}+v\frac{\partial H_q}{\partial z}=\mu vH_q+\varepsilon\left[(\Delta-\mu^2)H_q+2\mu e^{-\mu z}\frac{\partial H_z}{\partial q}\right],\\ &\frac{\partial H_z}{\partial t}+v\frac{\partial H_z}{\partial z}=\varepsilon\left(\Delta-2\mu\frac{\partial}{\partial z}\right)H_z.\end{aligned}\tag{6}$$

The equation for the z-component of the field separated, therefore asymptotically, for $t\to\infty$ the component H_z decays. Indeed, let us multiply the last equation by H_z and integrate over the volume contained between the planes $z=0$ and $z=1$. Recognizing that the integrals $\int H_z^2dpdq$ on these planes coincide, we obtain

$$\frac{d}{dt}\int H_z^2dpdqdz=-2\varepsilon\int(\nabla H)^2dpdqdz.$$

The negative character of the right-hand side of this equation proves the assertion. On the basis of this result we can for simplicity assume in the sequel that the component H_z of the field vanishes.

The equations for the p and q components differ only by the substitution $\mu\to-\mu$, it is therefore sufficient to consider only the q-component, which we denote by $H_q\equiv H$ with $\mu>0$:

$$\frac{\partial H}{\partial t}+v\frac{\partial H}{\partial z}=\mu vH+\varepsilon(\Delta-\mu^2)H.\tag{7}$$

Let us formulate the boundary conditions. This is simplest in the original coordinate system x, y, z. The symmetry (2) means that the function H is periodic in x, y:

$$H(x,y,z,t)=\sum_{n,m}H_{nm}(z,t)\exp[2\pi i(nx+my)]$$

$$=\sum_{\alpha,\beta}h_{\alpha\beta}(z,t)\exp[i(\alpha p+\beta q)],$$

where n, m are integers and α, β are related to $2\pi n$, $2\pi m$ by a linear transformation corresponding to the transition from the coordinates x, y to p, q [a rotation of the cartesian axes x, y by an angle $\arctan(2-\lambda_1)\simeq 72°$]. The symmetry with respect to a shift along the z axis:

$$H(x,y,z,t)=H(2x+y,\ x+y,\ z+1,\ t)$$

allows us to impose restrictions on the Fourier amplitudes. Substituting the last relation into Eq. (8), we obtain

$$H_{\mathbf{n}}(z+1)=H_{A'\mathbf{n}}(z),$$

where $\mathbf{n}=(n,m)$, and A' is the transpose of the matrix A;

in the case under discussion $A'=A$.

Thus, a shift along the z axis is equivalent to a transition from the Fourier amplitudes with indices $(n, m)=\mathbf{n}$ to the Fourier amplitudes with indices $A\mathbf{n}$. An exception is the case $n=m=0$, when the magnetic field does not depend on x, y, or p, q. In that case $A0=0$. In the general case, applying the matrix A to the vector $\mathbf{n}$ shifts the point n, m along a hyperbola in the (n, m) plane, see the figure.

If the function $H(x, y, z, t)$ is analytic its Fourier harmonics $h_{\alpha\beta}(z, t)$ must decay exponentially in α and β. This means, according to Eq. (2) that the function $h_{\alpha\beta}(z, t)$ must decrease not slower than a double exponential. If the function is k times differentiable, then the decay will be according to a power law, as is well known.

Thus, the solutions to Eq. (7) must be periodic in (x, y). If the solution does not depend on p, q then $H(z, t)$ is periodic in z. If there is a dependence on p, q, then the Fourier harmonics $h_{\alpha\beta}(z, t)$ of this solution must decrease rapidly with the increase of $|z|$.

We first consider the case $\varepsilon=0$. Then, going over into a Lagrangian reference frame, it is easy to obtain a solution to the Cauchy problem. Returning to Eulerian coordinates we have

$$H(t, p, q, z) = e^{\mu v t} H(0, p, q, z-vt). \quad (9)$$

When the initial field does not depend on p and q this solution can be represented as a superposition of the eigenfunctions $\exp(2\pi i m z)\exp(\gamma_m t)$ belonging to the eigenvalues

$$\gamma_m=\mu v-2\pi i m v,\ m=0, \pm 1, \pm 2, \ldots.$$

This is easily seen by expanding Eq. (9) into a Fourier series with respect to z. If the initial field depends on p and q (i.e., is a periodic function of x and y), then, as shown above, the Fourier harmonics of the expansion of (9) in terms of p and q, must decrease with the increase of $|z|$. Therefore the indicated set of functions does not describe solutions satisfying the boundary conditions. In fact this is related to the circumstances that the translation operator along the z axis has a continuous spectrum.

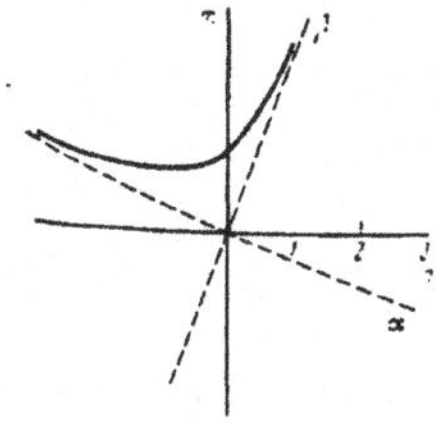

FIG. 1. The dashed axes indicate the directions of the eigenvectors of the matrix A corresponding to the eigenvalues $\lambda_1>1>\lambda_2$. Since $\lambda_1\lambda_2=1$, the product nm is conserved under the action of A on a vector with components (n, m), i.e., there occurs a shift along a hyperbola. One such hyperbola is shown in the figure.

We now go over to the general case $\varepsilon \neq 0$. As before, Eq. (7) has solutions periodic in z which are independent of p and q, with the set of eigenvalues

$$\gamma_m=\mu v+\varepsilon(4\pi^2 m^2-\mu^2)-2\pi i m v. \quad (10)$$

However, when the initial field depends on p and q, the character of the solution is completely different from (9). The shift $z-vt$ along the z axis is equivalent to a translation (along the hyperbola) of the labels of the harmonics $h_{\alpha\beta}(z, t)$ for fixed z. Therefore any given harmonic will shift with time into the region of large wave numbers, where dissipation becomes important, and asymptotically, for $t\to\infty$, will decay independently of the magnitude of ε. Let us describe this process.

We look for solutions of the form (8)

$$h_{\alpha\beta}(z, t)e^{i(\alpha p+\beta q)}.$$

Then the equation takes on the form

$$\frac{\partial h_{\alpha\beta}}{\partial t}+v\frac{\partial h_{\alpha\beta}}{\partial z}+(\varepsilon\mu^2-\mu v+\varepsilon\alpha^2 e^{2\mu z}+\varepsilon\beta^2 e^{-2\mu z})h_{\alpha\beta}=\varepsilon\frac{\partial^2 h_{\alpha\beta}}{\partial z^2}. \quad (11)$$

It is natural to assume that for $\varepsilon\to 0$ the leading role is played by terms containing the exponentials. Therefore we consider the reduced equation

$$\frac{\partial h_{\alpha\beta}}{\partial t}+v\frac{\partial h_{\alpha\beta}}{\partial z}+[\varepsilon(\alpha^2 e^{2\mu z}+\beta^2 e^{-2\mu z})-\mu v]h_{\alpha\beta}=0.$$

It has two first integrals

$$I_1=z-vt,$$

$$I_2=h_{\alpha\beta}(z, t)\exp\left[-\mu z+\frac{\varepsilon}{2\mu v}(\alpha^2 e^{2\mu z}-\beta^2 e^{-2\mu z})\right],$$

with the help of which it is easy to construct a solution of the Cauchy problem with the initial field $h_{\alpha\beta}(z, 0)$:

$$h_{\alpha\beta}(z, t)=h_{\alpha\beta}(z-vt, 0)\exp\left\{\mu v t-\frac{\varepsilon}{2\mu v}[\alpha^2 e^{2\mu z}(1-e^{-2\mu v t})-\beta^2 e^{-2\mu z}(1-e^{2\mu v t})]\right\}\approx h_{\alpha\beta}(z-vt, 0)\exp\left\{\mu v t-\frac{\varepsilon}{2\mu v}\beta^2 e^{-2\mu z+2\mu v t}\right\}. \quad (12)$$

We see that for a prescribed initial function which is bounded in z, $h_{\alpha\beta}(z, 0)$, each $\alpha\beta$-harmonic will first increase exponentially in proportion to $e^{\mu v t}$, being at the same time translated to the right along the z axis with velocity v, and then the growth is replaced by a sharp decay[1)]

$$\exp\left\{-\frac{\varepsilon\beta^2}{2\mu v}e^{-2\mu z+2\mu v t}\right\}$$

after the lapse of a characteristic time

$$t_*\approx\frac{z}{v}+\frac{1}{2\mu v}\ln\frac{1}{\varepsilon\beta^2}. \quad (13)$$

However, during this process the scale of the field in the z direction starts changing rapidly [approximately over the same time interval, only with $\ln(\varepsilon\beta^2)^{-1}$ replaced by $\ln(\varepsilon|\beta|)^{-1}$], i.e., the transition to the compact equation is no longer justified.

In order to find the asymptotic solution for $t\to\infty$ we go over from the equation (11) to an equation of the Schrödinger type:

$$\Delta\psi+(k-U)\psi=0;$$

$$\psi=H\exp\left(-\frac{v}{2\varepsilon}z-\gamma t\right),\quad k=-\gamma-\frac{v^2}{4\varepsilon}+\mu v-\mu^2\varepsilon,$$

$$U(z)=\alpha^2e^{2\mu z}+\beta^2e^{-2\mu z}=2|\alpha\beta|^{1/2}\,\mathrm{ch}\,\mu\tilde{z},\quad \tilde{z}=z+\frac{1}{2\mu}\ln\left|\frac{\alpha}{\beta}\right|.$$

The potential U has a minumum at the point $\tilde{z}=0$:

$$U_{min}=2|\alpha\beta|^{1/2}$$

and increases exponentially rapidly on both sides of the minumum. One can roughly estimate that the lowest "energy" level is of the order of U_{min}. This leads to

$$\gamma\approx-\frac{v^2}{4\varepsilon}+\mu v-\mu^2\varepsilon-U_{min}\underset{\varepsilon\to 0}{\longrightarrow}-\frac{v^2}{4\varepsilon}.$$

It can be shown that for $z\to\infty$ the corresponding eigenfunction has the form

$$h_{\alpha\beta}(z,t)\approx\exp\left[\left(-\frac{v^2}{4\varepsilon}+\mu v\right)t+\frac{v}{2\varepsilon}\tilde{z}-2|\alpha\beta|^{1/2}\,\mathrm{ch}\,\tilde{z}\right].$$

We thus come to the conclusion that asymptotically, for $t\to\infty$, only the solution which is independent of p and q survives.

We want to stress the fundamentally three-dimensional character of the problem. The expansion occurs in the p, q plane, and the velocity of the flow is along the z axis. It was the shift along the z axis which was responsible for the increasing factor $e^{\mu v t}$, which does not depend on p and q, in the field. On the other hand, the same shift along the z axis gives rise to the continuous spectrum, leading to the sharp decay of the harmonics in the expansion with respect to p and q.

1) Roughly speaking we have $k_{eff}=k_0e^{\mu vt}$, $h_{\alpha\beta}=h_{\alpha\beta}(0)e^{-ck^2}$, which leads to the exponential in the argument of the exponential.

[1] K. H. Moffatt, Generation of Magnetic Fields in a Conducting Medium (Russian Transl.), Mir, 1980.

[2] Ya. B. Zel'dovich and A. A. Rumaĭkin, Zh. Eksp. Teor. Fiz. 78, 980 (1980) [Sov. Phys. JETP 51, 493 (1980)].

[3] A. A. Ruzmaikin and D. D. Sokoloff, Geophys. Astrophys. Fluid Dyn. 16, 73 (1980).

[4] V. I. Arnol'd, Prikl. Matem. Mekh. 36, 255 (1972).

[5] A. P. Kazantsev, Zh. Eksp. Teor. Fiz. 53, 1806 (1967) [Sov. Phys. JETP 26, 1031 (1968)].

[6] S. I. Vaĭnshteĭn, Zh. Eksp. Teor. Fiz. 79, 2175 (1980) [Sov. Phys. JETP 52, 1099 (1980)].

[7] Ya. B. Zel'dovich, Zh. Eksp. Teor. Fiz. 31, 154 (1956) [Sov. Phys. JETP 4, 460 (1957)].

[8] P. G. Saffman, J. Fluid Mech. 16, 545 (1963).

Translated by Meinhard E. Mayer

ary conditions, for example $u(t, 0) = \varepsilon^2\mu_0$ [t, u(t, 0), p(t, 0) ε], where the function μ_0 is regular in ε and periodic in t; 2) for general systems of the gas-dynamic type that are reducible to Riemann invariants.

[1]O. V. Rudenko and S. I. Soluyan, Theoretical Foundations of Nonlinear Acoustics [in Russian], Nauka, Moscow (1975).
[2]N. N. Bogolyubov and Yu. A. Mitropol'skii, Asymptotic Methods in the Theory of Nonlinear Oscillations [in Russian], Nauka, Moscow (1974).
[3]A. L. Shtaras, Dokl. Akad. Nauk SSSR 237, 525-528 (1977).
[4]V. V. Kaner, O. V. Rudenko, and R. V. Khokhlov, Akust. Zh. 23, 756-765 (1977) [Sov. Phys. Acoust. 23, 432 (1977)].
[5]Yu. K. Éngel'brekht and U. K. Nigul, Nonlinear Deformation Waves [in Russian], Nauka, Moscow (1981).
[6]L. A. Ostrovskii, Nonlinear Acoustics, Preprint No. 32, Scientific-Research Institute of Radiophysics, Gorkii (1973).
[7]B. L. Rozhdestvenskii and N. N. Yanenko, Systems of Quasilinear Equations [in Russian], Nauka, Moscow (1978).

Translated by David G. Miller

Steady-state magnetic field in a periodic flow

V. I. Arnol'd, Academician Ya. B. Zel'dovich, A. A. Ruzmaikin, and D. D. Sokolov

M. V. Lomonosov State University, Moscow; M. V. Keldysh Institute of Applied Mathematics, Academy of Sciences of the USSR, Moscow

(Submitted June 9, 1982)

Dokl. Akad. Nauk SSSR 266, 1357–1361 (October 1982)

PACS numbers: 03.40.Gc

1. Of particular interest in the consideration of a magnetic field in a specified flow of a conducting fluid is the problem of the enhancement or maintenance of the initial magnetic field (kinematic dynamo). Steady-state fields (the maintenance case) usually arise with an increase in the magnetic Reynolds number, when the exponential attenuation changes to an exponential growth, which is finally limited by the nonlinearity arising because of the effect of the field on the motion. We shall indicate another type of steady-state magnetic field, which exists for any magnetic Reynolds number. The simplest example of such a field is a nonzero averaged field in infinite space. The attenuation mechanism related to dissipation leads to a finite system, as a rule, to an attenuation time τ proportional to the square L^2 of the characteristic size. As $L\to\infty$, $\tau\to\infty$ also. One can say that the attenuation disappears in the limit $L\to\infty$. However, one can construct an analog of this situation with finite dimensions. This is the subject of this article. Maintenance of steady-state unattenuated fields is possible in multiply connected (or periodic) velocity fields. Although steady-state fields of this topological type have been obtained in a number of examples (see, for example, Ref. 1), the fundamental difference between them and steady-state fields of the first type has not been discussed. We shall discuss the problem of the existence and number of such solutions. The interesting problem of the spatial structure of the steady-state field remains beyond the scope of this article. As the example of the convective flows investigated in Ref. 2 shows, the field distribution depends significantly on whether the flow is two- or three-dimensional.

2. Problem of the steady-state solution. Let us consider the steady-state flow $\mathbf{v}(\mathbf{r})$ of a fluid having constant conductivity. The evolution of the initially specified magnetic field $\mathbf{H}_0(\mathbf{r})$ is described by the induction equation, which in dimensionless variables has the form

$$\frac{\partial \mathbf{H}}{\partial t} = \operatorname{curl} \mathbf{v}\times\mathbf{H} + R_m^{-1}\Delta\mathbf{H},\quad \operatorname{div}\mathbf{H}=0, \tag{1}$$

where R_m is the magnetic Reynolds number, which is determined by the characteristic scale and velocity of the flow and the conductivity.

In the dynamo problem Eq. (1) is supplemented by boundary conditions corresponding to the absence of external emf's. When the velocity decreases at infinity, it is natural to distinguish the intrinsic field depending on the processes in the region where the fluid is moving. We shall ignore those fields which increase at infinity or tend to a constant and limit ourselves to fields satisfying the condition $\mathbf{H} = O(|\mathbf{r}|^{-3})$, $|\mathbf{r}|\to\infty$. In a periodic velocity field it is natural to require the same periodicity of the magnetic field, instead of the decrease at infinity.

For the formulated boundary conditions Eq. (1) conserves the integral $\int \mathbf{H}(t, \mathbf{r})d^3\mathbf{r}$. In periodic flow the integration is carried out over a periodicity cell, and in aperiodic flow the integration is carried out over all space. If $\mathbf{H} = O(|\mathbf{r}|^{-3-\varepsilon})$, $\varepsilon > 0$ at infinity, then the integral converges to zero. For $\varepsilon = 0$ the integral converges in the sense of the principal value and, within a factor of $8\pi/3$,[1] equals the magnetic dipole moment of the field configuration under consideration.

To prove the conservation of the integral, we integrate Eq. (1) over the volume. The integrals on the right-hand side reduce, according to the Stokes theorem, to surface integrals:

$$\frac{d}{dt}\int \mathbf{H}d^3r = \int d\mathbf{S}\times(\mathbf{v}\times\mathbf{H} - R_m^{-1}\operatorname{curl}\mathbf{H}). \tag{2}$$

In infinite space the integrals over an infinitely remote surface vanish by virtue of the decrease of the magnetic field. The periodic problem is equivalent to consideration of the vectors **v** and **H** on a compact manifold (which on extension gives the periodic pattern). For example, the fields in Cartesian coordinates x, y, z with periods l_1, l_2, l_3 are equivalent to the fields on a torus obtained from Euclidean space after the identifications $(x, y, z) = (x + nl_1, y + ml_2, z + pl_3)$, where n, m, p are integers. In this

0038-5689/82/100814-03$02.20

Originally published as Dokl. Akad. Nauk SSSR, 266:6, 1357-1361, © Russ. Acad. Sciences 1982
English transl.: Sov. Phys. Dok. 27:10, 814-816 © American Institute of Physics 1982
Reprint by Springer-Verlag Berlin Heidelberg 2014 with kind permission, DOI 10.1007/978-3-642-31031-7_35

case the integrals on the right-hand side of (2) vanish simply because of the absence of a boundary. For an individual periodicity cell one can speak of compensation of the contribution of the identified boundaries.

The result can be interpreted in the following manner. Let us specify an initial field with a nonzero integral $\int H_0 d^3r$. Then is the discrete spectrum of problem (1) is complete, it necessarily has a zero eigenvalue and corresponding eigenfunction, which appears with a nonzero coefficient in the expansion of the initial field. In the aperiodic case the zero point also belongs to the spectrum (in view of the conservation of the magnetic moment), but, as the examples of Ref. 1 showed, steady-state solutions in the aperiodic case are usually obtained only for fixed magnetic Reynolds numbers.

In the periodic case the zero point belongs to the discrete spectrum. This was first noted in the numerical experiments of E. I. Korkina. This eigenvalue is degenerate, i.e., it corresponds to several eigenfunctions. For example, there is a triple degeneracy in the case of the three-dimensional torus. This is related to the fact that the initial field distribution with nonzero magnetic moment can be specified in three different ways by directing the magnetic moment along the x, y, or z axes. In general the degree of degeneracy is no less than the number of topologically different closed curves on the manifold (the Betti number). In the aperiodic case the zero point, which in this case may have no corresponding eigenfunctions, is also no less than triply degenerate, in correspondence with the three linearly independent directions of the magnetic moment.

Let us emphasize the necessity of a constant component of the initial field. Formally such a field is periodic with any period. Without a constant component the integral $\int H d^3r$ vanishes in the periodic problem (1).

It is important to obtain an answer to the question of whether the solutions for which $\int H_0 d^3r \neq 0$ initially converge to the steady-state solution described above. The positive answer to this question is obvious for sufficiently small Reynolds numbers, when the increasing modes are not excited in the system. Of course, in the general case the answer to the question is negative, since examples of exponentially increasing solutions are known for sufficiently large magnetic Reynolds numbers.

These results are formulated mathematically in the following manner. Let us consider the evolution of a k-dimensional differential form on a compact Riemannian m-dimensional manifold without a boundary under the effect of transfer by the flux of the vector field (not necessarily solenoidal!) and diffusion. Then: 1) the dimensionality r of the space of steady-state solutions is no less than the k-dimensional Betti number of the manifold: $r \geq b_k$; 2) for high viscosity $r = b_k$ and the evolution always leads to a steady state; 3) in the two-dimensional case m = 2, k = 1 the emergence to a steady-state regime occurs for any viscosity (the exclusion theorem); 4) in the three-dimensional case m = 3, k = 2 examples exist with $r > b_2$, and there also exist flows in which the magnetic field with a unique vector potential increases exponentially.

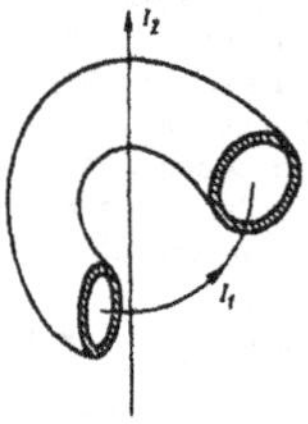

FIG. 1

At first glance, the result for the two-dimensional case contradicts the exclusion theorem forbidding a dynamo (even steady-state) for two-dimensional flows.[3] Actually, magnetic fields with boundary conditions corresponding to the aperiodic case were considered in this theorem. Let us generalize this theorem to the periodic case.

3. Exclusion theorem for the two-dimensional case. The problem of the behavior of a vector field lying on a surface, as is well known, leads to the problem of the transport of a scalar quantity, say, the temperature in the flow. We shall consider the truly two-dimensional problem (two-dimensional velocity field and two-dimensional vector field, independent of the third coordinate).

For a scalar quantity the transport equation

$$\frac{\partial T}{\partial t} + (\mathbf{v}\nabla)T = \chi \Delta T \tag{3}$$

for any (even three-dimensional) velocity $\mathbf{v}(\mathbf{r})$ and initial temperature leads with the assumption of incompressibility ($\mathrm{div}\,\mathbf{v} = 0$) to the dissipation theorem

$$\frac{d}{dt}\int (T-\bar{T})^2 d^2r = \frac{d}{dt}\int T^2 d^2r = -2\chi \int (\nabla T)^2 d^2r. \tag{4}$$

For an initial smooth $T_0(\mathbf{r})$ and small thermal diffusivity the temperature gradients increase at first; in particular, the integral $\int (\nabla T)^2 d^2r$ starts to increase. However, this integral necessarily changes in time to a decrease and the integral on the right-hand side of (4) vanishes.

The dissipation theorem can be generalized to the case of variable χ: $\chi \Delta T \to \mathrm{div}(\chi \nabla T)$ and even to the nonlinear case $\chi(T)$, as well as to the case of an arbitrary function of T:

$$\frac{d}{dt}\int f(T) d^2r = -2\int \chi \frac{d^2 f}{dT^2} (\nabla T)^2 d^2r.$$

Hence it follows that $\nabla T \to 0$ everywhere asymptotically as $t \to \infty$, i.e., the scalar field becomes uniform $T(t, \mathbf{r}) =$ const. In the periodic velocity field the solution T = const is formally the eigenfunction of Eq. (3) with eigenvalue $\gamma = 0$.

The two-dimensional magnetic field is the curl of a vector potential having a single nonzero component A, normal to the surface of the motion. With respect to transformations of the coordinates on the surface, A behaves like a scalar. In a locally Euclidean two-dimensional space,

$$H_x = \frac{\partial A}{\partial y},$$

$$H_y = -\frac{\partial A}{\partial x},$$

and the magnitude of the field is determined by the modulus of the two-dimensional gradient:

$$|H| = \sqrt{H_x^2 + H_y^2} = \sqrt{\left(\frac{\partial A}{\partial x}\right)^2 + \left(\frac{\partial A}{\partial y}\right)^2}.$$

But, in contrast to a scalar, the vector potential on a compact, multiply connected surface is not necessarily unique. For example, on the torus x, y(mod 1), where there are two families of lines not contractable to a point, the potential does not necessarily return to the initial value after one circuit of these lines, and one can obtain constant increments a and b. In the periodic problem corresponding to the torus the potential can be represented in the form

$$A = ax + by + A_1(x, y, t), \tag{5}$$

where A_1 is a periodic function of x and y. The magnetic field of this potential contains the constant component of $(H_x, H_y) = (b, -a)$, which is conserved in time by virtue of the conservation of $\int H d^2r$. Therefore, a and b do not depend on t.

The potential A satisfies an equation resulting from (1):

$$\frac{\partial A}{\partial t} + (v\nabla)A = R_m^{-1}\Delta A. \tag{6}$$

Multiplying by A and integrating over the periodicity cell, we obtain

$$\frac{d}{dt}\int A^2 d^2r = -2R_m^{-1}\int(\nabla A)^2 d^2r + R_m^{-1}\int\nabla(A^2)dl - \int A^2 v\, dl,$$

[compare with (4)]. The second integral over the boundary of the cell vanishes only for a unique potential. Therefore, attenuation occurs only in this case.

The steady state for the magnetic field must be established asymptotically as $t \to \infty$ in the general case. For specified constants a and b and d/dt = 0 we obtain from (6) the steady-state average value of the magnetic field deviating from the constant field (the deviation is described by the periodic potential A_1):

$$R_m^{-1}\int(\nabla A_1)^2 d^2r = -\int A_1(axv_x + byv_y)d^2r. \tag{7}$$

Hence, it is seen that a characteristic scale of the order of $\sim R_m^{-1/2}$ is formed. We notice that $A_1 = 0$ does not satisfy Eq. (6) with nonzero a, b, and v.

The origin of a nonunique vector potential can be interpreted physically by imagining a thin layer of a conductor (a liquid metal, e.g., mercury or sodium) between the walls of two nested tori (Fig. 1). An annular conductor with current I_1 is inserted within the torus, and a straight conductor with current I_2 penetrates the entire torus. The constants a and b are proportional to I_1 and I_2, respectively. Of course, the magnetic field is not two-dimensional in this case. There is a field component normal to the surface of the torus, and this component is transported by the motion without stretching but is coupled to other components through the diffusion term. However, the presence of this component does not change the conclusion about the nonuniqueness of the potential.

The motion of the fluid does not affect the currents I_1 and I_2, but generates a small-scale field. The energy of this small-scale magnetic field and the energy lost to the enhanced dissipation are derived from the forces causing the motion of the fluid. If the fluid is at rest, then A_1 vanishes [cf. Eq. (7)]; i.e., there are no currents in the conducting layer and, correspondingly, no energy dissipation.

We note in conclusion that a definite average value of a scalar with zero average gradient arises in a periodic field. In contrast to this, the two-dimensional magnetic situation with a nonunique vector potential is equivalent to the scalar problem with a specified average gradient, i.e., to the problem in which, with periodic conditions on the velocity, the solution does not satisfy the periodicity condition. Such a thermal problem can be realized in infinite space with periodicity cells in v, but with infinite temperature at spatial infinity. But this situation cannot be realized on the torus, where all the uniquely defined variables are strictly periodic. In the problem under consideration the dissipation alters the flow entering from the outside.

[1]H. K. Moffat, Magnetic Field Generation in Electrically Conducting Fluids, Cambridge Univ. Press (1978).
[2]E. M. Drobyshevskii and V. S. Yuferev, J. Fluid Mech. 65, 33 (1974).
[3]Ya. B. Zel'dovich, Zh. Eksp. Teor. Fiz. 31, 154 (1956) [Sov. Phys. JETP 4, 460 (1957)].

Translated by David G. Miller

SOME REMARKS ON THE ANTIDYNAMO THEOREM

V. I. Arnol'd

Vestnik Moskovskogo Universiteta. Matematika,
Vol. 37, No. 6, pp. 50-57, 1982

UDC 538.4

1. DESCRIPTION OF THE RESULTS

1. The antidynamic theorem is the assertion that a magnetic field is attenuated under the action of magnetic viscosity (despite its elongation by the motion of the medium, into which the field would be frozen if there were no viscosity). Cowling [1] and Zel'dovich [2] proved antidynamic theorems for a two-dimensional incompressible medium filling a simply connected region. From the general theorems proved below, there follow, in particular, generalizations of the antidynamic theorem for the nonsimply-connected case, when instead of attenuation of the field we observe the establishment of a stationary (constant with respect to time) field. The assumption that the medium is incompressible turns out to be superfluous (in the compressible case we need merely consider not the vector field but the vector density). The assumption that the evolving field is solenoidal is also superfluous: we shall see that the evolution automatically leads in the end to a solenoidal density, even if the initial condition was arbitrary. The assumption of two-dimensionality is extremely important, but some results can be proved even for motion in a field of arbitrary dimensionality m, for example, m = 3.

2. Thus, we consider the evolution of k-forms on a compact n-dimensional Riemannian manifold without boundary, which is acted upon by translation and diffusion (k = 0 corresponds to functions, k = m to densities, k = m - 1 to vector densities, etc.). For such an evolution the closed form remains closed. It is found that for any k the number of linear independent stationary (remaining unchanged during the evolution closed k-forms is not less than the k-dimensional Betti number on the manifold (the number b_k of independent k-dimensional cycles). In the case k = 0 or k = m, i.e., in the case of the evolution of functions and densities, we can assert the following: the number of linearly independent stationary solutions is exactly equal to the Betti number, i.e., is equal to a number of connectivity components of the manifold, and evolution with any initial condition leads in adjoint to a stationary distribution of the scalar in question.

Originally published in: Vestn. Mosk. Univ. Ser. I, Mat. Mekh. 6, 50-57, © Izd. Moscow State Univ. 1982
English transl.: Moscow Univ. Math. Bull. 37, 57-66, © Allerton Press, Inc., New York, 1982
Reprint by Springer-Verlag Berlin Heidelberg 2014 with kind permission, DOI 10.1007/978-3-642-31031-7_36

In the case k = 1 for closed 1-forms (and hence for the total differentials of multi-valued functions) the number of linearly independent stationary solutions is equal to the Betti number b_1, and evolution with any closed, i.e., vortex-free) initial condition leads to a stationary solution with the same periods (integrals over closed curves); this stationary solution is uniquely defined by the periods. For the case k = m - 1 (for vector densities) we establish that the evolution as time goes on makes the density devergence-free (even if the medium is compressible), but in this case we cannot assert that the evolution leads to a time-stationary vector density; only in the two-dimensional case (m = 2), as mentioned earlier, does the emergence to the stationary divergence-free density as for any initial condition.

3. In the case which is of greatest physical importance, namely, a magnetic field in three-dimensional space (k = 2, m = 3), we give examples showing that: 1) the number of linearly independent stationary solutions may exceed the Betti number b_2; 2) the evolution of a divergence-free field of general position may, as time goes on, lead not to a stationary field but to a field which grows exponentially (if the viscosity is sufficiently low). Our growing field is cohomologous to zero (it has a single-valued vector potential).

The increments of the growing field as the viscosity decreases will tend not to zero but to a positive limit ("half dynamo"). It seems to us that this effect is due not to the topology of the manifold in which the flow takes place but simply to the exponential elongation of the particles of liquid, and that it can take place for flows in limited regions of Euclidean three-dimensional space, but this has not been proved.

When the viscosity is sufficiently high, then for any k and m the dimension of the space of stationary solutions is equal to the Betti number b_k, and as time goes on, a stationary regime is established (unique in the class of cohomologies of initial closed form). It is probable that for any values of viscosity in each class of cohomologies there is a stationary solution, but this has not been proved.

2. EQUATIONS OF MOTION

We consider the equation of transfer and diffusion of k-forms

$$\partial\omega/\partial t = L_v\omega + D\Delta\omega \qquad (1)$$

on a compact m-dimensional manifold M. Here ω is a differential k-form on M evolving under the action of the mostness of field, given by the vector field v on M, and the diffusion with coefficients D > 0. The Lie differentiation operator L_v is defined by the condition that the form is frozen into the medium: the value of the form evolving under the action of Eq. (1), with D = 0, on the vectors drawn on the particles of the medium and of their images as the particles move (given by the velocity field v) does not change with time, although the particles move to a new place.

The linear operator L_v is expressed in terms of the operator i_v (substitution of the field into the form as the first argument) and the external-derivative operator d by the "homotopy formula" $L_v = i_v \circ d + d \circ i_v$. The Laplace operator Δ on the k-forms is defined by the formula $\Delta = d\delta + \delta d$, where δ is the operator conjugage to d in the sense of the Riemannian metric on M.

Example 1. Suppose that M is a three-dimensional Euclidean space (with the normal metric $ds^2 = dx^2 + dy^2 + dz^2$). We can specify a 2-form $\omega = P dy \wedge dz + Q dz \wedge dx + R dx \wedge dy$ by the vector field Ω with components P, Q, R, i.e., $\omega = i_\Omega \tau$, where $\tau = dx \wedge dy \wedge dz$ is an element of volume. Equation (1) for 2-forms, as we can readily convince ourselves, takes on the form of the following equation on the vector field Ω:

$$\partial\Omega/\partial t = \operatorname{rot}[v \times \Omega] + v \operatorname{div} \Omega + D(\operatorname{grad} \operatorname{div} \Omega - \operatorname{rot} \operatorname{rot} \Omega). \quad (2)$$

In particular, if the field Ω is solenoidal for $t = 0$ ($\operatorname{div} \Omega = 0$), then it remains solenoidal for all t and will evolve according to the law

$$\partial\Omega/\partial t = \operatorname{rot}[v \times \Omega] - D \operatorname{rot} \operatorname{rot} \Omega. \quad (3)$$

If, in addition, the medium is incompressible (div v = 0), then Eq. (3) takes the form

$$\dot{\Omega} = \{\Omega, v\} - D \operatorname{rot} \operatorname{rot} \Omega \quad (4)$$

(where {,} are the Poisson brackets) and for D = 0 the field Ω is frozen into liquid. Thus, Eq. (1) is a direct generalization of Eqs. (4), (3), and (2).

Example 2. For k = 0, i.e., for functions, Eq. (1) takes the form of the heat-conducting equation with translation:

$$\partial f/\partial t = \langle v, \operatorname{grad} f \rangle + D \Delta f \quad (5)$$

[which is also convenient in the Riemannian case if we take Δ as div grad and define the vector operations by the condition $\langle \xi, \operatorname{grad} f \rangle = df|\xi$ for any vector ξ (the scalar product is Riemannian); $\tau \operatorname{div} A = d i_A \tau$, where τ is the Riemannian element of the volume].

Example 3. For k = m, i.e., for ω = gτ (where τ is an element of volume), Eq. (1) takes the form of an equation on the scalar density g, namely,

$$\partial g/\partial t = \operatorname{div}(gv) + D\Delta g. \quad (6)$$

59

3. FORMULATION OF THE RESULTS

Definition. A closed k-form ω is called *stationary* if it satisfies the equation

$$L_v\omega + D\Delta\omega = 0. \qquad (7)$$

Theorem 1. The number of linearly independent stationary k-forms is not less than the k-dimensional Betti number of the manifold M.

Examples in which the first number is strictly larger than the second are given below.

Theorem 2. If the diffusion coefficient D is sufficiently large, then the numbers proved in Theorem 1 are equal, and: 1) in each cohomology class of closed k-forms there is a stationary form; 2) there is exactly one such form; 3) the evolution of Eq. (1) with any closed initial conditions leads in the limit to a stationary form belonging to the same cohomology class, i.e., with the same integrals with respect to k-dimensional cycles; 4) the evolution of Eq. (1) with any initial conditions leads in the limit to a closed form; 5) all solutions of Eq. (7) are closed.

Remark 1. Examples show that Assertions 2 and 3 are no longer true if the viscosity is sufficiently low (except in the cases k = 0, k = 1, and k = m, considered below). In the example proposed in [3], k = 2, m = 3. Equation (4) has an increasing solution $\Omega = e^{\gamma t} e_q$, where $\gamma = \mu - D\mu^2$ (in the notation of [3]). If $D = \mu^{-1}$, then this solution is stationary. It is cohomological to zero ($e_q = -\mu^{-1}\text{rot}\, e_p$). The manifold of the example considered here has $b_2 = 1$. For any D, Eq. (4) has the stationary harmonic solution $\Omega = v$ (in the notation of [3]). For $D = 1/\mu$ the dimension of the space of stationary solutions is not less than $2 > b_2 = 1$.

As another example with the same properties, we can consider the geodesic flow on a manifold of constant negative curvature (as e_p and e_q we have the velocity field of the oricyclic flows, and as v we have the field of geodesic flow).

Theorem 3. We consider the heat-conduction equation with transfer for scalars (5) on a connection compact Riemannian manifold. For any positive value of the diffusion coefficient: 1) every stationary solution of Eq. (5) is constant; 2) the solution of Eq. (5) with any initial condition tends to a constant as $t \to \infty$.

Theorem 4. We consider the heat-conduction equation with transfer for scalar densities (6) on a connected compact Riemannian manifold. For any positive value of the diffusion coefficient D: 1) the dimension of the space of stationary solutions of Eq. (6) is equal to 1; 2) there exists a stationary solution (and the solution is unique) with any value of the integral over the entire manifold; 3) the solution with any initial conditions tends as $t \to \infty$ to a stationary solution with the same integral; 4) in particular, the solution with initial conditions $g = \operatorname{div} \Omega$ tends to zero as $t \to \infty$, irrespective of the field Ω.

Theorem 5. We consider Eq. (1) for closed 1-forms on a compact Riemannian manifold. For any positive value of the diffusion coefficient:

1) the dimension of the space of stationary solutions is equal to the one-dimensional Betti number of the manifold;

2) there exists a stationary solution (and the solution in unique) with any values of the integrals over (independent) one-dimensional cycles;

3) the solution with any initial conditions tends as $t \to \infty$ to a stationary solution with the same integrals.

Theorem 6. We consider Eq. (1) for an $m - 1$-form on a compact m-dimensional Riemannian manifold, i.e., for vector densities.* Then as $t \to \infty$, the divergence of the evolving density tends to zero for any value of the diffusion coefficient $D > 0$. In particular, every stationary solution of Eq. (1) for $m - 1$-forms is closed.

Example 4. The solution of Eq. (2) with any initial conditions which is periodic in all three coordinates (for a field v of the same periodicity) becomes solenoidal as $t \to \infty$, i.e., $\operatorname{div} \Omega \to 0$.

In particular, every stationary solution of Eq. (2) is solenoidal.

Theorem 7. We consider Eq. (1) for 1-forms on a compact two-dimensional manifold. For every positive diffusion coefficient every solution tends as $t \to \infty$ to a stationary closed 1-form.

Example 5. We consider the system of equations

$$\dot{P} = (P_x + Q_y)u + (Pv - Qu)_y + D\Delta P,$$
$$\dot{Q} = (P_x + Q_y)v - (Pv - Qu)_x + D\Delta Q$$

in the unknown functions P and Q with periods (X, Y) with respect to the variables (x, y), where u, v are known functions with the same periods. By Theorem 7, the solution with any initial conditions tends as $t \to \infty$ to a stationary solution which is divergence-free $(P_x + Q_y \equiv 0)$. For any constant a and b, there exists a sta-

*For an $m - 1$-form ω with vector density Ω we have $\omega = i_\Omega \tau$ (τ is the Riemannian element of the volume); the law of evolution of the vector density Ω is

$$\dot{\Omega} = \{\Omega, v\} + \Omega \operatorname{div} v + D\Delta\Omega.$$

61

tionary solution (and the solution is unique) of the form

$$P = a + \psi_y, \quad Q = b - \psi_x,$$

where ψ is a function of period X in x and Y in y.

In particular, we consider the equation

$$\Omega^{\cdot} = \mathrm{rot}[w \times \Omega] + D\Delta\Omega \qquad (8)$$

with respect to a solenoidal field Ω which is periodic in x and y and is independent of z, with components (P, Q, 0), for a field w which is independent of z and periodic with respect to x and y and has components (u, v, 0).

Theorem 8. For every D > 0, Eq. (8) has: 1) a stationary solution with any predetermined periods a and b, and there is only one such solution; 2) a solution with any solenoidal initial conditions (periodic and independent of z) tends as $t \to \infty$ to a stationary solution with the same "flows" a and b as the initial condition.

4. PROOFS

Proof of Theorem 1 (according to E. I. Korkina). We consider on the closed k-forms the operator $A = L_v + D\Delta$. We denote by Ker A the set of solutions of the homogeneous equation $A\omega = 0$ and by Im A the image of A - the subspace of the space H of closed k-forms. The factor space H/Im A is called the cokernel (Coker A). The difference between the dimensions dim Ker A-dim Coker A is called the index of the operator. In the finite-dimensional case the index of the operator acting from a space into itself is always equal to zero. In the infinite-dimensional case this is not always so, but it is true for the Laplace operator, and since our operators differs from $D\Delta$ only in its higher-order terms (L_v is of first order), it follows that $\mathrm{ind}\, A = \mathrm{ind}\, \Delta$ (see, e.g., [4,5]). This means that dim $\mathrm{Ker}\, A = \dim \mathrm{Coker}\, A$. But $\mathrm{Im}\, A \subset \mathrm{Im}\, d$ (since $A\omega = d(i_v\omega + D\delta\omega)$ if $d\omega = 0$). Consequently,

$$\dim H/AH \geq \dim H/\{d\omega^{k-1}\} = b_k$$

(de Rama's theorem).

Proof of Theorem 2. A. In the evolution of Eq. (1) the cohomology class of the closed form does not change, since $A\omega = d(i_v\omega + D\delta\omega)$ is an exact form. B. On the orthogonal complement of the harmonic forms in closed forms, i.e., on the exact forms ω, the following equations hold:

$$\langle \omega, \omega \rangle \leq \alpha \langle \delta\omega, \delta\omega \rangle, \quad \langle \omega, L_v\omega \rangle \leq \beta \langle \delta\omega, \delta\omega \rangle,$$

where α and β are positive constants independent of ω. C. From A and B it follows that in the space of exact forms the evolution of Eq. (1) leads to a degeneracy if D is sufficiently large:

62

$$\langle\omega,\omega\rangle^{\cdot} = 2\langle\omega, L_v\omega\rangle - 2D\langle\omega, d\delta\omega\rangle \leqslant 2(\beta - D)\langle\delta\omega,\delta\omega\rangle \leqslant -2\gamma\langle\omega,\omega\rangle,$$

if $D \gg \beta + \alpha\gamma$. D. From A and B it follows that in an affine space of closed forms lying in one (any) cohomology class, the evolution of Eq. (1) defines the flow of *compressed* transformations (in the Hilbert metric) and consequently has a fixed point. We have thus proved assertions 1)-3). E. Both L_v and Δ commute with d, and therefore $d\omega$ satisfies an equation (1) of the same kind as ω. But the form $d\omega$ is exact, and therefore, in accordance with C, it tends exponentially to zero as $t \to \infty$. Consequently, the distance from $\omega(t)$ to the space of closed forms tends exponentially to zero as $t \to \infty$. Moreover, the same tendency to zero is found in metrics which take account of the derivative (if the diffusion coefficient D is sufficiently large; this is proved, as in C, by using inequalities of the type $<\Delta\omega, \Delta L_v\omega> \ll D<\Delta\omega, \Delta^2\omega>$ for the exact forms).

We now denote the orthogonal decomposition of the initial form ω into exact, harmonic, and co-exact (lying in the image of the operator δ) terms by $\omega = p + h + q$. Equation (1) takes the form of the system

$$\dot{p} = A_1 p + A_2 h + A_3 q,\ \dot{h} = A_4 q,\ \dot{q} = A_5 q$$

(since for $q(0) = 0$ the form remains closed, i.e., $q(t) \equiv 0$, and a closed form retains its cohomology class, i.e., $\dot{h} = 0$ for $q = 0$).

Now, since $q(t) \to 0$ (in metrics with derivative) exponentially, $h(t)$ tends to a finite limit (also in metrics with derivative). But in accordance with C, the transformation $\exp(A_1 t)$ is compressed, and therefore $p(t)$ also tends to a finite limit. Consequently, $\omega(t)$ converges to a finite limit $p(\infty)+h(\infty)$ which is a closed form. This proves assertions 4) and 5).

Proof of Theorem 3 (according to Yu. S. Il'yashenko and E. M. Landis). If the stationary solution were at any point larger than its minimum, it would immediately increase everywhere (since the heat is propagated instantaneously) and would not be stationary (the so-called strengthened maximum principle). Consequently, it must be everywhere equal to its minimum, i.e., it must be constant.

The same reasoning shows that a time-periodic solution of Eq. (5) must also be a constant. Consequently, the operator $A = L_v + D\Delta$ on functions has no pure imaginary eigenvalues and has a single eigenvector with eigenvalue zero. It is also impossible to have any Jordan cells with eigenvalue zero (by the maximum principle); this means that zero is an eigenvalue of multiplicity one, and all other eigenvalues lie strictly in the left-plane.

Since A is the sum of an elliptic operator $D\Delta$ and an operator L_v of lower order, from the information we have obtained about the spectrum we can derive by standard arguments the convergence of all solutions and constants (even in metrics with derivatives).

Proof of Theorem 4. The operator B on the right side of Eq. (6), which carries g into $\mathrm{div}(gv+D\Delta g)$, is conjugate to the operator $A=L_{-v}+D\Delta$ on function: $A = B^*$. The eigenvalues of the operators B and B* coincide, and therefore the dimensions of the spaces of stationary solutions of Eqs. (6) and (5) are identical; they are equal to 1, by Theorem 3. Assertions 2) and 3) of Theorem 4 follow from the information on the spectrum of the operator A which we obtained in proving Theorem 3. Assertion 4) follows from 3), since $\int(\mathrm{div}\,\Omega)\tau=0$.

Proof of Theorem 5. It should be noted that the operator $L_v+D\Delta$ commutes with d. From this it follows that the solution with initial condition $\omega_0 = df_0$ evolves under the implicit Eq. (1) in the same way as the derivative df of the solution f of Eq. (5) with initial condition f_0. From Theorem 3 it follows that $f \to$ const (with derivatives). This means that $df \to 0$, i.e., the exact 1-form degenerates as time goes on. Thus, the sole stationary solution which is an exact form is zero. But by Theorem 1, the dimension of the space of solutions of a stationary equation is not less than b_1, i.e., the codimension of the subspace of exact 1-forms in the space of closed forms. Since the space of stationary solutions intersects the subspace of exact forms only at zero, its dimension is exactly equal to the Betti number b_1, and its projection onto the space of adjacency classes of closed forms for exact forms is an isomorphism. This proves assertions 1) and 2). Assertion 3) follows from the fact that the exact 1-forms are degenerate.

Proof of Theorem 6. Since the operators d and $L_v+D\Delta$ are commutative, the m-form $d\omega = g\tau$ evolves according to the law (6). By Theorem 4, item 4, g tends to zero as $t \to \infty$ (the condition $d\omega = g\tau$ means that g is the divergence of the vector field which specifies the form ω).

Proof of Theorem 7. For m = 2, the 1-form ω is an m - 1-form. By Theorem 6, as $t \to \infty$, it becomes closed ($d\omega \to 0$). Here the tendency to zero is exponential (even in metrics with derivatives). Using the same reasoning as in the proof of Theorem 2, item E, and using Theorem 5 to investigate the behavior of the exact forms, we arrive at the conclusion that the limit of ω as $t \to \infty$ exists and is closed.

Proof of Theorem 8. This is a special case of Theorem 7: it must be applied to the 1-form $\omega=Pdy-Qdx$.

5. THE DISCRETE CASE

Suppose that $g:M \to M$ is a diffeomorphism of a compact Riemannian manifold, $g_1=g^{*-1}$ is its action on the forms (by forward translation), h_t is the evolution during time t under the action of the diffusion equation:

$$h_t=\exp(t\Delta), \quad f_t=h_t g_t .$$

We consider the action of g_t on the cohomologies, $G_t : H^*\to H^*$ $(H^* = H^*(M,R))$.

Theorem 9: 1. The cohomology class of the closed form $f_t\omega$ is obtained from the class of the closed form ω by the action of G_t. 2. If t is sufficiently large and G is the identity transformation, then

a) for any closed form ω the limit $\lim_{n\to\infty} f_t^n\omega$ exists;

b) this limit is a unique closed form, cohomologic to ω and fixed under the action of f_t;

c) if the form ω is exact, then $f_t^n\omega\to 0$ as $n\to\infty$;

d) for any form ω (not necessarily a closed form), as $n \to \infty$, the sequence of forms $f_t^n\omega$ is convergent and the limit is a closed form.

Theorem 10. Let M be a two-dimensional manifold; then assertions a)-d) of Theorem 9 are true for any $t > 0$ (and not only for large values of t).

These discrete analogs of Theorems 2 and 7 (and the analog of the other theorems) are proved in the same way as the theorems themselves.

The same method can be used to prove, for example, that if $M = T^2$ and $G_t=\begin{pmatrix}1&1\\0&1\end{pmatrix}$, then $f_t^n\omega$ increases not more rapidly than the first power of n.

Remark 2. The case $G_t=\begin{pmatrix}2&1\\1&1\end{pmatrix}$ is used in [3,6] for constructing a kinematic dynamo on a three-dimensional compact Riemannian manifold. The flow on a three-dimensional torus [3]

$$v=(A\sin y+B\cos z,\ B\sin z+C\cos x,\ C\sin x+A\cos y),$$

which creates an exponential elongation of the particles, is a natural candidate for the kinematic amplification of a magnetic field, i.e., for the existence of a growth increment of the closed 2-form ω in the evolution of (1) such that the increment remains positive as $D \to 0$.

E. I. Korkina, in an attempt to discover numerically the growth of the field, encounters a stably zero increment (for large values of D). This observation led successively to Theorems 2 and 1 and then the remaining results of this paper. For $D = 1/12$, $A = B = C = 1$, E. I. Korkina used the Galerkin method to discover the positive increment: $\lambda=0{,}0079\pm 6{,}2i$ (the two roots crossed the imaginary axis and entered the unstable half-plane).

The author is grateful for valuable comments to M. I. Vishin, Ya. B. Zel'-dovich, Yu. S. Il'yashenko, E. I. Korkina, E. M. Landis, A. G. Khovachkii, and M. A. Shubin.

REFERENCES

1. T. G. Cowling, "The magnetic field of sunspots," Mon. Notic. Roy. Astron. Soc., vol. 94, pp. 39-48, 1934.

2. Ya. B. Zel'dovich, "A magnetic field in connection with the two-dimensional motion of a conductive turbulent liquid," ZhETF, vol. 31, pp. 154-156, 1956.

3. V. I. Arnol'd, "Remarks on the behavior of flows of a three-dimensional ideal liquid for a small perturbation of the initial field of velocities," Prikl. Matem. i Mekhan., vol. 36, no. 2, pp. 255-262, 1972.

4. R. Pale, Seminar on the Atieh-Singer Index Theorem [Russian translation], Moscow, 1979.

5. Pham Ngoc Thao, "Natural differential operators on compact manifolds," Differents. Uravnen., vol. 5, no. 1, pp. 186-198, 1969.

6. V. I. Arnol'd, Ya. B. Zel'dovich, A. A. Ruzmaikin, and D. D. Sokolov, "A magnetic field in a stationary flow with elongation in Riemannian space," ZhETF, vol. 81, no. 6, pp. 2052-2058, 1981.

19 May 1982

Evolution of a magnetic field under the action of transfer and diffusion

V.I. Arnold
Translated by Gerald Gould

The equation of a (divergence-free) magnetic field H in the flow of an incompressible conducting fluid with velocity field v and coefficient of diffusion μ has the form

$$\dot{H} = \{v, H\} + \mu\Delta H,$$

where $\Delta = -\operatorname{rot}\operatorname{rot}$ and $\{\cdot,\cdot\}$ is the Poisson bracket. The fields H and v are assumed to be 2π-periodic in (x,y,z). *The flow*

$$v = (\cos y + \sin z, \cos z + \sin x, \cos x + \sin y)$$

exponentially stretches the fluid particles (the increment is of order 0.15, cf. [1]). The calculation of the eigenvalue of the operator $A = \mu^{-1}\{v,\cdot\} + \Delta$ with the largest real part was carried out by Korkin [2] for $\mu^{-1} \leq 18$ (about 20,000 harmonics were taken into account in the Galerkin approximation). The growth of H was discovered for $9 \leq \mu^{-1} \leq 17$, with the largest increment for $\mu^{-1} \approx 12$ being of order 0.01.

The increasing mode turned out to have no harmonics $e^{i(k\cdot x)}$ *for which* $k^2 = 3,4,7,12,15,16,23,28,31,39,47,48,55,60,63,64,71,\ldots$. The absence of $k^2 = 4^a(8b+7)$ is explained by the absence of representation by the sum of three squares. The remaining k^2 in the list are of type 4^a and $3\cdot 4^a$. The absence of all of them would be explained by the symmetry of the problem if the following conjecture could be proved.

Conjecture. *The square and the triple square of an odd prime admit a non-trivial representation as the sum of three squares.*

The group of symmetries of the operator A is isomorphic to the group of rotations of the cube. In the decomposition of the eigenspaces of the operator A into irreducible ones, for $k^2 = 2$ there is no trivial one-dimensional representation, while for

Talk at the Petrovsky seminar on differential equations and mathematical problems of physics, October 13, 1982

Originally published in: Usp. Mat. Nauk 38:2, 226-227, © Russ. Acad. Sciences 1983
Translated by G. Gould. Springer-Verlag Berlin Heidelberg 2014 with kind permission, DOI 10.1007/978-3-642-31031-7_37

$k^2 = 3$ both one-dimensional representations are absent. Therefore [3] *the increasing mode is transformed according to a non-trivial one-dimensional representation* and its Fourier series starts with

$$H_1 = (\cos y - \sin z, \cos z - \sin x, \cos x - \sin y).$$

As μ increases the eigenvalues of the operator A move into the left half-plane, for $\mu \approx 1$ they collide near $\lambda = -1.35$, and as $\mu \to +\infty$ they diverge to -1 and -2.

The latter results were also obtained by perturbation theory (taking account of symmetry, for $k^2 \leq 5$, only 5 harmonics survive out of 112). The specific nature of the field v enables one to find *in finite form* the initial part of any finite length of the Puiseux series of the eigenvalues. The corresponding diagram technique is applicable to any equation of "Mathieu type", for example, to the Hill equation, in which the constant coefficient has been modulated by a trigonometric polynomial. *The width of the forbidden zone in this case decreases with decreasing the depth of the modulation as a certain power of the depth; this power increases directly proportionally to the number of zones of the modulation and inversely proportionally to the degree of the polynomial.*

For the general equation $\dot{\omega} = L_r\omega + \mu A\omega$ of the evolution of a closed k-form ω on a compact Riemannian manifold (under the action of transfer by the flow of the divergence-free field v) it is known that *the dimension of the space of stationary solutions is not less than the kth Betti number of the manifold. It is not known whether there is a stationary solution in each cohomology class.* For large viscosity μ this is true, and even an initially non-closed form evolves to a stationary closed form as $t \to \infty$ [4].

References

[1] M. Henon, *Sur la topologie des lignes de courant dans un cas particulier*, C. R. Acad. Sci. Paris (1966), **262**.

[2] V. I. Arnol'd, E. I. Korkina, *Growth of a magnetic field in three-dimensional steady flow of an incompressible fluid*, Vestn. Moskov. Gos. Univ., Ser. Mat. Mekh., No. 3 , 43–46 (1983).

[3] V. I. Arnol'd, *Evolution of a magnetic field under the action of transfer and diffusion*, Collection of articles in memory of V. M. Alekseev, Moskov. Gos. Univ., Moscow (1983).

[4] V. I. Arnol'd, *Some remarks on the antidynamo theorem*, Vestn. Moskov. Gos. Univ., Ser. Mat. Mekh., No. 6 (1982).

THE GROWTH OF A MAGNETIC FIELD IN A THREE-DIMENSIONAL STEADY INCOMPRESSIBLE FLOW

V. I. Arnol'd and E. I. Korkina

Vestnik Moskovskogo Universiteta. Matematika,
Vol. 38, No. 3, pp. 43-46, 1983

UDC 538.4

We describe a numerical experiment for determining the growth of a magnetic field in a three-dimensional flow with exponential particle dilatation for Reynolds numbers of the order 60. In a certain range of Reynolds numbers, an exponential growth of the field is revealed, which changes to decay as the Reynolds numbers increase. The computations are carried out by the Galerkin method, taking into account approximately 20,000 harmonics.

The aim of this paper is the numerical investigation of the behavior of the solutions of an equation that describes the evolution of a magnetic field in a steady incompressible flow:

$$\partial H/\partial t = \{v, H\} + \varepsilon\Delta H. \tag{1}$$

Here v is a given field of divergence zero (the field of current speeds), H is a magnetic field, evolutionary in the course of time t, $\{v, H\} = v\nabla H - H\nabla v$ is the Poisson bracket, $\Delta = -\text{rot rot}$ is the Laplace operator, and $\varepsilon > 0$ is magnetic viscosity. We assume that v and H are defined in the whole of three-dimensional space and have period 2π in each of the three coordinates (x, y, z).

For the investigation we shall choose the special field of speeds

$$v = (\cos y + \sin z)\,\partial/\partial x + (\cos z + \sin x)\,\partial/\partial y + (\cos x + \sin y)\,\partial/\partial z.$$

For this field it has been established experimentally [1] that the particles of the liquid expand exponentially, and so exponential growth of the field H is observed when $\varepsilon = 0$.

The index of the exponent $(\lim_{t\to\infty} \ln \|H(t)\|/t)$ is defined as the largest characteristic Lyapunov index of the field v (roughly speaking, v has Lyapunov index λ at a given point if the particle emanating from this point increases with time according to the function $e^{\lambda t}$). The largest characteristic Lyapunov indices that we have observed have order of magnitude 0.15 (for example, for the initial point

Originally published in: Vestn. Mosk. Univ. Mat. 38:3, 43-46, © Izd. Moscow State Univ. 1983
English transl.: Moscow Univ. Math. Bull, Ser. I, Math. Mech. 3, 50-54, © Allerton Press, Inc., New York, 1983
Reprint by Springer-Verlag Berlin Heidelberg 2014 with kind permission, DOI 10.1007/978-3-642-31031-7_38

$x=0{,}1\cdot 2\pi$, $y=0{,}4\cdot 2\pi$, $z=0{,}51\cdot 2\pi$). On the other hand, the field v is very convenient for investigating the equation by means of Galerkin approximations, since it has small harmonics.

1. RESULTS OF THE COMPUTATIONS

We have computed the eigenvalue of the linear operator A_ε on the right hand side of (1) that has the largest real part. For a large magnetic viscosity ε this real part turns out to be exactly zero and does not depend on ε.

As the magnetic viscosity decreases to some critical value ε_1 the pair of conjugate complex numbers goes from the left-plane to the right. If the viscostiy decreases till further, the real parts of these eigenvalues increase until the viscosity attains a second critical value ε_2, after which they begin to decrease; for a third critical value of the viscosity ε_3 they vanish again, and then become negative.

We have thus observed the growth of the magnetic field ("kinematic dynamo") in the interval between the values ε_1 and ε_3 of the magnetic viscosity. We have not been able to carry out the computations for values of the viscosity significantly less than ε_3, and we do not know whether all the eigenvalues lie in the left half-plane as $\varepsilon \to 0$ or go over to the right.

The variation of the real part of the eigenvalue λ as ε decreases is shown in Fig. 1. The critical values of the viscosity are:

$$\varepsilon_1\approx 0{,}112,\quad \varepsilon_2\approx 0{,}083,\quad \varepsilon_3\approx 0{,}057.$$

In the table we give the positions of the pair of eigenvalues that goes into the right half-plane, for certain values of ε.

$1000\,\varepsilon$	435	250	143	111	100	90,9	83,4	71,5	55,6
1000 Re λ	−433	−151	−22,6	0,44	5,3	7,6	7,9	6,1	−0,67
$1000\,\lvert \mathrm{Im}\,\lambda\rvert$	548	602	625	607	625	623	618	604	580

For large values of ε it is convenient to regard the Poisson bracket in (1) as a perturbation, and so in this case we have computed the eigenvalues of the operator

$$H\to\Delta H+\eta\{v,H\},\qquad \eta=1/\varepsilon \tag{2}$$

(these numbers differ from those considered above by a factor η).

If $\eta = 0$, the eigenvalues of the operator (2) are equal to 0 (with multiplicity 3), -1 (with multiplicity 12), -2 (with multiplicity 24), and so on. The triple eigenvalue 0 is preserved for all values of η (the stationary solutions of (1) correspond to it). Three linearly independent stationary solutions correspond to the three components of the vector of the mean value of H. The mean value of H is preserved under the evolution (1). For sufficiently large viscosity (sufficiently small η) as $t \to \infty$ the field H tends to a stationary field with the same mean value as the original field. The eigenvalue -1 of the operator (2) is also preserved for all η (it corresponds to the eigenvector $H = v$).

We have investigated the behavior as η decreases, that is, as the viscosity increases, of the pair of conjugate complex eigenvalues of (2), which goes through the imaginary axis at the critical value $\eta_1 = 1/\varepsilon_1$. It turns out that as η decreases to a critical value η_0, for which the real part of the eigenvalues of the pair becomes -1, these eigenvalues have the largest real part among all the eigenvalues of (2) in the space of solenoidal fields with mean value zero. As η further decreases, the eigenvalues of the pair approach one another. They merge and go over to the real axis for the next critical value η_*, after which as $\eta \to 0$ one of the eigenvalues tends to -1 and the other to -2.

The behavior of the real parts of the eigenvalues λ as η varies is shown in Fig. 2. The critical values are:

$$\eta_1 \approx 8{,}93, \quad \eta_0 \approx 2{,}29, \quad \eta_* \approx 0{,}96.$$

The corresponding eigenvalues are:

$$\lambda_1 \approx 5{,}6, \quad \lambda_0 \approx -1 + 1{,}25\, i, \quad \lambda_* \approx -1{,}35.$$

The eigenvectors corresponding to these eigenvalues go into themselves under a cyclic permutation of the coordinates (x, y, z). The first harmonics of these vectors do not depend on the parameter (ε or η) and have the form

$$H_1 = (\cos z - \sin y)\, \partial/\partial x + (\cos x - \sin z)\, \partial/\partial y + (\cos y - \sin x)\, \partial/\partial z.$$

The distribution of the amplitudes of the Fourier coefficients of an eigenvector according to the numbers of the harmonics reveals distinctive regularities. We denote by $E(\vec{k}^2)$ the sum of the squares of the moduli of the Fourier coefficients of the harmonics $e^{i(\vec{k},\vec{x})}$ with wave vectors, the square of whose lengths is equal to $\vec{k}^2$. For certain values of $\vec{k}^2$ the computed value of $E(\vec{k}^2)$ turns out to be zero. These values are

$$\vec{k}^2 = 3, 4, 7, 12, 15, 16, 23, 28, 31, 39, 47, 48, 55, 60, 63, 64, 71, \ldots$$

Among them are, firstly, all numbers of the form $4^a(8b+7)$ that are not squares of lengths of wave vectors), and secondly, all numbers of the form 4^a and $3 \cdot 4^a$ (explained by the special symmetry of the eigenvector [3]). The symmetry reduces

52

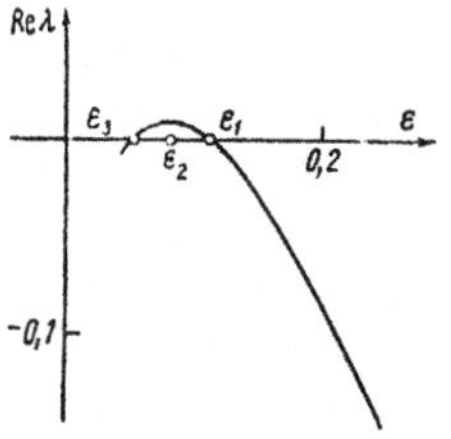

Fig. 1

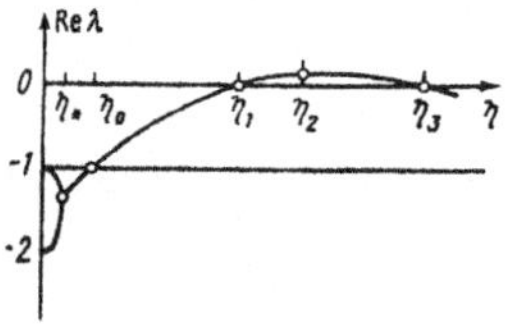

Fig. 2

the Reynolds magnetic number by 3 (to $lv\eta/3 \approx 2\pi\sqrt{3}\,\eta/3$).

2. THE TECHNIQUE OF COMPUTATION

To attain the necessary accuracy in the computation of the eigenvalues for our values of the viscosity (up to $\varepsilon \approx 1/20$) we need to find the eigenvalues of a matrix of order $\approx 20{,}000$, so we explain how we have computed them.

We have approximated the operator (2) by a Galerkin approximation, that is, we have considered trigonometric sums H with wave vectors $\vec{k}$ for which $|\vec{k}^2| < R^2$. We have replaced the Poisson bracket $\{v, H\}$ by the projection on this space, that is, we have rejected harmonics with large $|k^2|$. We have computed the eigenvalue of the resulting operator A in a space of finite dimension N $\left(N \approx 2 \cdot \frac{4\pi}{3} \cdot R^3\right)$ that has largest real part, by the iterations

$$H^{(n+1)} = (E + tA)\,H^{(n)},$$

where we have chosen t to be sufficiently small (usually half of $1/R^2$).

We have computed the approximate value of the eigenvalue with largest real part from the projections of the last three vectors $H^{(n)}$, $(E+TA)H^{(n)}$, $(E+TA)^2H^{(n)}$ on the plane of the harmonics with small wave vectors. The value of T is chosen large enough to distinguish the eigenvalue from unity, but small enough to nullify the influence of the negative eigenvalues of the Laplacian that are large in absolute value (in practice, we have chosen $T \approx 0.1$).

The iterations cease for values of n so large that the approximate eigenvalue ceases to change in the third decimal place.

By this method of computation we have managed to avoid storing the elements of the matrix A in the memory. It is very important that an eigenvalue is computed from the behavior of the projections onto the low-dimensional space of the smallest harmonics; we have not succeeded in carrying out the computations without this improvement in the iteration method.

We have chosen the parameter R (the radius of cut-off of the Galerkin approximation) as follows. Firstly, we have controlled the stability of our approximation to the eigenvalue in relation to the increase in R, and secondly the smallness of the amplitudes of the high harmonics. We have stopped at those values of R such that no more than 10^{-5} of the total energy of the eigenvector is concentrated in the harmonics with $|\vec{k}^2| = R^2$. It turns out that it is sufficient to choose R of order $3\sqrt{\eta}$.

The largest value of η attained was equal to 18, and sufficient accuracy is given by choosing R = 13, that is, the harmonics are calculated up to $|\vec{k}^2| = 169$.

The computations were carried out by E. I. Korkina on a VAX computer with the high speed of 300,000 operations per second. For η = 18 the computation of the eigenvalue took about 200 kilobytes of operating memory and 50 hours of machine time.

The authors are grateful to A. G. Khovanskii, who took an active part in discussing the paper.

REFERENCES

1. M. Henon, "Sur la topologie des lignes de courant dans un cas particulier," C.R. Acad. Sci. Paris, vol. 262, pp. 312-314, 1966.
2. V. I. Arnol'd, "Some remarks on the anitdynamo theorem," Vestn. Mosk. Univ. Mat., Mekh. [Moscow University Mathematics Bulletin], no. 6, pp. 50-57, 1982
3. V. I. Arnol'd, "On the evolution of a magnetic field under the action of translation and diffusion," in: Collection in Memory of V. M. Alekseev, Moscow, 1982.

26 November 1982

ОБ ЭВОЛЮЦИИ МАГНИТНОГО ПОЛЯ ПОД ДЕЙСТВИЕМ ПЕРЕНОСА И ДИФФУЗИИ

В.И.Арнольд

§ I. Проблема стационарного кинематического динамо

Уравнение эволюции магнитного поля H дивергенции 0 под действием переноса потоком с полем скоростей v дивергенции 0 и диффузии с коэффициентом диффузии μ имеет вид

$$\dot{H} = \{v, H\} + \mu \Delta H, \tag{I}$$

где $\Delta = -\mathrm{rot}\,\mathrm{rot}$ - лапласиан, $\{,\}$ - скобка Пуассона. Мы рассматриваем это уравнение с периодическими граничными условиями (H не меняется при сдвигах на кратные 2π вдоль любой из трех координатных осей). Поле скоростей v - фиксированное поле такой же периодичности, а именно:

$$v = (\cos y + \sin z)\partial/\partial x + \tag{2}$$
$$+ (\cos z + \sin x)\partial/\partial y + (\cos x + \sin y)\partial/\partial z .$$

Нас будет интересовать зависимость инкремента $\gamma = \mathrm{Re}\,\lambda$ наиболее быстро растущей моды $H = e^{\lambda t} H_0(x, y, z)$ от магнитного числа Рейнольдса $1/\mu$. Поле v называется динамо (при данном числе Рейнольдса), если существует растущая мода $(\gamma > 0)$. Динамо называется сильным, если инкремент остается ограниченным снизу положительной постоянной при уменьшении магнитной вязкости, т.е. если

$$\lim \gamma > 0 \qquad \text{при} \quad \mu \to 0 .$$

Существует ли сильное динамо - неизвестно. Двумерное течение не может быть сильным динамо, так как в таком течении нет экспоненциального растяжения частиц.

Специальное течение (2) было выбрано для исследования по следующим соображениям. При $\mu = 0$ уравнение (I) означает вмороженность поля H в жидкость. Если фазовый поток поля v экспоненциально растягивает частицы жидкости (т.е. если наибольший характеристический показатель Ляпу-

Originally published in: Some Problems in Modern Analysis, 8-21 (Russian), © Izd. MGU, Moscow 1984
Reprint by Springer-Verlag Berlin Heidelberg 2014 with kind permission, DOI 10.1007/978-3-642-31031-7_39

нова положителен на множестве положительной меры), то вмороженное поле H экспоненциально растет (растягивается потоком). Поле (2) является одним из простейших полей, обладающих экспоненциальным растяжением частиц (насколько можно судить по данным численных экспериментов, ср. [I]). Растяжение частиц течением "общего положения" происходит предположительно примерно так же, как в этом течении. Поскольку течение растягивает частицы неравномерно, растущее поле H со временем становится изрезанным. Вязкость быстро гасит высшие гармоники. Будет ли работать динамо (будет ли $\gamma>0$) зависит от того, какой процесс идет быстрее: рост вмороженного поля или вязкое затухание.

§ 2. Результаты численного эксперимента

Зависимость инкремента γ от магнитного числа Рейнольдса $1/\mu$ исследовалась численно при помощи галеркинских приближений Е.И.Коркиной. Искалось собственное число оператора A_R, $A_R H = R\{v, H\} + \Delta H$ с наибольшей вещественной частью. Собственный вектор H раскладывался в ряд Фурье и высшие гармоники отбрасывались. Вычисления были доведены до числа Рейнольдса $R=19$, что потребовало учета гармоник $e^{(k,r)}$ с $k^2 \leq 169$: таким образом, порядок матрицы, собственное число которой искалось, был около 20 000. Вычисления контролировались несколькими способами: изменением порядка учитываемых гармоник, сохранением бездивергентности эволюционирующего поля и выполнением условий симметрии, о которых подробно сказано ниже. Оказалось, что необходимо учитывать гармоники с $k^2 \lesssim 9R$.

Вычисления были начаты с малых чисел Рейнольдса, чтобы использовать найденный собственный вектор в качестве первого приближения при вычислениях с большими числами Рейнольдса. Однако оказалось, что при малых числах Рейнольдса R собственное число λ оператора A_R не зависит от числа Рейнольдса: оно в точности равно нулю.

Причина этого явления состоит в том, что при большой вязкости решение уравнения (I) выходит при $t\to\infty$ на стационарный режим, определяемый классом когомологий начального поля H (т.е. вектором средних значений поля). Выход на стационарный режим подробно обсуждается в [2] (в более общей ситуации эволюции k - форм на n - мерном многообразии). Чтобы избавиться от этого эффекта мы ограничились рассмотрением полей со средним нуль. Однако и в этом случае вычисления дали не зависящее от числа Рейнольдса собственное число $\lambda=-1$ оператора A_R, пока число Рейнольдса R остается меньшим критического значения $R_1\approx 2{,}3$.

Причина здесь состоит в том, что $\{v, v\}=0$, а в силу (2) $\Delta v=-v$, поэтому само наше поле v является собственным вектором оператора A_R с собственным числом $\lambda=-1$.

При увеличении числа Рейнольдса обнаруживается комплексная пара собственных чисел с $\gamma>-1$, которая движется вправо и пересекает границу динамо $\gamma=0$ при $R_2\approx 9{,}0$. Наибольшее значение γ на этой ветви достигается вблизи $R_3\approx 12{,}5$ и равно 0,096. Затем инкремент уменьшается и при $R_4\approx 17{,}5$ он снова становится отрицательным.

Таким образом, поле v является динамо при $R_2<R<R_4$. Вопрос о том, является ли это поле динамо при $R\to\infty$, остается открытым. Д.Галловей обнаружил динамо при $30<R<100$.

Распределение энергии найденного собственного вектора H по гармоникам обнаружило странную аномалию: при некоторых "пустых" значениях K^2 величина $\sum|H_K|^2$ точно равна нулю, а при некоторых "удивительных" значениях – с точностью вычислений. Пустыми оказались значения $K^2=7$, $15, 23, 28, 31, \ldots$, а удивительными – значения $K^2=3, 4, 12, 16, 48, 64, \ldots$. Пустые значения объясняются просто: это все числа вида $4^a(8b+7)$. Как известно, эти и только эти числа не допускают представлений суммой трех квадратов.

Удивительные значения имеют вид 4^a и $3\cdot 4^a$. Их появление объяснено ниже соображениями симметрии (они связаны с разложением представлений группы вращений куба в пространствах тригонометрических векторных многочленов на неприводимые). Эти соображения позволяют ускорить вычисления в десятки и даже сотни раз. Например, собственный вектор галеркинского приближения с учетом $K^2\le 5$ мы выписываем явно (без учета симметрии потребовалось бы рассмотреть матрицу порядка 112). Мы даже точно находим первую гармонику истинного собственного вектора изучаемой моды при любых числах Рейнольдса:

$$H_1=(\cos y-\sin z)\partial/\partial x+(\cos z-\sin x)\partial/\partial y+$$
$$+(\cos x-\sin y)\partial/\partial z.$$

10

Однако тот факт, что именно эта мода растет быстрее всех (имеет наибольший инкремент) получен только из численного эксперимента и только при $R \le 19$.

§ 3. Симметрии поля скоростей

Из формулы (2) видно, что поле скоростей v переходит в себя при циклической перестановке координат. Легко доказывается

Теорема. Группа симметрии поля v, сохраняющих Δ, содержит 24 элемента и изоморфна группе вращений куба.

Действительно, легко проверить, что поле v переходит в себя при преобразовании $g_4 : (x, y, z) \mapsto (x + \pi/2, z - \pi/2, \pi/2 - y)$. Преобразование g_4 имеет порядок 4 (т.е. $g_4^4 = id$). Вместе с циклической перестановкой $g_3 : (x, y, z) \mapsto (y, z, x)$ оно порождает группу G из 24 элементов, изоморфную группе вращений куба. Отметим еще сохраняющее Δ и меняющее знак v преобразование h : $(x, y, z) \longrightarrow (x + \pi, y + \pi, z + \pi)$.

Группа вращений куба имеет 5 неприводимых представлений:

Тривиальное одномерное (I);
нетривиальное одномерное (-I) (перестановка двух вписанных тетраэдров);
двумерное (2) (перестановка трех осей координат);
трехмерное (3) (действие на куб);
подкрученное трехмерное (-3) (тензорное произведение (-I) и (3)).

Эта группа G действует на пространстве векторных полей дивергенции 0 на торе.

Действие G коммутирует с действиями операторов $\Delta, \{v, \cdot\}$, а значит и A_R. Поэтому

I) G действует на собственных подпространствах оператора Δ.

2) Оператор A_R распадается в прямую сумму пяти операторов, действующих в пространствах, где представление группы G кратно каждому из пяти неприводимых.

3) Собственный вектор "почти наверное" принадлежит одному из 5 указанных пространств – ведь совпадение собственных чисел двух из 5 операторов прямой суммы "невероятно".

4) Операторы A_{-R} и A_R изоморфны (и следователь-

но, их собственные числа одинаковы, так что характеристическое уравнение четно относительно R).

Свойство 4) следует из того, что h переводит A_R в A_{-R} .

§ 4. Разложение представлений на неприводимые

Рассмотрим представление группы G вращений куба в собственном подпространстве оператора Лапласа, действующего на бездивергентные поля на торе со средним значением нуль. Собственное подпространство состоит из тригонометрических многочленов с векторными коэффициентами, ортогональными волновым векторам

$$\sum H_k e^{i(k,r)}, \quad (k, H_k)=0,$$

с фиксированной суммой квадратов компонент волнового вектора $k_1^2+k_2^2+k_3^2 = k^2 > 0$ (собственное число оператора Лапласа есть $-k^2$). Размерности этих пространств вдвое превосходят числа целых точек на соответствующих сферах, например,

$$d(1)=12, \quad d(2)=24, \quad d(3)=16, \quad d(4)=12,$$
$$d(5)=48, \quad d(6)=48, \quad d(7)=0, \quad d(14)=96.$$

Приведенные в § 3 формулы действия образующих g_3 и g_4 на торе позволяют найти характеры представления группы G в каждом собственном пространстве оператора Лапласа.

Зная характеры, мы раскладываем представление на неприводимые.

Результаты (довольно длинных) вычислений таковы.

Все целые точки на сфере $\sum k_i^2 = k^2$ разобьем на орбиты группы перестановок координат и изменений их знаков. Для каждого из 7 типов орбит, указанных в первой графе следующей таблицы, в следующих графах указаны размерность χ_1 соответствующего собственного подпространства оператора Лапласа (т.е. удвоенно число точек орбиты), затем остальные характеры, затем кратности вхождения неприводимых представлений и наконец, три первые значения k^2 , при которых встречается такая орбита:

орбита	χ_1	χ_{g_3}	χ_{g_4}	$\chi_{g_4^2}$	χ_{g_2}	(I)	(-I)	(2)	(3)	(-3)	K^2		
$(2r,0,0)$	I2	0	0	-4	0	0	0	0	2	2	4	I6	36
$(2r+1,0,0)$	I2	0	0	4	0	I	I	2	I	I	I	9	25
$(a,b,0)$	48	0	0	0	0	2	2	4	6	6	5	I0	I3
$(a,a,0)$	24	0	0	0	-4	0	2	2	4	2	2	8	I8
(a,b,c)	96	0	0	0	0	4	4	8	I2	I2	I4	2I	26
(a,a,c)	48	0	0	0	0	2	2	4	6	6	6	27	36
(a,a,a)	I6	-2	0	0	0	0	0	2	2	2	3	I2	27

Например, представления в собственных пространствах оператора Лапласа с малыми собственными числами разлагаются так:

$$K^2=1:(1)\oplus(-1)\oplus 2(2)\oplus(3)\oplus(-3)\ ,$$
$$K^2=2:2(-1)\oplus 2(2)\oplus 4(3)\oplus 2(-3)\ ,$$
$$K^2=3:2(2)\oplus 2(3)\oplus 2(-3),$$
$$K^2=4:2(3)\oplus 2(-3)\ ,$$
$$K^2=5:2(1)\oplus 2(-1)\oplus 4(2)\oplus 6(3)\oplus 6(-3).$$

Из приведенных таблиц вытекает

Теорема. Разложение представления в пространстве полей с данным K^2 на неприводимые имеет вид

$$A\left[(1)\oplus(-1)\oplus 2(2)\oplus 3(3)\oplus 3(-3)\right]\oplus B\left[(1)\oplus(-1)\oplus 2(2)\oplus (3)\oplus(-3)\right]\oplus C\left[(3)\oplus(-3)\right]\oplus D\left[(-1)\oplus(2)\oplus(3)\oplus 2(-3)\right]\oplus E\left[(2)\oplus(3)\oplus(-3)\right],$$

где коэффициенты выражаются через число N целых точек на сфере так:

K^2	A	B	C	D	E
$\neq n^2, 2n^2, 3n^2$	$N/12$	0	0	0	0
$(2r+1)^2$	$(N-6)/12$	I	0	0	0

13

$4n^2$	$(N-6)/12$	0	2	0	0
$2n^2$	$(N-12)/12$	0	0	2	0
$3n^2$	$(N-8)/12$	0	0	0	2

<u>Замечание</u>. А.Б.Гивенталь указал мне, что эту теорему можно получить короче при помощи двойственности Фробениуса, примененной к паре (G , стационарная подгруппа волнового вектора). Для нас существенны только приведенные выше таблицы разложений, поэтому мы не останавливаемся на этом подробнее.

§ 5. <u>Симметрия растущей моды</u>

Сопоставим "удивительные значения" K^2 из § 2, соображение 3 из § 3 и таблицу разложений из § 4. Очевидна

<u>Теорема</u>. Амплитуды гармоник искомой моды с данным K^2 будут нулевыми в тех случаях, когда представление, по которому мода преобразуется, не входит в разложение пространства полей с данным K^2 на неприводимые.

Таким образом, "удивительные значения" являются своеобразными отпечатками пальцев, по которым можно найти если не саму моду, то во всяком случае ее симметрию.

Из таблиц § 4 видно, что при $K^2 = 3, 4, 12, 16, 48, 64, \ldots$ в разложение не входит нетривиальное одномерное представление (и только оно). Следовательно, мода принадлежит пространству, преобразующемуся по этому представлению. Это пространство полей, удовлетворяющих условиям $g_3 H = H$, $g_4 H = -H$. Оно распадается в прямую сумму своих пересечений с собственными пространствами оператора Лапласа. Размерности этих пересечений, согласно таблицам § 4, таковы

K^2	I	2	3	4	5	6	8	9	I0	II	I2	I3	I4	I6
dim	I	2	0	0	2	2	2	3	2	2	0	2	4	0

14

При "удивительных" значениях $k^2 = 4^\alpha$ и $3 \cdot 4^\alpha$ размерность пересечения равна 0. При всех остальных значениях k^2 размерность положительна. Это вытекает из следующих утверждений теории чисел:

1. Квадрат каждого нечетного простого числа допускает нетривиальное представление суммой трех квадратов.

2. Утроенный квадрат каждого простого числа допускает представление суммой трех не равных всех вместе квадратов.

Эти утверждения доказаны уже Гауссом (автор благодарен Ж.-П.Серру за эту информацию).

§ 6. Алгебра четных и нечетных полей

Чтобы найти поля, преобразующиеся по представлению (-I), удобно поступить так. Назовем поле H четным, если $g_3 H = H = g_4 H$, и нечетным, если $g_3 H = H = -g_4 H$. Суммы четных и нечетных полей - инвариантные относительно g_3 и g_4^2 поля. Они образуют алгебру Ли.

Наше исходное поле $v = (\cos y + \sin z)\partial/\partial x$ + (циклические перестановки) четно. Нетрудно проверить, что поле

$$H_1 = (\cos y - \sin z)\,\partial/\partial x \quad + \text{(циклические перестановки)}$$

нечетно. Кратная скобка Пуассона четных и нечетных полей будет четной или нечетной в зависимости от того, четно или нечетно число участвующих в ней нечетных полей.*) Таким образом, все поля $\{v, H_1\}$, $\{v, \{v, H_1\}\}$, ... нечетны. Если поле четно (нечетно), то его проекция на каждое собственное подпространство оператора Лапласа также четна (нечетна). Мы будем обозначать проекцию поля H на подпространство с собственным числом $-k^2$ (т.е. сумму гармоник поля H с волновым вектором с квадратом длины k^2) знаком $(H)_{k^2}$. Например, $(v)_1 = v$. Комбинируя скобки Пуассона и проекции можно получить из v и H_1 много нечетных полей (может быть, все?). Возникающая алгебра довольно запутана. Начальный отрезок ее таков:

---

*) Иными словами, алгебра Ли сумм четных и нечетных полей $\mathbb{Z}_2$ - градуирована.

15

Теорема. Имеют место соотношения

$$\{v, H_1\} = -2H_2, \quad \{v, H_2\} = H_5, \quad (\{v, H_5\})_2 = -\frac{1}{2}H_2 + H_2', $$

$$\{v, H_2'\} = \frac{1}{2}H_1 + \frac{1}{2}H_5', \quad (\{v, H_5'\})_2 = -H_2',$$

где явные выражения базисных полей таковы:

$$H_2 = (cxcz + sxsy)\partial_x + \ldots,$$

$$H_2' = sycz\partial_x + \ldots,$$

$$H_5 = (cxs2y - sxs2z - \frac{1}{2}c2xcy - \frac{1}{2}c2xsz - $$
$$- \frac{1}{2}c2ysz - \frac{1}{2}cyc2z)\partial_x + \ldots,$$

$$H_5' = (cyc2z + c2ysz)\partial_x + \ldots .$$

Здесь c и s означают $\cos$ и $\sin$, а ... – циклическую перестановку.

§ 7. Представление колчана, порожденное коммутированием с v

Назовем векторное поле на торе полиномиальным, если его компоненты – тригонометрические многочлены. Назовем полиномиальное поле однородным, если оно является собственным для оператора Лапласа, т.е. если для всех его гармоник $\vec{H}e^{i(\vec{k},\vec{r})}$ величина $\vec{k}^2$ имеет одинаковое значение. Это общее значение k^2 мы назовем степенью однородного поля.

Пространства четных и нечетных полиномиальных полей являются прямыми суммами своих однородных составляющих. Оператор $\{v, \cdot\}$ скобки Пуассона с полем (2) перемешивает однородные составляющие не произвольным образом.

Назовем гармоники $e^{i(\vec{k},\vec{r})}$ и $e^{i(\vec{k}',\vec{r})}$ соседними, если целочисленные вектора $\vec{k}$ и $\vec{k}'$ отличаются лишь на I в одной из трех компонент (у каждой гармоники, таким образом, 6 соседей). Образуем теперь граф, вершинами которого являются степени нечетных однородных полей. Две вершины k^2 и k'^2 соединим ребром, если они являются степенями входящих в нечетные поля соседних гармоник.

16

Тогда скобка Пуассона нашего поля v с однородным полем степени K^2 имеет составляющими только поля соседних с K^2 в построенном графе степеней (ибо само поле однородно степени I).

Начальный отрезок графа имеет, согласно таблицам § 4, вид

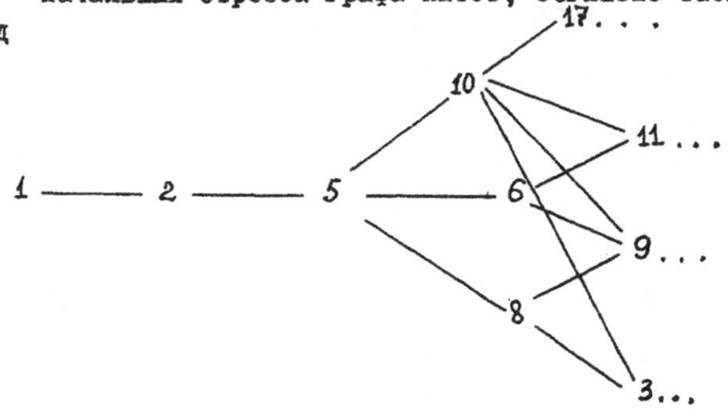

<u>Теорема</u>. Каждая четная вершина соединена только с нечетными, а нечетная только с четными. При этом никаких других отрезков, кроме указанных, к вершинам I,2,5,6,8,I0 не подходит. В некоторых случаях полезнее граф с большим числом вершин, соответствующих не степеням K^2, а орбитам действия группы перестановок координат волнового вектора и смен их знаков.

Ограничение A_i оператора $\{v,\cdot\}$ коммутирования с полем v на пространство $\mathcal{F}_i$ однородных нечетных полей степени i представляется в виде конечной суммы однородных слагаемых

$$A_i = \oplus A_{i,j} \quad , \quad A_{i,j} : \mathcal{F}_i \to \mathcal{F}_j ,$$

где j пробегает всех соседей i в построенном графе.

Если заменить каждое ребро нашего графа парой противоположно направленных стрелок, то получится (бесконечный) колчан. Выше построено представление этого колчана: вершине i сопоставлено конечномерное линейное пространство $\mathcal{F}_i$, стрелке $i \to j$ – линейный оператор $A_{i,j}$.

<u>Теорема</u>. Размерности первых трех пространств $\mathcal{F}_1$, $\mathcal{F}_2$

2 - I864

и $\mathcal{F}_5$ равны I,2,2 соответственно, базисы в них образуют приведенные в § 6 поля H_1, H_2 и H_2', H_5 и H_5'. Матрицы операторов $A_{i,j}$ в этих базисах следующие:

$$(A_{1,2})=\begin{pmatrix}-2\\0\end{pmatrix},\ (A_{2,1})=(0,1/2),\ (A_{2,5})=\begin{pmatrix}1&0\\0&1/2\end{pmatrix},$$

$$A_{5,2}=\begin{pmatrix}-1/2&0\\1&-1\end{pmatrix}.$$

Это – другая формулировка теоремы § 6.

§ 8. Галеркинская пятимодовая система

Отбрасывая в ряду Фурье $H=(H)_1+(H)_2+(H)_5+\ldots$ члены выше 5 степени, мы получаем галеркинскую систему из 5 линейных уравнений для компонент нечетного собственного вектора

$$(H)_1=a_1H_1,\quad (H)_2=a_2H_2+a_2'H_2',$$

$$(H)_5=a_5H_5+a_5'H_5'$$

оператора $A_R^{(5)}=P^{(5)}[R\{v,\ \}+\Delta]$, где $P^{(5)}$ – проекция на $\mathcal{F}_1\oplus\mathcal{F}_2\oplus\mathcal{F}_5$. Заметим при этом, что соотношения на уровне $\mathcal{F}_1$ и $\mathcal{F}_2$ получаются в точности такие же, как для полной (не галеркинской) системы, т.к. вершины I и 2 в графе соединены лишь с вершиной 5.

Теорема. Явный вид галерскинской системы пяти нечетных мод таков:

$$(R/2)a_2'=(\lambda+1)a_1,\quad -R(2a_1+a_5/2)=(\lambda+2)a_2,$$

$$R(a_5-a_5')=(\lambda+2)a_2',\quad Ra_2=(\lambda+5)a_5,$$

$$(R/2)a_2'=(\lambda+5)a_5'.$$

Доказательство формулы § 7.

Решая эту систему мы без труда находим характеристическое уравнение

$$4(\lambda+1)(\lambda+2)^2(\lambda+5)^2+4(\lambda+1)(\lambda+2)(\lambda+5)R^2+$$
$$+(5\lambda+21)R^4=0,$$

18

$$R^2 = \frac{-2(\lambda+2)(\lambda+5)\left[(\lambda+1) \pm 2\sqrt{-(\lambda+1)(\lambda+5)}\right]}{5\lambda+21}.$$

График этой функции изображен на рисунке

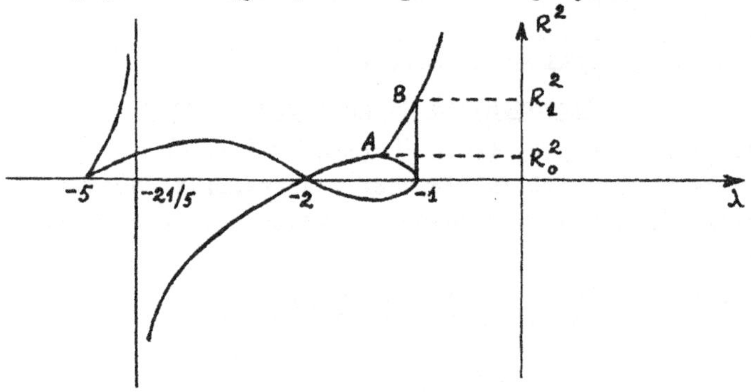

Если вязкость достаточно велика, то мы можем пренебречь высокими гармониками. Поэтому при малых значениях числа Рейнольдса R можно использовать полученный график для суждения о поведении инкремента быстрее всего растущей (точнее, медленнее всего затухающей) нечетной моды. Сравнение с численным счетом при больших числах Рейнольдса показывает хорошее согласие изучаемого пятимодового галеркинского приближения с точным решением при $R^2 \ll 10$.

При нулевом числе Рейнольдса $R=0$ спектр веществен $(\lambda=-1, -2, -2, -5, -5)$.

<u>Теорема</u>. В галеркинской пятимодовой системе с ростом R два собственных числа ($\lambda=-1$, $\lambda=-2$) начинают двигаться навстречу друг другу. При критическом числе $R_0 \approx 0,9324$ два вещественных собственных числа сталкиваются (у значения $\lambda_0 \approx -1,356$) и сходят в комплексную область. (Точка A на рис. I.) После этого вещественная часть γ родившейся пары комплексных собственных чисел растет с возрастанием числа Рейнольдса (линия AB на рис. I). Около $R_1 \approx 2,032$ (точка B) величина γ достигает значения -I, и нечетная мода обгоняет четную моду v (для которой $\gamma=-1$ при всех числах Рейнольдса). Начи-

19

ная с этого момента изучаемая мода становится ведущей (имеет наименьшее затухание). При дальнейшем росте числа Рейнольдса γ продолжает расти (асимптотически как $R\sqrt{1+\sqrt{5}}/2 \approx 0,9R$), переходя через 0 при $R\approx 4,32$.

Хотя ясно, что при таких больших числах Рейнольдса отбрасывание гармоник выше 5 степени недопустимо, поведение $\gamma(R)$ при $R<5$ удовлетворительно описывается пятимодовым приближением; это приближение выявляет происхождение растущей комплексной моды из столкнувшейся пары вещественных затухающих мод степеней I и 2 (в точной системе $R_0\approx 0,96$; $\lambda_0\approx -1,32$; $R_1\approx 2,286$).

§ 9. Диаграммная техника

В действительности, наши вычисления дают больше, чем галеркинские приближения: мы получаем также точные выражения для коэффициентов рядов Тейлора (или Пюизе) разложений собственных чисел полной (не галеркинской) системы (I) по степеням $R=1/\varepsilon$. Ответ дается в терминах построенного выше представления колчана: члены степени n в ряду Тейлора корня, родившегося из $-k^2$, отвечают петлям длины n с началом в вершине k^2 .

Для простоты мы начнем с однократного корня $\alpha_0=-1$.

Теорема. [3] Первый ненулевой коэффициент при R^n в разложении собственного числа $\lambda=\alpha_0+\alpha_1 R+\alpha_2 R^2+\dots$, $\alpha_0=-1$, оператора $\Delta+R\{v,\cdot\}$, дается суммой

$$\alpha_n=\sum\frac{A_{i_{n-1},i_0}\dots A_{i_1,i_2}A_{i_0,i_1}}{(\alpha_0-\lambda_{i_{n-1}})\dots(\alpha_0-\lambda_{i_2})(\alpha_0-\lambda_{i_1})}, \tag{3}$$

распространенной на все петли $i_0\to i_1\to\dots\to i_{n-1}\to i_0$ длины n , начинающиеся и кончающиеся в вершине i_0 .

В нашем случае $i_0=1$; операторы $A_{i,j}$ введены в § 7. Доказательство формулы (3) несложно и мы покажем лишь, как ею пользоваться.

I) Из формулы (3) следует, что α_1 и вообще все нечетные α_n равны нулю, поскольку длина любой петли в нашем колчане четна.

2) Единственная петля длины 2 с началом I есть

$1 \to 2 \to 1$. Согласно формулам § 7, $A_{2,1} A_{1,2} = 0$. Следовательно, $\alpha_2 = 0$.

3) Единственная петля длины 4 с началом I есть $1 \to 2 \to 5 \to 2 \to 1$. Перемножая матрицы, получаем

$$\alpha_4 = \frac{(0, 1/2)\begin{pmatrix} -1/2 & 0 \\ 1 & -1 \end{pmatrix}\begin{pmatrix} 1 & 0 \\ 0 & 1/2 \end{pmatrix}\begin{pmatrix} -2 \\ 0 \end{pmatrix}}{1 \cdot 4 \cdot 1} = -\frac{1}{4} .$$

Таким образом, для полной бесконечномерной системы $\lambda = -1 - R^4/4 + O(R^4)$, как и для галеркинского приближения § 8.

4) Мы заключаем также, что на α_n влияет только ограниченная часть колчана, и можем оценить какая.

В общем случае, когда корень α_0 кратный, формула (3) определяет не число, а линейный оператор $\alpha_n : \mathcal{F}_{i_0} \to \mathcal{F}_{i_0}$.

При расчете возмущения кратного собственного числа можно заменить бесконечномерное пространство полей конечномерным пространством $\mathcal{F}_{i_0}$ и весь оператор - матрицей $\alpha_0 E + R\alpha_1 + R^2\alpha_2 + \ldots$.

Например, в вершине 2 начинаются ровно две петли длины 2, $2 \to 1 \to 2$ и $2 \to 5 \to 2$. Им соответствуют слагаемые $\begin{pmatrix} -2 \\ 0 \end{pmatrix}(0, 1/2)/-1 = \begin{pmatrix} 0 & 1 \\ 0 & 0 \end{pmatrix}$, $\begin{pmatrix} -1/2 & 0 \\ 1 & -1 \end{pmatrix}\begin{pmatrix} 1 & 0 \\ 0 & 1/2 \end{pmatrix}/3 = \begin{pmatrix} -1/6 & 0 \\ 1/3 & -1/6 \end{pmatrix}$. Итак, $\alpha_0 E + R^2\alpha_2 = \begin{pmatrix} -2 - R^2/6 & R^2 \\ R^2/3 & -2 - R^2/6 \end{pmatrix}$; чтобы найти λ с погрешностью $O(R^2)$ нужно знать еще α_4 . Для галеркинской системы $\alpha_4 = 0$, $\lambda = -2 + R^2(-1/6 \mp \sqrt{1/3}) + O(R^4)$ в согласии с формулами § 8.

Литература

1. Henon M. Sur la topologie de lignes de courant dans un cas particulier. - C.R. Acad. Sci, Paris, 1966, v. 262, p. 312-314.
2. Арнольд В.И. Несколько замечаний об антидинамо теореме. - Вестник МГУ, сер. матем., мех., I982, № 6.
3. Арнольд В.И. Замечания о теории возмущений для задач типа Матье. - УМН, I983, т. 38, № 4, с. I89-208.
4. Арнольд В.И., Коркина Е.И. Рост магнитного поля в трёхмерном стационарном потоке несжимаемой жидкости. - Вестник МГУ, сер. матем., мех, I983, № 3 , с. 43-46.

364

Exponential scattering of trajectories and its hydrodynamical applications

V.I. Arnold
Translated by Gerald Gould

At the beginning of the 1960s in the theory of dynamical systems with finitely many degrees of freedom some remarkable events occurred: the stability of exponential divergence of trajectories in phase space or on a subset of it attracting neighbouring trajectories was discovered. The fundamental importance of these discoveries for the description of turbulence-type phenomena both in finite-dimensional and infinite-dimensional systems was understood fairly soon. However, this idea became widely accepted only in the seventies after Ruelle and Takens invented the term "attractor", which rapidly became fashionable, and took the whole area out of the restraining framework of rigorous theorems (of Anosov, Sinai, Smale, and others) into the vast terrain of numerical and physical experiments.

In this lecture, I will talk on three interrelated directions of rigorous mathematical investigation, resulting from the interaction of ideas in the modern theory of finite-dimensional dynamical systems and hydrodynamics: upper estimates of the dimension of attractors, applications of the geometry of infinite-dimensional Lie groups in the hydrodynamics of an ideal fluid, and certain questions relating to the problem of the stationary kinematic dynamo in magneto-hydrodynamics.

1. The dimension of strange attractors

Ten years ago A.N. Chetaev became interested in the following *problem of pursuit*. Suppose that in n-dimensional phase space $\mathbb{R}^n$ an m-dimensional submanifold M (the *target*) is given, and at each point of phase space it is known toward which target point to aim. In mathematical language this means that we are given a map $f : \mathbb{R}^n \to M^m$. The differential equation of pursuit has the form

$$\dot{x} = f(x) - x,$$

Translation of *N.E. Kochin and the Development of Mechanics*, Nauka, Moscow (1984), 185–193, 254.

Originally published in: N.E. Kochin and the Development of Mechanics, © Nauka, Moscow 1984
Translated by G. Gould. Springer-Verlag Berlin Heidelberg 2014 with kind permission, DOI 10.1007/978-3-642-31031-7_40

that is, we move directly towards the target (though evasive, since as our state changes our target alters).

Such a system, generally speaking, establishes some limiting régime (attractor), situated near the target manifold M^m, but, generally speaking distinct from M^m. Chetaev posed the question *of the dimension* of the limiting set of states of the process arising under arbitrary initial conditions, that is, the set

$$X = \bigcap_{t>0} (g^t B),$$

where g^t is the transformation of the phase flow and B is a sufficiently large ball in the phase space (it is easy to see that X does not depend on the choice of this ball, provided that it contains M). The set X is called the *global attractor* of the system.

An analysis of the dimension of the global attractor led A.N. Chetaev and Yu.S. Il'yashenko to the general theory of "k-compressing" systems, which is an interpolation between the theory of contracting maps ($k = 1$) and systems of negative divergence ($k = n$). While refereeing their work I noticed that the new theory can also be useful for the investigation of infinite-dimensional systems, for example, the Navier–Stokes system. At present this idea has been realized (Yu.S. Il'yashenko) for a two-dimensional Navier–Stokes system; here *for the dimension of the global attractor an upper estimate by a quantity of the order* Re^4 *was obtained.* (Strictly speaking, in [1] Yu.S. Il'yashenko considers: a) only Galerkin approximations and b) only periodic boundary conditions. The estimate of the dimension does not depend on the order of the approximation. Probably in the near future restrictions a), b) will be removed[1] but it is as yet unknown whether one can give a similar estimate in the 3-dimensional case.)

Here I merely want to explain the essence of the matter, namely, the concept of a *k-contraction* and its connection with the dimension of an attractor.

A map $g : V \to V$ of an n-dimensional Euclidean space (or a Riemannian manifold) into itself is said to be k-contractible if it uniformly decreases k-dimensional volumes of k-dimensional submanifolds.

To check whether a map is a k-contraction it suffices to consider the axes of an infinitesimal ellipsoid into which an infinitesimal ball centered at a point x is taken by the derivative of g at the point x. If r is the radius of the ball and $a_1 \geq \ldots \geq a_n$ are the semiaxes of the ellipsoid obtained from it then the *k-contractibility condition has the form* $a_1 \ldots a_n / r^k < \lambda < 1$ (the coefficient λ is the same for all points x).

Theorem. *Suppose that a subset M of V is invariant with respect to the map g, that is, $gM = M$. Then M has k-dimensional measure zero.*

This is easy to understand because

$$\mathrm{mes}_k M = \mathrm{mes}_k (gM) \leq \lambda \, \mathrm{mes}_k M, \quad \lambda < 1.$$

---

[1] The restriction a) was removed by Yu.S. Il'yashenko himself and b) by M.I. Vishik and by A.B. Babin, who also obtained estimates of the order $\exp(C\mathrm{Re}^4)$ on a two-dimensional manifold with boundary.

(Here mes_k is k-dimensional Hausdorff measure; it is defined for any set M, is equal to zero for manifolds of dimension less than k, is equal to infinity for manifolds of dimension greater than k, and is equal to the ordinary k-dimensional volume for k-dimensional manifolds.)

Suppose now that we are given a differential equation $\dot{x} = v(x)$, for example, the equation of pursuit or the Navier–Stokes system. In such a case the k-contractibility condition for the phase flow is easily verified by calculating the derivatives of v. For example, the n-contractibility condition ($n = \dim\{x\}$) simply has the simple form $\max \operatorname{div} v < 0$. In the general case one has to consider the quadratic form $(v_*\xi, \xi)$ (its matrix elements are $(\partial v_i/\partial x_j + \partial v_j/\partial x_i)/2$). Let $\lambda_1 \geq \ldots \geq \lambda_n$ be the eigenvalues of this matrix. The k-contractibility condition has the form $\max(\lambda_1 + \cdots + \lambda_k) < 0$.

In the case of the equation of pursuit such a condition (with k, of order of the dimension of the target, being independent of the dimension of the phase space) is satisfied due to the term $-x$ on the right-hand side. In the case of the Navier–Stokes equation the viscosity summand plays a similar role: the dissipation on the high harmonics is so strong that our quadratic form can have only finitely many positive eigenvalues and the sum of the negative ones rapidly diverges to $-\infty$.

The Hausdorff measure of attractors of the equations of hydrodynamics was also considered up to the work of Il'yashenko in a number of papers by Mallet-Paret, Foias and Ruelle. However, in these works the existence of a global attractor was not established, and only conditional upper estimates were obtained for the dimension of any attractor occurring in some function space or other. Il'yashenko's work guarantees also the convergence to the attractor of estimated dimension starting from any initial condition, but, unfortunately, it is this technically most difficult part of the work that depends the most on the two-dimensionality of the problem.

2. Curvature of the group of diffeomorphisms and poor predictability of a fluid motion

Although the passage of the fluid motion after loss of stability of a laminar flow to the established régime with exponential divergence of the trajectories in phase space appeared obvious already at the beginning of the 60s, a rigorous mathematical proof of these ideas is still missing. Here I shall discuss one of the first attempts along these lines.

It has been known (since the time of Hadamard) that the motion along geodesics on manifolds of *negative* curvature has the property of exponential instability. On the other hand, *the motion of an ideal* (incompressible inviscid) *fluid can be regarded as the motion along a geodesic on an* (infinite-dimensional) *manifold*, namely, on the group of diffeomorphisms of the flow domain that preserve the volume element. The kinetic energy defines on this infinite-dimensional Lie group a right-invariant Riemannian metric (strictly speaking, it is only a "weak metric" in view of the fact that the manifold is infinite-dimensional).

There arises the following question: what is the curvature for two-dimensional sections of this infinite-dimensional Riemannian manifold? The negativeness of the curvature is an indication of the flow instability and the curvature value affects the growth rate of the perturbations. It turned out that the curvature of the diffeomorphism group of a two-dimensional torus has both signs, but in the majority of the two-dimensional directions it is *negative* [2]. For example, the degree of instability of a "trade-wind flow" with the stream function $\sin y$ on a two-dimensional torus is such that the perturbations increase at least 20-fold for the period of one orbit of the quickest particles around the torus. This makes it impossible to give a long-range forecast of the motion of the fluid (if one regards the Earth as the torus and the atmosphere as an incompressible fluid, then the prediction error of the motion of the atmosphere over two months increases by a factor 10^5 as compared with the error in one's knowledge of the initial conditions.

In recent years A.M. Lukatskii has continued the study of the curvature of the diffeomorphism group in a number of papers. In particular, he has extended it to the cases of three-dimensional and n-dimensional flows [3], as well as to the case of flows on a two-dimensional sphere [4]. The results obtained are qualitatively (and even quantitatively) close to the results for a two-dimensional torus. I would like to note an entirely new advance in this area, which is also due to Lukatskii: the introduction and investigation of the *Ricci curvature* of infinite-dimensional manifolds.

By the Ricci curvature of an n-dimensional manifold in a fixed direction we mean the sum of its curvatures in the directions of two-dimensional planes drawn through the line of the chosen direction and through $n-1$ lines orthogonal to the chosen line and to each other.

For an infinite-dimensional manifold this sum is in general equal to infinity. But if one divides the Ricci curvature of an n-dimensional manifold by $n-1$, then one obtains the *mean value* of the curvature over all two-dimensional planes containing the chosen line. This *(normalized) Ricci curvature* for many infinite-dimensional manifolds is already finite (for example, for a sphere of radius 1 it is equal to 1).

Theorem (A.M. Lukatskii). *The Ricci curvature of the group of diffeomorphisms of an n-dimensional torus that preserve the volume element is negative for any $n > 2$ in the direction of any non-constant field.*[2]

Thus, although the curvature in some two-dimensional directions is positive, *in the mean* the curvature is negative; consequently, exponential divergence between the fluid flows with close initial conditions is more likely than power-type divergence or oscillation of one flow around another, which corresponds to positive curvature.

The study of the curvature of a group of diffeomorphisms was undertaken as an attempt to obtain a rigorous lower estimate of the dimension of an attractor for Navier–Stokes equations with low viscosity. Although rigorous theorems along these lines have not been proved so far, it seems to me that the exponential instability of the motion of an ideal fluid, which is indicated by negative curvature, is

[2] The curvature in a direction of a constant field is zero: $\operatorname{Ricc} v = -\mathrm{const} \cdot \|\sqrt{-\Delta v}\|^2$.

a basic mechanism involving an increase in the dimension of the attractor, an increase in the number of exponentially increasing independent perturbations and an increase in entropy (the sum of the exponents of the increasing perturbations) as the Reynolds number increases.

The geometry of infinte-dimensional Lie groups also manifests itself in other problems of hydrodynamics: for example, the fact that the justification of three-dimensional hydrodynamics lags a long way behind that of two-dimensional hydrodynamics is explained by the difference of the orbit geometry of the corresponding coadjoint representations.

Although much still remains to be done here, I recall a recent result of Shnirelmann.

Theorem ([5]). *The group of volume-preserving diffeomorphisms of a compact three-dimensional manifold has finite diameter, while for a two-dimensional manifold the diamater is infinite.*

In other words, any transformation of all the particles of a three-dimensional incompressible fluid to these new positions can be undertaken by a continuous path of bounded length (in the energy metric), whereas in the two-dimensional case the length of the required path can be arbitrarily large. This is related to the twist possible in the two-dimensional case: the path in the group of diffeomorphisms twisting the fluid particles in such a way that it completes many revolutions in one direction *in the two-dimensional case* and cannot be replaced by a short path bringing about the same final transformation of the particle positions.

3. The problem of the stationary kinematic dynamo

We consider the stationary flow[3] of an incompressible fluid given by a vector field v in three-dimensional space; $\operatorname{div} v = 0$.

We also consider a "magnetic" divergence-free vector field H "frozen" in the fluid (that is, transported by fluid particles in accordance with the rule

$$H(g^t x, t) = g^t_{*x} H(x, 0),$$

where $\{g^t\}$ is the phase flow of the field v and g^t_{*x} is its derivative at the point x). Suppose that the field H is further subjected to a weak diffusion; in other words, we consider the equation of magneto-hydrodynamics, which is linear with respect to the unknown magnetic field H,

$$\partial H/\partial t = \{v, H\} + \mu \Delta H, \quad \operatorname{div} H = 0,$$

with fixed stationary velocity field v of zero divergence and with small magnetic viscosity μ; here $\{v, H\} = \operatorname{rot}[v \times H]$ is the Poisson bracket of the field v and H.

[3] The velocity field does not depend on time.

The problem of the stationary kinematic dynamo (see, for example, [6]) is as follows. Suppose that fields v and H have compact support (equal to 0 outside a finite region) or satisfy periodic boundary conditions. Can H *exponentially grow with time* (with exponent not tending to zero as $\mu \to 0$)?

It was known for a long time that if the flow v is two-dimensional, then such a dynamo is impossible (this was established by Zel'dovich in 1956). The essence of the matter is that in a *two-dimensional* stationary flow of an incompressible fluid in a bounded region there does not occur *exponential* stretching of particles (the distance between originally close particles increases for most initial conditions only as the first power of the time). Therefore a magnetic field frozen in a two-dimensional fluid flow expands only linearly (rather than exponentially) with time.

In the *three-dimensional* case stationary flows of an incompressible fluid with exponential stretching of the majority of the particles in a bounded region are possible. One of the simplest examples of this kind was suggested by the author in 1965: the divergence-free velocity field is given by the components

$$(A\sin z + C\cos y, B\sin x + A\cos z, C\sin y + B\cos x) \tag{1}$$

(for this field $\operatorname{rot} v$ is proportional to v).

Although the exponential stretching of the majority of particles in such a field has so far not been rigorously proved, numerical experiments, carried out at the author's request by M. Henon [8] in 1966 and E.I. Korkina in 1981 clearly indicate exponential divergence of close trajectories, including the case $A = B = C$.

In the diagram there are shown returns of the trajectory of the field (1) on the torus $(x, y, z) \mod 2\pi$ in the plane $x = 0$. All the points belong to one and the same trajectory. Clearly this trajectory does not visit some regions of the plane, while it fills others densely. Successive intersection points of the trajectory with the plane $x = 0$ appear in weakly intercorrelated places due to the exponential scattering of the trajectories.

Thus, the exponential growth of a magnetic field with zero magnetic viscosity for this flow is experimentally established with sufficient certainty. The question whether this conclusion still holds for a small but finite magnetic viscosity is highly non-trivial, since the most rapidly increasing field for zero viscosity looks very discontinuous (it is concentrated on the ergodic component of the dynamical system defined by the field v, while this system is non-integrable and has complicated (in the set-theoretic sense) ergodic components similar to the zones of instability of the three-body problem and of other non-integrable problems of classical mechanics).

The fact that the most rapidly increasing field has numerous folds leads to its rapid decay due to the magnetic viscosity; therefore it is not clear how this struggle between the exponential growth (due to the stretching of the particles) and the dampening (due to the viscosity) will turn out.

I think that this question could be solved (if the exponential growth of H for our field v does not decay as $\mu \to 0$) by considering Galerkin approximations on a fairly powerful computer. Carrying out such computations would be very desirable, since the geometric question formulated above apparently does not yield to the rigorous methods of contemporary mathematics.

It is, however, possible to solve it by a certain idealization of the problem: it suffices to consider the motion of a fluid in a suitable three-dimensional compact Riemannian manifold, rather than in three-dimensional Euclidean space, in order to construct an example of exponential growth of the magnetic field with exponent bounded away from zero as the magnetic viscosity tends to zero. An example of such a compact manifold and flow was described by the author in 1972, see [9]; it was applied to the problem of the magnetic dynamo in a recent joint work [10] by the author, Zel'dovch, Ruzmaikin and Sokolov.

The manifold M is constructed from three-dimensional space by the following identifications:

$$(x,y,z) \sim (x+1,y,z) \sim (x,y+1,z) \sim (2x+y,x+y,z+1). \tag{2}$$

The Riemannian metric on this manifold is defined by the formula

$$ds^2 = e^{2\alpha z}dp^2 + e^{-2\alpha z}dq^2 + dz^2, \tag{3}$$

where p and q are the Cartesian coordinates in the (x,y) plane with axes directed along the eigenvectors of the linear transformation $A(x,y) = (2x+y,x+y)$ and α is the logarithm of the eigenvalue of this transformation.

The velocity field v is given in (x,y,z) coordinates by the components $(0,0,1)$. If one takes the magnetic field H to be directed along the contracting eigenvector of the transformation A and constant in magnitude, then $\Delta H = 0$ and hence such a field independently of the magnetic viscosity value is transformed as a frozen one. The frozen field directed this way exponentially increases because the phase flow of the field v exponentially stretches the length of the eigenvector of the contracting direction for A.

It is interesting to note that the fields H of other directions behave in the following manner for a finite viscosity μ: during time of order $\log\mu$ they increase up to a value of order $1/\mu$ and then rapidly decay.

All these results have the character of an exactly soluble model example (in a sense it is a question of the well-known Hubble flow, which is brought about by means of mathematical tricks on a compact manifold). Another exactly soluble example of this sort (not checked in detail as far as I know) is obtained from the geodesic flow on a surface of constant negative curvature (the exact solubility is guaranteed by the fact that the velocity field of a horocyclic flow is harmonic in the natural Riemannian metric of the phase space).[4]

By the way, the last example refutes the point of view that the conditions

$$B = \operatorname{rot} A, \quad (B, A) = 0, \quad A \neq 0 \tag{4}$$

imply that the flow B is *two-dimensional.* If for A one takes a negative horocyclic field, then the field $B = \operatorname{rot} A$ will be positive horocyclic and consequently the lines of the field B fill a *three-dimensional* manifold everywhere densely.

In fact, conditions (4) merely ensure that the lines of the field B lie on two-dimensional surfaces (orthogonal to trajectories of field A): this is Frobenius's theorem. But these surfaces can be non-closed and densely fill a three-dimensional region.

However, I do not know whether such a phenomenon is possible in a region of the three-dimensional Euclidean space.

One of the approaches to the seemingly analytically unapproachable problem of the growth of H in the field (1) can consist in considering small perturbations of the velocity field v or the metric (3) on the manifold (2) and investigating the stability of the exponential growth of H with respect to the perturbations of v and ds^2.

The point is that an exact solution in the example (2), (3) is guaranteed by a common eigenvector of the operators of shift of the frozen field and diffusion (that is, the operator ad_v of the Poisson bracket with v and the Laplace operator Δ). Under generic perturbations of the field v and the metric ds^2, the common eigenvector disappears and the problem becomes more similar to the problem for the field (1), while remaining, nevertheless, close to exactly soluble, and consequenly not so unapproachable.

To conclude this survey I should like to recall a number of geometric problems connected with so-called hellicity (that is, the Hopf invariant or the average linking number of the field trajectories). These problems were discussed in detail in [11] and, as far as I know, they remain unsolved.[5]

---

[4] Editors' note: This example was thoroughly studied in Section V.4 of the book "Topological methods in hydrodynamics" by V. Arnold and B. Khesin (1998), Springer.

[5] Editors' note: Some of these problems were solved in the papers by J.-M. Gambaudo and É. Ghys (1997), S. Müller and P. Spaeth (2011), and others.

References

[1] Yu. S. Il'yashenko, *Weakly contractible systems and attractors of Galerkin approximations of the Navier–Stokes equation*, Uspekhi Mat. Nauk **36**, No. 3, 243–244 (1981).

[2] V. I. Arnol'd, *Sur la courbure de Riemann des groupes des difféomorphismes*, C. R. Acad. Sci. Paris **260**, 5668–5671 (1965).

[3] A. M. Lukatskii, *On the curvature of measure-preserving diffeomorphisms of an n-dimensional torus*, Uspekhi Mat. Nauk **36**, No. 2, 187–188 (1981).

[4] A. M. Lukatskii, *On the curvature of measure-preserving diffeomorphisms of a two-dimensional sphere*, Funktsional. Anal. i Prilozhen. **13**, No. 3, 23–27 (1979).

[5] A. I. Shnirel'mann, *Geometry of the group of diffeomorphisms and the motion of an ideal incompressible fluid*, Uspekhi Mat. Nauk **35**, No. 5, 216 (1980).

[6] H. K. Moffatt, *Magnetic field generation in electrically conducting fluids*, Cambridge University Press, 1978.

[7] Ya. B. Zel'dovich, *The magnetic field for the two-dimensional motion of a conducting turbulent liquid*, JETP **31**, 154–155 (1956).

[8] M. Henon, *Sur la topologie des lignes de courant dans un cas particulier*, C. R. Acad. Sci. Paris (1966).

[9] V. I. Arnol'd, *Notes on the three-dimensional flow pattern of an ideal fluid under a small perturbation of the initial velocity field*, Prikl. Mat. Mekh., **36**, No. 2, 255–262 (1972).

[10] V. I. Arnol'd, Ya. V. Zel'dovich, A. A. Ruzmaikin and D. D. Sokolov *The magnetic field in a stationary flow with stretchings in a Riemannian space*, JETP **11**, No. 6, 2052–2058 (1981).

[11] V. I. Arnol'd, *The asymptotic Hopf invariant and its applications*, In the book: *Materials of the All-Union School on differential equations with infinitely many independent variables and on dynamical systems with infinitely many degrees of freedom*, Dilijan, Erevan, 1973; English translation: Selecta Math. Soviet. **5**, 327–345 (1986).

Kolmogorov's hydrodynamic attractors

By V. I. Arnol'd

Steklov Mathematical Institute, Vavilova 42, GSP-1, Moscow 117966, U.S.S.R.

This paper discusses briefly the content and the lasting influence of Kolmogorov's 1958 seminar on dynamical systems and hydrodynamic instability.

A. N. Kolmogorov considered his work on turbulence to be non-mathematical. He wanted to explain observed phenomena from first principles, i.e. from the Navier–Stokes equations. This dream was one of the sources of his interest in the theory of dynamical systems, in which he perceived a set of tools for the description of turbulent chaos.

In the fall of 1958, the programme of a new series of Kolmogorov seminars, 'Theory of dynamical systems and hydrodynamical instability', appeared on the announcement board of the MechMat (Mechanics and Mathematics) Faculty of Moscow Lomonosov University (Arnol'd & Mechalkin 1960). The text of the programme was:

1. Boundary problems for hyperbolic equations, whose solutions depend everywhere discontinuously on a parameter (see for instance the article of S. L. Sobolev, *Doklady* **109**, 1956, 707).

2. Problems of classical mechanics, whose eigenfunctions depend everywhere discontinuously on a parameter (see the survey of this topic in A. N. Kolmogorov's talk at the Amsterdam Congress, 1954).

3. Monogenous functions of Borel, and quasianalytical functions of Gontchar (in the hope of using these functions as tools for the study of problems (1) and (2)).

4. The appearance of high-frequency oscillations in systems with a small parameter multiplying the highest derivative term (for ordinary differential equations – see the works of Volosov and Lykova).

5. In the mathematical theory of partial differential equations containing a small parameter multiplying the term with the highest derivative; until now only such phenomena as boundary layers or interior layers (their limiting position for vanishing viscosity being surfaces of discontinuity of the limit solutions or of their derivatives) have been studied.

6. Stability problems for laminar flows. Asymptotically vanishing stability (at least for the same model equations). [The shrinking of the attracting domain of some stationary solutions to a point as the Reynolds number tends to infinity – V.A.]

7. The possibilities of application of the techniques of the metric theory of dynamical systems to real physical and mechanical problems. The problem of stability of different spectral types. The structurally stable system and the structurally stable properties (almost nothing is known in this domain for systems with more than one degree of freedom).

8. The discussion (at least at the level of models) of the conjecture, that in the presence of the phenomena described at the end of (5) the dynamical system becomes

Proc. R. Soc. Lond. A (1991) **434**, 19–22
Printed in Great Britain

Originally published in: Proc. Royal Soc. London A 434:1890, 19-22, © The Royal Society, London 1991
Reprint by Springer-Verlag Berlin Heidelberg 2014 with kind permission, DOI 10.1007/978-3-642-31031-7_41

in the limit a stochastic process (the conjecture of the practical impossibility of long-time weather prediction).

At that time Kolmogorov had in mind at least two scenarios for the transition from the laminar to turbulent flows. The first scenario was an extension of the Landau idea of bifurcations, transforming a steady flow into a period one, a periodic flow into a quasi-periodic one with two basic frequencies, described by a two-dimensional torus-attractor in the phase space, transforming a two-dimensional torus into a three-dimensional torus, and so on.

Kolmogorov posed the question as to whether an infinite sequence of such bifurcations can occur within a finite range of Reynolds numbers. He suggested that a system with a discrete spectrum generated by an infinite number of basic frequencies describing motion on an infinite-dimensional torus would seem very similar, for an experimentalist, to a mixing system whose spectrum is continuous.

It is known that such an infinite sequence of bifurcations is impossible in two-dimensional flows. Today most experts believe that it is also impossible in the three-dimensional case (but, as far as I know, this still has not been proved rigorously).

Kolmogorov related the poor predictability of flows with the influence of the higher-order harmonics on the basic, lower-order modes. Suppose, he would say, that the velocity field be changed in every cubic kilometre without changing its average in this cube. We have to study what the time interval is beyond which this change will crucially affect the weather. It is clear that dynamic weather prediction is impossible for longer periods (and will remain impossible in spite of all the future progress in computer techniques).

This reasoning, which Kolmogorov related to infinite-dimensional tori, is in fact independent of their conjectural existence: an attracting invariant torus of sufficiently high finite dimension, covered by quasi-periodic orbits, would lead to the same conclusions.

As far as I know, Kolmogorov's ideas on the experimentalist's difficulties in distinguishing between quasi-periodic systems with many basic frequencies and genuinely chaotic systems have not yet been formalized.

Suppose that some modes (of small amplitude) are neglected in an approximate description of a system whose attractor is filled by quasi-periodic orbits. Since some of the phases of the initial conditions are neglected in the approximation, the observed projection of the quasi-periodic motion on a torus of smaller dimension will seem to be sensitively dependent on the initial conditions, to have positive Ljápunov exponents and positive Kolmogorov–Sinai entropy. The quasi-periodicity becomes evident only much later, when we observe the orbit during a time period which grows exponentially with the number of neglected modes. The distinction between quasi-periodic attractors with many basic frequencies and really chaotic attractors with positive Ljápunov exponents is still a difficult experimental problem.

The alternative scenarios with finite-dimensional attractors carrying complicated dynamics under generic small perturbations was mentioned by Kolmogorov in his 1954 talk at the Amsterdam Congress of Mathematicians. He discussed there the motion along the geodesics on negatively curved manifolds and raised the question as to whether their ergodic properties should persist under generic hamiltonian perturbations (which, he believed, should be the case).

S. Smale's conjecture of the stability of this kind of 'chaotic' dynamics under *arbitrary* smooth perturbation appeared several years after the Kolmogorov 1958–59 seminar. The works of Sinai and Anosov on the structural stability and ergodic properties of these exponentially unstable motions appeared even later. Thus Kolmogorov's scenario was based on the complexity of the dynamics along the attractor rather than on its stability.

In a sense he was right. The attractors observed in real hydrodynamics are rather far from the structural stability paradigm, which makes the Sinai–Anosov theory so nice and which gives it its definitive character (making it at the same time so useless for application to the study of most of the attractors occurring in physics).

To isolate the main difficulty, Kolmogorov suggested the following model hydrodynamic problem: to study the flow of an incompressible viscous fluid along a torus (the periodic boundary conditions replacing the walls) under the action of a distributed exterior force field (replacing the exterior pressure or the boundary motion).

The viscosity and the exterior force are supposed to be small and proportional to the same small parameter – the scale of the resulting attractor has to be fixed while the Reynolds number grows to infinity.

The flow in this 'Kolmogorov model' is usually supposed to be two-dimensional and the exterior force field has usually the form of the first harmonic. If the Reynolds number is small, the steady laminar flow emerges: its velocity profile has a sinusoidal form. This flow loses its stability as the Reynolds number grows. The Kolmogorov 1958 programme was to study rigorously the bifurcations of the attractors in this system and their dependence on the growing Reynolds number. This programme was partially fulfilled in the work of Mechalkin & Sinai (1961). As far as I know, the bifurcations in such systems have not been completely understood even now, even in the two-dimensional case and even numerically, using the Galerkin approximation for *moderate* Reynolds number (but see Nicolaenko & She 1990).

It should be stressed that the Lorentz model, in which the chaotic motion along the attractor has been observed, is related to another mechanism of generation of oscillations.

The Kolmogorov system, written as a system of ordinary differential equations (using, say, the Galerkin approximation) takes the form

$$\dot{x}_i = \Sigma\, a^i_{jk}\, x_j\, x_k - b_i\, x_i + f_i.$$

The first nonlinear term corresponds to the inertial term, in the Euler equation. The linear term takes into account the viscosity, while the constant term describes the amplitude of the corresponding harmonics of the exterior force field. In the Kolmogorov model both the dissipation b and the force f are proportional to the same small parameter (to the viscosity).

The inertial term satisfies the incompressibility conditions $\Sigma_i\, a^i_{ik} = 0$ and the energy conservation condition

$$\Sigma\, a^i_{jk}\, x_i\, x_j\, x_k = 0.$$

The viscous linear term dissipates the energy, if the exterior force is absent:

$$\tfrac{1}{2}x^2 = \text{const.}, \quad \text{if } b = f = 0.$$

Taking into account the exterior force, we find that the energy is dissipated if the amplitudes are high (outside some ellipsoid in the phase space). Inside this ellipsoid the energy is injected by the exterior force field.

In the Lorentz model (Lorentz 1963) the constant term is absent, and the linear term is responsible both for energy dissipation and injection. The boundary separating the dissipation region from the injection region is a quadratic cone surface. This larger injection region enhances the instability of the trajectories along the attractor. Hence the Lorentz type systems are more sensitive to initial conditions than the Kolmogorov type systems, whose attractors have more relation to the Navier–Stokes equation attractors than those of the Lorentz model.

In any case, Kolmogorov's problem of the proof of the instability of the trajectories along the attractors of his system, formulated by him in the fifties, remains unsolved even now. In modern terms one of his problems reads:

> Does the minimum of the Hausdorff dimensions of the attractors grow with Reynolds number for the Navier–Stokes system on a compact two-dimensional manifold (for, say, the Kolmogorov model)?

References

Arnol'd, V. I. & Mechalkin, L. D. 1960 The Kolmogorov seminar on analysis (1958–1959). *Usp. mat. Nauk.* **15**, 247–250. (In Russian.)

Lorentz, E. 1963 Deterministic nonperiodic flow. *J. atmos. Sci.* **20**, 130–141.

Mechalkin, L. D. & Sinai, Ya. G. 1961 The study of the stability of a stationary solution of one system of equations of the plane motion of the incompressible viscous fluid. *Appl. Math. Mech.* **6**, 1140–1143.

Nicolaenko, B. & She, Z.-S. 1990 Coherent structures, homoclinic cycles and vorticity explosions in Navier–Stokes flows. In *Topological fluid mechanics* (ed. H. K. Moffatt & A. Tsinober), pp. 265–277. Cambridge University Press.

Proc. R. Soc. Lond. A (1991)

Annu. Rev. Fluid Mech. 1992. 24: 145–66

TOPOLOGICAL METHODS IN HYDRODYNAMICS

V. I. Arnold

Steklov Mathematical Institute, 42 Vavilova St., 117966 Moscow GSP-1, USSR

B. A. Khesin

Department of Mathematics, University of California, Berkeley, CA 94720

KEY WORDS: incompressible fluid, diffeomorphism group, invariants of motion, linking number

INTRODUCTION

A group theoretical approach to hydrodynamics considers hydrodynamics to be the differential geometry of diffeomorphism groups. The principle of least action implies that the motion of a fluid is described by the geodesics on the group in the right-invariant Riemannian metric given by the kinetic energy. Investigation of the geometry and structure of such groups turns out to be useful for describing the global behavior of fluids for large time intervals.

We begin with a survey of conservation laws for incompressible and barotropic fluid flows and superconductivity. These laws are determined by the infinitesimal structure of the corresponding diffeomorphism groups (i.e. the structure of their Lie algebras and coalgebras). For example, the equations of an inviscid incompressible fluid are Hamiltonian on the coadjoint orbits of the group of volume-preserving diffeomorphisms (Arnold 1966, 1969a, 1989). It is well known that for a two-dimensional flow there is an infinite number of enstrophy-type integrals $[\int f(\text{curl}\,\mathbf{v})\,d^2x]$, and in a three-dimensional case there is the total helicity integral $[\int (\text{curl}\,\mathbf{v}, \mathbf{v})\,d^3x]$. It turns out that these ideal hydrodynamics equations (as well as barotropic fluid and superconductivity equations) have an infinite

0066–4189/92/0115–0145$02.00

Originally published in: Annu. Rev. Fluid Mech. 24, 145-166, © Annual Reviews Inc., Palo Alto, California, USA 1992
Reprint by Springer-Verlag Berlin Heidelberg 2014 with kind permission, DOI 10.1007/978-3-642-31031-7_42

number of invariants for flows on an arbitrary even-dimensional manifold ("generalized enstrophies"). In an odd-dimensional case, they have at least one first integral ("generalized helicity") [we follow the paper of Serre (1984) and Khesin & Chekanov (1989)].

The second Section is devoted to the ergodic interpretation of hydrodynamical invariants. For three-dimensional manifolds, the helicity invariant coincides with the average linking number of the trajectories of corresponding curl vector fields (Arnold 1974). We present a survey of recent works extending this theme: new energy estimates for nontrivial linked fields (Freedman 1988; Freedman & He 1991a,b), ergodic interpretation of multidimensional hydrodynamic invariants as the average multilinking number of certain surfaces, and interpretation of Novikov invariants (Khesin 1991) and of Godbillon-Vey-type characteristic classes (Tabachnikov 1990).

Further results concern the geometry and curvatures of the different diffeomorphism groups themselves. Shnirelman's theorem (1985) states that the group of volume-preserving diffeomorphisms of a three- (or higher-) dimensional ball has a finite diameter if it is considered to be a Riemannian manifold. Contrary to this case, it was recently shown that the symplectomorphism group of any even-dimensional manifold has an infinite diameter (Eliashberg & Ratiu 1991b). In the conclusion, we discuss some results on the curvatures of diffeomorphism groups. The negativeness of these curvatures is responsible for hydrodynamical instability and for unreliable forecasts (Arnold 1966, 1989).

1. INVARIANTS OF MOTION FOR FLUID FLOWS

1.1 *Hydrodynamics on Riemannian Manifolds*

Let M^n denote a compact Riemannian manifold (without boundary) and μ a volume form [which, in general, has no relation to the volume form induced by the metric $(.,.)$]. The equation of the incompressible fluid on M is:

$$\dot{\mathbf{v}} = -(\mathbf{v}, \nabla)\mathbf{v} + \nabla p \qquad (1)$$

where $\mathbf{v}$ and p are a time-dependent vector field and a function on M, respectively, the flow of $\mathbf{v}$ preserves the form μ (i.e. $L_{\mathbf{v}}\mu = 0$ or $\operatorname{div}\mathbf{v} = 0$), and $(\mathbf{v}, \nabla)\mathbf{v}$ denotes the covariant derivative $\nabla_{\mathbf{v}}\mathbf{v}$ for the Riemannian connection.

Theorem 1 *Equation (1) has*
(i) the integral

$$I(\mathbf{v}) = \int_M u \wedge (du)^m \tag{2a}$$

in the case of an arbitrary odd-dimensional manifold M ($n = 2m+1$);

(ii) an infinite series of integrals

$$I_f(\mathbf{v}) = \int_M f\left(\frac{(du)^m}{\mu}\right)\mu \tag{2b}$$

in the case of an arbitrary even-dimensional manifold M ($n = 2m$), *where u is the 1-form induced from* $\mathbf{v}$ *by the "lifting of indices" defined by the metric* $u(\xi) = (\mathbf{v}, \xi)\ \forall\, \xi \in T_xM$, *and f is an arbitrary function of one variable.*

The statement above was settled for the standard $\mathbf{R}^n$ by explicit coordinate calculations by Tartar and Serre (Serre 1984). In our summary, we follow the works by Khesin & Chekanov (1989) and Ovsienko et al (1989) in which the general theorem was proved. This generalization differs considerably from that proposed by Dezin (1983) in which the odd-dimensional integral is obtained. Symmetries of the two- and three-dimensional equations were investigated by Olver (1982).

Proof This theorem is based primarily on the investigation of the structure of the Lie algebra $\mathscr{G}$ of divergence-free vector fields and its dual space.

There exists a natural isomorphism between the dual space $\mathscr{G}^*$ and the quotient space of differential 1-forms over M modulo differentials of functions: $\Omega_1/d\Omega_0$ (Marsden & Weinstein 1983). The corresponding pairing is $\langle [u], \mathbf{v}\rangle = \int_M u(\mathbf{v})\mu$ where $\mathbf{v} \in \mathscr{G}$ and the form $u \in \Omega_1$ is an arbitrary representative of $[u] \in \Omega_1/d\Omega_0$.

Let $G = \mathrm{SDiff}(M)$ be the group of all volume-preserving diffeomorphisms. Then the definition of the change of variables in the integral and the invariance of μ imply the coincidence of the coadjoint G-action with the G-action on the space of 1-forms.

The crucial point of the proof is the following

Proposition *The integrals* $I(u)$ *and* $I_f(u)$ *given by (2a,b) are well-defined functionals on* $\mathscr{G}^*$ *(i.e. they don't depend on the choice of the representative u in the class* $[u]$*) and are invariants of the coadjoint action.*

Proof Since the coadjoint action is push-forward, the statement follows from the change of variables formula and the coordinate-free definition of the corresponding integrals.

Let $(.,.)$ be a Riemannian metric on M (whose volume form differs, in general, from the given volume μ). It defines a nondegenerate scalar product on $\mathscr{G}$: $(\mathbf{v},\mathbf{w})_{\mathscr{G}}=\int(\mathbf{v},\mathbf{w})\mu$ and, hence, an invertible operator $A:\mathscr{G}\to\mathscr{G}^*$, called the inertia operator [see Arnold (1989)].

This operator takes Equation (1) on $\mathscr{G}$ into the following equation on $\mathscr{G}^*$: $[\dot{u}]=-L_{\mathbf{v}}[u]$, where $[u]=A\mathbf{v}$ and the operator $L_{\mathbf{v}}$ is the Lie derivative determined by the vector field $\mathbf{v}$ on M. [Indeed, for a particular representative u of this class $[u]$, this equation has a more recognizable form: $\dot{u}=-L_{\mathbf{v}}u+d\psi$, cf (1).] So, in fact, the equation for an ideal fluid is the equation on $\mathscr{G}^*$; moreover, this is Hamiltonian under the canonical Lie-Poisson linear structure on the coalgebra $\mathscr{G}^*$ [the Hamiltonian function is the kinetic energy $H(\mathbf{v})=\langle A\mathbf{v},\mathbf{v}\rangle$, see Arnold (1966, 1989)]. This implies that trajectories of the Euler equation are tangent to the orbits of the coadjoint G-action on $\mathscr{G}^*$, hence, $I_f([u])$ and $I([u])$ are its integrals [for details see Khesin & Chekanov (1989)].

Example: In the standard metric of $\mathbf{R}^3$ the integral (2a) coincides with $I(\mathbf{v})=\int(\mathbf{v},\operatorname{curl}\mathbf{v})\mu$, and for $\mathbf{R}^2$ the integral (2b) coincides with $I_f(\mathbf{v})=\int f(\operatorname{curl}\mathbf{v})\mu=\int f(\Delta h_{\mathbf{v}})\mu$, where $h_{\mathbf{v}}$ is the "stream function" of the vector field $\mathbf{v}$, i.e. the Hamiltonian for $\mathbf{v}$, relative to the symplectic form μ [see Arnold (1966)].

Remark 1 The invariant (2b) of the plane-parallel $2m$-dimensional flow induced by any $(2m-1)$-dimensional flow is trivial, since $(du)^m=0$. Therefore, the reduction of dimension gives no integrals different from (2a).

Remark 2 The integrals $I([u])$ and $I_f([u])$ do not form a complete set of continuous invariants of coadjoint orbits. By analogy with two- and three-dimensional cases (Arnold 1974), it is possible to construct parametrized families of orbits with equal values of these functionals. A precise description of the coadjoint orbits for the diffeomorphism groups still remains an unsolved and intriguing problem. In particular, it is unknown whether there exists a dense orbit in each level set of the integral $I(\mathbf{v})$ for $n=3$.

Remark 3 The manifold M may be multi-connected. In the nonsimply-connected case, the cohomological class of $[u]$ is also an invariant [compare Arnold (1969b)]. Other examples of (discrete) invariants of the Euler equation are the number of points on M where du is degenerate and the order of its degeneracy (here $[u]=A\mathbf{v}$).

Remark 4 The manifold M may be noncompact or may have a boundary (we may consider $M=\mathbf{R}^n$). In general, we should consider vector fields tangent to the boundary.

The rest of this section is devoted to the case of fluid motion on odd-dimensional manifolds. In this case, Theorem 1 provides the existence of one invariant. The geometrical approach to its proof allows us to obtain the following statements.

Corollary 1 *On an odd-dimensional manifold ($n = 2m+1$), define the curl vector field ξ by $i_\xi \mu = (du)^m$, where $i_\xi *$ denotes the inner product of the vector field ξ with a differential form. Then ξ is "frozen in the fluid."*

Proof The class $[u]$ [and, therefore, $(du)^m$] is Lie transported by the flow; the volume μ is invariant and, hence, the vector field ξ is also Lie transported geometrically, i.e., is "frozen in the fluid."

Corollary 2 [for $n = 3$ see Arnold (1974)]. *The incompressible fluid equations on an odd-dimensional manifold have a set of integrals $I_c(v) = \int_c u \wedge (du)^m$, where the integral is taken over any ergodic component c of the momentary curl vector field.*

The proof follows immediately from the Stokes formula and from the observation that the restriction of $(du)^m$ to the boundary of any ergodic component vanishes.

Remark 5 All the integrals in question are invariants of the coadjoint representation of the corresponding Lie groups (the so-called Casimir elements), i.e. they do not depend on the particular choice of the Hamiltonian. This opens the way to the investigation of the nonlinear stability problems by Routh or by the energy-momentum method (Arnold 1967, Holm et al 1985, Marsden et al 1989). We also think that information about the orbits can be of help in the study of the Cauchy problem of multi-dimensional hydrodynamics.

Note that the existence of an infinite series of integrals for the flow of an even-dimensional fluid has nothing to do with the complete integrability of the corresponding hydrodynamic equations. The invariants considered only define the coadjoint orbits (generally speaking, infinite-dimensional) on which the evolution takes place. For the equations on this orbit, there is a unique energy integral, while the integrability requires an infinite number of integrals.

Nevertheless, in the recent paper by Murometz & Razboynich (1990), it is shown that the hydrodynamics equations on a plane admit finite-dimensional truncations of arbitrarily large size that turn out to be Hamiltonian integrable systems. It would be very interesting to find nontrivial integrable subsystems for multidimensional hydrodynamics.

1.2 *Generalized Superconductivity and Barotropic Fluid Equations*

It turns out that multidimensional generalizations of superconductivity and barotropic fluid equations have conservation laws analogous to the hydrodynamic invariants. In both cases, the equations have an infinite series of integrals if the dimension n of the manifold is even, and at least one integral if this dimension is odd [see Holm & Kuperschmidt (1983) and Holm et al (1985) for $n = 3$ and Khesin & Chekanov (1989) for the general cases]. The reasons for these similarities are different.

(*a*) The relation of superconductivity to the equations of incompressible fluid is due to the fact that, at a high density, an electronic gas is similar to a fluid. Indeed, the repelling of the particles having equal charges in the electronic clusters makes the gas incompressible.

The equation of (nonrelativistic) superconductivity in $\mathbf{R}^3$ is

$$\dot{\mathbf{v}} = -(\mathbf{v}, \nabla)\mathbf{v} - \mathbf{v} \times \mathbf{B} + \nabla p, \tag{3}$$

where $\mathbf{v}$ denotes a divergence-free field of the electronic gas velocity, $\mathbf{B}$ is a constant external divergence-free magnetic field, and the symbol $\times$ is the vector product for the standard metric (Feynman 1972).

We define the analog of this equation on an arbitrary manifold. Let M^n be a manifold with volume μ and Riemannian metric g. We suppose that $\mathbf{v}$ is a divergence-free vector field with respect to μ, and $\mathbf{B}$ is a "strictly divergence-free" $(n-2)$-vector field with respect to the volume $\sqrt{g}$, i.e. the substitution of $\mathbf{B}$ in $\sqrt{g}$ is exact: $i_{\mathbf{B}}\sqrt{g} = d\alpha$ (for example, if $H_2(M) = 0$, the condition $di_{\mathbf{B}}\sqrt{g} = 0$ is sufficient). We define "the vector product" of the field $\mathbf{v}$ and $(n-2)$-vector field $\mathbf{B}$ in the standard way: $\mathbf{v} \times \mathbf{B} = *(\mathbf{v} \wedge \mathbf{B})$, where $*$ is the Hodge isomorphism of k- and $(n-k)$-vector fields induced by the metric g. We call (3) the generalized superconductivity equation.

Theorem 2 *The multidimensional superconductivity Equation (3) has the integrals $I(v)$ and $I_f(v)$ given by (2a,b) with the replacement of u by $u+\alpha$, where α is a 1-form satisfying the relation $d\alpha = i_{\mathbf{B}}\sqrt{g}$. The 1-form u is obtained by the "index lifting" action of the metric g on the vector field v.*

In fact, the Euler form of these equations differs from the standard hydrodynamics equation by a Coriolis force-type term. As in the ideal hydrodynamics case, the superconductivity equation is Hamiltonian on $\mathcal{G}^*$, but the corresponding Hamiltonian function differs from the quadratic form by the shift of the origin of $\mathcal{G}^*$.

(*b*) The configuration space of the barotropic fluid on a manifold is a semi-direct product of the diffeomorphism group and the space of all functions on the manifold considered; see Marsden et al (1984) for a derivation via reduction for the Lagrangian representation. The similarity of the barotropic fluid to the ideal one can be explained by its "incompressibility" in the coordinates connected with the density.

More precisely, a barotropic fluid (the pressure of which depends only on the density) on a manifold M with metric g is described by the following system of equations for the velocity $\mathbf{v}$ and density ρ:

$$\dot{\mathbf{v}} = -(\mathbf{v}, \nabla)\mathbf{v} + \nabla p(\rho)$$
$$\dot{\rho} + \operatorname{div}(\rho \mathbf{v}) = 0. \tag{4}$$

We use the n-form $\theta = \rho\mu \in \Omega(M)$, where ρ is the fluid density function and μ is the standard volume form induced by the metric, i.e. $\mu = \sqrt{g}$.

Theorem 3 *The barotropic fluid Equations (4) admit the integrals $I(\mathbf{v})$ and $I_f(\mathbf{v})$ given by (2a,b) with the replacement of n-form μ by $\theta = \rho\mu$.*

A heuristic proof of the theorem is based on the fact that the density ρ is transported by the flow, and the fluid is incompressible with respect to the new volume θ (depending on time and on the initial conditions). Thus, we can apply Theorem 1, the assumptions of which require no connection between the metric and the volume form.

1.3 *Topology of Steady Flows*

The investigation of steady flows is one of the most advanced areas in hydrodynamics. For the three-dimensional case the complete description of analytic steady ideal flows is given by the following theorem.

Theorem 4 (Arnold 1966) *Assume that the region D is bounded by a compact analytic surface and that the field of velocities is analytic and not everywhere collinear with its curl. Then the region of the flow can be partitioned by an analytic submanifold into a finite number of cells, in each of which the flow is constructed in a standard way. Namely, the cells are of two types: those fibered into tori invariant under the flow and those fibered into surfaces diffeomorphic to the annulus $\mathbf{R} \times S^1$, invariant under the flow. On each of these tori the flow lines are either all closed or all dense, and on each annulus all flow lines are closed.*

In this theorem, it is important that the velocity and vorticity vector fields are not collinear. Computer experiments conducted by Henon (1966) show a more complicated behavior for the flow lines for a velocity field satisfying curl $\mathbf{v} = \lambda\mathbf{v}$ (Beltrami flow) than described in the theorem for the general case. The results of Henon's calculations suggest that some flow lines densely fill a three-dimensional region [see further research in Dombre et al (1986)].

Another approach to the investigation of steady flows (or rather, another side of the same approach) is connected with the description of the flows as extremals of the energy functional $\int \mathbf{v}^2 d^n x$. For vector fields "frozen in the fluid" (for example, magnetic fields), the corresponding variational problem goes as follows: minimize the energy $\left(\int \mathbf{v}^2 d^n x\right)$ of a divergence-free vector field $\mathbf{v}$ by the action of volume-preserving diffeomorphisms on $\mathbf{v}$. The energy of such a frozen field is closely related to the topology of its flow lines (see Section 2), and the extremal field is stationary (Arnold 1974).

For the two-dimensional case, the problem is to minimize the Dirichlet integral, $\int (\nabla h)^2 \, d^2x$, among all functions h "equal-valued" with the given function h_0 (i.e. among all functions obtained from the given one by area-preserving diffeomorphisms). If the initial function on a disk has only one critical point, then the extremum in the problem is accessible on the symmetrized function, the value of which depends only on the distance to the center of the disk (Mostow 1968). This is the only case of a satisfactory description of an extremum. It should be mentioned that this extremal vector field not only has the energy minimum among all diffeomorphic fields, but also has the energy maximum among all isovortical fields (Kop'ev & Leont'ev 1988). The last result implies the acoustic instability of such rotations if any weak dissipation of energy exists.

So far the extremals are unknown if the topological type of the initial function is more complicated. Perhaps they should have some singularities (Moffatt 1986). Further investigations of this and related variational problems are contained in the recent papers by Laurence & Stredulinsky (1990) and Laurence & Avallaneda (1991).

In conclusion, we recall two results concerning the Dirichlet problem in 2-dimensional domains. For ideal and barotropic steady flows there is only one solution for the boundary problem in a rectangle in the analytic category, while there is an infinite number of solutions in the C^∞-category (Troshkin 1988). As far as we know, analogous results for higher dimensions have not been proved yet.

2. ERGODIC INTERPRETATION OF HYDRODYNAMIC INVARIANTS

The "frozenness" of a vorticity vector field into an ideal fluid and the helicity conservation law are crucial points of three-dimensional hydrodynamics. A description of the relation between these two concepts and the ergodic interpretation of total helicity as the average linking number of trajectories of a curl field (Arnold 1974) stimulated interest and recent progress in this subject.

2.1 *Main Definitions for the Three-Dimensional Case*

Let M^3 be a simply-connected manifold with volume μ, and ξ and η two divergence-free vector fields on M; let g^t_ξ and g^t_η denote their phase flows. Given two points $x, y \in M$, we define the "asymptotic linking number" of the trajectories of g^t_ξ and g^t_η starting at x and y, respectively. For this purpose, we first connect any two points on M by a "short" path Δ (the conditions imposed on the short paths will be described below and are satisfied by almost any choice of the "short" paths Δ).

We select two large numbers, T and S, and close the segment $g_\xi^t x$ $(0 \leq t \leq T)$ and $g_\eta^t y$ $(0 \leq t \leq S)$ of the trajectories issuing from x and y by short paths $\Delta(g_\xi^T x, x)$ and $\Delta(g_\eta^S y, y)$, so that we obtain two closed curves $\Gamma = \Gamma_T(x)$ and $\Gamma' = \Gamma_S(y)$. We assume that these curves are non-intersecting (this is true for almost all pairs of x and y, and for almost all T and S). Then, the linking number $N_{T,S}(x, y)$ of Γ and Γ' is well-defined.

Definition 1: The asymptotic linking number of the pair of trajectories $g_\xi^t x$ and $g_\eta^t y$ is defined as the limit

$$\lambda(x, y) = \lim_{T,S \to \infty} \frac{N_{T,S}(x, y)}{T \cdot S}$$

(T and S are to be varied so that Γ does not meet Γ'). It turns out that this limit exists almost everywhere and is independent of the system of short paths Δ [as an element of $L_1(M \times M)$].

Definition 2: The average linking number λ of two divergence-free vector fields ξ and η is

$$\lambda = \iint_{M \times M} \lambda(x_1, x_2) \mu_1 \mu_2.$$

Theorem 5 (Arnold 1974) *The average linking number λ of two divergence-free vector fields $\boldsymbol{\xi}$ and $\boldsymbol{\eta}$ on a simply connected three-dimensional manifold M with volume μ coincides with $\int_M i_\xi \mu \wedge d^{-1}(i_\eta \mu)$.*

The condition of the vanishing of the divergence for the vector field $\boldsymbol{\eta}$ on M is equivalent to the condition $di_\eta \mu = 0$ or $i_\eta \mu = dv$ by simple connectivity of M. Note that the integral $\int i_\xi \mu \wedge v$ evidently does not depend on the choice of $v \in \Omega^1(M)$ for a fixed dv.

Remark If $\boldsymbol{\xi} = \boldsymbol{\eta}$, the integral $\int_M i_\xi \mu \wedge d^{-1}(i_\xi \mu)$ coincides with the helicity invariant for vector field $\mathbf{v}$, the curl of which is equal to $\boldsymbol{\xi} = \operatorname{curl} \mathbf{v}$.

Indeed, by definition, $i_{\operatorname{curl} \mathbf{v}} \mu = du$, where u is a 1-form induced from $\mathbf{v}$ by "index lifting" defined by the metric (see Section 1.1); therefore, $\int_M i_\xi \mu \wedge d^{-1}(i_\xi \mu) = \int du \wedge u$.

2.2 *Linking Numbers in Magnetohydrodynamics*

In magnetohydrodynamics, we assume that the magnetic field $\mathbf{B}$ is "frozen" in the ideal fluid of infinite conductivity, filling a manifold M. The fluid flow preserves the volume μ on M induced by the metric g. The velocity field $\mathbf{v}$ and the frozen magnetic field $\mathbf{B}$ ($\operatorname{div} \mathbf{v} = \operatorname{div} \mathbf{B} = 0$) satisfy the so-called ideal magnetohydrodynamic equations:

$$\dot{\mathbf{v}} = -(\mathbf{v}, \nabla)\mathbf{v} + \operatorname{curl}\mathbf{B} \times \mathbf{B} + \nabla p$$

$$\dot{\mathbf{B}} = [\mathbf{v}, \mathbf{B}]. \qquad (5)$$

(The second equation is the definition of the "frozenness" of the field **B**, $[.,.]$ denotes the Jacobi-Lie bracket of two vector fields.)

Notice that the frozenness of the field **B** immediately implies the conservation law for the average linking number of the trajectories of this field. (This means that the integral $\int i_{\mathbf{B}}\mu \wedge d^{-1}(i_{\mathbf{B}}\mu)$ is the invariant of the motion by the above theorem.)

It turns out that there is also another invariant for these equations:

Theorem 6 (Vishik & Dolžanskii 1978) *The value of $\int (\boldsymbol{v}, \boldsymbol{B})\mu$ is preserved by the solutions of the magnetohydrodynamic Equations (5).*

Corollary 3 (Khesin & Chekanov 1989) *The magnetohydrodynamic invariant $\int (\boldsymbol{v}, \boldsymbol{B})\mu$ on a simply connected three-dimensional manifold coincides with the average linking number of vector fields curl $\boldsymbol{v}$ and $\boldsymbol{B}$.*

Proof Indeed, applying our theorem to vector fields $\boldsymbol{\xi} = \mathbf{B}$ and $\boldsymbol{\eta} = \operatorname{curl}\mathbf{v}$ (and using the relation $[u] = A\mathbf{v}$, $du = i_{\operatorname{curl}\mathbf{v}}\mu$), we obtain $\int i_{\mathbf{B}}\mu \wedge d^{-1}(i_{\operatorname{curl}\mathbf{v}}\mu) = \int i_{\mathbf{B}}\mu \wedge d^{-1}(du) = \int i_{\mathbf{B}}\mu \wedge u = \int u(\mathbf{B})\mu$. QED

Note that, in spite of the dependence of $\mathbf{v} = A^{-1}[u]$ on the choice of metric, the field curl **v** is defined unambiguously by $i_{\operatorname{curl}\mathbf{v}}\mu = d[u]$.

The ergodic interpretation of $\int (\mathbf{v}, \mathbf{B})\mu$ as an average linking number of curl **v** and **B** is somehow unexpected since curl **v** (in contrast to **B**) is not "frozen" [see Equations (5)]. The evolution changes the field $[u]$ (and, hence, $d[u]$ as well) by some additive summand, which depends on **B**, but it turns out that the average linking number of the kernel field curl **v** and the vector field **B** is preserved.

Notice that the Lie algebra $\tilde{\mathcal{G}}$ associated with these equations is the semidirect product of the Lie algebra $\mathcal{G}$ of all divergence-free vector fields on M (with volume μ) and of its dual space $\mathcal{G}^*$: $\tilde{\mathcal{G}} = \mathcal{G} \ltimes \mathcal{G}^*$, see Holm & Kuperschmidt (1983) and Marsden et al (1984).

Moffatt (1990a) reviews the applications of flow invariants for obtaining solutions of nontrivial topological structure for two- and three-dimensional magnetohydrodynamic equations.

2.3 *Estimates of Energy and Helicity of Vector Fields*

Natural magnetohydrodynamical systems (for example, stars) are usually nonconservative and dissipate magnetic energy $E(\mathbf{B}) = \int (\mathbf{B}, \mathbf{B})\mu$. This diffusion corresponds to an additional dissipative term $\nu\Delta\mathbf{v}$ in the first equation of (5). Nevertheless, the field **B** remains frozen in the fluid (due to the second equation) as long as the fluid's evolution follows this system.

The question is whether the topology of the divergence-free vector field **B** determines a lower boundary for its energy E.

A quadratic form of helicity $I(\mathbf{B}) = \int (\text{curl}^{-1}\mathbf{B}, \mathbf{B})\mu$ turns out to be a very useful tool in the solution of this problem. Unlike the magnetic energy, this form is invariant under the group of volume-preserving diffeomorphisms due to its coordinate-free definition as $I(\mathbf{B}) = \int d^{-1}(i_{\mathbf{B}}\mu) \wedge i_{\mathbf{B}}\mu$. It turns out that complicated topology of flow lines (and so nontrivial helicity) is an obstacle to the full dissipation of the energy. The precise result is as follows. Consider the curl operator on divergence-free vector fields homological to zero. This operator is invertible and the corresponding inverse operator curl^{-1} has a spectrum accumulating at zero on both sides.

Theorem 7 (Arnold 1974) *An eigenvector field of the operator* curl^{-1} *corresponding to the eigenvalue of largest modulus* λ *has minimum energy in the class of divergence-free fields obtained from the eigenfield under the action of volume-preserving diffeomorphisms.*

For example, the standard Hopf vector field on $S^3[\mathbf{v}_{\mathrm{H}}(x_1, x_2, x_3, x_4) = (-x_2, x_1, -x_4, x_3)$ on $S^3 = \{(x_1, x_2, x_3, x_4) \in \mathbf{R}^4, \Sigma x_i^2 = 1\}]$ corresponds to the maximal eigenvalue $(=1)$ on S^3 with the standard metric and has the minimal energy among the diffeomorphic fields [notice that $I(\mathbf{v}_{\mathrm{H}}) = 1$].

The existence of the vector field with minimal energy on the given orbit implies, of course, that the energy must be separated from zero on this orbit. If all trajectories of a vector field are closed and unlinked (for example, the rotation field of a 3-dimensional ball around a one-dimensional axis), this field can be deformed to make its energy arbitrarily small. This fact supposed by Zeldovich (see Arnold 1974) was recently proved by Freedman.

To the contrary, now let a divergence-free vector field **v** on a manifold M have a trajectory (or a set of trajectories) realizing an essential link L and, moreover, let this field **v** be "modeled on L" (i.e. there is a tubular neighborhood of $L \subset M$ that carries a foliation by circles that are integral curves of **v** near the link L). Freedman (1988) proved that in such a case there is a positive lower bound of the energy $E(g^*\mathbf{v})$ over the orbit, with $g \in S\,\text{Diff}(M)$. In fact he proved that there exist no diffeomorphisms that simultaneously transform all long linked trajectories together into trajectories that are short enough.

Moffatt (1990b) suggested using these lower boundaries of the energy as invariants of knots or links (or, more precisely, invariants of tubular neighborhoods of knots or links). Namely, for any knot, consider a satellite flux-tube of volume V carrying an "unlinked" vector field of flux Q (across any meridian section of the tube), and look at the associated energy of this vector field. This energy can be decreased by a diffeomorphism action, preserving both V and Q, to a topological accessible minimum. On dimensional grounds this minimal energy $E(\mathbf{v}) = mQ^2V^{-1/3}$, where $m = m(Q, V)$ is a positive real number depending on the knot topology. If for a given knot, different minima of energy states are accessible, then the sequence $\{m_0, m_1, \ldots, m_r\}$ of possible values could be reasonably described as the energy spectrum of the knot. The lowest number m_0 provides a possible natural measure of knot complexity.

Another point of view on this knot spectrum is the following. Introduce an "energy" function on the space of embeddings of a given knot, tending to infinity when the embedding becomes singular. The number of critical points can be estimated then by Morse theory and by Vassiliev's calculation of the Betti numbers of the space of embeddings (Vassiliev 1990). Perhaps the equilibrium states for the Moffat spectrum correspond to the critical points for a particularly chosen "energy," and a new energy function may define a new invariant.

In recent papers, Freedman & He (1991a,b) introduced an "asymptotic crossing number" of a vector field on a three-dimensional manifold and obtained new energy estimates by means of the topology of the vector field in these terms. For two closed curves in $\mathbf{R}^3$ the crossing number is equal to the integral of the modulus of the Gauss integrand for their linking number.

An asymptotic version of this crossing number $c(\mathbf{v})$ allows the authors to give the following energy estimate for an arbitrary divergence-free vector field $\mathbf{v}$ in $\mathbf{R}^3$:

$$E(\mathbf{v}) \geq \left(\frac{16}{\pi}\right)^{1/3} \frac{c(\mathbf{v})}{\mathrm{vol}\,(\mathrm{supp}\,\mathbf{v})^{1/3}}$$

where supp $\mathbf{v}$ denotes the support of $\mathbf{v}$. For particular cases of linked and knotted tori, combinatorial versions of this formula have been proved.

2.4 *Ergodic Meaning of Multidimensional Invariants*

We recall that the averaged asymptotic linking number of two divergence-free vector fields $\boldsymbol{\xi}$ and $\boldsymbol{\eta}$ on a simply-connected manifold M^3 turns out to be equal to the Hopf invariant $I(\alpha, \beta) = \int_M \alpha \wedge d^{-1}\beta$ of the corresponding closed two-forms α and β ($i_\xi\mu = \alpha$, $i_\eta\mu = \beta$, where μ is a volume form on M^3, see Section 2.2). For manifolds of higher dimensions, there are analogous integral invariants of Novikov (1984); we discuss here their ergodic interpretation (Khesin 1991).

It happens that, unlike the three-dimensional case where the asymptotic linking number is defined for almost every pair of trajectories of vector fields, for the multidimensional case there is no satisfactory definition of the system of short paths; hence, it is necessary to consider asymptotic linkage of any trajectory with the whole foliation of codimension 2. The interpretation presented here is a kind of dictionary for the translation of the facts about differential forms into the description of their kernel intersections.

More precisely, the closed 2-form α of rank ≤ 2 determines a (non-regular) foliation of codimension 2 on the manifold M^n. If M is equipped with the volume form μ, then this foliation as the field of $(n-2)$-vectors $\mathscr{A}$ (the kernel of α) is given by the relation $i_{\mathscr{A}}\mu = \alpha$. This field is completely integrable by the Frobenius criterion, as the kernal field of a closed 2-form.

Definition 3: The average linkage of a curve Γ and the foliation $\mathscr{A}$ is the flow of α through the arbitrary film $\partial^{-1}\Gamma$ bounded by Γ:

$$k(\Gamma, \mathscr{A}) = \int_{\partial^{-1}\Gamma} \alpha = \int_{\Gamma} d^{-1}\alpha.$$

Notice that for a foliation with compact nonsingular fibers, the number k is in fact the averaged linking number of the curve Γ with every fiber. Similar to the 3-dimensional case, we can now define an asymptotic linking $k(x, \mathbf{v}, \mathscr{A})$ of the trajectory of vector field $\mathbf{v}$ passing through the point $x \in M$ and the foliation $\mathscr{A}$. This is the time-average of the linking number of the long segment (for time T) of the trajectory $g_{\mathbf{v}}^{T}x$, starting in $x \in M$ and closed by some short path:

$$k(x, \mathbf{v}, \mathscr{A}) = \lim_{T\to\infty} \frac{1}{T} k(g_{\mathbf{v}}^{T}x, \mathscr{A}).$$

Then, the average linking number of the vector field $\mathbf{v}$ and the foliation $\mathscr{A}$ over M equipped with the volume form μ is $\bar{K}(\mathbf{v}, \mathscr{A}) = \int_M k(x, \mathbf{v}, \mathscr{A})\mu$.

Now, we are able to give an ergodic interpretation of Hopf-type integral invariants $I(\alpha, \ldots, \omega) = \int_M d^{-1}\alpha \wedge \beta \wedge \cdots \wedge \omega$ for a set of $m+1$ closed 2-forms $\alpha, \ldots, \omega$ on an odd-dimensional manifold M^{2m+1} with a volume form μ. Notice that $I(\alpha, \ldots, \omega)$ is symmetric under the permutation of $\alpha, \ldots, \omega$ and doesn't depend on the choice of the 1-form $d^{-1}\alpha$.

Theorem 8 *Let the rank of one of the forms (for example, α) be at most 2. Then $I(\alpha, \beta, \ldots, \omega) = \int_M d^{-1}\alpha \wedge \cdots \wedge \omega$ coincides with the average linking number of the vector field curl$(\beta, \ldots, \omega)$ and the foliation $\mathscr{A}$: $I(\alpha, \ldots, \omega) = \bar{K}[\mathrm{curl}(\beta, \ldots, \omega), \mathscr{A}]$, where the fields curl$(\beta, \ldots, \omega)$ and $\mathscr{A}$ are defined by $i_{\mathrm{curl}(\beta, \ldots, \omega)}\mu = \beta \wedge \cdots \wedge \omega$ and $i_{\mathscr{A}}\mu = \alpha$.*

Certainly, for the definition of the flow of α, and so in this theorem, rank α is not important: In the general case, we would consider the linkage with an $n-2$ vector field instead of an $n-2$ dimensional foliation. If, conversely, all these forms have rank ≤ 2 (of course, this is seldom the case), then one can interpret the number $I(\alpha, \ldots, \omega)$ as the multilinkage of the corresponding foliations.

Namely, by analogy with the usual linking number of two surfaces P^k and Q^l in M^n with $k+l = n-1$, $P \cap Q = \varnothing$ (i.e. the linking number is the bilinear form on the space of nonintersecting submanifolds of appropriate dimensions), we define the multilinking number as a multilinear form on the space of r-tuples of submanifolds $(P_1, \ldots, P_r)$ such that

$$\sum_{i=1}^{r} \operatorname{codim} P_i = n+1$$

and

$$\bigcap_{i=1}^{r} P_i = \varnothing.$$

By definition this form is equal to the number of the intersection points (with the corresponding signs) of a film bounded by one of these surfaces P_i with all the other submanifolds. If these submanifolds are equipped with transversal orientations then all films bounded by them and their intersections are also oriented, and so the signs of the intersection points are well-defined. For example, it is possible to link three circles on a plane or two spheres and one circle in 3-space.

We recall that every closed 2-form of rank ≤ 2 determines a foliation of codimension 2, and if the leaves would be compact, one could consider the multilinkage of these leaves for $(m+1)$ 2-forms in M^{2m+1}

$$\left[\sum_{i=1}^{m+1} (\text{codimension of leaves}) = 2m+2 = \dim M+1\right].$$

So, in these terms, the theorem above reads as "The Hopf-type invariant is equal to the average asymptotic multilinking number of leaves determined by the given 2-forms."

Remark In fact, this interpretation may be thought of as a sort of dictionary: instead of "closed 2-forms," we say their "kernel fields" (for forms of a rank ≤ 2, the kernel fields form foliations of codimension 2), so the operations d^{-1} and $\wedge$ correspond to the transfer from the surfaces to the films bounded by them and to their intersections, respectively. Finally, the integration $\int_M$ is the summation of the intersection points with the corresponding signs. This translation has precise meaning for singular

2-forms concentrated on separate $(n-2)$-surfaces. In the general case, it is necessary to average this amount.

So far, we have discussed invariants on odd-dimensional manifolds. Novikov (1984) suggested a set of invariants on M of arbitrary dimension, and here, for illustration, we consider the case of M^4.

Let α and β be closed 2-forms (on M) satisfying the following conditions: $\alpha \wedge \alpha = \alpha \wedge \beta = \beta \wedge \beta = 0$. These conditions ensure that the forms α and β determine certain foliations of dimension 2 on M^4 ($\mathrm{rk}\,\alpha, \mathrm{rk}\,\beta \leq 2$), that the intersections of their leaves form a 1-dimensional foliation ($\alpha \wedge \beta = 0$), and moreover, the distribution which is the sum of the kernels of α and β defines a 3-dimensional foliation. Then the integrals

$$J(\alpha, \alpha, \beta) = \int_M d^{-1}\alpha \wedge \alpha \wedge d^{-1}\beta$$

and

$$J(\alpha, \beta, \beta) = \int_M d^{-1}\alpha \wedge \beta \wedge d^{-1}\beta$$

don't depend on the choice of $d^{-1}\alpha$ and $d^{-1}\beta$.

To describe the ergodic meaning of these integrals, let's extend the concept of multilinking: We drop the codimension condition if it is compensated by new intersection conditions. For example, two circles S^1 and a sphere S^2 can't be linked in $\mathbf{R}^3$ (one can untie any configuration of them not passing through triple points). However, if these two circles are meridians of the same ball (and so their intersection consists of two points, i.e. S^0), the linkage may be nontrivial. In the definition of invariants J, the ($\alpha \wedge \beta = 0$)-type conditions provide the nongeneric intersections of corresponding leaves.

Theorem 9 *The invariant $J(\alpha, \alpha, \beta)$ $[J(a, \beta, \beta)]$ coincides with the average linking number of the foliation $\mathscr{A}$ of 2-form α (foliation $\mathscr{B}$ of β) with the vector field curl $d(d^{-1}\alpha \wedge d^{-1}\beta)$.*

Roughly speaking, these two amounts are the average linking numbers of the 1-dimensional foliation of the intersections $\mathscr{A}$ and $\mathscr{B}$, with each of the foliations $\mathscr{A}$ and $\mathscr{B}$ (determined by α and β).

Remark In concluding this section, we discuss the relation of the asymptotic generalized Hopf invariant to quantum anomalies, following Novikov (1984). Let's consider the space $\mathscr{L}$ of smooth mappings $f: S^q \to M^n$ homotopic to zero. It turns out that the invariants described above determine multivalued functions $F(f)$ on the space $\mathscr{L}$ (or closed 1-forms ∂F on $\mathscr{L}$), depending on a finite number of derivatives of f:

$$F_\theta(f) = \int_{f(D^{q+1})} \theta.$$

Here θ is a closed $(q+1)$–form on M (examples: $\alpha \wedge d^{-1}\alpha$ for $q+1 = 3$ or $d^{-1}\alpha \wedge \alpha \wedge d^{-1}\beta$ for $q+1 = 4$), f is extended to a mapping $D^{q+1} \to M^n$, and F depends just on $f|_{\partial D^{q+1} = S^q}$. Moreover, there is the integer lattice inside the space of $\{\theta\}$ consisting of "homotopy invariant elements." The meaning of this lattice is exactly equivalent to the role of the usual integer-valued Hopf invariant of mapping $S^3 \to S^2$ among all asymptotic linking invariants for arbitrary divergence-free vector fields on S^3. It is natural to call the last phenomenon a quantization condition (Novikov 1984).

2.5 *Interpretation of Godbillon-Vey-type Characteristic Classes*

Let $\mathscr{F}$ be a cooriented foliation of codimension 1 on the manifold M, and γ be a 1-form determining this foliation. Then $d\gamma = \gamma \wedge \eta$ for a certain 1-form η. The form $\eta \wedge d\eta$ turns out to be closed, its cohomology class does not depend on the choice of γ and η, and it is called the Godbillon-Vey class of the foliation $\mathscr{F}$.

Following Tabachnikov (1990), we describe the corresponding asymptotic Hopf invariant of $\eta \wedge d\eta$ on a 3-dimensional manifold M. Fix a vector field $\mathbf{v}$ such that $\gamma(\mathbf{v}) = 1$ and let g^t be the flow of $\mathbf{v}$. If μ is the volume form on M^3, we define the vector field

$$\xi = \operatorname{curl} \frac{d}{dt}(g^t\gamma \wedge \gamma)|_{t=0} \quad \left(\text{or } i_\xi\mu = \frac{d}{dt}(g^t\gamma \wedge \gamma)|_{t=0}\right).$$

This field ξ is tangent to $\mathscr{F}$ and determines the axis and the angular velocity of tangent spaces of $\mathscr{F}$ with respect to the flow g^t. Let the vector field ζ be the velocity of ξ along g^t:

$$i_\zeta\mu = \frac{d^2}{dt^2}(g^t\gamma \wedge \gamma)|_{t=0}.$$

Theorem 10 (Tabachnikov 1990) *The asymptotic Hopf invariant of the field ζ is equal to the Godbillon-Vey class of $\mathscr{F}$.*

The Godbillon-Vey class on manifolds of higher dimensions is an element of $H^3(M)$, hence it is determined by its values on 3-cycles. Any such value coincides with the asymptotic Hopf invariant of the corresponding field ζ, constructed for the induced foliation on this 3-cycle.

It would be interesting to consider whether similar techniques can be applied to generalize the Casson invariant and the Floer homology of homological 3-spheres to aspherical $4k-1$ manifolds with an additional structure (say, to contact manifolds).

3. DIFFERENTIAL GEOMETRY OF DIFFEOMORPHISM GROUPS

3.1 *Finiteness of the Diameter for the Group of Volume-Preserving Diffeomorphisms*

The group of volume-preserving diffeomorphisms SDiff(M) of a manifold M (with a volume form μ) is the configuration space of an ideal incompressible fluid on M. If the manifold M is Riemannian, it is natural to equip this diffeomorphism group with a right-invariant Riemannian metric. Namely, the tangent space T_eSDiff(M) to the group SDiff(M) at the identity element e is the space of all divergence-free vector fields on M. The scalar product of two such vector fields $\mathbf{v}, \mathbf{w}$ on M [$\mathbf{v}, \mathbf{w} \in T_e\mathrm{SDiff}(M)$] is defined as

$$(\mathbf{v}, \mathbf{w})_{T_e\mathrm{SDiff}(M)} = \int_M [\mathbf{v}(x), \mathbf{w}(x)]_M \mu.$$

Now, extending this metric from T_eSDiff(M) to the whole tangent bundle TSDiff(M) by right translation, make the group SDiff(M) a weak infinite-dimensional Riemannian manifold. It is hence natural to investigate its intrinsic differential-geometric properties.

We can compute the length of any curve on SDiff(M) and define the distance between any two diffeomorphisms, i.e. any two different points of the group. Namely, the distance between the diffeomorphisms ϕ_0 and ϕ_1 will be the minimal length of all paths connecting these two points: i.e.

$$\mathrm{dist}(\phi_0, \phi_1) = \inf_{(\gamma)} l(\gamma),$$

where γ is a curve on SDiff(M) connecting ϕ_0 and ϕ_1, and $l(\gamma)$ is the length of γ in the introduced metric.

We'll say that the group SDiff(M) has infinite diameter if there is a sequence of diffeomorphisms $\phi_1, \ldots, \phi_n, \ldots$ such that $\sup_n \mathrm{dist}(e, \phi_n) = +\infty$ and that SDiff(M) has finite diameter if there is a constant K such that for any $\phi \in \mathrm{SDiff}(M)$, $\mathrm{dist}(e, \phi) \le K$. This constant K depends only on M and on the metric on M. In other words, for the last case, one can connect each two points on SDiff by a short path. The theorem of Shnirelman (1985) states that the diameter of the group SDiff(M) is finite if M is a 3-dimensional cube. More precisely:

Theorem 11 (Shnirelman 1985) *Let $\Delta_\phi(x) = |x-\phi(x)|$ be the Euclid-*

ean distance between two points $x \in M$ and $\phi(x) \in M$ [$\phi \in SDiff(M)$] for 3-dimensional contractible $M \subset \mathbf{R}^3$. Then $dist(e, \phi) \leq c \| \Delta_\phi \|^{\alpha}_{L^2(M)}$ where $c > 0$, $\alpha > 0$ are constants depending only on M.

Though Shnirelman formulated his theorem only for a 3-dimensional cube, his proof can be modified for the case of the group of volume-preserving diffeomorphisms of any compact simply-connected Riemannian manifold of dimension greater than 2. However, the diameter can become infinite if the fundamental group is not trivial (Eliashberg & Ratiu 1991b).

Shnirelman also proved that there are configurations for a 3-dimensional fluid [i.e. diffeomorphisms ϕ_0, $\phi_1 \in \mathrm{SDiff}(M^3)$] such that it is impossible to connect them by a shortest path (the infimum of the distance between ϕ_0 and ϕ_1 isn't accessible). This means that the corresponding boundary-value problem for the hydrodynamical Euler equation has no classical solutions.

3.2 *Infinite Diameter of the Symplectomorphism Group*

It turns out that the situation is completely different for the group of symplectic diffeomorphisms Symp(M) (which plays a similar role in plasma physics as SDiff(M) plays for fluid dynamics). Eliashberg & Ratiu (1991a,b) proved that the diameter of this group for any compact exact sympletic manifold M (necessary with boundary: $\partial M \neq \emptyset$) is infinite. Unlike for the volume-preserving case, the infinite diameter of the group of symplectomorphisms has a local nature and does not depend on the topology of the underlying manifold. The distinction between the cases comes from the fact that the fundamental group of the group of linear symplectic transformations is infinite, while in the volume-preserving case it is finite in dimension >2.

In fact, Eliashberg & Ratiu (1991b) proved that the diameter is infinite in the metric on Symp(M) generated by the L^1 (rather than the L^2) norm on its Lie algebra. This implies the result of all L^p norms, $p \geq 1$. The same holds for the biinvariant metric recently introduced by Hofer (1989). The main tool of the proof is the use of the Calabi invariant, which measures the difference between the 1-forms defining the symplectic structure on M before and after the application of symplectomorphisms.

The proof is greatly simplified for the case of the group of exact symplectomorphisms fixed at the boundary of the manifold. In this case, the three main steps of the proof are: (a) an inequality between the length of a path and the value of the Calabi invariant of the final symplectomorphism, (b) the independence of the Calabi invariant from the choice of the path, and finally, (c) the existence of symplectomorphisms with arbitrarily large values of the Calabi invariant [see Eliashberg & Ratiu (1991a)].

As an example of an arbitrarily long path on the symplectomorphism group, one may consider a flow (within any ball) generated by an isotropic nonquadratic Hamiltonian for a long period of time. Notice, that although this path will be long in the class of "symplectic" curves, there is a short path connecting its initial and final points within the class of volume-preserving diffeomorphisms by Shnirelman's theorem—if the dimension of the manifold is more than 2 (see Section 3.1).

3.3 *Curvatures of Diffeomorphism Groups*

It is known that negative curvature of a Riemannian manifold implies exponential instability of the geodesics on it. In the work of Arnold (1966), explicit formulas were obtained for the connection and curvature of the right-invariant metric on the subgroup $S_0\mathrm{Diff}(T^2) \subset \mathrm{SDiff}(T^2)$ consisting of divergence-free diffeomorphisms that leave the center of mass of the torus fixed. Negativeness of these curvatures (under the assumption that the motion of the atmosphere looks like the flow of an ideal 2-dimensional fluid on a torus) implies that a small error ε at the initial moment in the state of the weather will give an error of prediction after n months of order $\approx 10^{2.5n} \cdot \varepsilon$. For example in order to predict the weather two months in advance, we must have initial data with five more digits of accuracy than the prediction accuracy. Practically, this means that calculating the weather for such a period is impossible.

Analogous results on the curvatures of the group of volume-preserving diffeomorphisms on S^2, T^n, were obtained by Lukatskii (1979, 1981). It turned out that these curvatures for $\mathrm{SDiff}(S^2)$ have the same order, and therefore the same conclusion holds as to the impossibility of long-term forecasts.

Later, the computations of curvatures were extended to the cases of arbitrary 2-dimensional compact manifolds (the answer is given in a form that includes the commutator of vector fields and the Laplace operator on the manifold), and of an arbitrary manifold with flat metric (Lukatskii 1988, and in preparation).

Let's recall now that the group of volume-preserving diffeomorphisms is a subgroup of the wider object: the group of all diffeomorphisms of the manifold. Just like its subgroup this large group is also equipped with a weak Riemannian metric: $\langle \phi^*\xi, \phi^*\eta \rangle = \int_M [g(\xi, \eta)]_{\phi(x)} \mu(x)$, where $\xi, \eta \in \mathrm{Vect}(M)$, $g(\xi, \eta)$ is the scalar product of ξ and η with respect to the metric g on M, and $\phi \in \mathrm{Diff}(M)$. Already this metric is not right-invariant on $\mathrm{Diff}(M)$. Bao & Ratiu (1990) have studied the total geodesic (or asymptotic) directions on the Riemannian submanifold $\mathrm{SDiff}(M) \subset$

Diff(M), i.e. such directions in the tangent space T_eSDiff(M) (or such divergence-free vector fields on M) for which the second fundamental form of SDiff(M) relative to Diff(M) vanishes. For an arbitrary manifold M^n they have obtained the explicit description of such directions $\phi^*\mathbf{v} \in T_\phi$SDiff($M$) in the form of a certain first-order nonlinear partial differential equation on **v**. For the 2-dimensional case this equation can be rewritten as an equation for the stream function. For easier formulation, let $\partial M = \varnothing$ and $H^1(M) = 0$.

Theorem 12 (Bao & Ratiu 1990) *For 2-dimensional M the divergence-free vector field $\phi^*v \in T_\phi SDiff(M)$ is a totally geodesic direction on SDiff(M) iff the stream function h of the field* **v** *satisfies the degenerate Monge-Ampere equation $2\cdot det(Hess\ h) = K\cdot g\cdot \|dh\|^2$ (here $g = det(g_{ij})$ in some coordinates $\{x_i\}$, $(Hess\ h)_{ij} = \partial^{ij}h/\partial x_i\partial x_j$, and K is the Gaussian curvature function on M).*

Their paper also contains examples of manifolds for which the Monge-Ampere equation has or has no solutions. Bao et al (1991) characterized all manifolds M for which SDiff(M) has harmonic asymptotic vectors in Diff(M).

Apparently the same Monge-Ampere equation describes the totally geodesic directions on Symp(M) $\subset$ Diff(M) for any symplectic manifold M (where h is the Hamiltonian function of the Hamiltonian field **v**).

ACKNOWLEDGMENTS

We are deeply grateful to V. Ginzburg, S. Tabachnikov, T. Ratiu, and A. Weinstein for their interest in our work, numerous valuable remarks, and fruitful discussions.

Literature Cited

Arnold, V. I. 1966. Sur la géométrie differentielle des groupes de Lie de dimension infinie et ses applications á l'hydrodynamique des fluides parfaits. *Ann. Inst. Fourier* 16: 316–61

Arnold, V. I. 1967. An a priori estimate in the theory of hydrodynamic stability. *Izv. Vyssh. Uchebn. Zaved. Mat.* 54(5): 3–5. Transl. 1969 in *AMS Transl.* 19: 267–69

Arnold, V. I. 1969a. The hamiltonian character of the Euler equation for the dynamics of a rigid body and an ideal fluid. *Usp. Mat. Nauk* 24(3): 225–26 (In Russian)

Arnold, V. I. 1969b. One-dimensional cohomologies of Lie algebras of divergence free vector fields and rotation numbers of dynamical systems. *Funct. Anal. Appl.* 3(4): 319–21

Arnold, V. I. 1974. The asymptotic Hopf invariant and its applications. *Proceedings of Conference on Differen. Equations, 1973, Erevan*; English translation 1986: *Sel. Math. Sov.* 5(4): 327–45

Arnold, V. I. 1989. *Mathematical Methods of Classical Mechanics*. Berlin: Springer-Verlag. 511 pp. 2nd ed.

Bao, D., Lafontaine, J., Ratiu, T. 1991. On the extrinsic geometry of the volume-preserving diffeomorphism group of a boundaryless manifold. *Pac. J. Math.* In press

Bao, D., Ratiu, T. 1990. On the geometrical origin and the solutions of a degenerate Monge-Ampere equation. *Proc. Symp.*

Pure Math. "Differential Geometry." Summer School, UCLA. In press
Dezin, A. A. 1983. Invariant forms and some structure properties of the Euler equations of hydrodynamics. *Zeit. Anal. Anwend.* 2: 401–9 (In Russian)
Dombre, T., Frisch, V., Greene, J. M., Henon, M., Mehr, A., Soward, A. M. 1986. Chaotic streamlines in the ABC flows. *J. Fluid Mech.* 167: 353–91
Eliashberg, Y., Ratiu, T. 1991a. On the diameter of the symplectomorphism group of the ball. *Seminaire Sud-Rhodanieu a Berkeley, MSRI Springer Series.* New York: Springer-Verlag
Eliashberg, Y., Ratiu, 1991b. The diameter of the symplectomorphism group is infinite. *Invent. Math.* 103: 327–40
Feynman, R. P. 1972. *Statistical Mechanics.* New York: Benjamin
Freedman, M. H. 1988. A note on topology and magnetic energy in incompressible perfectly conducting fluids. *J. Fluid Mech.* 194: 549–51
Freedman, M. H., He, Z.-X. 1991a. Divergence-free fields: energy and asymptotic crossing number. *Ann. Math.* 134: 189–229
Freedman, M. H., He, Z.-X. 1991b. Links of tori and the energy of incompressible flows. *Topology* 30(2): 283–87
Henon, M. 1966. Sur la topologie des lignes de couzant dans un cas particulier. *C. R. Acad. Sci. Paris* 262: 312–14
Hofer, H. 1989. On the topological properties of symplectic maps. Preprint
Holm, D. D., Kupershmidt, B. A. 1983. Poisson structures of super-conductors. *Phys. Lett. A* 93: 177–81
Holm, D. D., Marsden, J. E., Ratiu, T., Weinstein, A. 1985. Nonlinear stability of fluids and plasma equilibra. *Phys. Rep.* 123(1): 1–116
Khesin, B. A. 1991. Ergodic interpretation of integral hydrodynamic invariants. *Preprint PAM-522, Univ. Calif. Geometry and Physics.* In press
Khesin, B. A., Chekanov, Yu. V. 1989. Invariants of the Euler equations for ideal or barotropic hydrodynamics and superconductivity in D dimensions. *Physica D* 40: 119–31
Kop'ev, V. F., Leont'ev, E. A. 1988. Acoustic instability of plane vortex flows with circled flow lines. *Acousticheskii J.* 34(3): 475–80 (In Russian)
Laurence, P., Avallaneda, M. 1991. On Voltjer's variational principle for force free fields. *J. Math. Phys.* 32(5): 1240–53
Laurence, P., Stredulinsky, E. 1990. Convergence of a sequence of free boundary problems associated with the grad variational problem in plasma physics. *Comm. Pure Appl. Math.* 43: 547–73
Lukatskii, A. M. 1979. Curvature of groups of diffeomorphisms preserving the measure of the 2-sphere. *Funct. Anal. Appl.* 13(3): 174–78
Lukatskii, A. M. 1981. On the curvature of the group of measure-preserving diffeomorphisms of an *n*-dimensional torus. *Russ. Math. Surv.* 36(2): 179–80
Lukatskii, A. M. 1988. Structure of the curvature tensor of the group of measure-preserving diffeomorphisms of a compact 2-dimensional manifold. *Siber. Math. J.* 29(6): 947–51
Marsden, J. E., Simo, J. C., Levis, D., Posbergh, T. A. 1989. Block diagonalization and the energy-momentum method. *Contemp. Math.* 97: 297–313
Marsden, J., Ratiu, T., Weinstein, A. 1984. Semidirect product and reduction in mechanics. *Trans. Am. Math. Soc.* 281: 147–77
Marsden, J., Weinstein, A. 1983. Coadjoint orbits, vortices and Clebsh variables for incompressible fluids. *Physica D* 7: 305–23
Moffatt, H. K. 1986. On the existence of localized rotational disturbances which propagate without change of structure in an inviscid fluid. *J. Fluid Mech.* 173: 289–302
Moffatt, H. K. 1990a. Structure and stability of solutions of the Euler equations: a lagrangian approach. *Philos. Trans. R. Soc. London Ser. A* 333: 321–42
Moffatt, H. K. 1990b. The energy spectrum of knots and links. *Nature* 347(6291): 367–69
Mostow, G. D. 1968. Quasi-conformal mappings in *n*-space and the rigidity of hyperbolic space forms. *Publ. Math. IHES* 34: 53–104
Murometz, Y., Razboynick, S. 1990. Integrability of models of two-dimensional turbulance. In *Integrable and Superintegrable Systems,* ed. B. A. Kupershmidt, pp. 34–45. London: World Scientific
Novikov, S. P. 1984. The analytic generalized Hopf invariant: multivalued functionals. *Russ. Math. Surv.* 39(5): 113–24
Olver, P. J. 1982. A nonlinear Hamiltonian structure for the Euler equations. *J. Math. Anal. Appl.* 89: 233–50
Ovsienko, V. Yu., Khesin, B. A., Chekanov, Yu. V. 1989. Integrals of the Euler equations of multidimensional hydrodynamics and superconductivity. *Zap. LOMI* 172: 105–13 (English translation in *J. Sov. Math.* In press)
Serre, D. 1984. Invariants et degenerescence symplectique de l'equation d'Euler des

fluides parfaits incompressibles. *C. R. Acad. Sci. (Paris), Ser. A* 298: 349–52
Tabachnikov, S. L. 1990. Two remarks on asymptotic Hopf invariants. *Funct. Anal. Appl.* 24(1): 74–75
Shnirelman, A. I. 1985. On the geometry of the group of diffeomorphisms and the dynamics of an ideal incompressible fluid. *Mat. Sb.* 128(1). English translation *Math. USSR Sb.* 56(1): 79–105
Troshkin, O. V. 1988. On topological analysis of structure of hydrodynamical flows. *Russ. Math. Surv.* 43(4): 153–90
Vassiliev, V. 1990. Cohomology of knot space. *Adv. Sov. Math. Vol. 1: Theory of Singularities and Its Applications*, ed. V. I. Arnold. Providence: Am. Math. Soc.
Vishik, S. V., Dolžanskii, F. V. 1978. Analogs of the Euler-Lagrange equations and magnetohydrodynamics equations connected with Lie groups. *Sov. Math. Dokl.* 19: 149–53

Translator's preface to J. Milnor's book "Morse Theory"

V.I. Arnold
Translated by Gerald Gould

Milnor's book is a textbook on Morse theory. Starting with the simplest example and finishing with Bott's Periodicity theorem, the exposition remains geometrically intuitive but rigorous. It is modern but also elegant; it is extensive[1] but also self-contained: the necessary facts from differential geometry, calculus of variations, and so on, are derived in the book itself in a form required by the author.

Morse theory, that is, the study of critical points of functions and functionals "in the large", plays a significant role in modern topological research. "Morse surgeries" are constantly required as a flexible and adequate tool in work with differentiable manifolds; this tool is considerably more convenient and powerful than the combinatorial approach. The techniques developed here have already given a whole series of fundamental results. For example, the "Theorem on the exactness of inequalities" proved by Smale implies the Poincaré conjecture[2] in dimensions greater than 5, as well as the equivalence of the notions of h-homology and diffeomorphism, which is essential for the classification of differential structures on the sphere (Milnor and Kervaire).[3]

The theory of critical points of functionals obtained an interesting application in the works of Bott. At the time when Poincaré, Birkhoff, Morse, Shnirel'mann and Lusternik were applying topological methods to problems of variational calculus in

---

[1] We point out that Milnor's book does not claim to be a complete account of all the questions of the calculus of variations in the large. In particular, the classical problems on the geometrically distinct critical points and on non-self-intersecting closed geodesics are completely untouched.

[2] A manifold homotopy equivalent to an n-sphere S^n is combinatorially equivalent (and therefore homeomorphic) to the sphere.

[3] Let $|\theta(n)|$ be the number of differentiable manifolds homeomorphic to S^n and not diffeomorphic to each other. Then

n	1	2	3	4	5	6	7	8	9	10	11	12	13	14	...
$\|\theta(n)\|$	1	1	1	?	1	1	28	2	8	6	992	1	3	2	...

See M. Kervaire and J. Milnor, *On differential structures on spheres*, Ann. Math. **77**, No. 3, 504–537 (1963).

Originally published in: Translation of J. Milnor: Morse Theory, 5-7, © MIR, Moscow 1965
Translated by G. Gould. Springer-Verlag Berlin Heidelberg 2014 with kind permission, DOI 10.1007/978-3-642-31031-7_43

the large, Bott applied the methods of variational calculus in the large to topological problems. By considering minimal geodesics on classical Lie groups he found the "stable homotopy groups" of the latter. For example, let $\mathbb{O}(N)$ be the group of orthogonal matrices of order N, and $\pi_n(M)$ the nth homotopy group of the manifold M (that is, the group of homotopy classes of maps of the n-sphere into M). Then for $N \gg n$ we have and $\pi_{n+8}\mathbb{O}(N) = \pi_n\mathbb{O}(N)$ and

n	0	1	2	3	4	5	6	7	8	9	10	...
$\pi_n\mathbb{O}(N)$	$\mathbb{Z}_2$	$\mathbb{Z}_2$	0	$\mathbb{Z}$	0	0	0	$\mathbb{Z}$	$\mathbb{Z}_2$	$\mathbb{Z}_2$	0	...

Here $\mathbb{Z}$ is the group of integers and $\mathbb{Z}_2$ is the group of residues modulo 2 consisting of the elements 0, 1.

Bott's periodicity theorem ($\pi_{n+8}\mathbb{O} = \pi_n\mathbb{O}$) became the cornerstone of "K-theory", which presently has been intensively developed. As a result, there has been solved such a classical problem as the determination of the maximal number $k(n)$ of linearly independent vector fields on a sphere S^n of any dimension (by Adams).[4] Another example is the calculation of the index of elliptic differential equations in the multidimensional case (Atiyah and Singer).

It is assumed that the reader of the book has only basic knowledge in topology, some acquaintance with manifolds, homology, homotopy and fibre bundles.[5] The meaning of certain terms less known to the Russian reader is explained in the appendix written by D.V. Anosov.

One can hope that Milnor's book, which is not overburdened with algebraic formalism, can help Soviet readers to enter the realm of ideas and methods of modern differential topology.

---

[4] For n even $k(n) = 0$, while the first few values of $k(n)$ for odd values of n are given in the table below:

n	1	3	5	7	9	11	13	15	17	19	...	31	...
$k(n)$	1	3	1	7	1	3	1	7	1	3	...	9	...

[5] The required minimum is amply covered by the books of Seifert and Threlfall [11], Steenrod [34] and Hu Sy-Tsan [43].

Henri Poincaré: Selected Works in Three Volumes: Vol. I. New Methods of Celestial Mechanics

V.I. Arnold
Translated by Gerald Gould

This book contains the first two volumes of "New methods of celestial mechanics". The third volume is included in the second book of the present edition. This major work of the remarkable French mathematician and physicist is published in the Russian language for the first time.

In "New methods of celestial mechanics", H. Poincaré developed the theory of integral invariants, constructed the theory of asymptotical expansions, investigated periodic orbits, and brought a significant contribution to the solution of a number of other problems of applied mathematics, mechanics, and astronomy. This work, which has become a classic, has had a great influence on the development of the exact sciences, and even nowadays has not lost any of its value.

From the editorial board

Henri Poincaré (1854–1912) entered the scientific world as a mathematician and in the course of just over thirty years of his creative activity managed to enrich almost all areas of mathematics by results of the highest significance. At the same time, while remaining a mathematician and applying mathematics, he, in a series of courses and memoirs, encompassed the whole of the theoretical and mathematical physics of his time; with his own work, opened a new epoch in the history of celestial mechanics; took part in the creation of the theory of relativity and quantum theory, which were emerging at that time; embarked on general questions of science, and was its brilliant popularizer. He can rightly be called not only an extraordinary mathematician, but also a first-class engineer, physicist, and astronomer.

A substantial portion of Henri Poincaré's heritage remains a living component of modern science, and the tremendous volume of this heritage makes the task of selecting his work for the present edition quite challenging. Yet two decisions seemed

Originally published in: H. Poincaré: Selected Works in Three Volumes. Vol. I, New Methods of Celestial Mechanics, Preface, From the Editorial Board, Comments, 747-752, © Nauka, Moscow 1971
Translated by G. Gould. Springer-Verlag Berlin Heidelberg 2014 with kind permission, DOI 10.1007/978-3-642-31031-7_44

to be indisputable: to minimise the repetition in our collection of Poincaré's work existing in Russian translation, and to include the three volumes of his celebrated "New methods of celestial mechanics", which is strikingly abundant with results and ideas. They take up the first volume and part of the second volume of our edition. The rest of the second volume consists of the classical works of Henri Poincaré on topology and other memoirs adjacent to "New methods of celestial mechanics". The third volume consists of mathematical works, in particular, works on function theory, as well as works on physics and some articles on general problems of science.

The first and second volumes of "New methods of celestial mechanics" have been translated: the first by A.A. Bryandinskaya and I.V. Ioslovich under the editorship of V.I. Arnold, and the second by Yu. A. Danilov under the editorship of V.M. Alekseev. The overall editing of "New methods" was carried out by V.I. Arnold.

It has to be pointed out that in the preparation of this volume for printing a considerable amount of work of a technical nature was carried out and numerous mistakes in the formulae of the French original version were corrected.

Comments

Almost all the material of the first volume of "Méthodes Nouvelles" has become classical. The first chapter contains a compact account of the main theorems of analytical dynamics accompanied by examples of applications to the choice of suitable coordinates in various problems of celestial mechanics. Poincaré introduces and systematically uses the canonical action-angle variables (Delaunay variables), first in the restricted and later in the unrestricted three-body problem. He then introduces new canonical variables (Poincaré variables), which are convenient for the study of near-circular orbits.

The second chapter contains analytical preparatory material for further investigations. Here the method of Cauchy majorants is set forth in detail. Next, the implicit function theorem, algebroid singular points and Puiseux series are considered. Finally the conservation of parity of the number of real singular points under the deformation of a function is proved.

The third chapter is devoted to periodic solutions. Here the behaviour of periodic solutions under perturbations (bifurcation theory), and the "Poincaré method" of finding periodic solutions by a power-series expansion in a small parameter are investigated. The results are applied to the three-body problem, for which Poincaré found many interesting periodic solutions. Next, Poincaré finds periodic solutions for a general canonical system of differential equations that are close to integrable; the ideas of this section served as the basis of Birkhoff's theorems on the existence of infinitely many periodic solutions in a neighbourhood of a given periodic solution.

The fourth chapter, characteristic exponents, is in two parts. In the first part, the general thoery of linear equations with periodic coefficients is set forth; here the presence of first integrals and integral invariants is taken into account. In the second part, characteristic exponents of periodic solutions of the three-body problem are studied, and their expansions in a (generally fractional) power series of a small parameter are found.

The fifth chapter is devoted to the proof of non-integrability of the three-body problem. Here the non-existence of an analytic (and dependent analytically on the parameters) first integral that is independent of the classical ones and is single-valued in the action-angle variables is proved. The essence of Poincaré's idea is that the complicated behaviour of the solutions of a perturbed system (in particular, the birth of numerous non-degenerate periodic solutions) is incompatible with integrability: the presence of each additional first integral imposes on the behaviour of the solutions rather stringent restrictions. The opportunities afforded by this idea are far from exhausted, even today.

In the sixth chapter the analytical properties of the perturbation function are studied, that is, the interaction energy and the asymptotics of its high-order Fourier coefficients. From a mathematical point of view, this is a question of the investigation of the singular points of the integral of an analytic function of several variables, regarded as a function of parameters ("pinch points"). This investigation is then carried out in detail for the special case of a perturbation function of the three-body problem. The result is applied to the verification of the non-degeneracy of the three-body problem, required in the previous chapter for the proof of the non-existence of first integrals to go through. However, the value of these investigations by Poincaré goes far beyond the framework of celestial mechanics, since similar problems on asymptotics and integrals depending on a parameter are encountered in the most diverse areas.

The seventh chapter, asymptotical solutions, is devoted to invariant manifolds in connection with periodic solutions, stable and unstable "moustaches". The theory outlined by Poincaré was later developed by Hadamard and Perron. The chapter ends with the construction of the asymptotical expansions for invariant manifolds of near-integrable Hamiltonian systems.

2. (p. 16). The problem of the attraction of two fixed centres is worked out in detail in the book *Celestial mechanics* by Charlier (C. L. Charlier, Die Mechanik des Himmels, 1902-1907 (2nd edition 1927), Leipzig: Veit (2 volumes)). Inaccuracies in Charlier's qualitative investigations were later corrected in the works of Talquist (Acta Soc. Sci. Fennicae **1**, Nos. 3, 5 (1927) and Badalyan (Astron. J. II, No. 4 (1934); Comment. Phys.-Math. Soc. Sci. Fennicae, **8**, No. 2 (1935).

In recent years further attention has been given to the problem of two fixed centres. The potential of two centres well approximates the potential of a slightly elongated ellipsoid. If one places the centres at imaginary points, then one obtains a good approximation of the potential of a flattened ellipsoid. Therefore the solution of the problem about two fixed centres can be used as an approximation in the investigation of the motion of the satellites of oblate planets.

In this connection see, for example, the paper by Aksenov, Grebennikov and Demin in Astron. Zh. **40**, No. 2 (1963) and also W.T. Kyner, *Qualitative properties of orbits about an oblate planet*, Commun. Pure Appl. Math. **17**. No. 2, 227–236

(1964); C. C. Conley, *A disc mapping associated with satellite problems*, ibid., 237–243.

12. (p. 75). Taken literally, the expressed statement is refuted by the example $F_0 = x_1 + \sqrt{2}x_2, \quad F_1 = F_2 = \ldots = 0.$

Poincaré's assertion that in any neighbourhood of any point there is a closed trajectory can naturally be referred to systems "in general position". In such a form it is plausible, but is as yet unproven. Important results in this direction have recently been obtained by Pugh, *The Hamiltonian closing lemma*, at the International Mathematical Congress in Moscow 1966. Pugh showed that by means of a small (with first derivatives) change of the Hamiltonian vector field it can be arranged that a trajectory passing through a given point becomes closed. It is unknown whether this is achieved by a smoother deformation, and especially an analytic one, as Poincaré required.

17. (p. 111) In these words of Poincaré, the embryo of what is now called "Morse Theory" can be seen.

In fact, he is talking about the number of critical points of a smooth function on a torus.

Poincaré's argument was probably as follows.

In the memoirs "On curves defined by differential equations" (C. R. Acad. Sci. Paris **XCVII**, 287–289 (1884)) Poincaré essentially proved that for any smooth function on a manifold the number

$$\chi = M_0 - M_1 + M_2 \cdots,$$

where M_0 is the number of minima, M_1 is the number of saddle points with one negative square, M_2 with two, ..., is the same and coincides with the "Euler-Poincaré characteristic" of this manifold, $\chi = b_0 - b_1 + \cdots$, where b_i is the ith Betti number. In particular, the Euler characteristic of the torus is zero, therefore

$$M_0 - M_1 + M_2 = 0.$$

But since the number M_0 of minima and the number M_2 of maxima is at least 1, it follows that M_1 is at least 2. Hence Poincaré arrives at the conclusion that the total number of critical points of a function on a torus is at least 4. For a three-dimensional torus the Euler-Poincaré characteristic is equal to 0, therefore

$$M_0 - M_1 + M_2 - M_3 = 0.$$

Although M_0 and M_3 are at least 1, Poincaré was not able to derive from this any information about the existence of other critical points, hence the concluding part of his sentence on p. 111.

In fact, any smooth function on a manifold not homeomorphic to a sphere has critical points other than maxima and minima (J. Milnor, *Sommes de variétés différentiables des sphères*, Bull. Soc. Math. de France, **87**, 439–447 (1959)).

The relation between the number of critical points of a function on a manifold and the topology of the manifold was investigated in detail by M. Morse (see, for example, Milnor's book *Morse Theory*, Princeton Univ. Press, Princeton NJ 1963). It follows from Morse's inequalities that the total number of critical points is at least the sum of the Betti numbers.

$$M_0 + M_1 + M_2 + \cdots \geq b_0 + b_1 + b_2 + \cdots .$$

For an n-dimensional torus the sum of the Betti numbers is equal to 2^n. Therefore a function on a two-dimensional torus has at least four critical points, on a three-dimensional torus at least eight, and so on.

In accordance with this, under a perturbation of a three-dimensional torus filled with periodic trajectories there are at least four periodic solutions, of a four-dimensional torus at least eight and so on (cf. V. I. Arnold, C. R. Acad. Sci. Paris **261**, 3719–3722 (1965)).

Note that above by critical points we always meant to have in view non-degenerate ones, that is, those critical points at which the second differential of the function is non-degenerate; if, in fact, there are also degenerate critical points, then they need to be taken into account with their multiplicities.

The number of *geometrically distinct* critical points on a manifold can also be estimated via topological invariants of the manifold, such as the Lyusternik-Schnirelmann category (see, for example, Lyusternik and Schnirelmann *Topological methods in variational problems*, Moskov. Gos. Univ., 1930).

The category of a two-dimenional torus is 3, and on an n-dimenional torus $n+1$. Hence it follows that a function on a two-dimenional torus has at least three geometrically distinct critical points, and on an n-dimenional torus at least $n+1$.

These estimates are attained: there exist functions having exactly $n+1$ critical points.

In accordance with this, under a perturbation of an n-dimensional torus filled with periodic trajectories there arise at least *n geometrically distinct* periodic solutions.

21. (p. 205). The non-existence of a single-valued analytic integral in the three-body problem has so far not been proved with full rigour. In fact, Poincaré considers only integrals that are analytic not only with respect to the Delaunay variables, but also with respect to the parameter μ or $\sqrt{\mu}$. The first accurate proof of the non-integrability of a Hamiltonian system of sufficiently generic form is due to C. L. Siegel (Mathematische Annalen, **128** (1954), 144–170); see also J. Moser, *On the integrability of area preserving Cremona mapping near an elliptic fixed point*, Buletin de la Sociedad Matematica Mexicana (1961) and the literature cited therein.

It interesting to note that *non-analytic* integrals are possible in the problems considered above; their existence in problems with two degrees of freedom follows from a theorem of A. N. Kolmogorov (see A. N. Kolmogorov, Dokl. Akad. Nauk SSSR **48**, No. 4, 527–530 (1954) and V. I. Arnol'd, Uspekhi Mat. Nauk **18**, Nos. 5, 6 (1963)). By contrast, in the case when the number of degrees of freedom is greater than two, for systems of general form, most likely even continuous integrals are not possible. (see V. I. Arnol'd, Dokl. Akad. Nauk SSSR **156**, No. 1, 9–12 (1964)).

Comments on the paper "On a geometric theorem" by Henri Poincaré

V.I. Arnold
Translated by Gerald Gould

Associated with Poincaré's "geometric theorem", there are a number of proven and unproven propositions on fixed points of symplectic (or, as called by Poincaré, canonical) diffeomorphisms of symplectic manifolds more general than a circular annulus (or, even more generally, on intersections of so-called Lagrangian submanifolds of a symplectic manifold).

In the simplest of these generalizations the annulus is replaced by a torus. Consider a torus T with angular coordinates $(\phi_1, \phi_2) \mod 2\pi$ and the element of area $d\phi_1 d\phi_2$. The map $A : T^2 \to T^2$, which is homotopic to the identity transformation, is given by the formula

$$A(\phi_1, \phi_2) = (\psi_1, \psi_2) = (\phi_1 + f_1(\phi_1, \phi_2), (\phi_2 + f_2(\phi_1, \phi_2)),$$

where f_1 and f_2 are periodic functions with period 2π. We say that A is an *area-preserving* (or *symplectic*) *diffeomorphism of the torus* if the Jacobian of ψ with respect to ϕ is everywhere equal to 1. We say that A *keeps the centre of gravity of the torus fixed* if the mean values of the functions f_1 and f_2 are zero.

Poincaré's theorem then follows from the following proposition:

Every area-preserving diffeomorphism of the torus has at least four fixed points (counted with multiplicities); at least three of them are geometrically distinct.

(In order to derive Poincaré's theorem on the annulus from this, it is necessary to slightly extend the annulus along its boundaries and glue it to a second such annulus so as to obtain a torus.)

Unfortunately, the proof runs into the same difficulties as those encountered by Poincaré. Under certain additional restrictions a proof can be obtained from Morse theory (which must be applied to the generating function of the map A). For example, it suffices that the eigenvalues of the Jacoby matrix $\frac{\partial \psi}{\partial \phi}$ do not become equal to -1 at any point.

Originally published in: H. Poincaré: Selected Works in Three Volumes. Vol. II. Klassiki Nauki, 987-989, © Nauka, Moscow 1972
Translated by G. Gould. Springer-Verlag Berlin Heidelberg 2014 with kind permission, DOI 10.1007/978-3-642-31031-7_45

In order to carry over the formulation given above to more general manifolds one needs to analyse the condition concerning the centre of gravity.

Let (M, ω) be a compact *symplectic manifold*, that is, M is a smooth even-dimensional manifold and ω is a closed non-degenerate differential two-form on it (in Poincaré's terminology ω is the integral invariant $\sum dp_i dq_i$).

Since it is non-degenerate, the form ω gives an isomorphism between the tangent space and the cotangent space at each point. We denote this isomorphism by $i : TM \to T^*M$.

To each real function H on M, there corresponds the vector field $i^{-1}dH$ on M. This field is said to be Hamiltonian with the Hamilton function H. We say that a map $A : M \to M$ is *a symplectic diffeomorphism of M onto itself homologous to the identity*, if A can be joined with the identity transformation by a smooth homotopy, A_t, whose derivative at each instant of time t is a Hamiltonian vector field.

(It can be shown that Hamiltonian vector fields form the commutant of the Lie algebra of all preserving ω, that is, locally Hamiltonian, vector fields; symplectic diffeomorphisms homologous to the identity form the commutant of the connected component of the identity in the group of all symplectic, that is, ω-preserving, diffeomorphisms of M.)

In the special case of the torus, diffeomorphisms homologous to the identity are precisely those area-preserving maps that keep the centre of gravity fixed. Thus we arrive at the following generalization of Poincaré's theorem on the annulus.

A symplectic diffeomorphism of a compact symplectic manifold M onto itself that is homologous to the identity has at least as many fixed points, as has a smooth function on M critical points.

Under the hypothesis that the diffeomorphism A is not too far from the identity transformation the proof can again be obtained from Morse theory.

In particular, in the case when M is a surface of genus g, the number of expected fixed points is $2g+2$, and in the case when M is a torus of even dimension n, the number of expected fixed points is 2^n.

It is possible, however, that Poincaré's geometric theorem is connected with essentially two-dimensional effects similar to the following generalization of it, suggested as a conjecture by A.I. Schnirelmann.

Every area-preserving and orientation-preserving diffeomorphism of a 2*-sphere onto itself has at least two geometrically distinct fixed points.*

We point out that every diffeomorphism of a 2-sphere onto itself has a fixed point by purely topological considerations, and it is easy to construct a diffeomorphism not preserving areas, of course, that has only one fixed point.

The proof uses generating functions and the following elementary argument from two-dimensional topology.

The index of an isolated singular point of the gradient field of a smooth function on the plane does not exceed 1.

(In a space of three or more dimensions it is easy to construct a function with an isolated critical point at which the index of the gradient field is equal to any integer, and on the plane to any number from the series $1, 0, -1, -2, -3, \ldots$)

Acknowledgements

Springer-Verlag thanks the publishers of V.I. Arnold's papers for having granted permission to reprint them in these Collected Works.

The numbers given behind the original sources correspond to the numbers given in the Table of Contents.

We have made every effort to identify the copyright owners for all articles included in this book in order to obtain reprint permission. Some of our requests have however remained unanswered. The following list includes all sources, and owners where known.

Reprinted from Am. Math. Soc. Transl. (2)	6, 11
Reprinted from Annu. Rev. Fluid Mech.	42
Reprinted from C. R. Acad. Sc. Paris	3
Reprinted from Funct. Anal. Appl	9, 10, 24, 25
Reprinted from Izd. Inst. Akad. Nauk Ukrain. SSR, Kiev	30
Reprinted from J. Appl. Math. Mech.	1, 5, 31
Reprinted from Moscow Univ. Math. Bull.	21, 36, 38
Reprinted from Proc. Royal Soc. London	41
Reprinted from Russ. Math. Surv.	13, 28, 29
Reprinted from Selecta Math. Sov.	32
Reprinted from Some Problems in Modern Analysis, MGU Moscow	39
Reprinted from Sov. Math. Dokl.	4
Reprinted from Sov. Phys. JETP	34
Reprinted from Sov. Phys. Dok.	35
Reprinted from Transact. Math. Moscow Soc.	23
Reprinted from Usp. Mat. Nauk	22, 27

A.B. Givental et al. (Eds.), *Vladimir I. Arnold - Collected Works*,
DOI 10.1007/978-3-642-31031-7, © Springer-Verlag Berlin Heidelberg 2014

The manufacturer's authorised representative in the EU is Springer Nature Customer Service Centre GmbH, Europaplatz 3, 69115 Heidelberg, Germany. If you have any concerns regarding our products, please contact ProductSafety@springernature.com

Printed and bound by CPI Group (UK) Ltd, Croydon, CR0 4YY

30/09/2026

02225599-0020